COD AND HERRING

COD AND HERRING

THE ARCHAEOLOGY AND HISTORY OF MEDIEVAL SEA FISHING

Edited by

JAMES H. BARRETT AND DAVID C. ORTON

OXBOW | books

Oxford & Philadelphia

Published in the United Kingdom in 2016 by
OXBOW BOOKS
10 Hythe Bridge Street, Oxford OX1 2EW

and in the United States by
OXBOW BOOKS
1950 Lawrence Road, Havertown, PA 19083

Paperback Edition: ISBN 978-1-78570-239-6
Digital Edition: ISBN 978-1-78570-240-2 (epub)

A CIP record for this book is available from the British Library

Library of Congress Cataloging-in-Publication Data

Names: Barrett, James H. (James Harold), editor. | Orton, David C. (David Clive), editor.
Title: Cod and herring : the archaeology and history of medieval sea fishing
  / edited by James H. Barrett and David C. Orton.
Description: Oxford; Philadelphia: Oxbow Books, 2016. | Includes
  bibliographical references.
Identifiers: LCCN 2016013177| ISBN 9781785702396 (softcover) | ISBN
  9781785702402 (digital)
Subjects: LCSH: Fish remains (Archaeology) | Saltwater fishing–History–To
  1500.
Classification: LCC CC79.5.A5 C63 2016 | DDC 930.1–dc23 LC record available at http://lccn.loc.gov/2016013177

Printed in the United Kingdom by Latimer Trend

For a complete list of Oxbow titles, please contact:

UNITED KINGDOM
Oxbow Books
Telephone (01865) 241249, Fax (01865) 794449
Email: oxbow@oxbowbooks.com
www.oxbowbooks.com

UNITED STATES OF AMERICA
Oxbow Books
Telephone (800) 791-9354, Fax (610) 853-9146
Email: queries@casemateacademic.com
www.casemateacademic.com/oxbow

Oxbow Books is part of the Casemate Group

Front cover: Design by Katie Gabriel Allen using a woodcut image from *Historia de Gentibus Septentrionalibus* by Olaus Magnus
(used by permission of University of Glasgow Library, Special Collections)
Back cover: Medieval fish bones from Blue Bridge Lane, York (Photo: James Barrett)

Contents

Contributors

JAMES H. BARRETT, McDonald Institute for Archaeological Research, University of Cambridge, Cambridge, England

COLIN BREEN, School of Geography and Environmental Sciences, Ulster University, Coleraine, Northern Ireland

INGE BØDKER ENGHOFF, Natural History Museum of Denmark, University of Copenhagen, Copenhagen, Denmark

ANTON ERVYNCK, Flemish Heritage Agency, Brussels, Belgium

MARK GARDINER, Archaeology, School of Geography, Archaeology and Palaeoecology, Queen's University Belfast, Belfast, Northern Ireland

SHEILA HAMILTON-DYER, SH-D ArchaeoZoology, Southampton, England

JENNIFER HARLAND, Archaeology Institute, University of the Highlands and Islands, Orkney College, Kirkwall, Orkney, Scotland

POUL HOLM, School of Histories and Humanities, Trinity College Dublin, Dublin, Ireland

ANNE KARIN HUFTHAMMER, Department of Natural History, The University Museum, University of Bergen, Bergen, Norway

ANDREW K. G. JONES, Department of Archaeology, University of York, York, England

MARYANNE KOWALESKI, Department of History, Fordham University, Bronx, New York, USA

ALISON LOCKER, Escaldes-Engordany, Andorra

LEMBI LÕUGAS, Archaeological Research Collection, Tallinn University, Tallinn, Estonia

DANIEL MAKOWIECKI, Laboratory for Natural Environment Reconstruction, Institute of Archaeology, Nicolaus Copernicus University, Toruń, Poland

JAMES MORRIS, School of Forensic and Applied Sciences, University of Central Lancashire, Preston, Lancashire, England

GUNDULA MÜLDNER, Department of Archaeology, University of Reading, Reading, England

ARNVED NEDKVITNE, Department of Archaeology and History, University of Oslo, Oslo, Norway

ALF RAGNAR NIELSSEN, University of Nordland, Bodø, Norway

DAVID C. ORTON, BioArCh, Department of Archaeology, University of York, York, England

REBECCA REYNOLDS, Department of Archaeology, University of Nottingham, Nottingham, England

HELGE SØRHEIM, Museum of Archaeology, University of Stavanger, Stavanger, Norway

WIM VAN NEER, Royal Belgian Institute of Natural Sciences, Brussels, Belgium; Laboratory of Biodiversity and Evolutionary Genomics, University of Leuven, Leuven, Belgium

ORRI VÉSTEINSSON, Department of Archaeology, University of Iceland, Reykjavík, Iceland

Preface and Acknowledgements

The analysis of fish bones from archaeological sites is a highly specialised and painstaking task, requiring an abundance of the time that is so rarely available in either academic or commercial archaeology. Moreover, study of fish remains has seldom been at the top of archaeological research priorities. Nevertheless, over the last 40 years a few specialists across Europe have dedicated themselves to work of this kind, and thus to discovering the outlines of medieval fishing history around the North Atlantic, and the Irish, North and Baltic seas. Although mutually informed in terms of methodology, this fundamental research has often been carried out in the framework of national institutions and agendas. Concurrently, historians have independently striven to systematise and analyse complex corpora of textual evidence regarding medieval fishing and fish trade. Once again this work has sometimes occurred within national or regional schools of research. The results of these zooarchaeological and historical efforts have often proven surprising and important, revealing remarkable evidence of continuity and change. Archaeologists of medieval coastal settlements have also contributed much to our understanding of the relationship between people and the sea.

The present volume is an effort to enhance the value of this past work by crossing boundaries – between regions and between disciplines. It also emerges from a time when traditional zooarchaeology (the identification, quantification and interpretation of skeletal remains) has increasingly benefited from integration with biomolecular approaches, such as stable isotope analysis and the study of ancient DNA. These latter methods are not the main focus of the book – they are changing far too quickly for this to have been helpful. Nevertheless, they inform many of its chapters and Gundula Müldner has taken up the challenge of surveying the extant stable isotope evidence regarding human skeletal remains from medieval Britain.

Even in the fields of zooarchaeology and history it is recognised, even hoped, that this volume will quickly become outdated. It is our aspiration that the collaborative process of consolidating what is known and unknown may already have accelerated the pace of current research on medieval sea fishing.

The idea behind the book emerged from an interdisciplinary conference organised by one of us (JHB) in Westray, Orkney, Scotland, in June of 2008. It was several years, however, before the groundwork could be laid – including finishing the analysis of major collections and the synthesis of decades of fish-bone and historical research. The initial practicalities were skilfully managed by Cluny Johnstone, then a postdoctoral research fellow on the 'Medieval Origins of Commercial Sea Fishing' project funded by the Leverhulme Trust. After a period of maternity leave Cluny decided to be a full-time parent and editing became our responsibility. DCO began the process while a postdoctoral research fellow on the Leverhulme Trust project 'Ancient DNA, Cod and the Origins of Commercial Trade in Medieval Europe'. JHB was then able to see it through to completion. This book is also based upon work from the COST Action Oceans Past Platform, supported by COST (European Cooperation in Science and Technology).

We are grateful to Julie Gardiner of Oxbow Books for her helpfulness and patience during the book's long gestation. Jennifer Harland (also a postdoctoral research fellow on the 'Medieval Origins of Commercial Sea Fishing' project) and Christine Harcus assisted with the original conference in Orkney, which was funded by the Leverhulme Trust, the McDonald Institute for Archaeological Research and the History of Marine Animal Populations project (supported by the Alfred P. Sloan Foundation). Many thanks are owed to Suzanne Needs-Howarth, who copy-edited the volume and helped compile Appendix 1.1, and to the McDonald Institute for Archaeological Research, which contributed to the cost of her work. Dora Kemp also kindly assisted with copy-editing. The cover was designed by Katie Gabriel Allen using a woodcut image from *Historia de Gentibus Septentrionalibus* by Olaus Magnus (used by permission of University of Glasgow Library, Special Collections) and a photograph of medieval fish bones from York taken by JHB. Other image credits are given in the figure captions, and each chapter includes its own acknowledgements section when appropriate. Most importantly, we thank the contributors to this volume for the many years of careful research that their chapters represent.

James H. Barrett
University of Cambridge
David C. Orton
University of York

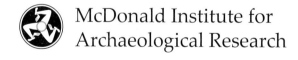

COST is supported by the EU Framework Programme Horizon 2020

COST (European Cooperation in Science and Technology) is a pan-European intergovernmental framework. Its mission is to enable break-through scientific and technological developments leading to new concepts and products and thereby contribute to strengthening Europe's research and innovation capacities. It allows researchers, engineers and scholars to jointly develop their own ideas and take new initiatives across all fields of science and technology, while promoting multi- and interdisciplinary approaches. COST aims at fostering a better integration of less research intensive countries to the knowledge hubs of the European Research Area. The COST Association, an International not-for-profit Association under Belgian Law, integrates all management, governing and administrative functions necessary for the operation of the framework. The COST Association has currently 36 Member Countries. www.cost.eu

Studying Medieval Sea Fishing and Fish Trade: How and Why

James H. Barrett

This book explores the changing human use of marine fish, mainly from AD 500 to 1550, around the North Atlantic Ocean and the Irish, North and Baltic seas. It does so by combining historical research with the study of fish bones from archaeological sites and, to a lesser degree, investigating the archaeology of fishing settlements and the chemistry of human bone. Its coverage is not comprehensive, but an effort has been made to gather and interpret sufficient evidence to frame working hypotheses regarding the chronology, causes and consequences of sea fishing in Europe's northern waters. Much of this evidence has never been brought together before, or in some cases even been published. Moreover, the combination of historical and archaeological evidence makes it possible to follow long-term trends in sea fishing that cross the source-based divide between a late and post-medieval Europe populated by fishers and fish merchants and an earlier world where they are difficult to perceive and even harder to understand.

In so doing, the book attempts to answer, in a preliminary way, a variety of questions that are important to economic, demographic, social and environmental archaeology and history. Was the growth of early medieval sea fishing a correlate of state formation and urbanisation? Were sea fisheries developed in response to social drivers, such as Christian fasting practices, elite demand and/or the cultural foodways of migrant communities? Was there an unprecedented sea-fishing revolution at the turn of the first to second millennia AD? Did sea fishing expand as a result of human impacts on freshwater ecosystems? Could marine fisheries of the High Middle Ages have overfished formerly superabundant species, such as cod (*Gadus morhua*) and herring (*Clupea harengus*)? When did the long-range trade of high-bulk and low-value staples, such as salted herring and dried cod, really begin? When did such trade expand to a pan-European scale? To what degree did demand for

sea fish influence the increasingly 'global' destinations of late-medieval mariners? In addressing these questions, the chapters in this book confirm some existing wisdom and provide new, sometimes counterintuitive, insights.

The available evidence is often preliminary in nature. A key limitation is that inter-regional comparison must remain qualitative in most cases. The comparability of much basic data is low. Exceptions exist where standard measures are possible. These range from historically based series of prices for dried cod and salted herring (Chapter 5) to stable isotope studies of cod bones (to infer local versus non-local catches) (see below) and human skeletal remains (to infer the contribution of marine protein to diet) (Chapter 20). Even in these instances there are source-critical issues (e.g. Hedges 2004; Hutchinson *et al.* 2015; Rigby 2005, xxxviii). Comparing zooarchaeological evidence, however, has always proven a very major challenge. The fundamental parameters of a bone assemblage – such as the species, sizes and anatomical parts of fish represented – are known to be influenced by preservation conditions, recovery methods and inter-analyst differences in recording protocols. It is difficult to control for all of these factors, and in some instances (such as the synthesis of old zooarchaeological reports) none of these variables may be within our power. Nevertheless, much can be gleaned even from the most fundamental data. To provide just one example, if herring bones are first found in inland Poland at elite centres of the tenth and eleventh centuries, it is reasonable to infer the possibility of some connection between Polish state formation and long-range fish trade (Chapter 12). Despite the imperfections of zooarchaeological evidence, fish bones are often our main (and sometimes our only) window on medieval sea fishing and fish trade.

Zooarchaeological and related methods merit a few additional words of explanation. Many fish can be identified to species based on the shape and surface features of some or most of their bones when examined by an experienced researcher using an appropriate reference collection. Some taxa, however, are extremely difficult to distinguish based on visual examination of their skeletal remains. An important example is flatfish of the family Pleuronectidae (Wouters *et al.* 2007). In particular, the common taxa plaice (*Pleuronectes platessa*) and flounder (*Platichthys flesus*) are extremely difficult to tell apart with certainty. Many past and present zooarchaeologists (including the present author) have thus often employed a combined category, such as plaice/flounder or plaice/flounder/dab (the latter being *Limanda limanda*). This is an important issue during interpretation because flounder can occupy both freshwater and marine environments, whereas plaice and dab are only found in salt water (Whitehead *et al.* 1986a, 1302–6). Similar combined groups are sometimes also used in other cases. Biomolecular methods based on proteins (e.g. Richter *et al.* 2011) and DNA (e.g. Ewonus *et al.* 2011) can provide definitive identifications, but these are not yet routinely applied on a large scale.

Turning to another fundamental methodological issue, there are five main approaches to identifying fish trade based on bones (Barrett 1997; Barrett *et al.* 2008; Dufour *et al.* 2007; Enghoff 1996; Hutchinson *et al.* 2015; Lõugas 2001; Perdikaris 1999). The first is biogeography, based on the movement of a species outside its natural range. The second is butchery patterning: the absence of some bones indicating processing elsewhere – typically head bones for gadids and hyoid, gill and appendicular elements for herring. These anatomical patterns may also be associated with distinctive cut marks. The third is fish-length estimates, which can imply specialisation in products of a particular size and/or suggest the import of fish that exceed the maximum size of a local fish population (when large cod are found in the eastern Baltic, for example). The fourth is stable isotope analysis, which can be used to identify fish that were likely caught in different aquatic ecosystems. The fifth is ancient DNA analysis, which can attribute specimens to different geographically constrained populations.

Ancient DNA is likely to have an increasing role in the future, but its application to the study of ancient fish trade remains is currently limited (e.g. Hutchinson *et al.* 2015). Conversely, the first four of the above-mentioned methods are used frequently in this book. Biogeography, anatomical patterning and fish-length estimates are sufficiently self-explanatory that further discussion can await individual chapters. Stable isotope provenancing of fish bone is a more developing technique. At present the most extensive work has involved cod (Barrett *et al.* 2008; 2011; Orton *et al.* 2011; Hutchinson *et al.* 2015), and this research helps to inform several chapters of the book. The basic premise is that the stable carbon ($\delta^{13}C$) and nitrogen ($\delta^{15}N$) isotope signatures of protein preserved in cod bones are determined by environmental parameters during the lifetime of the fish. These basic parameters vary by region. Examples include temperature, salinity, nutrient loading and food-web complexity. However, modern isotope values in aquatic ecosystems have been altered by such factors as pollution. Thus the signatures characteristic of each potential medieval fishing region must be established using data derived from archaeological cod remains. This can be done by using cranial bones as control samples to infer regional isotope signatures for different potential sources of traded cod – from the eastern Baltic to Iceland. This approach works because most (although not all) preserved cod products of the Middle Ages were decapitated during their preparation for trade by drying and/or salting (see below); skull bones therefore usually represent relatively local catches. One then compares the stable isotope values of archaeological vertebrae and appendicular elements (which typically remained in dried cod products) to these control data to see whether or not they match local values. When local values are not matched, implying the likelihood of imported cod, one can also make probability-based estimates regarding which potential source the fish may have derived from. This second step is more speculative, partly because there are not yet control data from every likely source and partly because there is some overlap in the isotope values of different regions.

The stable isotope method is continuously being refined as more control data (and additional isotopes, such as sulphur) are added (Hutchinson *et al.* 2015; Nehlich *et al.* 2013). Figure 1.1 provides an illustration of the control data published by Hutchinson *et al.* (2015), showing some of the strengths and limitations of the technique. Cod from some waters (e.g. the eastern Baltic Sea, or the Irish Sea and southern portions of the North Sea) have distinctive $\delta^{13}C$ and $\delta^{15}N$ signatures. Others (e.g. those from the waters of Iceland, northern Scotland and western Ireland) share similar values with neighbouring locations over extensive stretches of ocean. Nevertheless, this research has made important, if sometimes tentative, discoveries regarding trends in the chronology and geography of the medieval cod trade and/or long-range fisheries that will be cited where pertinent in the chapters to follow.

Other archaeological methods, such as dating the development of coastal fishing settlements, are often more straightforward. They can, however, be subject to their own biases, such as coastal erosion and submergence (Tys 2015, 131; Chapter 14). Historical methods are equally complex, given the wide range of source materials of wildly divergent detail and historicity. It is only in the fourteenth century, for example, that quantitative time-series exist for fish trade in northern Europe (Nedkvitne 2014). Even at this late date, the relevant customs accounts exist for only a few regions (particularly England and Germany), and these are highly variable in their level of detail, preservation and representativeness (e.g. Burkhardt

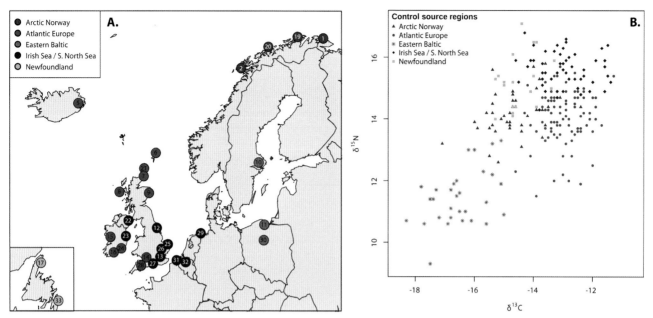

Figure 1.1. Spatial variability in the stable isotope values of protein from cod bones makes it possible, in some instances, to identify fish bones that have been transported by migrant fishers or trade: (a) archaeological sites from which cranial (control) bones of cod were sampled; (b) $\delta^{13}C$ and $\delta^{15}N$ values for the control samples grouped into five analytical regions (after Hutchinson et al. 2015, 5).

2013). Earlier textual sources are either one-offs (e.g. the Domesday Book), anecdotal (e.g. narrative sources, the most famous being the Icelandic sagas), or both. Not all regions are equally well documented in all periods. It is unrealistic to survey here the merits and weaknesses of all the relevant text-based evidence and of the skills that must be brought to their interpretation. For present purposes it is enough to know that the evidence base for our subject – archaeological and historical – is diverse, complex, unevenly distributed through time and generally low in both precision and accuracy.

Although this book is about sea fishing, it is instructive to understand contemporary developments in the exploitation of freshwater and migratory species to interpret trends in the use of marine fish (cf. Hoffmann 1996; 2002). Thus some chapters range widely in terms of taxonomy (see Appendix 1.1 for a list of common and scientific names of all the taxa mentioned). Of the fully marine fish, cod and herring arguably were of greatest socio-economic importance in medieval Europe (Sicking and Abreu-Ferreira 2009; Starkey *et al.* 2009). This is not to suggest that other marine taxa were insignificant. Flatfish such as plaice, for example, clearly played an important role for many around the coast of the southern North Sea (Chapters 13–14). One might also note the relevance of hake (*Merluccius merluccius*) fishing along the Atlantic coast of Ireland (Chapter 9; Chapter 19). These are just two of many examples. Catadromous and anadromous species, such as eel (e.g. Chapter 15) and salmon (e.g. Chapter 19), also played very important roles in specific times and places. Yet

fisheries for cod and herring were of especially widespread, albeit fluctuating, importance – in part due to their extensive natural distributions, high abundance and suitability for preservation (see below). However, cultural traditions also played a role in the popularity of these species, the consumption of which spread with migrants and merchants (Barrett *et al.* 2001; Orton *et al.* 2011; Chapters 11–12).

The word cod, as used in this book, refers specifically to *Gadus morhua* (Figure 1.2). More holistically, taxa of the cod family (Gadidae) are often considered together under the term gadids. The gadid taxa whiting (*Merlangius merlangus*), haddock (*Melanogrammus aeglefinus*), saithe (*Pollachius virens*) and ling (*Molva molva*) (the latter of which is attributed to the separate family Lotidae in some taxonomic schemes) were important targets of medieval fishers, as bycatches or as fisheries in their own right. Cod was among the most widely distributed and abundant members of the group, although it is now listed as vulnerable by the International Union for Conservation of Nature (Froese and Pauly 2015; IUCN 2015; Whitehead *et al.* 1986b, 686–7). The species can be found from the shoreline to depths of 600 m, with juveniles preferring shallow sublittoral environments and large, adult fish preferring deeper, colder waters. Its maximum total length varies by region, but can reach 2 m. Although shore-based fishing for small cod is known (e.g. Cerón-Carrasco 2011, 60; Enghoff 2011), in the Middle Ages large cod were usually caught from boats using hand lines with a single weight and hook (e.g. Chapter 6). Possible experimentation aside, long lines (fixed to the bottom and fitted with multiple hooks) and deep-water nets targeting gadids were post-medieval innovations, with the chronology of introduction varying by region (Starkey *et al.* 2009).

The flesh of cod and related gadids is low in fat and can thus be air dried without salt at high latitudes – in northern Norway and Iceland, for example, where the temperature remains sufficiently low during the fishing season (Cutting 1955, 173–4). In other contexts, such as the northern islands of Scotland, it can be air dried if protected from precipitation (Harland and Barrett 2012, 127). More typically, cod were

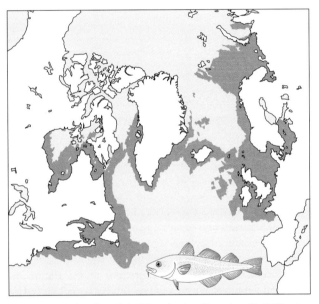

Figure 1.2. Cod and its global distribution (Drawing: Vicki Herring after Kaschner et al. *2013 and the Freshwater and Marine Image Bank, University of Washington).*

both dried and salted at lower latitudes when not sold fresh (e.g. Bennema and Rijnsdorp 2015; Cutting 1955, 157–82; Kowaleski 2003, 226). Gadid fish dried without salt are known as stockfish, regardless of how exactly they were butchered prior to preservation. Cod that were both dried and salted did not have a consistent name in the Middle Ages, although the post-medieval Norwegian term *klippfisk* (English: klipfish) can be employed for convenience (Barrett 1997, 619–20). Most varieties of stockfish and 'klipfish' were decapitated, but exactly which bones were or were not removed during processing varied based on local convention, the intended market and/or the size of the fish (cf. Barrett 1997, 619; Bennema and Rijnsdorp 2015). Six grades of stockfish, based on size and quality, were recognised in fourteenth-century England (Nedkvitne 2014, 500). In terms of what is archaeologically recognisable, however, the most important distinction is between what Norwegians call *rundfisk* (in which only the head was removed) and *råskjær* (in which both the head and anterior vertebrae were removed, creating a product that was partly split) (Chapter 18; see Figure 1.3). The 'shelf life' of dried cod could be very long. One estimate is five to seven years and a late fourteenth-century household management book even suggests that stockfish could keep for 10 to 12 years (Brereton and Ferrier 1981, 237; Wubs-Mrozewicz 2009, 188).

Herring are a small pelagic fish, reaching a maximum total length of *c.* 0.4 m. They have a wide distribution (Figure 1.4) and occur in dense shoals that rise to shallow water at night, bringing them within reach of pre-modern fishing methods. Most herring have seasonal spawning migrations into coastal waters, with the month and location differing by population (Froese and Pauly 2015; Whitehead *et al.* 1989, 273). They were thus both accessible to medieval

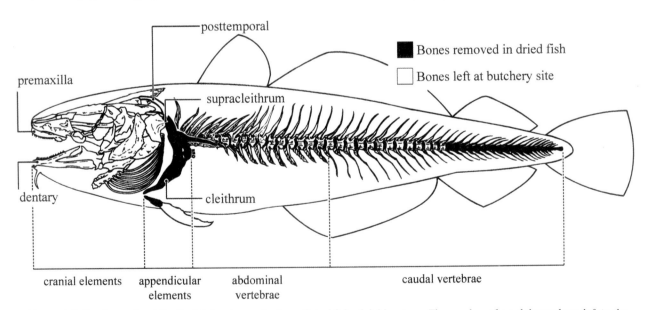

Figure 1.3. Butchery of cod family fish for the production of stockfish of råskjær *type. The number of caudal vertebrae left in the finished product varies (after Barrett* et al. *1999, 618).*

fishermen and conveniently staggered in the timing of their peak abundance from place to place. However, the local distributions of herring can change rapidly depending on ecological conditions. These fluctuations are more typical of some regions than others. For example, only isolated periods of large-scale herring abundance occurred along the Bohuslen (Bohuslän) coast of Sweden, whereas the Scanian fishery of the Middle Ages experienced only occasional poor years in a pattern of otherwise consistent availability (Chapter 2).

In this book the word herring refers to *Clupea harengus*. The related species sprat (*Sprattus sprattus*) was also targeted by some medieval fisheries (e.g. Bailey 1992, 16), and bones of the two are occasionally identified together as members of the family Clupeidae. Other clupeids, such as pilchard (also known as sardine, *Sardina pilchardus*) and shads (*Alosa* sp.), were regionally important at certain times (Kowaleski 2014, 47; Chapters 11–12), but do not play a major role in the research presented in this book.

Herring can be caught near the shore using pound or seine nets, especially in constricted waters, such as fjords (Starkey and Nielssen 2009, 39; Chapter 2). Gill nets set from boats in open water were also widely employed in medieval fisheries for this species, despite ambiguity regarding the date of their first introduction (cf. Chapter 2; Chapter 3). Barrelled herring that had been gutted and pickled in brine are one of the iconic trade goods of the Middle Ages, keeping up to two years (Jahnke 2009, 159). This cure was first produced in the Baltic Sea region, certainly by the twelfth–thirteenth centuries for catches off the island of Rügen (Chapter 2), but perhaps even earlier farther east based on Polish evidence

from Truso and Kołobrzeg-Budzistowo (Chapter 12). Preservation of gutted herring in barrels was only adopted around the North Sea in the late fourteenth and fifteenth centuries (Childs and Kowaleski 2000, 22; Rorke 2005; Unger 1978, 344), and its chronology in the Irish Sea is uncertain (Chapter 19). At other times and places, herring were preserved by various combinations of salting, drying and smoking. The resulting products could be the focus of surprisingly long-range trade. In the thirteenth and fourteenth centuries, for example, large quantities of smoked ('red') and salted ('white') herring from Yarmouth in eastern England were shipped to Gascony in exchange for wine (Kowaleski 2003, 186–7; Littler 1979, 203, 224–5). Chapter 3 describes a uniquely detailed account of herring curing on the manor of Wardley, near Newcastle, between AD 1289 and 1291. The herring were salted, hung on rods to dry and then packaged in baskets with straw. Thus the widespread use of herring for both trade and the payment of obligatory renders throughout much of the Middle Ages did not entail the barrelled and brined product that became ubiquitous in late medieval and early modern times.

The widespread distributions of cod and herring led to much local exploitation across medieval Europe. However, the abundance of these species was also concentrated during spawning in certain locations that were favourable for preservation. Thus the winter spawning migration of northeast Arctic cod brought it to the coasts of northern Norway, where conditions were ideal for stockfish production (Chapters 4–5). For herring, spawning migrations in the Baltic Sea coincided with local sources of salt (Chapter 2; Chapter 12). These serendipitous convergences provided opportunities on which fishermen, merchants and those who taxed them could draw as they endeavoured to survive, to prosper and to compete with one another. The result was long-range trade of specialised products.

A final issue meriting introduction is the degree to which the interpretations offered in the different chapters of this book converge or diverge. Differences of chronological interpretation exist among the authors – regarding the introduction of commercial fishing for cod and related gadids in the North Sea, for example (cf. Chapters 3, 14–17, and 21). There are also different points of view regarding whether or not medieval sea fishing expanded partly as a response to human impacts on freshwater ecosystems. These and other disagreements emerge from the use of different source material, differing allowances for the gap between a phenomenon and its first historical documentation and/or differing emphases on specific examples versus general trends. Chapter 21 represents one attempt to reconcile these alternative perspectives, but every interpretation must be seen as tentative in the light of our imperfect evidence base (see above).

The relationship between humans and marine resources was important in the past and remains so in the present (Rick and Erlandson 2008; Roberts 2007; Schwerdtner Máñez

Figure 1.4. Herring and its global distribution (Drawing: Vicki Herring after Kaschner et al. *2013 and the Freshwater and Marine Image Bank, University of Washington).*

et al. 2014). Herring represented one of Europe's largest commodity trades around AD 1400 (Chapter 2). Stockfish linked the far north with urban Europe (Chapters 4–8). Changes in the fortunes of the herring and cod trades had major economic, social and political ramifications (Chapter 21). Today, when *c.* 17% of the world's animal protein supply is provided by fish, and per capita fish consumption is growing (FAO 2014, 66), cod are already vulnerable to overfishing (IUCN 2015). Research on the history of fishing and the archaeology of fish bones is thus far from esoteric. Yet there is a need for better temporal and geographical coverage, improved chronological resolution and greater comparability between data – combined with continuing methodological development and a humanist's sense of the contingencies of history. These are achievable objectives. They must also be priorities if we are to understand the past in its own terms, without the back-projection of teleological thinking, and if we are to accurately comprehend how much the Anthropocene differs from conditions of the (not so distant) Middle Ages.

Acknowledgements

I thank the contributors to this volume for all their patient efforts towards a common goal, and especially my co-editor David Orton for keeping the project on track at a number of critical stages, especially during my stints as Acting Director of the McDonald Institute for Archaeological Research. The Leverhulme Trust has supported the research leading to this chapter and book and this article is also based upon work from the COST Action Oceans Past Platform, supported by COST (European Cooperation in Science and Technology).

References

Bailey, M. (ed.) 1992. *The Bailiffs' Minute Book of Dunwich, 1404–1430*. Woodbridge: Boydell Press/Suffolk Records Society.

Barrett, J. H. 1997. Fish trade in Norse Orkney and Caithness: a zooarchaeological approach. *Antiquity* 71: 616–38.

Barrett, J. H., R. P. Beukens & R. A. Nicholson. 2001. Diet and ethnicity during the Viking colonization of northern Scotland: evidence from fish bones and stable carbon isotopes. *Antiquity* 75: 145–54.

Barrett, J. H., R. A. Nicholson, & R. Cerón-Carrasco. 1999. Archaeo-ichthyological evidence for long-term socioeconomic trends in northern Scotland: 3500 BC to AD 1500. *Journal of Archaeological Science* 26: 353–88.

Barrett, J. H., C. Johnstone, J. Harland, W. Van Neer, A. Ervynck, D. Makowiecki, D. Heinrich, A. K. Hufthammer, I. B. Enghoff, C. Amundsen, J. S. Christiansen, A. K. G. Jones, A. Locker, S. Hamilton-Dyer, L. Jonsson, L. Lõugas, C. Roberts & M. Richards. 2008. Detecting the medieval cod trade: a new method and first results. *Journal of Archaeological Science* 35(4): 850–61.

Barrett, J. H., D. Orton, C. Johnstone, J. Harland, W. Van Neer, A. Ervynck, C. Roberts, A. Locker, C. Amundsen, I. B. Enghoff, S. Hamilton-Dyer, D. Heinrich, A. K. Hufthammer, A. K. G. Jones, L. Jonsson, D. Makowiecki, P. Pope, T. C. O'Connell, T. de Roo & M. Richards. 2011. Interpreting the expansion of sea fishing in medieval Europe using stable isotope analysis of archaeological cod bones. *Journal of Archaeological Science* 38: 1516–24.

Bennema, F. P. & A. D. Rijnsdorp. 2015. Fish abundance, fisheries, fish trade and consumption in sixteenth-century Netherlands as described by Adriaen Coenen. *Fisheries Research* 161: 384–99.

Brereton, G. E. & J. M. Ferrier (eds) 1981. *Le Menagier de Paris*. Oxford: Oxford University Press.

Burkhardt, M. 2013. Business as usual? A critical investigation of the Hanseatic pound toll lists, in J. Wubs-Mrozewicz & S. Jenks (eds) *The Hanse in Medieval and Early Modern Europe*: 215–38. Leiden: Brill.

Cerón-Carrasco, R. 2011. The ethnography of fishing in Scotland and its contribution to icthyoarchaeological analysis in this region, in U. Albarella & A. Trentacoste (eds) *Ethnozooarchaeology: the present and past of human–animal relationships*: 58–72. Oxford: Oxbow Books.

Childs, W. & M. Kowaleski. 2000. Fishing and fisheries in the Middle Ages, in D. J. Starkey, C. Reid & N. Ashcroft (eds) *England's Sea Fisheries: the commercial sea fisheries of England and Wales since 1300*: 19–28. London: Chatham.

Cutting, C. L. 1955. *Fish Saving*. London: Leonard Hill.

Dufour, E., C. Holmden, W. Van Neer, A. Zazzo, W. P. Patterson, P. Degryse & E. Keppens. 2007. Oxygen and strontium isotopes as provenance indicators of fish at archaeological sites: the case study of Sagalassos, SW Turkey. *Journal of Archaeological Science* 34: 1226–39.

Enghoff, I. B. 1996. A medieval herring industry in Denmark and the importance of herring in eastern Denmark. *Archaeofauna* 5: 43–7.

Enghoff, I. B. 2011. *Regionality and Biotope Exploitation in Danish Ertebølle and Adjoining Periods*. Copenhagen: Det Kongelige Danske Videnskabernes Selskab.

Eschmeyer, W. N., R. Fricke & R. van der Laan. 2015. *Catalog of Fishes Online Database*. San Francisco: California Academy of Sciences. Available at: www.calacademy.org/scientists/catalog-of-fishes-classification.

Ewonus, P. A., A. Cannon & D. Y. Yang. 2011. Addressing seasonal site use through ancient DNA species identification of Pacific salmon at Dionisio Point, Galiano Island, British Columbia. *Journal of Archaeological Science* 38: 2536–46.

FAO Food and Agriculture Organization of the United Nations. 2014. *The State of World Fisheries and Aquaculture 2014: opportunities and challenges*. Rome: Food and Agriculture Organization of the United Nations.

Froese, R. & D. Pauly (eds) 2015. *FishBase*. Available at: www.fishbase.org.

Harland, J. F. & J. H. Barrett. 2012. The maritime economy: fish bone, in J. H. Barrett (ed.) *Being an Islander: production and identity at Quoygrew, Orkney, ad 900–1600*: 115–38 (McDonald Institute Monograph). Cambridge: McDonald Institute for Archaeological Research.

Hedges, R. E. M. 2004. Isotopes and red herrings: comments on Milner *et al.* and Lidén *et al. Antiquity* 78: 34–7.

Hoffmann, R. C. 1996. Economic development and aquatic ecosystems in medieval Europe. *American Historical Review* 101: 631–69.

Hoffmann, R. C. 2002. Carp, cods, connections: new fisheries in the medieval European economy and environment, in M. J. Henninger-Voss (ed.) *Animals in Human Histories: the mirror of nature and culture*: 3–55. Rochester NY: University of Rochester Press.

Hutchinson, W. F., M. Culling, D. C. Orton, B. Hänfling, L. Lawson Handley, S. Hamilton-Dyer, T. C. O'Connell, M. P. Richards & J. H. Barrett. 2015. The globalization of naval provisioning: ancient DNA and stable isotope analyses of stored cod from the wreck of the *Mary Rose*, AD 1545. *Royal Society Open Science* 2: 150199.

IUCN International Union for Conservation of Nature. 2015. The IUCN red list of threatened species. Version 2015-3. www.iucnredlist.org.

Jahnke, C. 2009. The medieval herring fishery in the western Baltic, in L. Sicking & D. Abreu-Ferreira (eds) *Beyond the Catch: fisheries of the North Atlantic, the North Sea and the Baltic, 900–1850*: 157–86. Leiden: Brill.

Kaschner, K., J. Rius-Barile, K. Kesner-Reyes, C. Garilao, S. O. Kullander, T. Rees, & R. Froese. 2013. *AquaMaps: predicted range maps for aquatic species*. World wide web electronic publication, www.aquamaps.org, version 08/2013.

Kowaleski, M. 2000. The expansion of the south-western fisheries in late medieval England. *Economic History Review* New Series 53: 429–54.

Kowaleski, M. 2003. The commercialization of the sea fisheries in medieval England and Wales. *International Journal of Maritime History* 15(2): 177–231.

Kowaleski, M. 2014. Coastal communities in medieval Cornwall, in H. Doe, A. Kennerly & P. Payton (eds) *A Maritime History of Cornwall*: 43–59. Exeter: University of Exeter Press.

Littler, A. S. 1979. Fish in English Economy and Society Down to the Reformation. Unpublished PhD dissertation, University College Swansea.

Lõugas, L. 2001. Development of fishery during the 1st and 2nd millennia AD in the Baltic region. *Journal of Estonian Archaeology* 5: 128–47.

Nedkvitne, A. 2014. *The German Hansa and Bergen 1100–1600*. Köln: Böhlau.

Nehlich, O., J. H. Barrett & M. P. Richards. 2013. Spatial variability in sulphur isotope values of archaeological and modern cod (*Gadus morhua*). *Rapid Communications in Mass Spectrometry* 27: 2255–62.

Orton, D. C., D. Makowiecki, T. de Roo, C. Johnstone, J. Harland, L. Jonsson, D. Heinrich, I. B. Enghoff, L. Lõugas, W. Van Neer, A. Ervynck, A. K. Hufthammer, C. Amundsen, A. K. G. Jones, A. Locker, S. Hamilton-Dyer, P. Pope, B. R. MacKenzie, M. Richards, T. C. O'Connell & J. H. Barrett. 2011. Stable isotope evidence for late medieval (14th–15th C) origins of the eastern Baltic cod (*Gadus morhua*) fishery. *PLoS ONE* 6(11): e27568.

Perdikaris, S. 1999. From chiefly provisioning to commercial fishery: long-term economic change in Arctic Norway. *World Archaeology* 30: 388–402.

Richter, K. K., J. Wilson, A. K. G. Jones, M. Buckley, N. van Doorn, & M. J. Collins. 2011. Fish 'n chips: ZooMS peptide mass fingerprinting in a 96 well plate format to identify fish bone fragments. *Journal of Archaeological Science* 38: 1502–10.

Rick, T. C. & J. M. Erlandson (eds) 2008. *Human Impacts on Ancient Marine Ecosystems: a global perspective*, Berkeley: University of California Press.

Rigby, S. H. 2005. *The Overseas Trade of Boston in the Reign of Richard II*. Woodbridge: Boydell Press.

Roberts, C. 2007. *The Unnatural History of the Sea*. London: Gaia.

Rorke, M. 2005. The Scottish herring trade, 1470–1600. *The Scottish Historical Review* 84: 149–65.

Schwerdtner Máñez, K., P. Holm, L. Blight, M. Coll, A. MacDiarmid, H. Ojaveer, B. Poulsen & M. Tull. 2014. The future of the oceans past: towards a global marine historical research initiative. *PLoS One* 9: e101466.

Sicking, L. & D. Abreu-Ferreira (eds) *Beyond the Catch: fisheries of the North Atlantic, the North Sea and the Baltic, 900–1850*: 61–90. Leiden: Brill.

Starkey, D. J. & A. R. Nielssen. 2009. Environment and ecology, in D. J. Starkey, J. T. Thór & I. Heidbrink (eds) *A History of the North Atlantic Fisheries. Volume 1: from early times to the mid-nineteenth century*: 38–40. (Deutsche maritime Studien/ German Maritime Studies 6). Bremen: Hauschild.

Starkey, D. J., J. T. Thór & I. Heidbrink (eds) 2009. *A History of the North Atlantic Fisheries. Volume 1: from early times to the mid-nineteenth century*. (Deutsche maritime Studien/German Maritime Studies 6). Bremen: Hauschild.

Tys, D. 2015. Maritime environment and social identities in medieval coastal Flanders: the management of water and environment and its consequences for the local community and the landscape, in J. H. Barrett & S. J. Gibbon (eds) *Maritime Societies of the Viking and Medieval World*: 122–37. Leeds: Maney/Society for Medieval Archaeology.

Unger, R. W. 1978. The Netherlands herring fishery in the late Middle Ages: the false legend of Willem Beukels of Biervliet. *Viator* 9: 335–56.

Whitehead, P. J. P., M. L. Bauchot, J. C. Hureau, J. Nielsen & E. Tortonese (eds) 1986a. *Fishes of the Northeastern Atlantic and the Mediterranean. Volume 3*. Paris: United Nations Educational, Scientific and Cultural Organization.

Whitehead, P. J. P., M. L. Bauchot, J. C. Hureau, J. Nielsen & E. Tortonese (eds) 1986b. *Fishes of the Northeastern Atlantic and the Mediterranean. Volume 2*. Paris: United Nations Educational, Scientific and Cultural Organization.

Whitehead, P. J. P., M. L. Bauchot, J. C. Hureau, J. Nielsen & E. Tortonese (eds) 1989. *Fishes of the Northeastern Atlantic and the Mediterranean. Volume 1*. Paris: United Nations Educational, Scientific and Cultural Organization.

Wouters, W., L. Muylaert & W. Van Neer. 2007. The distinction of isolated bones from plaice (*Pleuronectes platessa*), flounder (*Platichthys flesus*) and dab (*Limanda limanda*): a description of the diagnostic characters. *Archaeofauna* 16: 33–95.

Wubs-Mrozewicz, J. 2009. Fish, stock and barrel. Changes in the stockfish trade in northern Europe, c. 1360–1560, in L. Sicking & D. Abreu-Ferreira (eds) *Beyond the Catch*: 187–208. (Northern World 41). Leiden: Brill.

Appendix 1.1 A list of common and scientific names of the taxa mentioned (after Eschmeyer et al. 2015; Froese and Pauly 2015; alternative common names have been included if used in the book).

Common name	Scientific name	Family
sea lamprey	Petromyzon marinus	Petromyzontidae
smooth-hound	Mustelus sp.	Triakidae
dogfish family		Squalidae
picked dogfish/spurdog	Squalus acanthias	Squalidae
ray family		Rajidae
(blue) skate	Dipturus batis	Rajidae
thornback ray	Raja clavata	Rajidae
common stingray	Dasyatis pastinaca	Dasyatidae
sturgeon	Acipenser sp.	Acipenseridae
(European) eel	Anguilla anguilla	Anguillidae
(European) conger	Conger conger	Congridae
herring family		Clupeidae
(Atlantic/Baltic) herring	Clupea harengus	Clupeidae
(European) sprat	Sprattus sprattus	Clupeidae
shads	Alosa sp.	Clupeidae
allis shad/twaite shad	Alosa alosa/ A. fallax	Clupeidae
allis shad	Alosa alosa	Clupeidae
twaite shad	Alosa fallax	Clupeidae
(European) pilchard	Sardina pilchardus	Clupeidae
(European) anchovy	Engraulis encrasicolus	Engraulidae
carp family/cyprinids		Cyprinidae
bleak	Alburnus alburnus	Cyprinidae
barbel	Barbus barbus	Cyprinidae
gudgeon	Gobio gobio	Cyprinidae
common dace	Leuciscus leuciscus	Cyprinidae
ide	Leuciscus idus	Cyprinidae
(freshwater/ common) bream	Abramis brama	Cyprinidae
roach	Rutilus rutilus	Cyprinidae
rudd	Scardinius erythrophthalmus	Cyprinidae
chub	Squalius cephalus	Cyprinidae
vimba bream	Vimba vimba	Cyprinidae
tench	Tinca tinca	Cyprinidae
wels catfish	Siluris glanus	Siluridae
pike	Esox lucius	Esocidae
(European) smelt	Osmerus eperlanus	Osmeridae
salmon & trout family/salmonids		Salmonidae
salmon/trout	Salmo salar/ S. trutta	Salmonidae
(Atlantic) salmon	Salmo salar	Salmonidae
(sea/brown) trout	Salmo trutta	Salmonidae
Arctic char	Salvelinus alpinus	Salmonidae
whitefish	Coregonu ssp.	Salmonidae

(Continued)

Appendix 1.1 A list of common and scientific names of the taxa mentioned (after Eschmeyer et al. *2015; Froese and Pauly 2015; alternative common names have been included if used in the book). (Continued)*

Common name	Scientific name	Family
grayling	*Thymallus thymallus*	Salmonidae
cod family/gadids		Gadidae (including Lotidae in this vol.)
(Atlantic/Baltic) cod	*Gadus morhua*	Gadidae
whiting	*Merlangius merlangus*	Gadidae
bib/pouting	*Trisopterus luscus*	Gadidae
pollack	*Pollachius pollachius*	Gadidae
saithe	*Pollachius virens*	Gadidae
haddock	*Melanogrammus aeglefinus*	Gadidae
burbot	*Lota lota*	Lotidae (previously Gadidae)
tusk/cusk/torsk	*Brosme brosme*	Lotidae (previously Gadidae)
ling	*Molva molva*	Lotidae (previously Gadidae)
(European) hake	*Merluccius merluccius*	Merlucciidae
garfish/garpike	*Belone belone*	Belonidae
sand smelt	*Atherina presbyter*	Atherinidae
three-spined stickleback	*Gasterosteus aculeatus*	Gasterosteidae
gurnard family		Triglidae
grey gurnard	*Eutrigla gurnardus*	Triglidae
fourhorn sculpin	*Myoxocephalus quadricornis*	Cottidae
shorthorn sculpin/bullrout	*Myoxocephalus scorpius*	Cottidae
(European) seabass	*Dicentrarchus labrax*	Moronidae
(Hapuku) wreckfish	*Polyprion oxygeneios*	Polyprionidae
pike-perch	*Sander lucioperca*	Percidae
(European) perch	*Perca fluviatilis*	Percidae
Atlantic horse mackerel/scad	*Trachurus trachurus*	Carangidae
sea bream family		Sparidae
blackspot/red sea bream	*Pagellus bogaraveo*	Sparidae
black sea bream	*Spondyliosoma cantharus*	Sparidae
mullet families		Mugilidae/Mullidae
grey mullet family		Mugilidae
wrasse family		Labridae
(viviparous) eelpout	*Zoarces viviparus*	Zoarcidae
sand-eel family		Ammodytidae
greater weever	*Trachinus draco*	Trachinidae
goby family/gobiids		Gobiidae
black goby	*Gobius niger*	Gobiidae
(Atlantic) mackerel	*Scomber scombrus*	Scombridae
tuna	*Katsuwonus/Sarda/Thunnus* sp.	Scombridae
flatfish order		Pleuronectiformes
turbot family		Scophthalmidae
turbot/brill	*Scophthalmus maximus/ S. rhombus*	Scophthalmidae
turbot	*Scophthalmus maximus*	Scophthalmidae

(Continued)

Appendix 1.1 A list of common and scientific names of the taxa mentioned (after Eschmeyer et al. 2015; Froese and Pauly 2015; alternative common names have been included if used in the book). (Continued)

Common name	Scientific name	Family
righteye flatfish/halibut family		Pleuronectidae
(European) plaice	*Pleuronectes platessa*	Pleuronectidae
(Atlantic) halibut	*Hippoglossus hippoglossus*	Pleuronectidae
(common) dab	*Limanda limanda*	Pleuronectidae
(European) flounder	*Platichthys flesus*	Pleuronectidae
(common) sole	*Solea solea*	Soleidae

Part I

Perspectives from History and Settlement Archaeology

Commercial Sea Fisheries in the Baltic Region
c. AD 1000–1600

Poul Holm

Introduction

The largest commercial fishery in medieval Europe developed in the western Baltic. The aim of this paper is to assess traded and landed amounts of the main target species, Atlantic herring (*Clupea harengus*), and to discuss the slow development of cod (*Gadus morhua*) and other fisheries in the eastern and northern parts of the Baltic.

I shall primarily consider the supply side, namely, the catch and procurement of fish, for which we have a growing body of evidence. The demand side of consumer preferences, in both local and international markets, remains under-researched, but in recent years some information has become available – primarily from the analysis of archaeological fish bones – that points to potential breakthroughs in the future. The commercial, cultural and political significance of the Baltic fisheries is currently poorly understood, and this paper will identify some possible future lines of investigation.

Methodology

A wide range of parameters determines the development of fisheries: natural causes, including both abiotic (e.g. climate, currents, salinity) and biotic (e.g. species interactions and food webs) conditions, as well as human variables. Ecologists have proposed important hypotheses for our understanding of the variability of Baltic fish populations from the evidence of climate, currents and salinity (MacKenzie *et al.* 2002); so far, however, ecological studies have concerned themselves mainly with the past hundred years, while the investigation of the past environment of the Baltic is still at an exploratory stage (Eero *et al.* 2011; Grupe *et al.* 2009). There is a need to understand the interaction of the Baltic aquatic environment with major

climatic change, such as the medieval warming period and the early modern cooling period. Sea-bottom sampling might provide direct evidence of past fish distribution and abundance, but such evidence is not yet available.

Currently, we have two main sources of knowledge for medieval fisheries: fish-bone remains, from both terrestrial and marine sediments, and documentary records. The archaeology of fish-bone remains has developed rapidly in recent decades, and we are likely to see important if not dramatic improvements in our knowledge of the historic Baltic fisheries in the next few years. In 1999, Enghoff presented an overview of the literature on Baltic archaeo-ichthyology. Since then, such methods as stable isotope analysis and genetic studies have introduced additional ways of interrogating the known record, but only a few results are available so far (e.g. Barrett *et al.* 2008; Orton *et al.* 2011; see Chapters 11–12).

Documentary records are preserved in archives in all of the Baltic countries. A recent survey by MacKenzie *et al.* (2007) identified a great many hitherto untapped or little-known records, and it is likely that full studies of these records will vastly improve the state of our knowledge. Documentary evidence provides direct information about extractions and, potentially, indirect information about past variability of marine resources. In this survey I shall critically examine the literature, with a view to evaluating the extractions effected by, the manpower involved in, and the economic significance of the Baltic commercial fisheries.

The natural environment

The Baltic emerged after the last Ice Age as an inland lake and has been open only to westerly inflows of saline water in the past 10,000 years. The sea is a brackish environment conditioned by the confluence of oceanic water through the bottleneck of the Danish Kattegat in the west (where haline conditions may be as high as 25‰ salinity) and sweet water from rivers running off into the Baltic. Surface salinity is as low as 6‰ between Stockholm and Tallinn; farther east and north, in the Gulf of Bothnia and the Bay of Finland, saline concentrations drop even lower. Strong westerly winds may bring heavy Atlantic salt water into the basin, while conditions of prevailing easterlies will force a reduction in haline conditions. Modern measurements show considerable changes year on year and even decade on decade (MacKenzie *et al.* 2002).

The stressful brackish-water environment in the Baltic provides for low marine biodiversity. The two major commercial species in the Middle Ages were Atlantic herring and cod, and these are the species on which I shall concentrate. The small and lean subspecies Baltic herring (*Clupea harengus membras*), also known as *strömming*, was considered less suited for trade, while such species as sturgeon (*Acipenser* sp.), Atlantic salmon (*Salmo salar*), brown trout (*Salmo trutta*), whitefish (*Coregonus* sp.), vimba bream (*Vimba vimba*), smelt (*Osmerus eperlanus*), eel (*Anguilla anguilla*), lamprey (*Petromyzon marinus*) and indeed seals (Phocidae) were only caught in relatively small quantities.

Fish in medieval diet

Medieval diets and food cultures were probably the most important driver of the Baltic fisheries, but they remain woefully understudied. It is often assumed that Christianisation and especially the practice of abstaining from meat during Lent would have promoted the eating of fish, but hard and fast evidence is limited, and it is not safe to infer general patterns. In addition to religious imperatives, such questions as taste, cooking preparation, elite preferences and gender roles may have influenced diets to degrees that are still little known (Woolgar 2010).

Diet in medieval Denmark seems to have changed little over time. A study of stable isotopes in finds of human bone detected little evidence of variation in the proportions of marine and terrestrial foods consumed at three coastal and inland sites dating from the eleventh to the sixteenth centuries. Only one site, the coastal city of Ribe, provided evidence of a reduction in the consumption of fish, in the years spanning the height of the Black Death. This reduction is possibly related to a belief that the plague was caused by 'bad air' from the fish. However, at the other two sites, no such change in diet was evident (Yoder 2010). A study of the Swedish city of Sigtuna shows that variation in diet over time was linked more to social stratification than to changing practices in general and that, although some marine fish bones have been found, for this inland city

the nearby lake remained the main source of fish through the period (Kjellström *et al.* 2009). In Poland, preliminary evidence suggests that the medieval diet was very low in marine foods (Reitsemaa *et al.* 2010).

Coastal and aristocratic peoples had a higher consumption of marine foods. Human bones from the coastal settlement of Ridanäs on the island of Gotland show a persistent high consumption of marine protein through the Viking and early medieval periods, from the ninth–twelfth centuries. Christianisation had no observable impact on local diet because fish was already a preferred foodstuff (Kosiba *et al.* 2007). A study of the skeletal remains of the Viking-period cemetery at Birka, Sweden, documents a marine emphasis in the diet among men buried with weapons. This food may have been consumed in connection with overseas travels (Linderholm *et al.* 2008).

In sum, the evidence for marine food in the medieval diet of Baltic people is as yet slim and does not indicate that major changes took place that would have driven the rise of the commercial fisheries. In some regions, marine fish was a preferred food even in the pre-Christian era, but in most places the food sources were overwhelmingly terrestrial. Clear differences in consumption were, however, in place within different levels on the social ladder. The economic and social elites around the Baltic demanded a variety of food sources, including marine, and would have supported a small but growing market for commercial fisheries.

The other major driver for the Baltic fisheries is likely to have been long-distance trade into Continental Europe. There is strong evidence that the rise of the Hanseatic League increased availability of marine fish throughout Continental Europe (Jahnke 2000) and indeed gave rise to an international fish market, which by the sixteenth century had developed links across the Atlantic, with as yet poorly understood consequences for the Baltic fisheries. In conclusion, the evidence suggests a steady home market with an increasing demand by the elites for traded fish products, as well as a rise in the international demand for Baltic marine products.

The western Baltic herring fisheries

The primary species targeted by the fishers of the western Baltic was Atlantic herring, caught as they were spawning in the Danish Sound and in the western Baltic. The small and less fatty Baltic herring spawns in the central Baltic, and while it is locally important for consumption, it does not seem to have been the basis for anything other than regional trade in the medieval period. Commercial herring fisheries in the medieval period were therefore overwhelmingly concentrated in the western Baltic.

In the sixth and seventh centuries AD and again in the eleventh century, abundant catches of herring are documented by archaeological finds from the island of

Bornholm. The catches seem to have been due to advances in fishing technology, namely, the introduction of nets, as indicated by finds of net floats. The herring bones are accompanied by bones of small cod, which may represent a secondary catch. The finds are the result of extensive sieving at a number of sites on the island, and they may therefore be considered representative as indications of natural fluctuations of local herring populations, since herring is known to be prone to huge natural variability. Tenth–thirteenth-century finds at northern German and Polish sites also seem to attest to extensive herring fishing with nets (Enghoff 1999, 78–9).

An early commercial fishery to feed an urban settlement is indicated by abundant ninth–twelfth-century finds of herring bone at the settlements of Haithabu and Schleswig, probably originating from weir fishing in the Schlei Fjord (Jahnke 1998). Other western Baltic coasts saw a similar rise in the eleventh and twelfth centuries of early commercial herring fisheries in sheltered coastal and estuarine environments. In the Roskilde Fjord, archaeological study of thirteenth–fourteenth-century fish bones detected the use of the method of removing the gills by twisting a small knife behind the head of the herring (see Chapter 13, which updates Enghoff 1996). The fish would then have been immediately salted, while the blood was still running and thus would have enabled the salt to penetrate into the soft tissues. Late medieval legend claimed that the method was a Dutch invention (Unger 1978), but in fact it seems to have been introduced in the Baltic centuries earlier. Viking Age herring finds from the same site by Roskilde Fjord do not show use of this method (Chapter 13).

Whilst these were commercial fisheries for a local market, perhaps the earliest-known fishery to serve a wider market developed in the twelfth century, off the island of Rügen. The fishery benefitted from proximity to salt mines in Lüneburg and the rapidly developing trade network of the Hanseatic city of Lübeck. The merchants developed institutional controls of the landing site, such as the right to maintain their own law within delimited areas known as *Vitte*, and strict quality controls on the curing and barrelling of herring. Exports reached deep into continental Europe by the thirteenth century. Nevertheless, by AD 1290, the fishery was clearly subordinated by the much larger fishery that had developed in the Danish Sound, particularly from the coast of Scania (Jahnke 2000, 15–38).

The Scanian herring fishery

Saxo Grammaticus (writing *c.* AD 1200) maintained that herring were so densely packed in the Sound that it was hard to row a boat, making it possible to catch the fish with one's bare hands. He observed that there were numerous fishing booths by the southwestern Scanian beach (Friis-Jensen and Zeeberg 2005). Arnold of Lübeck, writing around the same time, ascribed Denmark's wealth to the herring (Lappenberg

1868), and archaeological evidence supports the early development of the fishery (Ersgård 1988).

The market developed originally in two neighbouring places in southwestern Scania, Skanør and Falsterbo, which were protected by the Danish king. By the thirteenth century, Lübeck merchants obtained extensive privileges to bring in their own sheriffs and keep *Vitte*, and similar rights were later obtained by a string of German and Dutch towns (Friedland 1991, 68). Archaeological finds show that the *Vitte* were permanently settled by the early fifteenth century (Christensen 1994). By that time, such cities as Malmö, Helsingør, Copenhagen and Rønne (the latter located on Bornholm) had developed their own fish markets. The Scanian markets continued to be of overwhelming importance for the fish export trade and had developed into an international fair, serving as an exchange link for Baltic and west European commodities. The power of the Hansa rested on their control of the essential salt supplies from the salt mines at Lüneburg and Oldesloe (Friedland 1991), although Danish fishermen and merchants did have access to so-called green salt from Friesland, made by burning intertidal peat, which was cheap and of poorer quality, and to sea salt from the island of Læsø, made by boiling highly saline sea water (Hansen 2010). By the latter half of the fourteenth century, large ships were able to undertake the full journey into the Baltic markets, and the general fair diminished in importance. However, the fish market continued to be of international importance until the early seventeenth century.

The fishing technology used was nets, either allowed to drift attached to a buoy through the night or fixed to the sea bottom by means of poles. Drift nets might be several hundred metres in length. The boats needed one or two experienced men assisted by four or five additional men at the oars, who would also help draw the nets (Ventegodt 1990, 9). A small resident population fished throughout the year, but the season for the great herring shoals was more or less consistent with the official Scanian market, from 15 August to 9 October.

The herrings were cured in salt brine in wooden barrels, where they could keep for as long as two years, and were therefore eminently suitable for long-distance trade (Jahnke 2000, 222–3). The content weight of the barrel would have been regulated, probably according to a Rostock standard of 117 kg. Typically, one barrel of salt was used to preserve three barrels of herring. The herring would have been gutted before salting, using the gilling method, and we may therefore assume that one barrel represented about 100 kg of wet fish (Holm and Bager 2001). The size of the barrels remained constant, but the size of the herring fluctuated, so that in the fourteenth century a barrel contained around 900 herrings, while in 1524 the count was as many as 1200 (Knudsen 1836, 313).

The barrels would be transported all over northern and central Europe, in particular to the large German market

(Jahnke 2000). The market preference was clearly for herring sourced in the Sound, while herring landed in the western Baltic area by the island of Bornholm fetched considerably lower prices. Thus, for instance, the convent of Marienburg (Malbork) supplemented its provisions of Scanian herring with Bornholm herring at two-thirds of the price (DD 1402, 368; 1403, 117). The reason for this preference may have been that the Scanian market had strict quality controls in place, which encouraged consumer confidence in the product. Branded barrels could be sourced back to the provider in case the contents had gone off, and extensive legislation had been enacted to regulate the market. Similar branding and control was likely not in place in other Baltic markets, and prices were therefore typically only half those of the quality product.

The fishery attracted thousands of people. Around AD 1400, a total of 17,000 men attended the fishery, with an estimated additional 8000 engaged in related trades (Ventegodt 1990). In AD 1524, according to a witnessed declaration by the German superintendent of the Scanian market, no fewer than 38,515 fishermen paid dues of 240 herring each (Knudsen 1836, 313). The fishers came not only from the Sound region of Scania and Zealand, but also from all the towns 'inside the Scaw', meaning the Danish Baltic waters (as distinct from the North Sea coast of Jutland, which was referred to as 'past the Scaw'). There is evidence that German, Dutch and English fishermen participated as well, certainly in the fourteenth century (Hørby 1966, 245–72; 1971, 68–77; Tuck 1972, 75–88; Weibull 1966, 75). Danish noblemen invested heavily in the fishery, in terms of both capital and labour from their estates. In the first half of the sixteenth century, the most active towns, such as Stege, were almost depopulated during the herring season according to contemporary reports (Arup 1932, 417; Stoklund 1959, 101–22).

Total landings fluctuated, but we may obtain fairly good estimates of extremes thanks to a few well-recorded years (note that the following revises estimates in Holm 1996). The 'pound toll' records of the city of Lübeck document large trade volumes in AD 1368, when it imported around 76,000 barrels from the Sound, and again in the years AD 1398–1400, when the town imported almost 70,000 barrels each year (Lechner 1935; Weibull 1966). If Lübeck's share of the pound toll is an indication of its share of the total herring trade, then total imports of herring to the Baltic towns from the Scanian market were three times the amount of the Lübeck imports, or around 225,000 barrels. To this amount should be added the unknown exports to the North Sea Hansa towns, probably no fewer than 50,000 barrels. Again we have no complete information, but the Flemish port of Sluis handled an annual average of 20,000 barrels from AD 1374 to 1380. Herring was an equally important staple at Damme (Degryse 1973, 202–6). By the late fourteenth century, total exports, therefore, may have

been as many as 300,000 barrels, or around 30,000 metric tonnes, according to the Rostock standard.

To estimate total catches, as opposed to total exports, we must add home consumption to traded volume. According to the 'Hansa mote book', which regulated the market, each fisherman was allowed to salt six barrels for his own consumption (Holm and Bager 2001, 108–9). Because we know how many participated in the fishery, we can add 100,000 barrels to the total in AD 1400 if all fishermen brought home their allowed quota. In AD 1368, a landed total of 400,000 barrels would indicate a production of around 40,000 tonnes live weight.

In the early sixteenth century, landings may have been even higher. Unfortunately, export figures are not known for AD 1524, but in that year, as stated above, more than double as many men were active in the fishery than in the fourteenth century. If each man brought home six barrels for own consumption, this alone would have amounted to 230,000 barrels. Total landings are therefore likely to have been in excess of 50,000 metric tonnes.

Troughs followed peaks in the herring fishery, and contemporary chroniclers reported the herring as being virtually absent in AD 1402, 1425, 1469 and 1474–5. In AD 1494 and again in AD 1537, Danish exports from the Sound and the other major fishery, in the Limfjord, were only about 10,000 tonnes (96,000 barrels) (Holm and Bager 2001). Such variability is typical of pelagic fisheries and will have created periods of affluence followed by abject poverty. The likely driver of herring abundance is the availability of prey, as determined by changes in weather and currents for which we have, as yet, no scientific information.

We have no precise information on the size of the Sound fishery proper after AD 1540. The meeting of the Danish estates in AD 1558 still identified the fishery as one of the most important economic activities and confirmed the right of noblemen to engage in the herring trade (CCD no. 32). Many Danish towns now sought and obtained privileges to participate in the herring market (Jahnke 2000, 115). However, indications are that the fishery declined at the end of the century. We know that one of the fishing hamlets in North Zealand, Gilleleje, experienced a severe decline in tax revenues from fishing boats (Frandsen and Jarrum 1992, 105–39). The villagers who were professional fishermen paid an average of 340 shillings in the decade AD 1585–94. In AD 1610–19, they paid an average of only 35 shillings, but the amount rose again to 218 shillings in AD 1620–26, when contemporary anecdotal evidence indicates a temporary rise in the Sound fishery. After that, revenues averaged no more than 20–60 shillings throughout the seventeenth century. The cities of Stege, Odense, and Copenhagen reported that the herring trade had severely diminished and asked for the king's permission to engage in trade for cod in Bergen. Traders from the Hansa towns continued to travel to the herring market in Falsterbo, but the Hansa towns had lost their grip on the market. In AD 1605, King Christian IV even prohibited the Germans from bringing their own executioner to the market, which is an indication that they had lost their privileged status (Holm 1998).

How do the medieval production levels compare with modern production levels? In the beginning of the twentieth century, Danish landings of herring from the Sound fluctuated between a low of 100 tonnes and a high of 10,000 tonnes (Holm and Bager 2001, 108–10). Swedish catches would probably have had similar dimensions. Thus, the landings in the best years of the medieval fishery may have been two to three times larger than those in the best years around AD 1900. The drift-net technology probably remained more or less the same throughout the centuries, but fishing pressure, as represented by number of men and boats, was much higher in the Middle Ages than in later years, probably because the price of herring was much higher in relative terms than it was at the beginning of the twentieth century. Hence, while herring may have been more abundant in the Middle Ages, they need not have been vastly more abundant than in the early twentieth century. The difference in landings seems, instead, to reflect declining fishing effort. The causes of the decline of the medieval fishery remain obscure. We may, however, rule out the assumption by some historians that the demise of the Scanian fishery was caused by a changed migration pattern of the herring. Modern ecological knowledge would not give credence to a sudden change of spawning habitat unless one could point to a severe environmental stress factor.

The Limfjord herring fisheries

The fisheries in the Limfjord in northern Denmark became increasingly important during the fifteenth century. The wealth of the monasteries situated around this large estuary was based on a combination of rich pasture and good fishing. Likewise the town of Ribe on the North Sea coast prospered from trade in livestock and fish. For the supply of fish to the Danish population, the Jutland fisheries were perhaps of as much importance as the Sound fisheries, because the latter were controlled by Hanseatic merchants for export. In contrast, the Jutland fisheries were largely controlled by Danish merchants, and if the quality was lower, so was the price, making the Jutland fish affordable to the domestic consumer (Kjærsgaard 1978, 61). Viking Age deposits of herring bones show no use of the gilling method necessary for long-term conservation (Enghoff 1996; Chapter 13), but by the late medieval period the method had clearly been introduced, because barrels of salted Limfjord herring were being exported to the Baltic.

We know that exports of salted herring from the main commercial town in the Limfjord, Aalborg, ranged between 5000 and 50,000 barrels per year in the eighteenth century (Poulsen *et al.* 2007). We have no information on sixteenth-century production, but we know that the pound-net fishing technology was the same, and that the royal prerogative on buying herring on certain weekdays rendered the same amounts in the sixteenth century as it did in the seventeenth century (Holm 1999). Total output may therefore well have been at a similar or even higher level. The Aalborg merchants succeeded in securing the herring trade for themselves during the sixteenth century. Membership lists of the parrot guild in Aalborg show a significant German contingent by AD 1500, but by the middle of the century the local merchants were almost totally dominant. Royal privileges further enhanced the role of local interests at the expense of foreign merchants, who were only allowed to trade on certain days and might not own fishing gear themselves. If, by the late sixteenth century, the Sound fisheries had declined, the fjord fishery compensated for part of the loss as far as domestic consumption was concerned (Rasmussen 1968, 33–45).

The Bohuslen fisheries

The demise of the Scanian fishery opened the Baltic markets to imports from the Netherlands. Large catches were made in the North Sea from the late fifteenth century onwards thanks to technological innovation using sea-going vessels. When, starting in AD 1562, the Sound toll registers allow a quantification of the goods being sent into the Baltic, they show clearances of 30,000 barrels of herring per year from Dutch ports in the 1560s. Evidently the Dutch had by then made incursions into a market that had earlier been filled by exports from the Sound. Soon, however, the Dutch domination of the trade in herring to the Baltic was challenged. Their fisheries in the North Sea faltered in the AD 1570s during the War of Independence, and Dutch herring exports into the Baltic fell to an average of 5700 barrels per year (Holm 1996).

At this juncture Norwegian exports took over the Baltic market. The Norwegian fishery was carried out off the coast of Bohuslen, (Bohuslän) the area just north of present-day Gothenburg, which was only later to become Swedish territory. In geographical terms the Skagerrak and Kattegat waters are a transitional area between the North Sea and the Baltic, but in economic terms the region was by then thoroughly linked to the Baltic region. Like the Sound herring, the Bohuslen herring were of the Atlantic species, that is, invaders from the North Sea, as has been proven by investigations of fish bones (Höglund 1972). This was not the first time that herring were caught in huge quantities off Bohuslen. Similar herring adventures took place certainly in the early twelfth century, in the first half of the thirteenth century and in the first half of the fourteenth century. The Bohuslen herring periods were likely controlled by periods of negative North Atlantic Oscillation with reduced westerly winds, and of cold water in the North Sea and the Skagerrak (Alheit and Hagen 1997; Corten 2001, 204–5).

The herring shoals started to come close to shore in AD 1556 and were unusually large, so that the merchants counted only 640 herring to the barrel (Storm 1881, 94, 97). For the next two decades, Bohuslen was the main provider of herring for the Baltic market. In the 1560s, Bohuslen clearances were less than a third of the Dutch clearances (8400 barrels per year), but in the 1570s, Bohuslen exported 34,800 barrels. Dutch traders tried to compensate for the loss of the North Sea fisheries by calling at Bohuslen to fill their empty holds. In the 1580s, Bohuslen exported 39,180 barrels on yearly average, while the Dutch were slowly improving their North Sea fisheries, which supplied 13,080 barrels cleared from Dutch ports. Bohuslen exports peaked in AD 1585, at 75,600 barrels (Holm 1996).

The limiting factor would not have been the ability to extract the fish but, rather, to cure and market it. During the eighteenth-century herring boom, huge amounts of fish were wasted or used as train (fish) oil because curing and barrelling demanded heavy capital investments (Holm 1991). In the sixteenth century and before, lack of capital may have restricted the fishery severely. Considering the export amounts mentioned above and allowing for quantities for local consumption and reducing for train oil, the total Bohuslen production in the sixteenth century is likely to have been no more than in the region of 15,000–25,000 tonnes, or perhaps half the Scanian production figures in good years.

The actual fishery took place from a number of more or less permanent settlements in the Bohuslen archipelago, which attracted thousands of people from inland. The Danish merchants also brought their own fishermen. A tax list of AD 1589 shows that merchants from such towns as Fredrikstad, Aalborg, Flensborg, Sønderborg and Copenhagen owned booths in almost every major fishing settlement. Other booths were owned by local merchants from Marstrand, and also by merchants from Oslo and Tønsberg to the north; Lödöse and Varberg to the east; and Helsingør, Kalundborg, Odense, Kerteminde and Skælskør to the south (Pettersson 1953, 115–16). Even from the North Sea city of Ribe, fishermen went all the way around the Scaw to participate in the Norwegian fisheries. The Crown provided naval vessels to protect fishermen and merchants travelling from Denmark to Norway during the Nordic Seven Years' War (AD 1563–70), and so demonstrated the importance attached to the new fisheries. While dried cod and haddock from the west coast of Jutland were reserved for the navy as cheap and durable food, the herring fisheries were granted a general export licence during the war to provide the kingdom with much-needed foreign currency (Holm 1998).

The Bohuslen fishery was controlled by the Danish king, but regulations and trade patterns resembled those of the Scanian trade (Jahnke 2000, 281). Customs stations were established to obtain taxes from the fisheries, and in AD 1561, the Marstrand law-books were published, containing rules and regulations for the Bohuslen fisheries. The city of Marstrand, because of the wealth it had acquired through the herring fisheries, had to pay the same amount of war taxes as the large city of Bergen, very much to the dismay of its population. Foreigners were allowed to participate in the fish trade. Ships arrived from Denmark, England, France, Germany, Holland and Scotland at Bohuslen to transport salted herring to other European countries. Thousands – probably tens of thousands – of people from neighbouring countries settled with their families on the Bohuslen coast (Holm 1991; Storm 1881, 96).

Although of considerable size, the fishery of the Bohuslen region never gave rise to an international fair like that in Scania. It was always just a single-commodity market. The herring adventures occurred only in a centennial pattern, and they lasted only two or three decades – with the exception of the immensely rich eighteenth-century adventure, which lasted almost half a century. Capital was ploughed into the city of Marstrand, but not into the rest of Bohuslen region. The fishery benefited the many players who remained concentrated in the rich towns and cities of the western Baltic. In contrast, migrant workers swiftly left when a herring adventure came to an end, leaving the Bohuslen region essentially rural and dependent on its timber exports (Holm 1991).

The Baltic cod fishery

Successful cod recruitment depends on the existence of the right saline conditions for the hatching of eggs. Studies have documented huge variability in Baltic spawning stock biomass, with long-term knock-on effects for stock abundance. Spawner biomass was low during most of the twentieth century, until large saline intrusions following powerful storms created favourable conditions for hatching and spawner biomass increased ten-fold (Eero *et al.* 2011). It is likely that similar variations occurred in the past.

Cod was caught for local consumption using hook and line in the Iron Age and earlier periods. It was also line-caught to serve early-medieval markets, such as the Danish cities of Århus and Lund (Enghoff 1999, 79). However, archaeological evidence indicates that large-scale commercial cod fisheries in the Baltic did not develop until the late medieval period. Stable isotope signatures in fish bones from eight thirteenth- and fourteenth-century archaeological sites in Poland and Estonia, plus Uppsala and Lübeck, show strong North Atlantic signals, implying that cod was imported from an outside market, probably Norway. Late fourteenth–sixteenth-century cod bones, on the other hand, show clear Baltic signatures, indicating that local cod fishing had developed to supply markets in present-day Poland and the Baltic states (Orton *et al.* 2011). There is also historical evidence for the development of a dried cod and dried pike industry in the eastern Baltic in the course of the fourteenth century (e.g. Hoffmann 2009). Moreover, medieval Danish written records show a substantial inland trade from North Sea suppliers of haddock and cod to the perambulating royal household, which primarily was located in the eastern Baltic parts of the realm (Holm 1998).

Documentary evidence shows strong local cod fisheries in the western Baltic by the mid-sixteenth century, which seemingly taper off towards the eastern and northern stretches of the sea. Most of the fishery was conducted from the island of Bornholm and from some of the Scanian counties (Holm and Bager 2001). The full extent of the fishery is not known, but it was sufficient for the Danish king to tax the cod fishing effort in the form of a specified amount of dried (pre-1610) or barrelled (post-1610) salted cod. The amount of dried and barrelled cod supplied to the Danish king as a tax has been estimated for the period AD 1597–1660 (Holm and Bager 2001). This was a commercially important fishery of regional significance that paid handsome revenues to the king. Not all fish was consumed locally. Exports from one small town, Rönneby in Scania, to towns in northern Germany was estimated around 5–10 tonnes per year. The tax revenue evidence indicates that the cod fishery declined after AD 1613.

Swedish and Finnish data provide evidence that even the farthest stretches of the Baltic saw significant cod landings where today there would be hardly any. Cod fishing was at a high point during the AD 1550s. Stockholm received 5–15 tonnes per year from local fishing settlements. In southwestern Finland a cod fishery took place from at least the mid-1550s to the 1630s. Annual exports to Stockholm from the Finnish archipelago between AD 1561 and 1578 were an average of 10 tonnes per year, but they evidently decreased later. MacKenzie *et al.* (2007) observe that these landing levels were similar to the officially reported Swedish landings in the same area during AD 1998–2005, while Finnish landings have not exceeded 3 tonnes since AD 1996. As local consumption was not included for the sixteenth-century data and technology certainly was less developed relative to today, this is evidence of a considerable fishing effort at the time. Estonian, Latvian and Russian archives have revealed very limited cod landings, although information for other species, such as Baltic herring and salmon, has been found (Gaumiga *et al.* 2007; Lajus *et al.* 2007a; 2007b). Archival research in Germany or Poland may, however, reveal more information (cf. Hoffmann 2009).

Smaller commercial fisheries in the Baltic

Overall, fishing pressure was quite low in the inner parts of the Baltic. An Estonian study has revealed the wide extent of local fisheries, which were of some commercial importance. During the Little Ice Age of the late seventeenth century, cold-water marine fish (herring, flounder [*Platichthys flesus*]

and eelpout [*Zoarces viviparus*]) were of importance, and the fishing season for the major pelagic fish was substantially later in the year than it is during the much warmer conditions of today (Gaumiga *et al.* 2007). During the late seventeenth century, landings from the Gulf of Riga were at least 200 times less than at the end of the twentieth century. Most seventeenth century fisheries concentrated on the rivers.

In the eastern Baltic Sea area, migratory and sweet-water species were the predominant sources of fish. The change from riverine to marine fisheries, which took place in western Europe as early as the eleventh century (Barrett *et al.* 2004; Hoffmann 1996), seems not to have occurred until the nineteenth century in the Gulf of Finland region. Seventeenth-century evidence from the fish market of Narva shows that freshwater species dominated, while the major marine product, Baltic herring, constituted only about 20–30 metric tonnes, or about 10% of total landings. The herring was almost exclusively consumed locally, while the main export species was European eel for Stockholm (Verliin *et al.* 2013, 5). During the medieval period, Russian river fisheries profited from the availability of migratory fish species, such as sturgeon, Atlantic salmon, brown trout, whitefish, vimba bream, smelt, eel and lamprey. In the Russian fisheries, Baltic herring, and thus marine operations, did not increase in importance until the nineteenth century (Lajus *et al.* 2013).

Discussion

A full understanding of this evidence necessitates further research into the biases of the historical information, such as market demand and production supply, settlement patterns, cultural preferences and political conflicts and aims.

Herring and cod have been fished extensively in the Baltic since the Mesolithic, but commercial fisheries for these species developed only in the Middle Ages. The herring fishery gained international significance as a highly regulated, branded export industry for consumers across Europe. It was a fishery of enormous economic and political importance to Scandinavian and Hanseatic history. Historians agree that the European herring trade was one of the major trades of the Middle Ages, but just how important was it? Richard W. Unger (1978, 337 n. 3, 354) estimated that the wine trade from Bordeaux to England was about 11,000 tonnes annually between AD 1400 and 1440, while the Dutch North Sea herring trade was eight to ten times less. He concluded that herring was not used for mass consumption. However, he did not consider the herring trade from Scania, which was much more developed than the North Sea trade at the time. Around AD 1400, the estimated annual export trade, around 35,000 tonnes, was almost three times the volume of the Bordeaux wine trade. In terms of importance, the herring trade was likely only surpassed by the trade in grain and the trade in textiles.

We may use the figures for traded and locally consumed herring to estimate per capita consumption. The Danish home market is likely to have taken most of the herring that the fishermen were allowed to keep for themselves (that is, six barrels each). By 1368, the Danish population probably numbered no more than half a million, who might consume 20 kg per capita, or half a herring per day per person. By the middle of the eighteenth century, the yearly ration of herring for a Zealand agricultural worker was reckoned as one quarter of a barrel (Heckscher 1949, xxv), or around 200–300 herring, depending on the count of herring to the barrel. Such a level of consumption is likely to have been established as a norm in the medieval period. In the Baltic German towns and Dutch and Frisian areas, consumption levels may also have been very high, as they did send their own fishing vessels. But the traded volume as it spread around Europe made for very low per capita provisions. Total traded production was around 30,000 tonnes in 1368, and in the wake of the Black Death the European population may have numbered no more than around 40 million. The trade really only reached northern and central Europe, or perhaps a third of the European population. Average per capita consumption of Sound herring in northern and central Europe would then have been not much more than 2 kg per annum, or 20 herrings in a year.

Herring from the Sound was a well-defined quality product that sold at premium prices relative to competing local products. In that sense it may not have been for mass consumption, but it was certainly highly popular among broadly defined rural and urban elites in northern and central Europe, including monasteries and other public institutions, which bought herring as a staple product for a varied diet through the year and particularly for Lent (Jahnke 2000, 270). In terms of population health, there seems little doubt that the availability of herring would have contributed to social and regional disparities.

The decline of the Scanian fishery by the end of the sixteenth century presents questions of relevance to the general historical interpretation of the period. The demise of the largest commercial fishery of medieval Europe must have had political ramifications. We may ask if the decline of the Sound fishery was the unintended result of the fact that the Danish crown gained control of the herring market in the second half of the sixteenth century. The decline of Hanseatic control of the fish market after the defeat of Lübeck in the Danish Civil War of 1535 also meant the dissolution of branding controls and marketing. We may equally ask if the decline was the result of Dutch competition as the North Sea fisheries built up during the sixteenth century. In the seventeenth century, the preferred herring even among the Danish nobility was no longer the Baltic herring but imported Dutch North Sea fare, which commanded a mark-up on the price amounting to one-fourth on the Danish commodity (Holm 1998). These are questions

that require research over and beyond what is possible within the scope of this paper.

We may identify as a key driver for the rise and decline of the Sound fishery the ability to provide well-branded, quality-controlled products to the consumer in an emerging European market. As a perishable good, fish could only be turned into an attractive commodity by well-capitalised merchants. Otherwise it would remain for local or regional consumption only. The market conditions prevailing in the Baltic between the thirteenth and the late sixteenth centuries are key to understanding the rise of large-scale fisheries elsewhere.

We know much less about the Baltic cod fishery, which only emerges in the sources by the fourteenth century and did not become substantial until the sixteenth century. Today, cod rarely venture into the very brackish northern Baltic waters between Stockholm and the Gulf of Riga. The presence of a large cod fishery in these waters in the late sixteenth and early seventeenth centuries indicates that cod abundance in the Baltic must have been very great. The abundance is all the more remarkable because the population of top predators, such as seals, would have been much greater than today. MacKenzie *et al.* (2007) review possible explanations for the high cod abundance in the sixteenth century – such as the impacts of temperature; salinity; and, possibly, oxygen on cod reproduction and biomass – but conclude that the evidence is currently inconclusive.

The Little Ice Age is an attractive explanation for increased abundance of a cold-water species such as cod in the Baltic. However, cod was, in fact, present and was caught in considerable quantities throughout prehistory, even at times when the waters were warmer than today (Enghoff *et al.* 2007). Increased storms may have brought more saline waters into the Baltic, which would have increased egg survival and cod abundance. However, the Baltic, if anything, is thought to have been more brackish in the sixteenth century than it is today. In conclusion, MacKenzie *et al.* 2007 observe that the strong landings of cod from sixteenth-century northern Baltic waters may simply have been due to a new fishery being carried out on a pristine fish population of moderate abundance.

Technological innovation is a second supporting factor to explain sixteenth-century landings. While the single hook-and-line (hand line) was used throughout prehistory for cod fishing, we know that the long line, with a hundred hooks or more, was introduced in northern Europe by the late fifteenth century and in the next century or so spread to become the preferred fishing tool for demersal species (Holm 1991). Vastly increased landings of cod may have resulted from this innovation.

Finally, politics are a third supporting factor to explain increased cod fishing efforts. By the time of the Nordic Seven Years' War the Danish navy demanded large supplies of cod as cheap and nourishing food for its naval crews. Archaeological and archival evidence show that such Baltic fishing settlements as Sandhagen and Gilleleje grew rapidly at this time in response to increased demand (Holm 1998).

Conclusion

It is possible to estimate the total traded and landed volumes of herring caught in the Baltic in the medieval period, while similar estimates are not available for cod. We can establish regional and temporal patterns which may be related to as yet poorly understood changes in natural conditions. Overall, the main driver for the development of the fisheries was the establishment of market controls, which made herring one of the single largest commodities in the northern European market. The decline of the market for fish in the seventeenth century was conditioned by a number of political, economic and cultural factors which need further research.

Acknowledgements

The author is grateful for the support provided by consecutive fellowships at the Rachel Carson Center for Environment and Society, Munich, and the Swedish Collegium for Advanced Study. He wishes to thank, in particular, Dr Henn Ojaveer, Professor Anthony Lappin and Dr James Barrett for comments on the draft.

References

Alheit, J. & E. Hagen. 1997. Long-term climate forcing of European herring and sardine populations. *Fisheries Oceanography* 6(2): 130–9.

Arup, E. 1932. *Danmarks historie II: stænderne i herrevælde 1282–1624*. Copenhagen: H. Hagerup.

Barrett, J. H., A. M. Locker & C. M. Roberts. 2004. 'Dark Age Economics' revisited: the English fish bone evidence AD 600–1600. *Antiquity* 78: 618–36.

Barrett, J. H., C. Johnstone, J. Harland, W. Van Neer, A. Ervynck, D. Makowiecki, D. Heinrich, A. K. Hufthammer, I. B. Enghoff, C. Amundsen, J. S. Christiansen, A. K. G. Jones, A. Locker, S. Hamilton-Dyer, L. Jonsson, L. Lõugas, C. Roberts & M. Richards. 2008. Detecting the medieval cod trade: a new method and first results. *Journal of Archaeological Science* 35(4): 850–61.

CCD Corpus Constitutionum Daniæ: Forordninger, Recesser og andre kongelige Breve, Danmarks Lovgivning vedkommende 1558–1660. (Koldingske reces 32). Copenhagen.

Christensen, T. P. 1994. Kystbebyggelser i senmiddelalder og renæssance: en komparativ status for Østdanmark. *Hikuin* 21: 99–124.

Corten, A. A. H. M. 2001. *Herring and Climate: changes in the distribution of North Sea herring due to climate fluctuations*. Groningen: University of Groningen.

DD Diplomatarium Danicum. 2003. Copenhagen: Danske sprog- og litteraturselskab. Available at: http://diplomatarium.dk/

Degryse, R. 1973. De Vlaamse westvaart en de Engelse represailles omstreeks 1378. *Handelingen der Maatschappij voor Geschiedenis en Oudheidkunde te Gent* 27: 193–239.

Eero, M., B. R. MacKenzie, F. W. Köster & H. Gislason. 2011. Multi-decadal responses of a cod (*Gadus morhua*) population to human-induced trophic changes, fishing, and climate. *Ecological Applications* 21(1): 214–26.

Enghoff, I. B. 1996. A medieval herring industry in Denmark and the importance of herring in eastern Denmark. *Archaeofauna* 5: 43–7.

Enghoff, I. B. 1999. Fishing in the Baltic region from the 5th century BC to the 16th century AD: evidence from fish bones. *Archaeofauna* 8: 41–85.

Enghoff, I. B., B. R. MacKenzie & E. E. Nielsen. 2007. The Danish fish fauna during the warm Atlantic period (ca. 7000–3900 BC): forerunner of future changes? *Fisheries Research* 87: 167–80.

Ersgård, L. 1988. '*Vår marknad i Skåne': bebyggelse, handel och urbanisering i Skanör och Falsterbo under medeltiden*. Stockholm: Almqvist and Wiksell.

Frandsen, S. & E. A. Jarrum. 1992. Sæsonfiskelejer, åresild og helårsfiskerlejer ved Sjællands nordkyst. *Gilleleje Museum Årbog* 29: 105–16.

Friedland, K. 1991. *Die Hanse*. Stuttgart: Kohlhammer.

Friis-Jensen, K. & P. Zeeberg. (eds) 2005. *Saxo Grammaticus,Gesta Danorum*. Copenhagen: Gad.

Gaumiga, R., G. Karlsons, D. Uzars & H. Ojaveer. 2007. Gulf of Riga (Baltic Sea) fisheries in the late 17th century. *Fisheries Research* 87: 120–5.

Grupe, G., D. Heinrich & J. Peters. 2009. A brackish water aquatic foodweb: trophic levels and salinity gradients in the Schlei fjord, northern Germany, in Viking and medieval times. *Journal of Archaeological Science* 36: 2125–44.

Hansen, J. M. 2010. The salt industry on the Danish Kattegat island of Læsø (1150–1652): hypersaline source, climatic dependence, and environmental impact. *Geografisk Tidsskrift/ Danish Journal of Geography* 110: 1–24.

Heckscher, E. 1949. *Sveriges ekonomiska historia från Gustav Vasa*. Stockholm: Bonniers.

Hoffmann, R. C. 1996. Economic development and aquatic ecosystems in medieval Europe. *American Historical Review* 101: 631–69.

Hoffmann, R. C. 2009. *Strekfusz*; a fish dish links Jagiellonian Kraców to distant waters, in P. Górecki & N. Van Deusen (eds) *Central and eastern Europe in the Middle Ages: a cultural history*: 116–24. London: I. B. Tauris.

Höglund, H. 1972. *On the Bohuslän Herring During the Great Herring Fishery*. Lysekil: Institute of Marine Research.

Holm, P. 1991. *Kystfolk: Kontakter og sammenhænge over Kattegat og Skagerrak ca. 1550–1914*. Esbjerg: Fiskeri- og Søfartsmuseets Forlag.

Holm, P. 1996. Catches and manpower in the Danish fisheries, c. 1200–1995, in P. Holm & D. J. Starkey (eds) *The North Atlantic Fisheries, 1100–1976: national perspectives on a common resource*: 177–206. Esbjerg: Fiskeri- og Søfartsmuseets Forlag.

Holm, P. 1998. Fiskeriets økonomiske betydning i Danmark 1350–1650. *Sjæk'len: Årbog for Fiskeri- og Søfartsmuseet* 1998: 8–42.

Holm, P. 1999. *Aalborg og Søfarten gennem 500 år*. Aalborg: Selskabet for Aalborgs Historie.

Holm, P. & M. Bager. 2001. The Danish fisheries, c. 1450: medieval and early modern sources and their potential for marine environmental history, in P. Holm, T. D. Smith &

D. J. Starkey (eds) *The Exploited Seas: new directions for marine environmental history*: 97–122. St John's: International Maritime Economic History Association.

Hørby, K. 1966. Øresundstolden og den skånske skibstold: Spørgsmålet om kontinuitet, in T. E. Christiansen, S. Ellehøj & E. L. Petersen (eds) *Middelalderstudier: Tilegnede Aksel E. Christensen på tresårsdagen 11. September 1966*: 245–72. Copenhagen: Munksgaard.

Hørby, K. 1971. Skånemarkedet. *Kulturhistorisk Leksikon for Nordisk Middelalder* 16: 68–77.

Jahnke, C. 1998. Regulating the medieval fisheries of the Baltic region: the 'urban' and 'international market' compared, in P. Holm & D. J. Starkey (ed.) *The North Atlantic Fisheries: markets and modernisation*: 177–206. Esbjerg: Fiskeri- og Søfartsmuseets Forlag.

Jahnke, C. 2000. *Das Silber des Meeres: Fang und Vertrieb von Ostseehering zwischen Norwegen und Italien (12.–16. Jahrhundert)*. Cologne: Böhlau.

Kjellström, A., J. Storå, G. Possnert & A. Linderholm. 2009. Dietary patterns and social structures in medieval Sigtuna, Sweden, as reflected in stable isotope values in human skeletal remains. *Journal of Archaeological Science* 36: 2689–99.

Kjærsgaard, E. 1978. *Mad og øl i Danmarks middelalder*. Copenhagen: København Nationalmuseet.

Knudsen, H. (ed.) 1836. Cantsler Claus Giordsens Optegnelser, især om de danske og norske Lehn paa Kong Frederik den Førstes Tid. *Nye Danske Magazin* 6: 270–87.

Kosiba, S. B., R. H. Tykot & D. Carlsson. 2007. Stable isotopes as indicators of change in the food procurement and food preference of Viking Age and Early Christian populations on Gotland (Sweden). *Journal of Anthropological Archaeology* 26: 394–411.

Lajus, D., Z. Dmitrieva, A. Kraikovski, J. Lajus & D. Alexandrov. 2007a. Historical records of the 17th–18th century fisheries for Atlantic salmon in northern Russia: methodology and case studies of population dynamics. *Fisheries Research* 87: 240–54.

Lajus, D., Y. Alekseeva & J. Lajus. 2007b. Herring fisheries in the White Sea in the 18th–beginning of the 20th centuries: spatial and temporal patterns and factors affecting the catch fluctuations. *Fisheries Research* 87: 255–9.

Lajus, J., A. Kraikovski & D. Lajus. 2013. Coastal fisheries in the eastern Baltic Sea (Gulf of Finland) and its basin from the 15th to the early 20th centuries. *PloS ONE* 8(10): e77059. doi:10.1371/journal.pone.007059

Lappenberg, I. M. (ed.) 1868. *Arnoldi chronica Slavorum: scriptores rerum Germanicarum in usum scholarum recusi*. Available at: https://archive.org/stream/arnoldichronicas00arnouoft#page/n3/mode/2up

Lechner, G. 1935. *Die hansischen Pfundzollisten des Jahres 1368, 18. März 1368 bis 10. März 1369*. Lübeck: Verlag des Hansischen Geschichtsvereins.

Linderholm, A., C. H. Jonson, O. Svensk & K. Lidén. 2008. Diet and status in Birka: stable isotopes and grave goods compared. *Antiquity* 82: 446–61.

MacKenzie, B. R., J. Alheit, D. J. Conley, P. Holm & C. C. Kinze. 2002. Ecological hypotheses for a historical reconstruction of upper trophic level biomass in the Baltic Sea and Skagerrak. *Canadian Journal of Fisheries and Aquatic Sciences* 59: 173–90.

MacKenzie, B. R., M. Bager, H. Ojaveer, K. Awebro, U. Heino, P. Holm & A. Must. 2007. Multi-decadal scale variability in the eastern Baltic cod fishery 1550–1860–evidence and causes. *Fisheries Research* 87: 106–19.

Orton, D. C., D. Makowiecki, T. de Roo, C. Johnstone, J. Harland, L. Jonsson, D. Heinrich, I. B. Enghoff, L. Lõugas, W. van Neer, A. Ervynck, A. K. Hufthammer, C. Amundsen, A. K. G. Jones, A. Locker, S. Hamilton-Dyer, P. Pope, B. R. MacKenzie, M. Richards, T. C. O'Connell & J. H. Barrett. 2011. Stable isotope evidence for late medieval (14th–15th C) origins of the eastern Baltic cod (*Gadus morhua*) fishery. *PLoS ONE* 6: e27568. doi: 10.1371/journal.pone.0027568

Pettersson, J. 1953. *Den svenska skagerrakkustens fiskebebyggelse: en etnologisk studie*. Lund: C. W. K. Gleerups Förlag.

Poulsen, B., P. Holm & B. R. MacKenzie. 2007. A long-term (1667–1860) perspective on impacts of fishing and environmental variability on fisheries for herring, eel and whitefish in the Limfjord, Denmark. *Fisheries Research* 87: 181–95.

Rasmussen, H. 1968. *Limfjordsfiskeriet før 1825*. Copenhagen: København Nationalmuseet.

Reitsemaa, L. J., D. E. Crews & M. Polcyn. 2010. Preliminary evidence for medieval Polish diet from carbon and nitrogen stable isotopes. *Journal of Archaeological Science* 37: 1413–23.

Stoklund, B. 1959. Bonde og fisker: lidt om det middelalderlige sildefiskeri og dets udøvere. *Handels- og Søfartsmuseets Årbog 1959*: 101–22. Kronborg: Handels- og Søfartsmuseet.

Storm, G. (ed.) 1881. *Peder Claussøn Friis, samlede skrifter*. Kristiania: Brøgger.

Tuck, A. 1972. Some evidence for Anglo–Scandinavian relations at the end of the fourteenth century. *Mediaeval Scandinavia* 5: 75–88.

Unger, R. W. 1978. The Netherlands herring fishery in the late Middle Ages: the false legend of Willem Beukelsof Biervliet. *Viator* 9: 335–56.

Ventegodt, O. 1990. Skånemarkedets sild. *Maritim Kontakt* 14: 3–19.

Verliin, A., H. Ojaveer, K. Kaju & E. Tammiksaar. 2013. Quantification of the early small-scale fishery in the north-eastern Baltic Sea in the late 17th century. *PLoS ONE* 8(7): e68513. doi:10.1371/journal.pone.0068513

Weibull, C. 1922. *Lübeck och Skånemarknaden: Studier i Lübecks pundtullsböcker och pundtullskvitton 1368–1369 och 1398–1400*. Lund: Berlingska Boktryckeriet.

Weibull, C. 1966. Lübecks sjöfart och handel på de nordiska rikena 1368 och 1398–1400. *Scandia* 32: 1–123.

Woolgar, C. M. 2010. Food and the Middle Ages. *Journal of Medieval History* 36: 1–19.

Yoder, C. 2010. Diet in medieval Denmark: a regional and temporal comparison. *Journal of Archaeological Science* 37: 2224–36.

The Early Documentary Evidence for the Commercialisation of the Sea Fisheries in Medieval Britain

Maryanne Kowaleski

Historians have thus far not contributed as much to our understanding of the origins of commercial sea fishing in medieval Britain as have archaeologists (Barrett *et al.* 2004a; 2004b; 2011; Orton *et al.* 2014). Richard Hoffman (2000; 2002) has profitably examined the impact of commercialised fishing on the medieval environment, but he has focused on continental Europe, not Britain, and on inland rather than marine fisheries. Kowaleski (2003) has treated many aspects of the commercialisation of marine fishing in England and Wales, but says little about the documentary evidence for the period before the thirteenth century, an oversight that this contribution is meant to redress. The best study to date of the sea fisheries in medieval England is an unpublished PhD thesis by Alison Littler (1979), although she, too, does not consider the chronology or reasons behind the commercialisation of the fisheries in any depth. Starting in the late 1960s, but particularly in the past fifteen years, an excellent series of articles has appeared on the productivity and catches of particular sea fisheries during the later Middle Ages. Although they offer little about the earlier history of these regional fisheries, these articles do lay out reasons for their growth (or decline) from the late thirteenth through early sixteenth centuries (Bailey 1990; Childs 1995; 2000; Dulley 1969; Fox 1996a; 1996b; 2001; Gardiner 1996; Heath 1968; Hybel 1996; Kowaleski 2000a; 2000b; Middleton-Stewart 1996; Saul 1975; 1981; Vasey 1978). In addition to outlining the organisation and trade in fish in specific marine fisheries and emphasising the impact of marine fishing on the landscape and on trade, these essays have tracked the different trajectories and developmental pace of the eastern and western fisheries and the growing involvement of the British in long-distance fisheries in Ireland and Iceland. All of these works are extremely useful, but they are relatively few, and they focus largely on the later Middle Ages – although more recently Mark Gardiner (1997; 1998), James Campbell (2002), Hirokazu Tsurushima (2006), James Barrett (2007; 2012) and Vicki Szabo (2008) have considered particular types of documentary evidence for the eleventh century.

This study examines what the written primary source evidence for the tenth through thirteenth centuries can tell us about the earliest signs of commercialisation in the medieval sea fisheries of Britain, especially in England. Several sources shed light on the chronology, location and pace of commercialisation, while offering some hints that we need to think more about the role of technological developments in catching and curing marine fish. The herring (*Clupea harengus*) fisheries off the eastern coast of England were clearly producing for the market by the tenth and early eleventh centuries, but the pace of commercialisation quickened considerably over the course of the eleventh and early twelfth centuries. Consumer demand, particularly from urbanisation, was the prime mover behind this commercialisation, though the availability of salt for curing close to the rich fishing grounds near the coast of eastern England was a crucial factor in the ability to respond to this demand. There is little historical evidence that the English cod (*Gadus morhua*) fisheries were commercialised before the late twelfth and early thirteenth centuries

How does the archaeological evidence compare with the early documentary record? Analysis of assemblages of fish bones in England dating from the seventh through sixteenth centuries reveals that those from the seventh to late tenth centuries were dominated by freshwater and migratory species, especially eel (*Anguilla anguilla*) and cyprinids (Cyprinidae) (Barrett *et al.* 2004a; 2004b). Herring was present, especially in coastal and riverine trading settlements, but its importance increased fourfold in the eleventh and twelfth centuries. Analysis of stable carbon and nitrogen isotope ratios in human bones reinforces this dating in showing that even coastal residents ate little marine protein until the late ninth to eleventh centuries. From this period – termed the 'fish event horizon' by James Barrett and his colleagues – onwards, the diets of first male and then female residents of coastal communities began to include a significant portion of marine protein, a trend that reached inland communities a bit later (Barrett and Richards 2004; Barrett *et al.* 2004a; 2004b; Müldner and Richards 2006; Chapter 20). Cod and other gadids, which were not widely consumed by the English before the eleventh century, became more significant over the next two centuries, although it is unlikely that cod consumption ever outpaced that of herring in the medieval English diet. Most of the cod bones dating from the twelfth and thirteenth centuries probably originated in the southern North Sea off the British coast, but imported cod from more distant waters increased significantly in importance in the late thirteenth to fourteenth centuries. By the fifteenth and sixteenth centuries, imported cod dominated cod consumption, at least in large cities, such as London (Barrett *et al.* 2011; Orton *et al.* 2014). There is also some evidence that cod declined in importance in the later Middle Ages as the proportion of related species – such as haddock (*Melanogrammus aeglefinus*), ling (*Molva molva*), saithe (*Pollachius virens*) and hake (*Merluccius merluccius*) – rose. The later evidence also shows significant regional variations, with cod more abundant in eastern England, hake more common in western England, and conger eel (*Conger conger*) especially prominent in the south-coast counties of Devon, Dorset and Hampshire (Serjeantson and Woolgar 2006).

As other scholars have observed, the Anglo-Saxon documentary sources before the eleventh century say hardly anything about marine fishing, thus matching the archaeological findings of little consumption of marine protein before *c.* AD 1000. Anglo-Saxon sources that mention the abundance of fish never refer to marine fish (Hagen 1995, 160–1). The Old English word for 'herring' appears fewer than ten times in the Anglo-Saxon corpus: five times as part of a Latin/Old English glossary, twice as a place-name in charter boundaries, and once in a note about the herring season, although there are also several references to herring rents (DOE). In the very early eleventh century, Ely Abbey handed over to Thorney Abbey 2000 *hæringes* it had purchased for 40 pence, but they were dwarfed by the more than 26,000 eels recorded as rents from two of Ely's fen properties later in the document (Robertson 1956, 252–7, 505). The Latin for herring, *allecium*, is also not widely used until the late twelfth century, while all but one of the appearances of the Latin word for cod (*mulvellus*) occur in the thirteenth century (DML, vol. 1, 64; vol. 6, 1858). Anglo-Norman versions of the word for cod (*morue*, *mulvel*) also do not emerge before the thirteenth century

(AND). The earliest use of any Middle English word for a member of the gadid family was in 1228 in reference to ling and *milvil*, and in 1289 to codling; in its dried form as stockfish or haberdine, it occurs in the third quarter of the thirteenth century (MED; OED), which is clearly later than its appearance in the English archaeological record. This linguistic evidence suggests that cod, a deep-water fish, was not widely known and certainly not regularly traded in Britain much before the thirteenth century. Herring was referenced in a wider variety of contexts by the eleventh century, but its relative absence in the linguistic record until the late twelfth and thirteenth centuries suggests it, too, was not highly commercialised until later. The linguistic record thus seems to lag behind the archaeological data that suggests a 'fish event horizon' around AD 1000 in which the consumption and therefore the catch of marine fish accelerated significantly (Barrett *et al.* 2004a; 2004b). However, relying on published dictionary evidence of first citations is risky, given the non-systematic inclusion of Latin and vernacular forms (compare, for instance, the many Latin forms of *allecium* in Domesday Book and other sources before AD 1100 in Table 3.1 with those in the DML, vol. 1, 64). There are also currently a variety of debates on the etymology of the word cod and its variants (Fix 1986; Lockwood 2006; Sayers 2002, 24–5) that need to be resolved. Further attention to Old English glosses of Latin words for fish (and *vice versa*) could also prove helpful (Napier 1906, 278–9).

The most explicit Anglo-Saxon and Anglo-Latin references to sea fishing come in the dialogue composed by Abbot Aelfric around AD 1000 to help students learn Latin. One of his conversations is between a teaching master and a fisherman, which was meant to enlarge students' vocabulary by providing them with many fish names (Garmonsway 1939; Swanton 1993, 171–2).

Master: How do you catch the fish?

Fisherman: I board my boat and cast my net into the river; and throw in a hook and bait and baskets; and whatever I catch I take ...

Master: Where do you sell your fish?

Fisherman: In the city.

Master: Who buys them?

Fisherman: The citizens. I can't catch as many as I can sell.

Master: Which fish do you catch?

Fisherman: Eels and pike, minnows and turbot, trout and lampreys and whatever swims in the water. Small fish.

Master: Why don't you fish in the sea?

Fisherman: Sometimes I do, but rarely, because it is a lot of rowing for me to the sea.

Master: What do you catch in the sea?

Fisherman: Herring and salmon, porpoises and sturgeon, oysters and crabs, mussels, winkles, cockles, plaice and flounder and lobsters, and many similar things.

Master: Would you like to catch a whale?

Table 3.1. Early herring rents and fisheries in England (to c. AD 1130).

Place	Herring render	Rent or value and lord	Date (AD)	Source
		Essex		
Beaumont-cum-Moze	3M dried herring	Granted to Westminster Abbey by Geoffrey de Mandeville	1085–1086	Mason 1988, 297
		Norfolk		
Thorpe St Andrew	Then 2M, now 0	Then worth £12 and 1 sester of honey and 2M herring, now £30	1066–86	Morris 1975–92, vol. 33:1, 1/126
Norwich	1 last	From church of St Laurence to Bury St Edmunds Abbey	1038–1066	Hart 1966, 82
Repps	10M	Confirmation of grant to monks of Longueville Priory previously made by Earl Walter Giffard	1146–1155	Harper-Bill 1990, 96; Salter 1921, 2, 6
		Suffolk		
Blythburgh	Then 10M, now 3M	Then 10M herring to king, now 50s and 3M herring	1066–86	Morris 1975–92, vol. 34:1, 1/12
Dunwich	60M	Then £10, now held by Rob. Malet and worth £50 and 60M herring as a gift	1066–86	Morris 1975–92, vol. 34:1, 6/84
Dunwich	Herring tithes & 10M	Granted by Rob. Malet to Priory of Eye; by 1134–55, 10M herring to St Mary at Bernay (Normandy) before Advent	c. 1087	Brown 1992, 12–19
Dunwich	8M	Gilbert Blunt holds 80 men from Rob. Malet, and pays £4 and 8M herring	1086	Morris 1975–92, vol. 34:1, 6/84
Dunwich	30M	30M herring to monks of Ely	1110–1131	Karn 2005, 5–6; Miller 1951, 282–3
Beccles	Then 30M, now 60M	Then 30M herring to monks of Bury St Edmunds, now 60M	1066–86	Morris 1975–92, vol. 34:2, 14/20
Southwold	Then 20M, now 25M	Then 20M herring to monks of Bury St Edmunds, now 25M	1066–86	Morris 1975–92, vol. 34:2, 14/163
Stoven	1C	Then and now worth 7½s and 1C herring, to Hugh de Montfort	1066–86	Morris 1975–92, vol. 34:2, 31/7
Weston	4C	Now worth 5s and 4C herring, to Hugh de Montfort	1066–86	Morris 1975–92, vol. 34:2, 31/22
Willingham	1C	Now worth 8d and 1C herring, to Hugh de Montfort	1066–86	Morris 1975–92, vol. 34:2, 31/23
Willingham	Then 0, now 3M	Then 60s, now 30s30d and 3M herring, to Hugh de Montfort	1066–86	Morris 1975–92, vol. 34:2, 31/21
Hatheburgafelda	Then 0, now 9C	Then worth 10s, now 11s8d and 9C herring, to Hugh de Montfort	1066–86	Morris 1975–92, vol. 34:2, 31/24
Worlingham	1M	Then and now worth 10s6d and 1M herring, to Hugh de Montfort	1066–86	Morris 1975–92, vol. 34:2, 31/25
Beketun	1.5M	Then worth 4s, now 21s4d and 1.5M herring, to Hugh de Montfort	1066–86	Morris 1975–92, vol. 34:2, 31/26
Kessingland	Then 0, now 1M	Then worth 10s, now 22s and 1M herring, to Hugh de Montfort	1066–86	Morris 1975–92, vol. 34:2, 31/27
Pakefield	Then 0, now 6C	Then worth 5s, now 9s and 6C herring	1066–86	Morris 1975–92, vol. 34:2, 31/28
Wimundahala	Then 0, now 5C	Then worth 2s, now 3s and 5C herring, to Hugh de Montfort	1066–86	Morris 1975–92, vol. 34:2, 31/29
Gisleham	Then 2C, now 3C	Then 2 free men worth 2s6d and 2C herring, now 1 free man worth 5s and 3C herring, to Hugh de Montfort	1066–86	Morris 1975–92, vol. 34:2, 31/30
Hornes	1.6C	Worth 3s and 1.6C herring, to Hugh de Montfort	1066–86	Morris 1975–92, vol. 34:2, 31/31
Carlton Colville	4C + 3C	2 free men worth 3s and 4C herring; 1 free man then and now worth 5s and 3C herring, to Hugh de Montfort	1066–86	Morris 1975–92, vol. 34:2, 31/32
Kirkley	0 then, now 2C	Then worth 2s, now 3s and 2C herring, to Hugh de Montfort	1066–86	Morris 1975–92, vol. 34:2, 31/33
Rushmere	3C	Worth 5s and 3C herring, to Hugh de Montfort	1066–86	Morris 1975–92, vol. 34:2, 31/34

(*Continued*)

Place	Herring render	Rent or value and lord	Date (*AD*)	Source
Desning in Gazeley	5M	Grant of Gilbert de Clare to Stoke by Clare Priory of 5M herring taken at *Deseninges*	c. 1080	Harper-Bill and Mortimer 1984, no. 137
Kelsale	20M	Confirmation and grant by Wm. Bigod to Thetford Priory	c. 1107	Dugdale 1665, 665
Unknown places, but probably in Suffolk	3M, 1M, 1M, 1M	Donations in alms by men of Roger or Wm. Bigod to Thetford Priory: Bernard de Berneham; ... Rinaldi, dapifer; Bristrieus de Longelond; and Robert de Rekelund	c. 1107	Dugdale 1665, 665
Kent				
Sandwich	40M	Owed £40 and 40M herring for sustenance of the monks at Canterbury; now £50 and herrings as before	1066–86	Morris 1975–92, vol. 1, 2/2
Sandwich	4M	4M herring or 10s owed from 30 messuages to monks of St Augustine's Canterbury	c. 1124	Ballard 1920, 20
Luddenham	3C	One-half fishery renders 3C herring	1066–86	Morris 1975–92, vol. 1, 5/160
Sussex				
Southease	38.5M	Rendered by villeins to St Peter's, Winchester, and £4 for porpoises	1066–86	Morris 1975–92, vol. 2, 7/1
Ilford	16M	16M herring from demesne (with 26 burgesses of Lewes of William de Warenne)	1066–86	Morris 1975–92, vol. 2, 12/3
Brighton	4M	4M herring from rents owed to Ralph who holds of William de Warenne	1066–86	Morris 1975–92, vol. 2, 12/13
Lewes	4M	22s and 4M herring from 44 closes	1066–86	Morris 1975–92, vol. 2, 12/4
Bulverhythe	2M	Confirmation of grants of grandfather and father by Henry, count of Eu, to church and college of St Mary's in Hastings; prebend of W. Fitzallec receives annual rent of 2M herring and other fish dues	1090–1094	Gardiner 1989; Maxwell-Lyte 1900, no. 1073
Hastings	2M	Grant of Robert son of Ralf to his daughter and abbey of St Amand in Rouen of yearly rent of 2M herring	? 1067	Round 1899, 25–6
Surrey				
Southwark and London	2M	15 dwellings in Southwark and London render 6s and 2M herring at Walkingstead	1066–86	Morris 1975–92, vol. 3, 15/2
Southwark	5C	1 dwelling in Southwark pays 5C herring at Long Ditton	1066–86	Morris 1975–92, vol. 3, 19/21
Berkshire & Oxfordshire				
Abingdon	1C from each boat	1C herring or suitable price for them owed each year to cellarer from each boat of city of Oxford travelling on River Thames past Abingdon	1052–1066	Hudson 2002, 174–5, 218–19
Gloucestershire				
Tidenham	30M	Additional rent to Bath Abbey owed after life lease includes 1 mark, 30M herring and 6 porpoises per year	956–1060	Robertson 1956, 216–19
Lancashire and Cheshire				
Lands between Ribble & Mersey & in Wirral	3 M shad [a] each	Rent paid during shad season to Burton Abbey by two heirs	1002–1004	Sawyer 1979, xv–xix, 53–6

Notes: C=hundred and M=thousand, but note that most fish were measured by the long hundred (120) not the short hundred (100). [a]Shad are fish of the herring family that

Fisherman: *Not me!*

Master: *Why?*

Fisherman: *Because it is a risky business catching a whale. It's safer for me to go on the river with my boat, than to go hunting with many boats.*

Master: *Why so?*

Fisherman: *Because I prefer to catch a fish I can kill, rather than a fish that can sink or kill not only me but also my companions with a single blow.*

Master: *Nevertheless, many catch whales and escape danger, and make great profit by it.*

Fisherman: *You are right, but I dare not because of my timid spirit.*

In answering the questions posed by the Master, the Fisherman makes clear a number of salient points. One is that he sells his fish in the city and cannot catch enough to satisfy demand, implying that urbanisation was a driving force behind the intensification of fishing that was happening around this time. Second, his catch is mainly small river fish because he finds sea fishing too much effort. He particularly avoids fishing for whale because it is risky and requires more investment in boats and men. The sea fish he does mention are herring (but not cod), salmon (i.e. *Salmo salar*), sturgeon (i.e. *Acipenser* sp.), plaice (i.e. *Pleuronectes platessa*) and flounder (i.e. *Platichthys flesus*). Porpoise is included in the list and he also mentions lobster and other shellfish, such as oysters, mussels, crabs, winkles and cockles, all of which were regularly being marketed in Britain by the late twelfth and early thirteenth centuries. In other words, this source nicely represents the transitional point between consumers' reliance on freshwater fish and their growing taste for marine fish, and between conservative subsistence fishing and riskier commercial fishing for a growing urban market. It reveals an industry in the early stages of commercialisation in which supply has not yet caught up with demand.

The river fishing that Aelfric's Fisherman practises is amply documented in Domesday Book, that massive compilation listing the lands, tenants, rents and obligations of tenants-in-chief before and after the Norman conquest in AD 1066. The many hundreds of fisheries recorded in Domesday were primarily weirs or traps fixed in rivers or mill-ponds, almost all of them associated with freshwater fish (Darby 1977, 279–85). Few of the references specified the type of fish being caught, though eels (which spawn in salt water but spend most of their life in, and are usually caught in, rivers) and to a lesser extent salmon (which spawn in fresh water and are often captured in rivers as they are returning to the sea to feed) were singled out most often. Eel renders were especially frequent and occurred in sometimes massive amounts, such as the more than 100,000 owed from just four manors in the Cambridgeshire fens (Darby 1971, 306). Domesday also recorded fishing boats, nets and almost one hundred fishermen, most of whom

must have been river fishermen like Aelfric's Fisherman, since they resided at inland locations (Darby 1977, 282–3; Tsurushima 2006, 194).

The small number of references to sea-based fishing in Domesday Book occurs in references to sea weirs, marine fishermen and herring renders. Fisheries at coastal locations such as Southwold (with its *heia maris*, or sea-hedge) and Blytheburgh in Suffolk must have been foreshore or sea weirs (Darby 1977, 285), while many recorded in tidal waters would have also captured sea fish (Moore and Moore 1903, 3–4). Yet none of the early and mid-Saxon sea weirs that we know about from archaeological excavations (for example, in Norfolk, Suffolk, Essex and Somerset: Allen *et al*. 1997; Gilman 1998; Murphy 2010, 216–20; O'Sullivan 2003) or from documentary references (for example, in Calshot, Hampshire: Pierquin 1912, 582–3) were recorded in Domesday, an indication of how selective the Domesday Book references to fisheries can be. The early Essex sea weirs (often called 'kiddles' in eastern England and the River Thames), particularly those around Foulness, are also well documented later in the Middle Ages (Crump and Wallis 1992), so it is reasonable to speculate that other clusters of sea weirs known in later medieval documentation, such as the stone weirs off the north Devon and Somerset coasts, first noted in a late eleventh-century grant and frequently referenced from the thirteenth century onwards, could well have been operating in the Anglo-Saxon period if not earlier (Hancock 1903, 31–3, 171–2; Maxwell-Lyte 1909, vol. 1, 97, 296; McDonnell 1979, 134–5). The *goredi* off the Welsh coast are also in this category in that they are well attested in the archaeological context (Jones 1983) and are referred to in several early, albeit hard-to-date texts, including later copies of the tenth-century laws of Hywel Dda (Bond 1988, 78) and late medieval copies of *Hanes Taliesin*, a bardic poem partly written in an archaic Welsh that some date to the tenth century and purporting to describe events in the sixth century (Lewes 1924, 398). The cost of constructing and maintaining these sea weirs is one indication of their commercial potential, as is the care with which their annual value and output in terms of fish were recorded in Domesday and in charters. In Essex, each tide left about a cart-load of fish to be collected (Crump and Wallis 1992, 41), so places with two strong tides collected twice as much, a significant haul that speaks to the marketable surplus that hundreds of such structures along stretches of suitable coastline could yield.

Domesday Book also recorded estuarine fisheries or fish traps that captured sea fish, along with eel and salmon. Very large numbers of such fish traps were recorded in the River Severn, with 65 at Tidenham alone (Darby and Terrett 1954, 38; Hooke 2007, 48–9). The tide comes into the Severn twice a day at very great speeds, and archaeological evidence has confirmed weirs here since the ninth century and perhaps earlier (Godbold and Turner 1994, 36). Other documentary evidence also shows the early development of estuarine weirs in the Severn. Many fish traps – most probably basket and hackle weirs – are noted in Anglo-Saxon charters and in a mid-eleventh-century survey for Bath Abbey of its properties at Tidenham (Robertson 1956, 204–7; Seebohm 1890, 150–5). When the abbey leased the manor to the archbishop sometime between the mid-tenth and mid-eleventh centuries, the abbot expected a cash rent, as well as 30,000 herring and six porpoises each year (Robertson 1956, 216–19), which points not only to what the weirs were capturing, but also to the commercial value of the catch. Tidenham's villein tenants also profited from estuarine weirs, although they had to give every other fish they took to the lord, as well as every rare fish 'of value', which included herring, porpoise, sturgeon and other sea fish (Robertson 1956, 204–7). When the lord was on the estate, no one could sell any fish for money, an indication that a good proportion of the weirs' catch was still going to sustain a large religious household.

The hackle weirs on the Severn appear to have functioned best when fishermen working from boats stretched a net across the eddy in which the fish were channelled by the wattled hedges or walls of the weir (Seebohm 1890, 153). As demand for marine fish rose in the eleventh century, the occupation of sea fisherman must have become more widespread in coastal communities. Domesday Book notes several such groups, including 24 fishermen at Little Yarmouth in Suffolk, an unspecified number at Lyme Regis, and four at Bridge just east of Lyme Regis, although the 20 fishermen at Wisbech on the Wash in Cambridgeshire probably concentrated on eel fishing (Darby 1977, 285; Tsurushima 2006, 194). No fishers are mentioned at Great Yarmouth in Norfolk, even though we know that East Anglia was already a centre for the herring fishery in this period, an omission that casts some doubt on the extent to which Domesday Book represents anything like an accurate picture of the fishing industry in AD 1086. Indeed, it is clear that the Domesday survey focused primarily on recording river weirs and traps held by great lords, probably because they were fixed, tangible structures that could be owned and rented out for profit. It is likely that smaller weirs owned or operated by those lower down the social ladder were rarely included in the survey. Also completely excluded from Domesday were coastal fishing grounds exploited by local fishermen with nets, lines and hooks, which helps explain why Domesday did not record even one fishery in Cornwall, the county with the longest coastline and most diverse fishing grounds in England. A significant amount of inshore fishing must have been going on by the eleventh century in Cornwall given the evidence that fishing tolls collected as part of the 18 'port farms' of the earldom of Cornwall in the twelfth and thirteenth centuries had probably replaced the earls' rights to a share of the catch as the industry became more commercialised (Kowaleski 2001, 19–20).

Although Domesday did not directly record sea fisheries, we can get some idea of the geographic distribution and relative value of the herring fisheries by examining references to herring rents in Domesday and in Anglo-Saxon charters (Table 3.1). The vast majority of the herring renders were in East Anglia and the southeastern counties of Sussex and Kent. The two Surrey references were to fisheries on the Thames; if the Domesday return for London were extant, it would undoubtedly include additional fish weirs or kiddles. Quite substantial fish weirs dating from the eighth through tenth centuries have been excavated in the London stretch of the Thames (Milne 2003, 39). These weirs had proliferated to such an extent by the twelfth century that they had to be banned, since they were hindering the course of navigation down the Thames (Keene 2001, 166, 168). A glimpse of the Thames sea fisheries is also visible in the dues collected by the Constable of the Tower of London in the thirteenth century. They included tolls on 17 kiddles in the Thames and on boats taking 'small fish' with nets, on stall boats fishing for sprot (i.e. sprat (*Sprattus sprattus*)) from the Tower to the sea, and on boats carrying herring in from the sea (TNA, E101/4/10, 4/24, 16/31, 18/4, 531/19). The Constable appears to have held rights over these weirs for some time. Although their catch is not specified, other early weirs in tidal rivers such as the Thames and the Humber not recorded in Domesday would also have caught herring, salmon and other sea fish (Brett and Gribbin 2004, 27; Campbell 1973, 19; van Caenegem 1990–1, 218–19).

The largest number of herring fisheries recorded for the eleventh century was in Suffolk, which boasted more than 20 places with herring rents or renders (Table 3.1). Of the 12 Suffolk entries that compared profits or herring rents in AD 1066 with AD 1086, nine saw an increase, suggesting a healthy late eleventh-century demand for herring and easy readjustment to post-invasion conditions (Campbell 2002). Beccles, for instance, rendered 30,000 herring in AD 1066, but at the time of the great survey in AD 1086, this amount had doubled, to 60,000. In Carlton Colville, there used to be two free men worth 3 shillings and 400 herring. By AD 1086, there was only one free man, but he was worth 5 shillings and 300 herring, a significant per capita increase. In three places, the larger herring renders occurred in tandem with a reduction of ploughs, which suggests a growth in the labour directed towards fishing. Moreover, in seven places the herring renders may have only been assessed after AD 1066, another indication of an expanding industry. Most of these places were in the lordship of one man, Hugh de Montfort. Was he or his reeve quicker to take advantage of the growing by-employment that herring fishing was offering to his tenants, or was his reeve simply more careful than neighbouring reeves in reporting all the rents owed by his lord's properties? Either scenario is possible, but together this evidence – the minor Suffolk herring renders owed by only one or two freemen, or villeins, on quite small acreages, the increase in many of these herring renders over the 20 years covered by the Domesday survey, and the associated reduction in ploughs on some of these lands – suggests that coastal tenants were beginning to take greater advantage of the autumn herring fishery off the East Anglian coast in the late eleventh century (Campbell 2002).

Like Domesday's record of fishermen, its listing of herring rents is incomplete. It fails to record 4000 herrings in rent at Sandwich, 4000 at Hastings in Kent, and 5000 at Desinges in Suffolk, to name only a few examples (Table 3.1). What is significant here is not so much that once again Domesday Book provides only a partial record, but that herring (and eel) renders were so common in the coastal manors of eastern England, from Norfolk in the north to Sussex in the south. References to these rents in Domesday Book, as well as in charters, rentals and other records into the thirteenth century, show that both eel and herring were clearly being used as a type of currency to pay rents, tithes, debts, labour services and even marriage gifts (Darby 1940, 31; Jones 2000, 102; Lees 1935, 8–9, 154–5; Scargill-Bird 1887, xiii–xv, 8–14). Many were ancient rents that can almost certainly be traced back to the tenth century if not earlier. They were not a new phenomenon, for instance, when they first appeared in eleventh-century charters recording fish renders to large monasteries. Herring rents also extended far down the social ladder, appearing as a customary payment from villeins usually recorded by their Old English names, such as *fishsilver*, *heringsilver* or *heringlode*, suggesting an older history (Neilson 1910, 33–4). This usage implies that when the surge in herring fishing occurred in the early eleventh century, it was building upon a tradition of fishing that had already gone beyond subsistence fishing to produce a surplus for estate owners and villein tenants who were using part of their catch as a type of currency to pay rents and other obligations.

Although herring renders are typically interpreted as for the convenience of large, landowning households, particularly monks who imposed them to ensure a steady supply of fish to their table on the many meatless days of the Christian calendar, the sheer size of the renders and their geographic distribution exactly where the English herring fishery first intensified production for the market suggests that they represent the initial stage of commercialisation, and help explain why the transition to production for the market occurred so quickly. James Campbell (2002) has calculated that if herring renders in Domesday Book alone represented a modest one-twentieth of all herring caught annually, the total English catch would have been at least 3,298,000 herring in the late eleventh century. He noticed too the unusual concentration of herring renders in Suffolk in Domesday and the signs that herring rents were actually increasing in this region from AD 1066 to 1086. These factors led him to argue that the very dense population and prosperity of East Anglian coastal regions could be attributed not only to the region's agricultural fertility and exploitation of recently drained lands suitable for sheep husbandry, but also to the income being provided by three inter-connected industries: the concentration of salt works in southeast Norfolk, the ready availability of fuel from the peat mines of the Norfolk Broads to make salt, and the easy access to inshore herring during the fall months.

There is highly suggestive evidence that the herring fisheries were a viable commercial enterprise even earlier than the eleventh century. Anglo-Saxon charters of the late seventh–early ninth centuries refer to rights in fisheries on the Isle of Thanet and at Graveney in Kent (Hooke 2007, 45–6), and to a fishery and fish houses – probably for the storage of fishing equipment, such as lines, nets, hooks and boats – at the mouth of the River *Limen* near New Romney (Brooks 1988, 98; Brooks and Kelly 2013, nos 11, 11A). Another charter of 962 records a toll associated with one fishing boat at two coastal locations in Hampshire (Kelly 2000, 377–80). By AD 1038–40, the east coast herring fishery was significant enough that the confiscation of Sandwich by King Harold Harefoot was measured in terms of 'two whole herring seasons', probably to emphasise the financial loss that this valuable port represented (Robertson 1956, 175). It was Dunwich, however, that was probably the most important Anglo-Saxon fishing port, since it owed by far the largest herring renders in the eleventh century (Table 3.1) and ranked among the top six towns in terms of population and wealth (Campbell 2002, 14). Yarmouth was smaller and appears to have started as a seasonal fishing settlement, since the first definitive documentary reference is to a chapel, probably established sometime in the eleventh century on the sands of Yarmouth near to fishing huts for visiting fishermen (Saunders and Millican 1939, 31–2).

Domesday Book records Yarmouth as a borough with 70 burgesses and no fishermen, although it is likely that the 24 fishermen at Little Yarmouth in Suffolk were associated with the local fishing industry (Darby 1971, 141–2). The first documentary reference to the famous Yarmouth herring fair is not until AD 1208 (Letters 2007), but a large-scale fishing industry and trade was clearly operating in the area by the eleventh century. The Domesday entry for Dover, for example, refers to the king's truce that prevailed from 29 September to 11 November, which corresponds exactly to the time of the Yarmouth herring fair (Morris 1975–92, vol. 1, 1/1) and brings to mind the measurement of time at Sandwich by 'two whole herring seasons' during Harold Harefoot's reign. Dover's Domesday entry also alludes to the ship service it owed to the king, an obligation noted for Sandwich as well in the eleventh century (Ballard 1920, 20) and eventually recognised as one of the chief characteristics of the Cinque Ports confederation, which firmly controlled the Yarmouth herring fair by the thirteenth century. One prominent scholar even attributes Harold's puzzling discharge of the Anglo-Saxon fleet right before the

Norman invasion not to the stated shortage of provisions, but to the eagerness of the impressed sailors to participate in the mackerel (*Scomber scombrus*) fishery in the Channel and, presumably, the herring fishery that started off the East Anglian coast in late September (Oppenheim 1907, 128). The rights of the Cinque Ports to dry their nets and land on the shore at Yarmouth during the fishing season were not enshrined by charter until the thirteenth century, but the Yarmouth fishing activities of the Portsmen are evident by the early twelfth century (Saunders and Millican 1939, 32). Indeed, the foremost scholar of the Cinque Ports argues that the economic interests of the Ports in the annual herring fishery at Yarmouth – which appear to date from before the Conquest – were the prime motive behind the formation of the confederation (Murray 1935, 16–19). The Yarmouth herring fishery was attracting hundreds and perhaps thousands of foreign fishermen from at least the early twelfth century, as is evident from the reference in a saint's life to a fleet of forty vessels from the Low Countries at Yarmouth sometime between AD 1107 and 1127 (Bethell 1970, 94). In AD 1208, the burgesses of Yarmouth gave the king more than 110,000 herring so that they could farm their own borough (PR, vol. 61, 14), an indication of how enormous the catch was by that period.

The early commercialisation of the east-coast herring industry is also evident in the dates of fairs established in coastal communities, which were often timed to coincide with the migration of herring (Kowaleski 2010, 120–2). Thus the port towns of Northumberland and Durham had a group of August fairs, while farther south in Yorkshire, including the ports of Whitby, Scarborough, Filey and Bridlington, the fairs usually began later in August or in September. The shoals generally reached southern Yorkshire and northern Lincolnshire by September, which reflects the timing of fairs in Ravenser, Grimsby and Saltfleetby. Many of these fairs were established very early; the Whitby fair was operating by AD 1122, Bridlington's was recorded by AD 1200, Hartlepool's by AD 1201 and Yarmouth's by AD 1208 (but probably commenced much earlier); it is likely that the charters founding these fairs were really sanctioning well-established markets rather than initiating entirely new fairs. The herring fairs at Scarborough and Yarmouth, both of which lasted about six weeks, were also among the busiest and longest fairs in medieval Britain, another indication of the international market for herring caught off the English coast. The twelfth-century *Life of St Kentigern* and a charter of Malcolm IV (AD 1141–65) also note the presence of Scots, English, French and Flemings in the Scottish fisheries around the Isle of May at this time (Stuart 1868, vii).

The extant sources for the twelfth and thirteenth centuries are far better than those for the eleventh century, so we can see more clearly the commercial intensification of sea fishing. Tithe grants and disputes, for instance, offer insights into how much more lords of the foreshore thought

was at stake in this period. These cases also capture a fully developed commercial industry, suggesting that commercialisation had occurred in the eleventh century and possibly earlier. In the 1120s, for example, a dispute over tithes on the Yorkshire coast, between the abbot of Whitby and the prior of Bridlington was settled by laying out the obligation for tithe depending on where fishers landed their catch. Problems flared up again a few decades later, in the 1190s, however, and now involved not only the monks of Whitby infringing on the fish tithes that Bridlington claimed, but also canons from Grimsby who were trying to get a share of the profits coming in from the Filey fishermen, much to the distress of Bridlington priory (Farrer 1914–16, vol. 2, 222–3, 466–8). The fish tithes of Whitby itself were first noted in the 1090s, when they were granted by William Percy to the priory and monks of Whitby (Clay 1963, 20). A tithe dispute at Alnmouth in AD 1249 gives valuable details on fishers who borrowed boats, nets or tackle to fish in the 'Doggedrawe', probably an early reference to fishing for cod in the Dogger Bank (Bateson *et al.* 1893–1940, 470–1), a fishery first documented in AD 1189 in reference to Scarborough fishermen (Farrer 1914–16, vol. 1, 286). Indeed, although the proof is less explicit than for the Alnmouth case, there is strong evidence that fishermen were paying tithes on this fishery (which must have included cod) in Hornsea by the tenth century, and by the twelfth century in Alnmouth, Filey, Scarborough and Whitby as well as Berwick (Farrer 1914–16, vol. 2, 222–3, 466–8; vol. 3, 26–7; Littler 1979, 46). Tithe grants in the 1260s for the vicarage of Berwick mention 'white fish coming from boats of the sea', almost certainly a reference to cod fishing (Barlow 1945, 156–7).

A variation on tithes, the 'shares' that seigneurial lords (and, later, towns) claimed from their tenant-fishermen, are also mentioned more frequently in the twelfth and thirteenth centuries, when they become (like the port farms in Cornwall) paid in cash in lieu of kind, another indication of the marketable surpluses increasingly being produced. Fécamp Abbey was collecting shares on its sea fisheries at Rye from at least the first half of the twelfth century, when the local industry was already very well organised and diversified, with large boats of 26 oars (probably employed in offshore fishing at Yarmouth and elsewhere) and four other types of smaller boats likely oriented towards inshore fishing (Round 1898, 79–80). When the accounts survive, as they do for the late thirteenth century in Rye and Winchelsea, we can get a very good idea of the relative profitability of different fisheries (Kowaleski 2003, 187–9). In the AD 1281–8 accounts for Rye, for example, herring shares accounted for 55% of the profits; plaice, 29%; mackerel, 10%; and sprat, 6% (Dulley 1969, 38–9).

Toll lists and royal purchases of fish also show the growing market for marine fish, as well as developing regional specialisations, although this evidence tends to

Table 3.2. Early tolls on fish

Late eleventh- to mid-twelfth-century London tolls at Billingsgate (Robertson 1925, 72–3)
½d from fish coming by boat to the bridge, 1d if coming in a larger ship
6s owed by a large ship of the men of Rouen with wine and whale/porpoise [*craspisce*] and a twentieth part of the fish

AD **1052 × 1066 tolls at Abingdon (inland river port)** (Blair 2007, 258–9; Hudson 2002, 174–5, 218–19)
1C herring or suitable price for them owed each year to cellarer of Abingdon Abbey from each boat of city of Oxford travelling on River Thames past Abingdon

AD **1100 × 1135 Newcastle tolls** (Martin 1911, 334–6)
1d from each merchant carrying fish by horse
1d from 1M herring
Salt and herring brought to port must be sold on ship (Ballard 1913, 168)

AD **1141 × 1153 Chester** (Rees 1985, 60)
Haughmond Abbey is granted boat on River Dee and annual right to purchase 6M herring free of toll at Chester

AD **1147 × 1183 Cardiff tolls** (Ballard 1913, 177–8)
½d on cart loaded with corn or fish, buyer to pay 1d

AD **1180s Bury St Edmunds** (Butler 1949, 75–7)
15d toll on carts of herring of London citizens coming from Yarmouth

AD **1180 × 1189 Wearmouth** (Snape 2002, 133–5)
Merchandise brought by sea to be landed, except salt and herring, which can be sold on the ship or in the borough
The bishop retains custom on fish bought at Wearmouth, as Robert de Bruce has from his men at Hartlepool

AD **1194 × 1202 Thetford** (Ransford 1989, 412)
Canons of Waltham Abbey granted freedom from toll every year on 20 carts going through Thetford to and from Norwich with herring

AD **1225 murage toll for 3 years at Scarborough** (CPR 1216–25, 508–9)
12d from each great ship landing corn, salt, fish, herring or other saleable things
6d from each *sornecca* and 2d from each boat landing with same saleable goods
1d from each cart carrying fish or other saleable things into the town

AD **1228 Torksey tolls (an inland river port)** (Gras 1918, 156)
4d from 1 last herring
4d from 1C great ling
2d from 1C cod [*milvel*]
1d from 1C of cropling
2d from 1C of hard fish [stockfish]
½d from 1 salmon

AD **1232 murage toll for 2 years at Bristol** (CPR 1225–32, 483)
¼d from 1M herring coming to be sold
1d from each 1C salmon, congers and cod [mulwell] coming to be sold
¼d from each 1C hake coming to be sold

c. AD **1240 Exeter tolls** (Schopp 1925, 31, 36–8)
½d for 4 sticks eels for sale if carried by horseback
1 stick of eels (¼C) out of 7 if for sale and carried from the ship or boat
Custom owed to go to sea [nearby port of Topsham] to buy fish; if the merchant buys nothing, he can buy in town and pay nothing; if he buys half a load at sea, he can buy another half in town and pay nothing
½d a seam if fish in a ship or boat is landed; no custom if not landed
1d for 1M red herring coming to port and then carried by horseback
3½d per cart for red herring coming to port and then carried by cart
1C red herring for bailiff and 1M red herring for the castle when coming to port
1d per 1M red herring purchased in town
3½d per cart for red herring purchased in town

AD **1252 tolls at St Ives fair** (Moore 1985, 197)
4d for each … last of dried herring
10d from each marks' worth of dried fish

(Continued)

Table 3.2. Early tolls on fish (Continued)

AD 1216 × 1272 prisage granted to friars at Bristol (Bickley 1900: 89–90; CIM, vol. 1, 366)
2 conger from each boat carrying fresh conger
4 cod from each boat carrying fresh cod [*mulvellum*]
8 hake from each boat carrying fresh hake
8 haddock from each boat carrying fresh haddock
8 plaice from each boat carrying fresh plaice
4 ray from each boat carrying fresh ray [*retias*]

c. **AD 1266 London tolls** (Riley 1861, 204, 205, 206, 207)
2d per cart bringing fish or poultry into Westchepe
½d per cart bringing cod [mulwell], herring or other fish
London Bridge tolls (freemen exempt):
2 cod from a vessel bringing only cod [mulwell]
1 cod and 1 ray from vessel bringing cod [mulwell] and rays
1C herring from vessel with fresh or salted herring
Only 2d strandage from vessel bringing sea-bass, conger, surmullet, turbot, shad or eels
26 mackerel from vessel bringing mackerel; same for haddock
½d and 1 pannier from boat bringing 5 panniers of whelks
2d per 1C cod [mulwell], 1d per ½C and ½d per ¼C of cod [mulwell] carried into town by a stranger and out of town for resale
½d per 1M white herring purchased by a stranger
¼d for 1M red herring purchased by a stranger
1 fish per dorser brought by a dealer on a horse and ½d for the horse, except for cod [mulwell] and ray
¼d per horse-load to be paid by stranger dealers who buy fish to carry it out of city for resale
1 salmon (the second-best) from cart bringing salmon belonging to foreigner, 2d for cart
5 herrings from carts bringing white or red herring to city and 2d for the cart
5 mackerel from carts bringing mackerel to city and 2d for the cart
1 cod from carts bringing cod [mulwell] to city and 2d for the cart
Carts bringing eels pay nothing for the fish but 2d for the cart

AD 1291 Ipswich tolls (Twiss 1873, 192–3)
4d per 1ast of red herring; ½d per 1M red herring
4d per last of fresh or salt herring, unless seller caught it himself
2d per 1C any kind of hard fish
½d per salmon
4d per 1 quintal whale meat
4d from each float of nets dried on dry ground
2d on every barrel of sturgeon
2d on every cart bringing fish or herring to market for sale
½d on every horse-load bringing fish or herring to market for sale
¼d on every man's load or a barrow-load of fish or herring brought to market for sale
1d on every porpoise
¼d on every salmon

AD 1303 tolls at Ipswich (Gras 1918, 159–63)
1d from nets for taking herring
½d from nets for taking mackerel
4d from 1C whale
4d from 1C of turbot
4d from 1C cod [mulwell] if bought
1d from a cart with fish
½d from a horse-load of fish
10d from 1 last herring going to the River Thames
4d from 1 last herring going to France
10d from 1M mackerel going to the River Thames
4d from 1M mackerel going overseas
1d per oar from each fishing ship sold
½d from ling

Note: The list includes all known tolls on fish recorded before AD 1250. Only relevant fish tolls have been selected from what are often long lists of commodities paying customs. See also Masschaele 2007.

lag chronologically behind the archaeological evidence for the consumption of particular species at specific places. The selection of fish references in the toll lists noted here (Table 3.2) do, however, suggest that further analysis of the many lists that survive could add much to our understanding of not only the types of fish available locally and from overseas, but also the relative value of different types of fish, the means of transport adopted to get them to urban markets, and even the methods employed to catch them. They also open up the issue of the transactions costs involved in getting sea fish from the fishermen to the market to the consumer, as well as the relative pricing of fresh as opposed to preserved fish, questions that deserve more extended treatment in understanding the commercialisation of sea fishing (Kowaleski 2000c, 31). Such toll lists also support arguments (such as Lampen 2000) that town dwellers were the main purchasers of fish, particularly cured fish. The early appearance of occupational surnames like 'Harengarius' in towns, such as eleventh-century Winchester, also reinforces this argument (Barlow 1976, 39, 88).

Although tolls were regularly being assessed at English ports by the eighth century (Kelly 1992), they were not enumerated by commodity. Only in the mid–late twelfth century and increasingly in the thirteenth do we find fish and other goods being singled out for specific tolls, though in a context that implies that tolls on fish had already been collected for a long time (Burton 2004, 54; Ransford 1989, 412; Table 3.2). The earliest tolls on most goods were charged in kind, but in-kind tolls on fish in particular seem to have lasted longer (Table 3.2) – well into the thirteenth century in many cases – probably because it was easy to enumerate and take away fish and because of the value, especially during fasting periods, of an immediate supply of fish. Except for imported whale or porpoise, herring was the first type of fish to be singled out in toll lists. It took more than a century for other types of fish to be specified in tolls, but from the late 1220s through 1260s, many different types of sea fish began to be enumerated, indicating the growing profitability of the trade in marine fish throughout England, albeit especially in port towns.

Large royal purchases, which reveal the amounts and types of fish, and often where they were acquired, are recorded from the very early thirteenth century in the Pipe Rolls, royal wardrobe accounts, Liberate Rolls, Patent Rolls and Close Rolls, with herring from Norfolk and Suffolk among the first purchases (PR, vol. 51, 23, 235). By the 1230s and 1240s, the king's purchases reflected the increasing specialisation of the sea fisheries; he bought most of his merling (i.e. whiting [*Merlangius merlangus*]) and plaice from Winchelsea in Sussex and by the 1250s was acquiring herring, sole (*Solea solea*), haddock and conger on a regular basis (CCR, vol. 3, 402; vol. 6, 54, 429–30; vol. 7, 68; vol. 10, 153; CLR, vol. 2, 12, 7). Congers came mostly from Southampton, eels from Cambridge and

herring from Norfolk in these accounts (CLR, vol. 2, 10, vol. 12, 91, 114, 127, 148, 166, 168, 184, 215, 266). One of the earliest references to cod was a royal order in 1247 to the sheriff of Kent to send for Christmas provisions 200 large and 'well-powdered' cod ('*cc mulewellis grossis et bene pudratis*') (CCR, vol. 6, 96), which likely refers to cod from British waters preserved by lightly salting, not dried stockfish from more distant, Scandinavian waters. The importance of regional market identification was evident in the work of an anonymous thirteenth-century poem associating products with places; the poem contained more references to fish than any other single item, a strong indication of how commercialised the fish trade had become by this period (Rothwell 1975, 881–4). What is especially significant is the extent to which particular types of fish were associated with particular ports by this period. This regional specialisation is reflected in the poem's references to eels of Cambridge, herring of Yarmouth, plaice of Winchelsea, merling of Rye and mullet (Mugilidae) of Dengemarsh, all of which can also be seen in the wholesale purchase of fish for the royal household. Such market specialisation is the hallmark of full commercialisation.

What does this documentary evidence tell us about what prompted the commercialisation of marine fishing in medieval Britain? What clues does it offer to the particular timing, pace, geographic extent and reasons for the 'fish-event horizon'? Current theories (particularly from archaeologists) have focused on a variety of factors, including rising consumer demand from the increasing rate of urbanisation, the exhaustion of freshwater fisheries in the face of growing consumer demand, the claims for rent/tribute/taxes from state formation, the Christian dietary restrictions on eating meat, and the introduction of new tastes and fishing and curing methods from the Scandinavians (cf. Chapter 21). Expanding consumer demand was the most likely impetus behind the 'fish event horizon', particularly given what we know about the prosperity of eleventh-century England and its healthy seaborne trade, as well as the rapid pace of urbanisation over the eleventh to early thirteenth centuries (Barrett *et al.* 2011; Gardiner 1999; Sawyer 1965). The most comprehensive study to date of the fish trade in the medieval German Empire (widely defined) also attributes the eleventh-century rise in fish consumption to urban demand and downplays the influence of Christian fasting rules (Lampen 2000). The sentiments of Aelfric's Fisherman support this view: that marine fishing was known and practised before the eleventh century, but not widely carried out because of the costs in manpower and labour compared with river fishing, whether by boats, nets, lines or weirs. Rising consumer demand helped to spur greater capital investment, which in turn augmented the pace and extent of the commercialisation of marine fishing from the eleventh century on. One recent argument posits that fish eating was a mark of social status in the eleventh century, implying that wealthy fish eaters were willing to invest in sea fishing in order to assure a steady supply of this high-status food (Fleming 2000).

This intensified consumer demand would certainly have outstripped the supply of freshwater fish, but to claim that sea fishing was stimulated *c*. AD 1000 'as a result of a sharp decline in large freshwater fish', as popular simplification of archaeological results has done (Kinver 2009), probably goes too far, as do interpretations that see the rise of marine fishing as a response to an ecological crisis in the supply of freshwater fish (Roberts 2007). Although there is slight archaeological evidence that the length of some freshwater fish may have declined over time (Chapter 15), freshwater fish were always expensive, relatively scarce, and not widely available to most medieval people (Dyer 1988). Indeed, it was so expensive in the High Middle Ages that only the aristocracy could afford to eat freshwater fish on a regular basis, so it was not a mainstay of the peasant or artisan's diet. The nobility were the main force behind the (very costly) development of fish ponds, particularly in the thirteenth century. Rising consumer demand also stimulated the construction of weirs to catch more freshwater fish from the twelfth through thirteenth centuries, but the increasing regulation imposed on weirs in this period reflected concerns about their effect on navigation more than worries about over-fishing (Flower 1915; Hooke 2007, 53–42). During the late Middle Ages, regulation of weirs did proliferate and included strictly enforced rules about the size and placement of nets and traps and seasons to fish in order to protect small fry (for example, Riley 1868, 214–20, 483–6, 500–9; Wright 1997), but sea fishing was fully commercialised by this point. As a result, the chronology that associates the rise of sea fishing with over-fishing of freshwater fish is flawed, particularly when the significant expansion of weirs, freshwater fish ponds and cyprinid farming from the late twelfth century on is taken into consideration.

Nor do some of the other hypotheses prove very convincing. State formation does not represent a reasonable explanation of what may have stimulated sea fishing in medieval Britain, since the Anglo-Saxon state had been efficiently extracting taxation for centuries by the year AD 1000. The influence of the Scandinavians deserves more sustained attention, but their impact was likely strongest in northern Scotland, where a lasting Scandinavian community developed and Norwegian methods of preserving cod by air drying could be practised (Barrett 1997; 2012). Scandinavians were certainly not responsible for bringing their techniques of drying cod to eastern England, since the climate there was unsuitable for this curing process. With the exception of Norfolk and Suffolk, where the herring grounds also happen to be most prolific, thus making it difficult to ascertain the impact of Scandinavian settlers as opposed to other residents, there is not a very strong geographic correlation between the areas where the herring

industry first developed – including Essex, Kent, and Suffolk (Table 3.1) – and the regions of strongest Scandinavian influence, from Yorkshire down through Lincolnshire.

There are, however, two additional factors that deserve more attention if we are to understand the medieval origins of the commercialisation of sea fish. One concerns fishing methods, about which there is a lamentable lack of documentary information before the late thirteenth century. Scholars have nonetheless posited, albeit with no concrete evidence, that the introduction of drift nets around AD 1000 helped to increase significantly the herring catch (Ervynck *et al.* 2004, 232; Hagen 1995, 202; Hodges 1982, 143; Jones 1982, 84). Seine nets, in use for thousands of years, could be operated from the shore or small boats with as few as two men. Drift or gill nets could yield larger and more various catches because they were generally deeper and far longer than seine nets, which means they also needed more labour to work (Davis 1958, 54–61, 64–9; Steane and Foreman 1988, 156–9; Von Brandt 1984, 283–307, 355–61), thus reflecting Aelfric's Fisherman's comment that sea fishing required too much effort. The leads, weights and sinkers used on nets can be found in dateable archaeological contexts, as can the hooks used in line fishing, but these finds by themselves do not yield information on what the tackle caught, nor do they reveal much about the type of net used. It is perhaps instructive, however, to consider the story that Bede tells of the ignorance of the late seventh-century Saxons of Sussex, who knew nothing of the art of sea fishing until Bishop Wilfrid showed them how effective throwing eel nets into the sea could be (Colgrave and Mynors 1969, 374–5). It is nonsense to think that coastal residents were unfamiliar with net fishing even in this early period, but worth considering the idea that medieval people could realise the benefits of new fishing methods, particularly the innovation of using several shorter nets by stringing them together. Probably more important here, however, is not the invention or adaptation of fishing methods, but where the capital came from to finance more and larger nets, boats and lines, as well as the labour required to work them. One-third of the catch taken by the bishop's men, for instance, went to those who had supplied the nets, an early example of the share system in practice.

On occasion the documentary evidence is suggestive, but not conclusive. In the late eleventh or early twelfth centuries, for instance, St Werburgh Abbey in Chester was granted the tithes of fisheries in Rhuddlan and Anglesey, along with the right to have a ship of 10 nets, which sounds much more like drift nets than seine nets (Tait 1920, 14, 17, 21). The *c*. AD 1130 agreement on fishing shares between Fécamp Abbey and its tenants in Rye distinguished ships (rowed by 8 to 12 men) called *heccheres*, which probably refers to heak nets, which have been variously identified as herring nets, cod nets or trammel nets (Dulley 1969, 37; Round 1898, 79–80; *s.v. haking* in OED). Their reference within the context of shares, however, does point to the importance of pooling capital in the fishing industry, a risk-reduction strategy that allowed humble fishers contributing an oar, cords or only their labour to collect part of the fishing venture's profit, although those supplying capital equipment, particularly the boat or nets, took far larger percentages of the catch (Kowaleski 2006, 917–22). An echo of how capital investment in fishing equipment could be managed is also evident in a thirteenth-century manorial custumal from Sussex, which required boat owner–tenants of the bishop of Chichester to keep nets of 200 meshes with their cords at Yarmouth for the profit of the bishop, who agreed to pay 6d for this right, along with necessary repairs. Boat owners were also obliged to maintain a mackerel net of 150 meshes with two cords, with the bishop paying 4d and necessary repairs (Scargill-Bird 1887, 84). The capital investment made by the landlord here would have augmented that of the boat owners and other fishermen and may help explain the quickness with which the English moved sea fishing to a highly commercialised state in the eleventh and twelfth centuries.

Early statute legislation on fishing, primarily concerned with weirs, also makes reference to different types of nets, including a contraption called a wondrychoun, in AD 1376, which has been interpreted as the first documentary reference to trawling (Bailey 1990, 110; Moore and Moore 1903, 171–5). By the late fourteenth and the fifteenth centuries references to different types of nets and fishing methods had multiplied ten-fold. The wills of the small fishing port of Hythe (Kent), for instance, note flew nets (the classic herring drift nets of eastern England), seine nets, trammel nets, shot nets, nets called *depings*, sprat nets, herring nets, mackerel nets, small hooks, *herbewe* hooks, great ropes and fishing lines (Hussey 1937–8). Scholars need to do a lot more work to track when these different types of fishing gear came into use, and whether the increasingly varied and precise vocabulary of fishing tackle in the late Middle Ages reflects new technology or simply the enhanced terminology of a more literate society.

Even more important than a possible change in fishing methods were potential transformations in curing. The rapid spoilage rate of fresh fish means that it could not have been commercialised without being preserved. Fresh fish can be transported inland quite impressive distances, but the costs of this effort put such fish out of the reach of most consumers. Although great strides have been made by archaeologists in understanding the medieval curing processes that fish could have undergone, documentary historians have had little new to say about curing techniques in medieval Britain. Most studies rely on a 1955 book by a food chemist (Cutting 1955), who thoroughly described the chemical processes that curing involves, but who depended rather uncritically on a few printed sources and who treats the Iron Age to the Industrial Era as one period. There is, in fact, a fair amount of archival evidence available

that has yet to be systematically brought together for the medieval period. Some evidence has been assembled for the later Middle Ages, when we can more easily date new methods that involved producing a more commercially viable product by gutting fish and salting it at sea (Littler 1979, 137–48; Unger 1978). The early commercialisation of the herring industry in particular, however, can only be understood in the context of the curing processes that this fatty fish – which begins to spoil only a few hours after capture – underwent to enable it to be traded inland or overseas. Also important to remember here is that all early herring cures involved salt in some way, and that the British climate dictates that salt can only be gathered by heating it to high temperatures, which requires plentiful fuel. This relationship – as James Campbell (2002) has recognised in discussing the confluence of fuel sources in the Norfolk Broads, with salt pans and a rich supply of herring in Norfolk and Suffolk – is crucial if we are to track the factors that helped sea fishing commercialise.

There are several key questions that need to be resolved in assessing the role of curing in the commercialisation of the sea fisheries. One is tracing the source of salt before higher-quality Biscay salt began to be imported in the late thirteenth and early fourteenth centuries, as well as the processes employed and costs associated with making it in Britain in the crucial period between the tenth and early thirteenth centuries. We actually know a fair bit about the source of the salt (Keen 1988), but much less about how much it cost to make or its suitability for curing fish. Transport costs for the fishing industry, however, were probably very low because salt works were located on the foreshore, not far from where herring boats would have drawn up. Indeed, it is likely that the earliest cures consisted of little more than pouring salt on heaps of herring thrown up on the beach.

A second question concerns the types and costs of the cure most often used in the early and High Middle Ages. In the English climate, salting was probably the oldest curing process. Salt pans and fisheries were often associated in early charters; two eighth-century charters for coastal Kent grant not only a piece of land suitable for boiling salt, but also 120 full wagon-loads of wood for boiling the salt, and nearby there were fishermen's sheds (Brooks and Kelly 2013, nos 10, 11, 22). 'White' herring were salted, but there is considerable difference of opinion about whether they were fresh, lightly salted, heavily salted or packed in brine, and about how long the cure lasted. It is likely that the Anglo-Saxon herring renders (Table 3.1) were salted herring, which would hold up for several months. Accounts of the herringry in the Durham priory manor of Wardley (southeast of Newcastle) from 1289 to 1291 reflect the commercial curing process as it had probably existed for centuries (Britnell 2014). The purchase of herring accounted for 75% of the total costs in both years of the account, one of which produced about 9 lasts of cured herring and the other

15 lasts (there were 12,000 fish to each last). Salt was the second-largest expense, at about 14%, followed by the costs of curing (7%) which included salting, washing, putting on rods and hanging the herring, followed by the costs of binding and packing the herring with straw into baskets. The actual cost of labour (as opposed to the purchase of tubs, rods, baskets and straw) ranged from 4% to just over 6%, with the lower costs coming in the year that curing was farmed out to one man for a set rate of 10s. The remaining expenses were for cutting and carrying wood (40 cartloads the year that 9 lasts were cured), transporting and landing the herring and salt, and repair of the herringry, which was primarily constructed of wood, with a roof (probably thatch) and three doors with locks, surrounded by a woven fence with a gate. These expenses make clear that curing on a large scale required capital above all, since well over 90% of the total outlay was spent on purchasing and carrying herring and salt and repairing the herringry. The process for salting other 'fish', such as *Crassum piscis*, which refers to whale or porpoise, documented by the mid-twelfth century (PR, vol. 2, 51; vol. 60, 21), was probably similar.

Although salting fish was by far the preferred cure in early fishing, other techniques were also known. Drying fish by the heat of the sun or by fire was another ancient technique in Britain, although better suited to white-fleshed fish than to oily fish such as herring or mackerel. Herring could be dried, but like most dried fish, it also underwent an initial salt cure to prevent putrefaction. In AD 1185, the peasant tenants of Temple Ewell near Dover were required to preserve herring by salting and drying it; the lord undertook to find them the salt, rods for hanging the fish and wood for drying, implying that they were using heat to expedite the drying process (Lees 1935, 23). These provisions also point to the capital investment of the lord in curing fish. Less-fatty white fish could possibly be dried without artificial heat, which is what is suggested in the AD 1240 lease in Porthoustock (Cornwall) of a house and adjoining plot of land 10 ft by 20 ft (3.0 × 6.1 m) for drying fish, a facility from which the Gascons were excluded (Hockey 1974). Salted and dried hake and conger were a big enough industry in early thirteenth-century Cornwall to attract the attention of southern French merchants, who paid for the right to salt and export fish (Kowaleski 2003, 206). Smoking was a variation on drying, and most well known for the 'red' herring of Yarmouth, which were salted and smoked for at least one week and often longer. References to smoke-houses can be found as early as the mid-thirteenth century (Gardiner 1954, 42), but they are more common by the early fourteenth century (Saul 1981, 35). With further research, we will likely be able to date the use of smoke-houses more precisely. Wood, including thorn and hazel, was used to effect the smoked cure (Cutting 1955, 71; Vasey 1978, 20)

A third question concerns when herring began to be gutted before curing, since gutting provides a more long-lasting cure, one that considerably prolongs the shelf life and quality of the preserved herring. There has been considerable difference of opinion on this issue, although the documentary evidence does not allow us to trace this innovation much earlier than the mid-thirteenth century in Scania (Unger 1978; see also Chapter 13; Chapter 15). Gutting and salting the fish at sea was also a crucial innovation, especially when combined with further curing on shore, which produced a superior and exportable product, although it is unlikely this process occurred much before the late fourteenth century (Childs 2001, 23; Kowaleski 2003, 211).

Lastly, a word about cod. The archaeological evidence from London shows that the remains of traded cod do not really begin to appear in significant numbers until the mid-thirteenth century (Orton *et al.* 2014), a chronology supported by the documentary evidence presented here. It appears that the British cod industry (except for northern regions of Scotland, including Orkney and Shetland) underwent its greatest period of expansion in the late rather than the High Middle Ages. The same was true of the fish import trade. Herring dominated the fourteenth-century imports, but dried and often also salted cod became more plentiful in the fifteenth century, when the varieties of imported fish – some probably caught by English fishers in distant waters – proliferated (Kowaleski 2000a; 2003; Littler 1979, 286–313). New techniques developed by archaeologists have vastly increased our understanding of the development and direction of the medieval trade in cod, but they should not obscure the fact that herring commercialised more than a hundred years before cod and was probably consumed by more people and in larger amounts than was cod in medieval England. The herring industry also employed far more fishers and boats in England, and relied less on imports than did the cod trade. Scholars studying the commercialisation of the British sea fisheries need, therefore, to take into account the very different trajectories of the herring and cod fisheries and be wary of assuming that the same factors were responsible for the transformation of both industries in all regions.

Acknowledgements

I thank Martin Chase, David McDougall and especially Ian McDougall for help with the linguistic evidence; Lesley Abrams and Susan Kelly for guidance on Anglo-Saxon charters; Morgan Kay for advice on the Welsh-language sources; and Elizabeth Ewan for answering my queries on Scottish sources. Special thanks are due to the late Richard Britnell for allowing me to see his transcription of the section of the Durham account rolls on the herringry at Wardley before publication.

Archival references

TNA The National Archives, Exchequer Accounts Various, E101/4/10, 4/24, 16/31, 18/4, 531/19.

References

Allen, J. R. L., R. J. Bradley, M. G. Fulford, S. J. Mithen, S. J. Rippon & H. J. Tyson. 1997. The archaeological resource: chronological overview, in M. G. Fulford, T. Champion & A. J. Long (eds) *England's Coastal Heritage: a survey for English Heritage and the RCHME*: 103–53. London: English Heritage and Royal Commission on the Historical Monuments of England.

AND Anglo-Norman online hub: the Anglo-Norman dictionary. Aberystwyth: University of Aberystwyth and University of Swansea. Available at: http://www.anglo-norman.net/.

Bailey, M. 1990. Coastal fishing off south east Suffolk in the century after the Black Death. *Proceedings of the Suffolk Institute of Archaeology and History* 37: 102–14.

Ballard, A. (ed.) 1913. *British Borough Charters 1042–1216*. Cambridge: Cambridge University Press.

Ballard, A. 1920. *An Eleventh-century Inquisition of St Augustine's Canterbury*. London: Oxford University Press for the British Academy.

Barlow, F. (ed.) 1945. *Durham Annals and Documents of the Thirteenth Century*. Vol. 155. Durham: Surtees Society.

Barlow, F. 1976. The Winton Domesday, in M. Biddle (ed.) *Winchester in the Early Middle Ages*: 1–141. Oxford: Clarendon Press.

Barrett, J. H. 1997. Fish trade in Norse Orkney and Caithness: a zooarchaeological approach. *Antiquity* 71: 616–38.

Barrett, J. H. 2007. The pirate fishermen: the political economy of a medieval maritime society, in B. Ballin Smith, S. Taylor & G. Williams (eds) *West Over Sea: studies in Scandinavian sea-borne expansion and settlement before 1300*: 299–340. Leiden: Brill.

Barrett, J. H. (ed.) 2012. *Being an Islander: production and identity at Quoygrew, Orkney, AD 900– 1600* (McDonald Institute Monograph). Cambridge: McDonald Institute for Archaeological Research.

Barrett, J. H. & M. P. Richards. 2004. Identity, gender, religion and economy: new isotope and radiocarbon evidence for marine resource intensification in early historic Orkney, Scotland, UK. *European Journal of Archaeology* 7: 249–71.

Barrett, J. H., A. M. Locker & C. M. Roberts. 2004a. 'Dark Age economics' revisited: the English fish bone evidence AD 600–1600. *Antiquity* 8: 618–36.

Barrett, J. H., A. M. Locker & C. M. Roberts. 2004b. The origins of intensive marine fishing in medieval Europe: the English evidence. *Proceedings of the Royal Society B* 271(1556): 2417–21. doi: 10.1098/rspb.2004.2885

Barrett, J. H., D. Orton, C. Johnstone, J. Harland, W. Van Neer, A. Ervynck, C. Roberts, A. Locker, C. Amundsen, I. B. Enghoff, S. Hamilton-Dyer, D. Heinrich, A. K. Hufthammer, A. K. G. Jones, L. Jonsson, D. Makowiecki, P. Pope, T. C. O'Connell, T. de Roo & M. Richards. 2011. Interpreting the expansion of sea fishing in medieval Europe using stable isotope analysis of archaeological cod bones. *Journal of Archaeological Science* 38: 1516–24.

Bateson, E., H. H. E. Craster, M. H. Dodds, A. B. Hinds, J. Hodgson & K. H. Vickers (eds). 1893–1940. *A History of Northumberland, Issued under the Direction of the Northumberland County History Committee*. Vol. 2. Newcastle-upon-Tyne: A. Reid/London: Simpkin, Marshall, Hamilton, Kent & Co.

Bethel, D. 1970. The lives of St Osyth of Essex and St Osyth of Aylesbury. *Analecta Bollandiana* 88: 75–127.

Bickley, F. B. (ed.) 1900. *The Little Red Book of Bristol*. Bristol: W Crofton Hemmons/London: Henry Sotheran.

Blair, J. 2007. Transport and canal building on the upper Thames, 1000–1300, in J. Blair (ed.) *Waterways and Canal-building in Medieval England*: 254–86. Oxford: Oxford University Press.

Bond, C. J. 1988. Monastic fisheries, in M. Aston (ed.) *Medieval Fish, Fisheries and Fishponds in England*. (British Archaeological Report 182): 69–112. Oxford: Archaeopress.

Brett, M. & J. Gribbin (eds) 2004. *English Episcopal Acta. Volume 28: Canterbury 1070–1136*. Oxford: Oxford University Press for the British Academy.

Britnell, R. H. (ed.) 2014. *Durham Priory Manorial Accounts, 1277–1310* (Publications of the Surtees Society). Woodbridge: Boydell and Brewer.

Brooks, N. P. 1988. Romney Marsh in the early middle ages, in J. Eddison (ed.) *Romney Marsh: evolution, occupation, reclamation*. Oxford: Oxford University School of Archaeology.

Brooks, N. P. & S. E. Kelly (eds) 2013. *Charters of Christ Church, Canterbury*. (Anglo-Saxon Charters 17). Oxford: Oxford University Press for the British Academy.

Brown, V. (ed.) 1992. *Eye Priory Cartulary and Charters. Part I*. (Suffolk Charters 13). Woodbridge: Boydell Press for Suffolk Records Society.

Burton, J. (ed.) 2004. *The Cartulary of Byland Abbey*. (Publications of the Surtees Society 208). Woodbridge: Boydell Press for the Surtees Society.

Butler, H. E. (ed. & trans.) 1949. *The Chronicle of Jocelin of Brakelond: concerning the acts of Samson, abbot of the Monastery of St. Edmund*. New York: Oxford University Press.

CCR Calendar of Close Rolls, Henry III. 1902–75. Vol 3, 6, 7, 10. London: HMSO.

CIM Calendar of Inquisitions Miscellaneous. Vol. 1. 1916. London: HMSO.

CLR Calendar of Liberate Rolls, Henry III. 1917–64. Vol. 2. London: HMSO.

CPR Calendar of Patent Rolls 1216–25. 1901. London: HMSO.

CPR Calendar of Patent Rolls 1225–32. 1909. London: HMSO.

Campbell, A. (ed.) 1973. *Charters of Rochester*. London: Oxbow Books.

Campbell, J. 2002. Domesday herrings, in C. Harper-Bill, C. Rawcliffe & R. G. Wilson (eds) *East Anglia's History: studies in honour of Norman Scarfe*: 5–17. Woodbridge: Boydell Press.

Childs, W. R. 1995. England's Icelandic trade in the fifteenth century: the role of the port of Hull, in P. Holm, O. U. Janzen, O. Uwe & J. Thor (eds) *Northern Seas Yearbook, 1995* (Fiskeri- og sofartsmuseets studieserie 5): 11–31. Esbjerg: Fiskeri- og Søfartsmuseet.

Childs, W. R. 2000. Fishing and fisheries in the Middle Ages: the eastern fisheries, in D. J. Starkey, C. Reid & N. Ashcroft (eds) *England's Sea Fisheries: the commercial sea fisheries of England and Wales since 1300*: 19–23. London: Chatham.

Childs, W. R. 2001. Mercantile Scarborough, in D. Crouch & T. Pearson (eds) *Medieval Scarborough: studies in trade and civic life*: 15–32. (Yorkshire Archaeological Society Occasional Paper 1). Scarborough: Yorkshire Archaeological Society/ Scarborough Archaeological & Historical Society.

Clay, C. T. (ed.) 1963. *Early Yorkshire Charters. Volume 11: the Percy fee* (Record series, extra series 9). Wakefield: Yorkshire Archaeological Society.

Colgrave, B. & R. A. B. Mynors (eds) 1969. *Bede's Ecclesiastical History of the English People*. Oxford: Clarendon Press.

Crump, B. & S. Wallis. 1992. Kiddles and the Foulness fishing industry. *Essex Journal* 27(3): 38–42.

Cutting, C. L. 1955. *Fish Saving: a history of fish processing from ancient to modern times*. London: Leonard Hill.

Darby, H. C. 1940. *The Medieval Fenland*. Cambridge: Cambridge University Press.

Darby, H. C. 1971. *The Domesday Geography of Eastern England*. Cambridge: Cambridge University Press.

Darby, H. C. 1977. *Domesday England*. Cambridge: Cambridge University Press.

Darby, H. C. & I. B. Terrett (eds) 1954. *The Domesday Geography of Midland England*. Cambridge: Cambridge University Press.

Davis, F. J. 1958. *An account of fishing gear of England and Wales* (Fishery Investigations series 2, 21(8)). London: HMSO.

DML Dictionary of Medieval Latin from British Sources. R. D. Latham & D. R. Howlett (ed.) 1975–2013. London: Oxford University Press for the British Academy.

DOE Dictionary of Old English: Old English Corpus. Ann Arbor: University of Michigan Digital Library Production Service. Available at: http://ets.umdl.umich.edu/o/oec/.

Dugdale, W. 1665. *Monasticon Anglicanum, sive pandectæ coenobiorum Benedictorum, Cluniacensium, Cisterciensium, Carthusianorum*. Vol, 1. London: Typis Richardi Hodgkinsonne.

Dulley, A. J. F. 1969. The early history of the Rye fishing industry. *Sussex Archaeological Collections* 108: 36–64.

Dyer, C. C. 1988. The consumption of fresh-water fish in medieval England, in M. Aston (ed.) *Medieval Fish, Fisheries and Fishponds in England* (British Archaeological Report 182): 27–38. Oxford: Archaeopress.

MED Electronic Middle English Dictionary. Ann Arbor: University of Michigan Digital Library Production Service. Available at: http://quod.lib.umich.edu/m/med/.

Ervynck, A., W. Van Neer & M. Pieters. 2004. How the North was won (and lost again): historical and archaeological data on the exploitation of the North Atlantic by the Flemish fishery, in R. A. Housley & G. Coles (eds) *Atlantic Connections and Adaptations: economies, environments and subsistence in lands bordering the North Atlantic* (Symposia of the Association for Environmental Archaeology 21): 230–9. Oxford: Oxbow Books.

Farrer, W. (ed.) 1914–16. *Early Yorkshire charters*. Vols 1–3. Edinburgh: Ballantyne, Hanson and Co.

Fix, H. 1986. Dorsch, in H. Beck, D. Geuenich & H. Steuer (eds) *Reallexikon der germanischen Altertumskunde*. Vol. 6: 119–21. Berlin & New York: Walter de Gruyter.

Fleming, R. 2000. The new wealth, the new rich and the new political style in late Anglo-Saxon England. *Anglo-Norman Studies* 23: 1–22.

Flower, C. T. (ed.) 1915. *Public Works in Medieval Law*. Vol. 1. (Selden Society Publication 32). London: Selden Society.

Fox, H. S. A. 1996a. Cellar settlements along the south Devon coastline, in H. S. A. Fox (ed.) *Seasonal Settlement*: 66–9. (Vaughn Paper 39). Leicester: University of Leicester Press.

Fox, H. S. A. 1996b. Fishing in Cockington documents, in T. Gray (ed.) *Devon Documents in Honour of Mrs Margery Rowe*: 76–82. Tiverton: Special issue of Devon and Cornwall Notes & Queries.

Fox, H. S. A. 2001. *The Evolution of the Fishing Village: landscape and society along the South Devon coast, 1086–1550*. Oxford: Leopard's Head Press.

Gardiner, D. K. 1954. *Historic Haven: the story of Sandwich*. Derby: Pilgrim Press.

Gardiner, M. 1989. Some lost Anglo-Saxon charters and the endowment of Hastings College. *Sussex Archaeological Collections* 127: 39–48.

Gardiner, M. 1996. A seasonal fishermen's settlement at Dungeness, Kent. *Annual Report of the Medieval Settlement Research Group* 1996: 18–20.

Gardiner, M. 1997. The exploitation of sea-mammals in medieval England: bones and their social context. *Archaeological Journal* 154: 173–95.

Gardiner, M. 1998. Anglo-Saxon whale exploitation: some evidence from Dengemarsh, Lydd, Kent. *Medieval Archaeology* 52: 96–101.

Gardiner, M. 1999. Shipping and trade between England and the Continent during the eleventh century. *Anglo-Norman Studies* 22: 71–93.

Garmonsway, G. N. (ed.) 1939. *Ælfric's Colloquy*. London: Methuen.

Gilman, P. J. 1998. Essex fish traps and fisheries: an integrated approach to survey, recording, and management, in K. Bernick (ed.) *Hidden Dimensions: the cultural significance of wetland archaeology*: 273–89. Vancouver: University of British Columbia Press.

Godbold, S. & R. C. Turner 1994. Medieval fishtraps in the Severn estuary. *Medieval Archaeology* 38: 19–54.

Gras, N. S. B. 1918. *The Early English Customs System*. Cambridge, MA: Harvard University Press.

Hagen, A. 1995. *A Second Handbook of Anglo-Saxon Food and Drink: production and distribution*. Hockwold cum Wilton: Anglo-Saxon Books.

Hancock, F. 1903. *Minehead in the County of Somerset: a history of the parish, the manor and the port*. Taunton: Barnicott & Pearce.

Harper-Bill, C. (ed.) 1990. *English Episcopal Acta. Volume 6: Norwich 1070–1214*. Oxford: Oxbow Books.

Harper-Bill, C. & R. Mortimer (ed.) 1984. *Stoke by Clare Cartulary*. Part 3. Woodbridge: Boydell and Brewer for Suffolk Records Society.

Hart, C. R. 1966. *The Early Charters of Eastern England*. Leicester: Leicester University Press.

Heath, P. 1968. North Sea fishing in the fifteenth century: the Scarborough fleet. *Northern History* 3: 53–69.

Hockey, S. F. (ed.) 1974. *The Beaulieu Cartulary*. (Southampton Records Series 17). Southampton: Southampton University Press.

Hodges, R. 1982. *Dark Age Economics: the origins of towns and trade ad 600–1000*. New York: St Martin's Press.

Hoffmann, R. C. 2000. Medieval fishing, in P. Squatriti (ed.) *Working with Water in Medieval Europe: technology and resource-use*: 331–93. Leiden: Brill.

Hoffmann, R. C. 2002. Carp, cods, connections: new fisheries in the medieval European economy and environment, in M. J. Henninger-Voss (ed.) *Animals in Human Histories: the mirror of nature and culture*: 3–55. Rochester: University of Rochester Press.

Hooke, D. 2007. Uses of waterways in Anglo-Saxon England, in J. Blair (ed.) *Waterways and Canalbuilding in Medieval England*: 37–54. Oxford: Oxford University Press.

Hudson, J. (ed.) 2002. *Historia Ecclesiae Abbendonensis: the history of the church of Abingdon*. Vol. 2. Oxford: Oxford University Press.

Hussey, A. 1937–8. Hythe wills. *Archaeologia Cantiana* 49: 127–56; 50: 87–121; 51: 27–65.

Hybel, N. 1996. Sildehandel og sildefiskeri i den nordvestlige Nordsø i begyndelsen af det 14. århundrede, in H. Jeppesen (ed.) *Søfart, politik, identitet: tilegnet ole feldbæk*: 27–42. Cophenhagen: Falcon.

Jones, A. K. G. 1982. Bulk-sieving and the recovery of fish remains from urban archaeological sites, in A. R. Hall & H. K. Kenward (ed.). *Environmental archaeology in the urban context* (Council for British Archaeology Research Report 43). London: Council for British Archaeology.

Jones, C. 1983. Walls in the sea – the goradau of Menai: some marine antiquities of the Menai Straits. *International Journal of Nautical Archaeology & Underwater Exploration* 12(1): 27–40.

Jones, E. D. 2000. Some Spalding Priory vagabonds of the twelve-sixties. *Historical Research* 73: 93–104.

Karn, N. (ed.) 2005. *English Episcopal Act. Volume 31: Ely 1109–1197*. Oxford: Oxford University Press for the British Academy.

Keen, L. 1988. Coastal salt production in Norman England. *Anglo-Norman Studies* 11: 133–80.

Keene, D. 2001. Issues of water in medieval London to *c.* 1300. *Urban History* 28: 161–79.

Kelly, S. 1992. Trading privileges from eighth-century England. *Early Medieval Europe* 1: 3–28.

Kelly, S. (ed.) 2000. *Charters of Abingdon Abbey*. Vol. 2. London: Oxbow Books.

Kinver, M. 2009. Study unlocks history of the seas. *BBC News*. Available at: http://news.bbc.co.uk/2/hi/science/nature/8058351.stm.

Kowaleski, M. 2000a. The expansion of the south-western fisheries in late medieval England. *Economic History Review* (2nd series) 53: 429–54.

Kowaleski, M. 2000b. Fishing and fisheries in the Middle Ages: the western fisheries, in D. J. Starkey, C. Reid & N. Ashcroft (eds) *England's Sea Fisheries: the commercial sea fisheries of England and Wales since 1300*: 23–28. London: Chatham.

Kowaleski, M. 2000c. The internal and international fish trades of medieval England and Wales, in D. J. Starkey, C. Reid & N. Ashcroft (eds) *England's Sea Fisheries: the commercial sea fisheries of England and Wales since 1300*: 29–32. London: Chatham.

Kowaleski, M. (ed.) 2001. *The Havener's Accounts of the Earldom and Duchy of Cornwall, 1287–1356*. (Devon & Cornwall Record Society new series 44). Exeter: Devon & Cornwall Record Society.

Kowaleski, M. 2003. The commercialization of the sea fisheries in medieval England and Wales. *International Journal of Maritime History* 15(2): 177–231.

Kowaleski, M. 2006. Working at sea: maritime recruitment and remuneration in medieval England, in S. Caviocchi (ed.) *Richezza del mare, ricchezza dal mare, secc. XIII–XVIII*: 907–35. (Atti delle 'Settimane di Studi' e altri convegni, Istituto Internazionale di Sotria Economica 'F. Datini' Prato). Florence: Le Monnier.

Kowaleski, M. 2010. The seasonality of fishing in medieval Britain, in S. G. Bruce (ed.) *Ecologies and Economies in Medieval and Early Modern Europe: studies in environmental history for Richard C. Hoffmann*: 117–47. Leiden: Brill.

Lampen, A. 2000. *Fischerei und Fischhandel im Mittelater: Wirtschafts- und sozialgeschichtliche Untersuchungen nach urkundlichen und archäologischen Quellen des 6. bis 14. Jahrhunderts im Gebiet des Deutschen Reiches*. Husum: Matthiesen.

Lees, B. A. (ed.) 1935. *Records of the Templars in England in the Twelfth Century: the inquest of 1185 with illustrative charters and documents*. London: Oxford University Press for the British Academy.

Letters, S. (ed.) 2007. *Online Gazetteer of Markets and Fairs in England and Wales to 1516*. London: Centre for Metropolitan History. Available at: http://www.history.ac.uk/cmh/gaz/gazweb2.html.

Lewes, E. 1924. The goredi near Llanddewi, Aberarth, Cardiganshire. *Archaeologia Cambrensis* 7(4): 397–400.

Littler, A. 1979. Fish in English economy and society down to the Reformation. Unpublished PhD dissertation, University College of Swansea.

Lockwood, W. B. 2006. On the philology of *cod* and *stag*. *Transactions of the Philological Society* 104(1): 13–15.

Martin, M. T. (ed.) 1911. *The Percy Chartulary*. (Surtees Society Publication 117). Durham: Surtees Society.

Mason, E. (ed.) 1988. *Westminster Abbey Charters 1066–c. 1214*. (London Record Society 25). London: London Record Society.

Masschaele, J. 2007. Tolls and trade in medieval England, in L. Armstrong, I. Elbl & M. M. Elbl (eds) *Money, Markets and Trade in Late Medieval Europe: essays in honour of John H. A. Munro*: 146–86. Leiden: Brill.

Maxwell-Lyte, H. C. (ed.) 1900. *A Descriptive Calendar of Ancient Deeds*. Vol. 3. London: HMSO.

Maxwell-Lyte, H. C. 1909. *A History of Dunster and of the Families of Mohun and Luttrell*. Volumes 1–2. London: St Catherine Press.

McDonnell, R. 1979. Tidal fish weirs, west Somerset. *Proceedings of the Somerset Archaeology & Natural History Society* 124: 134–5.

Middleton-Stewart, J. 1996. 'Down to the sea in ships': decline and fall on the Suffolk coast, in C. Rawcliffe, R. Virgoe & R. G. Wilson (eds) *Counties and Communities: essays on East Anglian history presented to Hassell Smith*: 69–83. Norwich: Centre for East Anglian Studies, University of East Anglia.

Miller, E. 1951. *The Abbey and Bishopric of Ely: the social history of an ecclesiastical estate from the tenth century to the early fourteenth century*. Cambridge: Cambridge University Press.

Milne, G. 2003. *The Medieval Port of London*. Stroud: Tempus.

Moore, E. W. 1985. *The Fairs of Medieval England: an introductory study*. Toronto: Pontifical Institute of Mediaeval Studies.

Moore, S. A. & H. S. Moore. 1903. *History and Law of Fisheries*. London: Stevens and Haynes.

Morris, J. (ed.) 1975–92. *Domesday Book*. Vols 1, 2, 3, 33, 34. Chichester: Phillimore.

Müldner, G. & M. P. Richards. 2006. Diet in medieval England: the evidence from stable isotopes, in C. M. Woolgar, D. Serjeantson & T. Waldron (ed.) *Food in medieval England: diet and nutrition*: 228–38. Oxford: Oxford University Press.

Murphy, P. 2010. The landscape and economy of the Anglo-Saxon coast: new archaeological evidence, in N. J. Higham & M. J. Ryan (eds) *The Landscape Archaeology of Anglo-Saxon England*: 211–21. Woodbridge: Boydell Press.

Murray, K. M. E. 1935. *The Constitutional History of the Cinque Ports*. Manchester: Manchester University Press.

Napier, A. S. 1906. Old English lexicography. *Transactions of the Philological Society* 25: 265–335.

Neilson, N. 1910. *Customary Rents*. Oxford: Clarendon Press.

OED Oxford English Dictionary Online. Oxford: Oxford University Press. Available at: http://dictionary.oed.com/.

Oppenheim, M. 1907. Maritime history, in W. Page (ed.) *The Victoria History of the County of Sussex*. Vol. 2: 125–68. London: Archibald Constable.

Orton, D. C., J. Morris, A. Locker & J. H. Barrett. 2014. Fish for the city: meta-analysis of archaeological cod remains and the growth of London's northern trade. *Antiquity* 88: 516–30.

O'Sullivan, A. 2003. Place, memory and identity among estuarine fishing communities: interpreting the archaeology of early medieval fish weirs. *World Archaeology* 35: 449–68.

Pierquin, H. (ed.) 1912. *Recueil général des chartes anglo-saxonnes: les Saxons en Angleterre (604–1061)*. Paris: Alphonse Picard.

PR Pipe Roll Society. 1884–2007. *Publications of the Pipe Roll Society*. Vols 2, 51, 60, 61 London: Pipe Roll Society.

Ransford, R. (ed.) 1989. *The Early Charters of the Augustinian Canons of Waltham Abbey, Essex, 1062–1230*. Woodbridge: Boydell and Brewer.

Rees, U. (ed.) 1985. *The Cartulary of Haughmond Abbey*. Cardiff: Shropshire Archaeological Society: University of Wales Press.

Riley, H. T. (trans.) 1861. *Liber Albus: the white book of the City of London; compiled AD 1419, by J. Carpenter, Common Clerk, R. Whitington, Mayor*. London: Richard Griffin.

Riley, H. T. 1868. *Memorials of London and London life, in the XIIIth, XIVth, and XVth centuries*. London: Longman and Green.

Roberts, C. 2007. *The Unnatural History of the Sea*. Washington, DC: Ocean Island Press and Shearwater Books.

Robertson, A. J. (ed.) 1925. *The Laws of the Kings of England from Edmund to Henry I*. Cambridge: Cambridge University Press.

Robertson, A. J. 1956. *Anglo-Saxon Charters* (2nd edition). Cambridge: Cambridge University Press.

Rothwell, H. C. (ed.) 1975. *English Historical Documents. Volume 3: 1189–1327*. London: Routledge.

Round, J. H. 1898. Some early Sussex charters. *Sussex Archaeological Collections* 42: 78–80.

Round, J. H. (ed.) 1899. *Calendar of Documents Preserved in France. Volume 1: 918–1206*. London: HMSO.

Salter, H. E. (ed.) 1921. *Newton Longeville Charters*. (Oxford Record Society 3). Oxford: Oxford Record Society.

Saul, A. 1975. Great Yarmouth in the Fourteenth Century: a study in trade, politics and society. Unpublished PhD dissertation, University of Oxford.

Saul, A. 1981. The herring industry at Great Yarmouth, *c*. 1280–*c*. 1400. *Norfolk Archaeology* 38: 33–43.

Saunders, H. W. (trans.) & P. Millican (ed.) 1939. *The First Register of the Norwich Cathedral Priory*. (Norfolk Record Society 11). Norwich: Norfolk Record Society.

Sawyer, P. H. 1965. The wealth of England in the eleventh century. *Transactions of the Royal Historical Society* (5th series) 15: 145–64.

Sawyer, P. H. 1979. *Charters of Burton Abbey*. Oxford: Oxford University Press for the British Academy.

Sayers, W. 2002. Some fishy etymologies: Eng. *cod*, Norse *Þorskr*, Du. *kabeljauw*, Sp. *bacalao*. *Northwestern European Language Evolution: NOWELE* 41: 17–30.

Scargill-Bird, S. R. (ed.) 1887. *Custumals of Battle Abbey in the Reigns of Edward I and Edward II, 1283– 1312*. (Camden Society Publications new series 41). Westminster: Camden Society.

Schopp, J. W. (ed.) 1925. *The Anglo-Norman Custumal of Exeter*. Oxford: Oxford University Press.

Seebohm, F. 1890. *The English Village Community* (4th edition). London and New York: Longmans, Green.

Sergeantson, D. & C. M. Woolgar. 2006. Fish consumption in medieval England, in C. M. Woolgar, D. Sergeantson & T. Waldron (eds) *Food in Medieval England: diet and nutrition*: 102–30. Oxford: Oxford University Press.

Snape, M. G. (ed.) 2002. *English Episcopal Acta. Volume 24: Durham 1153–1195*. Oxford: Oxford University Press for the British Academy.

Steane, J. M. & M. Foreman. 1988. Medieval fishing tackle, in M. Aston (ed.) *Medieval Fish, Fisheries and Fishponds in England*. (British Archaeological Report 182): 137–86. Oxford: Archaeopress.

Stuart, J. (ed.) 1868. *Records of the Priory of the Isle of May*. Edinburgh: R. Clark for the Society of Antiquaries of Scotland.

Swanton, M. (trans.) 1993. *Anglo-Saxon Prose*. London: J. M. Dent.

Szabo, V. E. 2008. *Monstrous Fishes and the Mead-Dark Sea: whaling in the medieval North Atlantic*. Leiden: Brill.

Tait, J. (ed.) 1920. *Chartulary or Register of the Abbey of St Werburgh. Volume 1* (Chetham Society Publications, new series 79). Manchester: Chetham Society.

Tsurushima, H. 2006. The eleventh century in England through fish-eyes: salmon, herring, oysters, and 1066. *Anglo-Norman Studies* 29: 193–213.

Twiss, T. (ed.) 1873. *Monumenta Juridica: the black book of the Admiralty*. Vol. 2 (Rolls Series 55). London: Longman.

Unger, R. W. 1978. The Netherlands herring fishery in the late middle ages: the false legend of Willem Beukels of Biervliet. *Viator* 9: 335–56.

van Caenegem, R. C. (ed.) 1990–91. *English Lawsuits from William I to Richard I*. (Selden Society Publications 106–7). London: Selden Society.

Vasey, P. G. 1978. The later medieval herring industry in Scarborough. *Transactions of the Scarborough Archaeological & Historical Society* 21: 17–23.

Von Brandt, A. 1984. *Fish Catching Methods of the World*. (3rd edition). Farnham: Fishing News Books.

Wright, L. 1997. Medieval Latin, Anglo-Norman and Middle English in a civic London text: an inquisition on the River Thames, 1421, in S. Gregory & D. A. Trotter (eds) *'De mot en mot': aspects of medieval linguistics: essays in honour of William Rothwell*: 224–60. Cardiff: University of Wales Press.

Early Commercial Fisheries and the Interplay Among Farm, Fishing Station and Fishing Village in North Norway

Alf Ragnar Nielssen

Introduction

For decades Norwegian historians and archaeologists have been fascinated by the question of how far back in time the commercial cod (*Gadus morhua*) fisheries in northern Norway extended, and, connected with this, of the impact of this new industry on the settlement pattern in the north of the country. Traditionally, historians have tended to postulate a main shift from farm settlement to fishing village during the High and late Middle Ages (twelfth–thirteenth and fourteenth–fifteenth centuries, respectively), but more recent studies have shown that the development was far more complicated than this. As for the question of the age of the commercial fisheries, the breakthrough traditionally has been dated to the late eleventh century, based on written evidence, but new archaeological evidence may point to both earlier (i.e. tenth–eleventh-century) and later (i.e. thirteenth-century) episodes of economic intensification.

The first part of this chapter will scrutinise the most relevant of the early written sources for possible indications that the commercial cod fisheries of Norway pre-date the late eleventh century. The second part of the chapter will discuss the impact of commercial fisheries on the medieval settlement pattern of northern Norway. Here developments in the interplay among farm, fishing station and fishing village are of central concern.

The introduction of commercial cod fisheries for export

The first unambiguous written evidence for the existence of a commercial cod fishery refers to AD 1103–7, when later medieval sources say that a royal tax was levied on individuals fishing at Vågan, in Lofoten. More controversially, it has been contended that this duty originated in the AD 1030s

(Bjørgo 1982, 46). Farther north, in Finnmark, commercial fisheries seem to have developed during the twelfth century, when the resident Sami and Norwegian fishermen from farther south worked side by side from seasonal fishing stations (Bratrein 1998, 117–21; Hansen 2009, 69–70). As the fisheries developed, the actual fishing remained largely in the hands of the fishermen themselves. Farmers and peasants, for whom fishing was a part-time activity within the household economy, owned most of the fishing boats and the gear. The boats were small combined rowing and sailing vessels, carrying three to four men, and the only gear used in these cod fisheries was the hand line.

In Norway, society was still dominated by powerful chieftains until the early eleventh century. The Viking Age, had, however, brought more communication among regions in northern Europe. Moreover, the development from the early, sail-rigged Viking ships of the eighth–ninth centuries to the *knarr* – which is known from the mid-tenth century as a special ship type adapted for cargo – opened new opportunities for trade. However, even these ships typically had a low capacity compared with later vessels of the High Middle Ages. The Hanseatic *cog*, which was developed in the twelfth century (Crumlin-Pedersen 2003, 264), became of special importance for the growth of a trade in stockfish for grain between Bergen and other markets in northern Europe. The development of more sophisticated trade organisations in the High Middle Ages was certainly also of importance.

The fundamental basis of an emerging stockfish market in Europe was the 'commercial revolution', the beginning of which is usually dated to the eleventh century. It went hand in hand with population growth and resulted in the development of larger towns, especially in the North Sea area. All were prerequisites for extensive new trade in

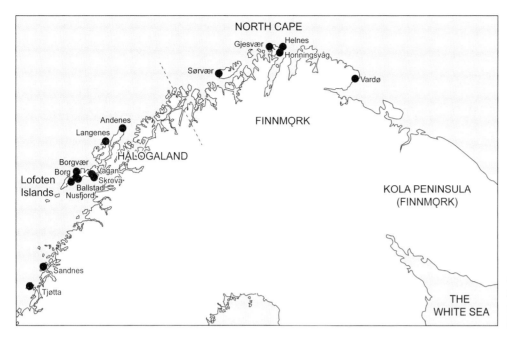

Figure 4.1. North Norway and settlements mentioned in the text (Drawing: Alf Ragnar Nielssen and Vicki Herring).

stockfish, salted herring and grain products. The other element that created a special demand for fish was the medieval church, which allowed fish to be eaten during Lent and other periods of fasting.

From the eleventh century onwards, population growth was rapid also in Norway. On the other hand, arable land was limited. Thus an emerging need for grain import becomes apparent in the historical sources in the thirteenth century at the latest (Helle 1982, 314–16). This trade was most important in the north, where climatic conditions made grain crops especially insecure. In the Viking Age the summer climate had been warm, while in the thirteenth and fourteenth centuries it probably entered a cooling phase (e.g. Young *et al.* 2012). These developments were probably the most important impetuses for the farmers of the far north to turn to commercial cod fisheries and a trade of fish for grain as a regular part of their economic adaptation.

Earlier historical references?

A few written sources of varying date have relevance to the discussion regarding whether or not the Norwegian stockfish trade pre-dates the late eleventh century. The earliest is the story of the chieftain Ohthere from northern Norway, who, during a trading visit to England around AD 890, was called to the court of Alfred of Wessex to relate conditions in his own country (Bately and Englert 2007). Ohthere told the king that the most important trading products in his country, and which also secured the wealth of the chieftains, were furs from reindeer, bears and otters, bird feathers and down, seal skin, sea-mammal skin (to be used as ship ropes) and

walrus ivory. These products were bought, or taken as a tax, from the Sami, who were the professional hunters of the region (Hansen and Olsen 2014). The chieftains were intermediaries in the trade. They exported these products, which originated both from land and from sea hunting, on their ships to the North Sea area. The traded articles were expensive and clearly meant for the upper classes. The text also notes that Ohthere brought walrus ivory to the king. Ohthere said nothing about return commodities, but hundreds of Iron Age grave finds in northern Norway document imported weapons and jewellery (Sjøvold 1974). Finer textiles, wheat and wine may also have been imported. Thus this source records a luxury trade, distinctly different from the bulk exchange of commodities such as stockfish and grain that is known from later sources.

Ohthere's relation, the only contemporary written source about northern Norway from the Viking Age, fails to mention stockfish or any other fish product, despite including a long list of other trade goods. Ohthere did observe Sami fishermen in Finnmark and on the Kola Peninsula during his voyage to the White Sea, but there is nothing in the text that connects these observations with trade. On the contrary, it is explicit that the voyage to the White Sea area was motivated by a desire for walrus tusks and sea-mammal skin.

The Icelandic 'family sagas' were written down from the twelfth century onwards, and they thus have limited credibility with regard to society in the Viking Age. They often look back to the Norwegian origin of their literary heroes. In *Egils saga*, the father of Egill Skallagrímsson emigrated (to Iceland) after the Norwegian king, Harald

Fairhair, had killed his brother, Torolv Kveldulvsson, who is said to have lived at the chieftain's residence Sandnes, in Helgeland, northern Norway. The events are nominally staged in the AD 870s, but the source itself probably dates to the thirteenth century (Byock 2004). One passage is particularly relevant to discussion of the early stockfish trade:

> Torolv owned a large ship which was meant for ocean voyages … He let his men carry on board stockfish and hides and ermine; he also sent a lot of squirrel fur and other fur that he had brought with him from the mountain area; and this was in abundance. He let Torgils Gjallande sail west to England to buy cloth and other things he needed. They sailed south along the coast and later across the sea, and arrived in England. They made a good bargain, loaded the ship with wheat and honey, and in the autumn they returned home. (Heggstad 1978, 38)

Most of the passage is about furs, in accordance with Ohthere and our general knowledge of Viking Age trade. The mention of cod as the first item, however, is a reflection of exchange patterns at the time of the saga's composition or redaction.

Another passage in *Egils saga* has received less attention, but is equally relevant: 'In spring Torolv returned to his farm. He sent people to the cod fishery at Vágar [Vågan], and some on herring fishery, and collected all kinds of products to his farm' (Heggstad 1978, 38).

The chiefly economy of the Viking Age was probably organised in a redistributive system (Hansen 1990, 129–32), in which the economic activity of each chieftain's region was organised to bring all kinds of food and other resources to the centre and then redistribute it to client subjects. Perhaps the pre-eleventh-century cod fishery of Lofoten can be better understood in such a context, where the intention was to bring extensive food supplies to chieftains' farms, rather than to produce a surplus for export trade. The potential importance of stockfish production for local consumption can be observed in the contemporary Icelandic situation, where dried cod served domestic demand from the time of settlement (*c.* AD 870) until the end of the thirteenth century (Perdikaris and McGovern 2009; Thór 2008, 329; see Chapter 7).

In the somewhat more historical 'king's sagas', evidence also exists for the ownership of fishing stations by northern Norwegian chieftains. Hårek of Tjøtta is mentioned by Snorri Sturluson in connection with a fishing and hunting station, about which he was in conflict with another local chieftain, Grankjell, nominally in the early eleventh century: 'There is a station toward the ocean, where one can hunt both seal and fowl, and which is also an egg and fishing station' (Schjøtt 1963, 175–6).

The only information regarding how these resources were used is that Hårek sent 10–12 of his men to the station and loaded a large rowing boat with all kinds of products. The context shows that hunting stations were valuable for chieftains. In this case the question of ownership was referred to the king himself, who decided in favour of Grankjell.

Another well-known example of a hunting station belonging to a chieftain's domain is Borgvær, in Lofoten, which was under the ownership of the farm Borg, where the largest Iron Age longhouse in Norway (*c.* 80 m) has been located (Narmo and Wickler 2008; Stamsø Munch *et al.* 2003; Wickler and Narmo 2014). Borgvær is situated towards the open sea and gives access to a variety of resources. In addition to winter and summer fisheries, conditions were excellent for seal and seabird hunting, and probably also for whaling. In an area where agriculture was limited for climatic reasons, chieftains had to turn to alternative resources to secure their income and maintain their power, such as those presented in Ohtheres' account (see above). However, nothing indicates that fishing had priority over sealing and fowling. The multiple use of the sea's resources was preferred.

Early archaeological evidence?

In contrast with the historical chronology argued above, it has been proposed by some Norwegian archaeologists that a substantial export of stockfish also could have taken place as early as the Iron Age (e.g. Narmo and Larsen 2004; Nilsen 1997). Nilsen's evidence is based on her research on the remains of boathouse sites on Vestvågøy, in Lofoten, where one important observation was that between AD 400 and AD 600 the size of the fishing boats was already the same as in the High Middle Ages and later (Nilsen 1997, 87–8). This led to the conclusion that winter fisheries for cod could already have been organised for an export trade in this period. On the other hand, her research does not explain where the European market for stockfish could have been located at this very early time, how the trade was organised or by which ship types a bulk cargo would have been transported. Narmo and Larsen's evidence is of a different character. They have excavated remains of turf huts at the famous fishing village of Nusfjord, in Lofoten, where the basal layers have radiocarbon dates centred around AD 600 (Narmo and Larsen 2004, 23–4). On this basis, they, too, have concluded that stockfish export could have taken place at this early time.

The main problem with arguments of this kind is that the potential markets for such an early export have not been identified. The above-mentioned studies are based on local archaeological material. On their own, they do not contain evidence that can support a thesis of a very early export trade. In northern Europe more generally, a market for mass trade is unlikely to have existed before the eleventh century, as mentioned above. What these studies do indicate, however, is that an organised cod fishery of some significance existed in Lofoten centuries before export to international markets became of importance. The earliest possible evidence of export to markets may be a few cod bones, probably originating from Arctic waters, that have been identified from the Danish Viking Age town of Hedeby (now German Haithabu), located on the Baltic coast of southern Jutland, in modern Germany (Barrett *et al.* 2008, 858; Orton *et al.* 2011). These finds are of broad ninth- to eleventh-century date and thus need not indicate the emergence of a market for stockfish much earlier than has traditionally been assumed. Another reservation regarding the interpretation of this material is that stockfish must also have been used as travelling provision by Scandinavians on their longer voyages (cf. Heinrich 2006; Lepiksaar and Heinrich 1977). Thus until larger amounts of Viking Age cod bones originating from fish caught in the Arctic have been identified in European consumer sites, it is difficult to treat archaeological evidence as indicative of a need to revise the traditional chronology.

The socioeconomic transformation of the eleventh and twelfth centuries

As noted above, the earliest references to Norwegian fishing stations are found in the medieval sagas. In *Morkinskinna*, the earliest extensive chronicle of the kings of Norway, it is related that Øystein (AD 1103–23) erected fishermen's shanties at Vågan, in Lofoten (Andersson and Gade 2000, 346). Owing to the seasonal character of Vågan's fisheries, there can be little doubt that they were intended for farmers participating in the winter fishery for spawning cod. Such investment was probably designed to increase the Crown's taxation revenue by centralising both the fishermen and the fish trade at Vågan, which, during the twelfth century, became the principal market for northern Norway. During the first half of the eleventh century, the rule of chieftains had been replaced by that of the kings, in co-operation with the church. In Norway, as in the rest of northern Europe, towns began to grow and traders developed as a professional group. Bergen became of the utmost importance for the fish trade, but the archbishopric in Nidaros (Trondheim) and the monasteries also participated.

People in northern Norway probably adapted easily to the new economic opportunities of the commercial revolution in Europe. For centuries they had been involved in trading with both the British Isles and continental Europe. The old trade had been predominately in luxuries. Furs, walrus tusks, sea mammal hides, train oil and perhaps *some* stockfish were exported, while imports were exclusively of weaponry, jewellery, wine and textiles. Trade routes and connections already existed when the commercial cod fisheries provided new opportunities for export.

Another important factor is that hunting, also with the use of boats, had long been an important element in the economy of this society. The boat technology already existed, and people knew how to handle different waters and difficult weather conditions. The only cod fishing gear was the hand line, which was not new at all, as it had been part of the equipment used on the combined hunting and fishing expeditions of earlier centuries. In the High Middle Ages, written sources indicate that both Norwegian farmers and Sami hunters took part in the commercial fisheries (Hansen 2009, 69–70). By the end of this period, many of the coastal Sami had also begun to settle on farms and take up a living that was inspired by their Norwegian neighbours. According to the source *Passio Olavi*, which is supposed to have been written between AD 1160 and 1188, the seasonal migration to Lofoten was probably being supplemented by a fishery for younger cod in Finnmark from late April till early June, starting as early as the twelfth century (Bratrein 1998, 117–21; Hansen 2009, 69–70). The main pattern that came to dominate the economic activity of northern Norway until the early twentieth century had been laid.

A third factor that smoothed the transition to commercialisation was the complementarity of fisheries that took place in the period late January–mid-April and late April–mid-June (with the first season in Lofoten and the second in Finnmark), at a time of year when the farming season had not yet started. It was only after the cod fishery in Finnmark finished in mid-June that the spring ploughing started. Cod fishing was similarly convenient for the coastal Sami, who could easily combine it with other economic activities, especially since the hunting of seals and whales took place in summer.

Farmers resided in fishing stations, especially in Lofoten, during the main seasonal cod fishery. When the commercial fishery was weak, such as occurred periodically in the seventeenth century – because of fluctuations in price, trade and/or the abundance of cod – they could withdraw to farming and household fishing only. Because of the absence of the men for almost half the year, animal husbandry became mainly a task for the women of the household, a division of labour that actually lasted as long as the fisherman–farmer adaptation – that is, until the first half of the twentieth century.

Vågan in Lofoten is, as mentioned above, the first known Norwegian station for commercial cod fishing. It soon also developed into the one and only town in northern Norway (Bertelsen 1994, 120–5; Bjørgo 1982). The written evidence about other fishing stations in Lofoten in the medieval period is very sparse, but in AD 1239 the well-known fishing station of Skrova is mentioned in the contemporary saga of Håkon Håkonsson (Nielssen 2011, 295). In the cadastre of archbishop Aslak Bolt, *c.* AD 1432, we also hear that about 200 men were fishing from each of the two stations of

Ballstad and Nusfjord, farther west in Lofoten (Jørgensen 1997, 159–62; Nielssen 2009, 84–5).

The essential requirement for a fishing station was proximity to good fishing grounds. Many were located where conditions did not favour permanent settlement and where, for example, even small-scale animal husbandry was impractical. However, if terrestrial resources were more plentiful, stations could and did develop into villages. Other types of fishing stations included diminutive villages and permanently settled farms near prolific fishing grounds. In Lofoten, fishermen swarmed to such places during the fishing season, renting rooms or shanties from residents.

The fishing village: a new element in the settlement structure

Between the late eleventh and mid-thirteenth centuries, commercial cod fisheries in northern Norway were solely conducted by Norwegian farmers and coastal Sami hunters who visited the fishing stations periodically throughout the year. The resulting product was purchased by Norwegian traders who in summer attended the market at Vågan – where all stockfish was to be sold according to a royal resolution (Bagge *et al.* 1973, 364–5). This system had little impact on the established settlement pattern. From the middle of the thirteenth century onwards, however, the first evidence of permanent settlement in fishing villages can be traced archaeologically and, shortly afterwards, also in the written material. The background may be that the German Hansa began trading in Norway at this time. The first peace and trade agreement between the town of Lübeck and the Norwegian king was signed in AD 1250 (Helle 1982, 379–88, 730–50), and by *c.* AD 1360 the merchants of Lübeck had established both their powerful *Kontor* in Bergen and a credit system for the fishermen. It was in this period that the commercial flow of grain to coastal areas in Norway became so secure that permanently settled fishing villages could flourish. In Finnmark and northern Troms, they soon came to dominate the settlement pattern, but to the south the interplay between farming and seasonal fishing from stations prevailed. The participation of farmers in seasonal fisheries must also have undergone a substantial rise in this period, when fish prices were especially high (Nedkvitne 1988, 26; see Chapter 5). One observable effect of the extended integration of commercial fisheries in the economy is the fact that stockfish took over from butter as a measure of value all over northern Norway in this period (Sandnes 1977, 176–8).

For decades Norwegian historians have undertaken research on questions concerning the structure and development of northern Norwegian settlement from the late Middle Ages onwards, when the impact of commercial fisheries was clearly significant. It has been common to characterise this period as the time when year-round settlement in fishing villages began. More recent research has shown, however, that this was only one aspect of commercial fishing, one that was mainly relevant for the northernmost parts of the area (that is, Troms and Finnmark). In most of northern Norway, conversely, traditional farm settlements continued to be dominant. Nevertheless, farmers started participating in the seasonal cod fisheries to a much greater extent than before, and commercial fishing became more important in the household economy of the farmers.

Ragnhild Høgsæt (Aarsæther) has made a distinction between the fishing village coast of the far north and the fisherman–farmer coast of the more southern areas of northern Norway, with the border between the two main areas having been a little north of Lofoten (Aarsæther 1980, 412–15). Arnved Nedkvitne has also identified different economic adaptations along the coast of western and northern Norway, pointing out characteristic differences between three main regions in the balance between fishing and farming (Nedkvitne 1988). In this context it is important to emphasise that the fisherman–farmer population of northern Norway, both in the late Middle Ages and after, was much larger than that of the fishing villages. The fishing village was introduced as a new kind of settlement in the thirteenth century, and became of great importance in the far north, but it never came to replace the farm as the main settlement form in the more heavily populated parts of northern Norway.

Little documentary evidence of fishing villages has survived for the period before AD 1500, but what does exist can be used in conjunction with archaeological findings to offer an impression of their development as settlement units. The most extensive archaeological work has been undertaken in Finnmark and northern Troms, where Norwegian and Sami populations overlapped. The oldest traces of fishing villages in the archaeological material in this area originate from the thirteenth century (Andreassen and Bratrein 1990; Bratrein 1989, 214).

The earliest community mentioned in written sources is Geirsvær (now Gjesvær), a few kilometres west of North Cape. The name appears in Snorri Sturluson's thirteenth-century *Heimskringla* in relation to events that are supposed to have taken place in the AD 1020s (Schjøtt 1963, 159). The text mentions a landing stage (quay) that was the first landfall encountered from the northeast by Hálogaland chieftains who had been trading and plundering in the White Sea. The implication is that a station had been established to facilitate commerce with, and probably the collection of taxes from, the Sami. This same station *may* also have served as a bridgehead for the foundation of fishing villages 200 years or so later. The first fishing village to be mentioned explicitly in written sources is Vardø, where, in the year AD 1307, the archbishop of Nidaros himself made the long voyage to consecrate a newly built church, and where a royal castle was also erected in the early AD 1300s (Nielssen 2009, 97). We can thus be certain that Norwegian settlement had expanded to eastern Finnmark by this time. The earliest fishing villages in western and mid-Finnmark mentioned in the written sources are Sørvær, Helnes and Honningsvåg in a document from AD 1385, but they must have been founded before that date (Nielssen 1994, 25).

The character of the fishing villages

The first historical sources that give an overall view of the size and character of the fishing villages are tax registers from the years AD 1520 and AD 1567 (Nielssen 1994, 25). The sources give a clear picture, especially for Finnmark, as they register the settlements one by one from the west to the east. The largest settlement was Vardø, in the northeast, with *c.* 50 taxpayers, indicating a total population of *c.* 250–300 persons. Many other fishing villages in Finnmark have lists of 20–40 taxpayers, and there were many large fishing villages also in Troms and Vesterålen. The southern counterparts to Vardø were Andenes and Langenes in Vesterålen, which both had a population close in size to that of Vardø.

While grain was not cultivated in fishing villages, subsistence agriculture was invariably undertaken, with families utilising the sparse grass supply, complemented by other types of animal feed, to engage in a small but necessary animal husbandry. Normally, one or two cows and a few sheep and goats were owned by each family (Nielssen 1984, 97–100). Women bore the primary responsibility for the small farming activity in fishing villages, while men concentrated on fishing. Land was seldom cultivated, but grass and other sources of fodder were collected where they grew. Grass was plentiful in many places during the summer, so resources were only truly meagre in a handful of fishing villages. People also transported their animals to nearby islands or to Sami areas on the mainland for grazing. Obtaining an adequate supply of winter feed posed the biggest challenge. In most places marine resources were thus an important component of winter fodder. Seaweed and fish waste (heads, entrails and bones), cooked together, were regularly fed to domestic animals. Marine resources were most important in late winter and spring, by which time most of the stored hay had been consumed and conditions for harvesting seaweed were favourable (Bratrein 1974, 17–32). The use of marine resources to feed domestic animals was a normal part of the life cycle, with at least 50% of all winter feed for animals in coastal districts of northern Norway coming from the sea (Bratrein 1974, 17–32). Nineteenth-century reports by state officers indicate that cows fed on such a diet were healthy and yielded excellent milk (Drivenes 1975, 206–10).

Immigration to the Norwegian fishing districts from southern Norway and abroad was of great importance from the late Middle Ages until the beginning of the seventeenth century. Through trading contacts, many foreigners were recruited and settled permanently in fishing villages.

Most of those who settled were attracted by the economic opportunities that the fisheries offered, and transients blended easily into this comparatively cosmopolitan society. The evidence of surnames indicates that the non-Norwegians came from Denmark (especially Jutland), Germany, Sweden, Holland, Scotland, Shetland and the Faroe Islands (Nielssen 2009, 87). The majority moved to the fishing villages via Bergen, where trading ships, which often carried passengers, arrived from all over northwestern Europe. Sometimes these new arrivals remained for just a year, but many got married, raised families and settled there permanently.

From the mid-thirteenth century to *c.* AD 1600, trade with Bergen and the German Hansa generated so much wealth in coastal regions that (in the sixteenth century, for example) fishing communities paid considerably more in taxes than agricultural districts (Holmsen 1976, 26). Tangible proof of this prosperity can be found in the decor of late medieval churches in northern Norway, which feature elaborate altars and wooden sculptures made by leading German and Dutch craftsmen that were donated by wealthy merchants.

During the AD 1620s, however, the decline in the cod fisheries signalled by a fall in prices precipitated a full-blown economic crisis. In Finnmark, negative signs were already apparent before AD 1620. While the number of Norwegian taxpayers in Finnmark ranged between 563 and 583 a year in the period AD 1593–1601, it fell to between 460 and 520 in the period AD 1610–1625. In AD 1629 it fell below 400, and in AD 1640 it stood at a mere 299 individuals (Nielssen 1986, 80–6). Several of the large fishing villages disappeared or shrank drastically during these years. People moved from fishing villages to fjord districts, where they could better combine farming with seasonal fisheries (Nielssen 1986, 87–90). From their new homes farther inland people could still migrate back to the coast for the fishing season. Some of the largest sixteenth-century fishing villages became fishing stations, while others survived as permanent settlements. In essence, the pattern that had been characteristic of Lofoten and the adjacent area to the south generally became prevalent along the entire coast. Even though there would be a return to profitable fisheries over the course of the seventeenth and eighteenth centuries, the era of the large permanently settled fishing village had passed. It was not until the mid-nineteenth century that the fishing village regained a position in the settlement pattern comparable to that evident in the late Middle Ages.

Conclusions

There is no evidence in the written sources from the Viking Age or the period after which supports the hypothesis that a developed and organised stockfish trade took place earlier than the eleventh century. Neither Ohthere's account nor the Icelandic saga sources contain evidence of an international trade of this character. Moreover, they do not indicate that cod fishing had priority over sealing, whaling and fowling. On the other hand, archaeological evidence from Lofoten confirms that organised cod fishing does go far back into the Iron Age. This early activity is best interpreted as a way of collecting food resources for the domestic population, possibly organised by chieftains. Other arguments against such an early export of stockfish are the lack of an identifiable stockfish market in Europe, of a developed trading organisation and of cargo ships suitable for the transport of bulk commodities.

The earliest documented fishing station meant for a commercial cod fishery is still Vågan in Lofoten, *c.* AD 1100. In its earliest phase the commercial fishery was probably concentrated in Lofoten and was conducted for only a few months during the winter by local farmers using their own small combined rowing and sailing boats. In consequence, the settlement structure was not much affected by the fishery at this time. In the thirteenth century, as the commercial fisheries grew in importance, permanently settled fishing villages developed. They became especially important in the far north, where farming was quite limited and grain could not be cultivated. In most of northern Norway, however, fisheries continued to be conducted by farmers who visited fishing stations seasonally. In the late Middle Ages, and even in the sixteenth century, the economic importance of commercial fisheries continued to grow due to the high price of stockfish compared with that of grain and due to the sophisticated trading system developed by Hanseatic merchants in Bergen.

In the seventeenth century, however, the market for stockfish declined. This considerably weakened the fishing village as a settlement unit, and most of the inhabitants moved to fjord districts, where they could take up farming in combination with frequenting fishing stations seasonally – in the same way that the vast majority of the population of northern Norway had done since the eleventh and twelfth centuries.

References

Aarsæther, R. 1980. Værmenn, fiskere, bønder og fiskerbønder. *Heimen* 3: 406–26.

Andersson, T. M. & K. E. Gade. (ed.) 2000. *Morkinskinna: the earliest Icelandic chronicle of the Norwegian kings (1030–1157).* London: Cornell University Press.

Andreassen, R. L. & H. D. Bratrein. 1990. Magerøy i middelalderen. (Tromura–Tromsø Museums Skriftserie, Kulturhistorie 14). Tromsø: Tromsø Museum/University of Tromsø.

Bagge, S., S. H. Smedsdal & K. Helle. 1973. *Norske middelalderdokumenter i utvalg.* Bergen: Universitetsforlaget.

Barrett, J. H., C. Johnstone, J. Harland, W. Van Neer, A. Ervynck, D. Makowiecki, D. Heinrich, A. K. Hufthammer, I. B. Enghoff, C. Amundsen, J. S. Christiansen, A. K. G. Jones, A. Locker, S. Hamilton-Dyer, L. Jonsson, L. Lõugas, C. Roberts & M. Richards. 2008. Detecting the medieval cod trade: a new method and first results. *Journal of Archaeological Science* 35(4): 850–61.

Bately, J. & A. Englert. (ed.) 2007. *Ohthere's Voyages: a late 9th-century account of voyages along the coasts of Norway and Denmark and its cultural context.* Roskilde: Viking Ships Museum.

Bertelsen, R. 1994. Helgelendingene og Vágar i Lofoten, in B. Berglund (ed.) *Helgelands Historie* Vol. 2: 113–32. Trondheim: Helgeland Historielag.

Bjørgo, N. 1982. Vågastemna i mellomalderen, in S. Imsen & G. Sandvik (eds) *Hamarspor: eit festskrift til Lars Hamre. 1912–23. januar–1982*: 45–60. Oslo: Universitetsforlaget.

Bratrein, H. D. 1974. Tradisjonell utnytting av tang og tare i Nord-Norge. *Ottar*: 17–32. Tromsø: Tromsø Museum/University of Tromsø.

Bratrein, H. D. 1989. *Karlsøy og Helgøy bygdebok. Volume 1.* Karlsøy: Karlsøy Kommune.

Bratrein, H. D. 1998. 'Passio Olavi' – et kildested om finnmarksfisket på 1100-tallet. *Håløygminne* 79: 117–21.

Byock, J. L. 2004. Social memory and the sagas: the case of *Egils saga. Scandinavian Studies* 76: 299–316.

Crumlin-Pedersen, O. 2003. Die Bremer Kogge – ein Schlüssel zur Geschichte des Schiffbaus im Mittelalter, in G. Hoffmann & U. Schnall (eds) *Die Kogge*: 256–71. Hamburg: Convent Verlag.

Drivenes, E. A. 1975. Åkerbruk og fehold i Altafjorden på 1800-tallet. Unpublished Master's dissertation. University of Tromsø.

Hansen, L. I. 1990. *Samisk fangstsamfunn og norsk høvdingøkonomi.* Oslo: Novus.

Hansen, L. I. 2009. Sami fishing in the pre-modern era: household sustenance and market relations, in D. J. Starkey, J. Th. Thór & I. Heidbrink (eds) *A History of the North Atlantic Fisheries. Volume 1: from early times to the mid-nineteenth century* (Deutsche maritime Studien/German Maritime Studies 6): 65–82. Bremen: Hauschild.

Hansen, L. I. & B. Olsen. 2014. *Hunters in Transition: an outline of early Sámi history.* Leiden: Brill.

Heggstad, L. 1978. *Egilssoga.* Oslo: Det norske samlaget.

Heinrich, D. 2006. Die Fischreste aus dem Hafen von Haithabu – handaufgelesene funde, in U. Schmölcke, K. Schietzel, D. Heinrich, H. Hüster-Plogmann & K. J. Hüser, *Untersuchungen an Skelettresten von Tieren aus dem Hafen von Haithabu* (Berichte über die Ausgrabungen in Haithabu 35): 157–93. Neumünster: Wachholtz Verlag.

Helle, K. 1982. *Bergen bys historie. Volume 1: Kongesete og kjøpstad: fra opphavet til 1536.* Bergen, Oslo & Tromsø: Universitetsforlaget.

Holmsen, A. 1976. *Nye studier i gammel historie.* Oslo: Universitetsforlaget.

Jørgensen, J. G. 1997. *Aslak Bolts jordebok.* Oslo: Riksarkivet.

Lepiksaar, J. & D. Heinrich. 1977. *Untersuchungen an Fischresten aus der frühmittelalterlichen Siedlung, Haithabu* (Berichte über die Ausgrabungen in Haithabu 10). Neumünster: Wachholtz Verlag.

Narmo, L. E. & K. T. Larsen. 2004. *Rapport fra utvidet arkeologisk registrering i Nusfjord.* Bodø: Nordland Fylkeskommune.

Narmo, L. E. & S. Wickler. 2008. *Arkeologiske rapporter fra Borgvær.* Borg: Lofotr Viking Museum.

Nedkvitne, A. 1988. *'Mens Bønderne seilte og Jægterne for': nordnorsk og vestnorsk kystøkonomi 1500– 1730.* Oslo: Universitetsforlaget.

Nielssen, A. R. 1984. Animal husbandry among the Norwegian population in Finnmark *c*. 1685–1705. *Acta Borealia* 1: 81–112.

Nielssen, A. R. 1986. Fra storvær til småbruk: den geografiske ekspansjonen i den norske bosettinga i Finnmark *ca*. 1570–1700. *Heimen* 2(22): 79–95.

Nielssen, A. R. 1994. The importance of the Hanseatic trade for the Norwegian settlement in Finnmark, in V. Henn & A. Nedkvitne (eds) *Norwegen und die Hanse*: 19–30. (Kieler Werkstücke 11). Pieterlen: Peter Lang.

Nielssen, A. R. 2009. Norwegian fisheries, *c*. 1100–1850, in D. J. Starkey, J. Th. Thór & I. Heidbrink (eds) *A History of the North Atlantic Fisheries. Volume 1: from early times to the mid-nineteenth century*: 83–118. (Deutsche maritime Studien/German Maritime Studies 6). Bremen: Hauschild.

Nielssen, A. R. 2011. Fiskeværsfenomenet i Lofoten – noen lange linjer. *Heimen* 4: 291–309.

Nilsen, G. 1997. Jernaldernaust på Vestvågøy i Lofoten. Unpublished Master's dissertation, University of Tromsø.

Orton, D. C., D. Makowiecki, T. de Roo, C. Johnstone, J. Harland, L. Jonsson, D. Heinrich, I. B. Enghoff, L. Lõugas, W. van Neer, A. Ervynck, A. K. Hufthammer, C. Amundsen, A. K. G. Jones, A. Locker, S. Hamilton-Dyer, P. Pope, B. R. MacKenzie, M. Richards, T. C. O'Connell & J. H. Barrett. 2011. Stable isotope evidence for late medieval (14th–15th C) origins of the eastern Baltic cod (*Gadus morhua*) fishery. *PLoS ONE* 6(11): e27568. doi:10.1371/journal.pone.0027568.

Perdikaris, S. & T. H. McGovern. 2009. Viking Age economics and the origins of commercial cod fisheries in the North Atlantic, in L. Sicking & D. Abreu-Ferreira (eds) *Beyond the Catch: fisheries of the North Atlantic, the North Sea and the Baltic, 900–1850*: 61–90. Leiden: Brill.

Sandnes, J. 1977. Mannedauen og de overlevende, in S. Imsen and J. Sandnes (eds) *Avfolkning og union*. (Cappelens Norgeshistorie 4). Oslo: Cappelen Forlag.

Schjøtt, S. 1963. *Snorre Sturlason Kongesoger: soga om Olav den heilage*. Oslo: Det Norske Samlaget.

Sjøvold, T. 1974. *The Iron Age Arctic Settlement of Northern Norway: a study in the expansion of European Iron Age culture within the Arctic Circle. Volume 2: late Iron Age (Merovingian and Viking periods)*. Tromsø: Norwegian University Press.

Stamsø Munch, G., O. S. Johansen & E. Roesdahl (eds) 2003. *Borg in Lofoten: a chieftain's farm in north Norway*. (Arkeologisk Skriftserie 1). Borg: Lofotr Vikingmuseum.

Thór, J. Th. 2008. Icelandic fisheries, *c*. 900–1900, in D. J. Starkey, J. Th. Thór & I. Heidbrink (eds) *A History of the North Atlantic Fisheries. Volume 1: from early times to the mid-nineteenth century*: 323–49. (Deutsche maritime Studien/German Maritime Studies 6). Bremen: Hauschild.

Wickler, S. & L. Narmo. 2014. Tracing the development of fishing settlement from the Iron Age to the Modern Period in northern Norway: a case study from Borgvær in the Lofoten Islands. *Journal of Island & Coastal Archaeology* 9: 72–87.

Young, G. H. F, D. McCarroll, N. J. Loader, M. H. Gagen, A. J. Kirchhefer & J. C. Demmler. 2012. Changes in atmospheric circulation and the Arctic Oscillation preserved within a millennial length reconstruction of summer cloud cover from northern Fennoscandia. *Climate Dynamics* 39: 495–507.

The Development of the Norwegian Long-distance Stockfish Trade

Arnved Nedkvitne

Introduction

A thorough examination of all the available written sources on the origins of the Norwegian stockfish trade suggests that the long-range or international (that is, beyond the borders of the medieval kingdom of Norway) part of it originated around AD 1100 (Nedkvitne 1983; 2014, 25–31; see also Chapter 4). In this chapter I will consider the written evidence on which this dating is based, and how the trade of dried cod (*Gadus morhua*) developed through the Middle Ages and into the post-medieval period. I will also consider how the Norwegian stockfish trade related to the other major commercial fishery of medieval Europe, that of herring (*Clupea harengus*). Lastly, I will consider whether the scale of medieval cod and herring fishing was sustainable in ecological terms, using comparison with the catches of later centuries as a benchmark.

Preserving cod and herring

A basic precondition for a particular species of fish being drawn into long-distance trade was that it had to be capable of being preserved in a manner that made the end product attractive to buyers weeks or months after the fish had been caught. It was expensive to transport fish over long distances, meaning that the product must ultimately fetch a sufficiently high price. In Norway, large shoals of both cod and herring reached the coast in winter. Cod can be dried without salt into highly marketable stockfish if the climate is cold, dry and windy, and this is the case at the Norwegian coast north of Stad. Drying without salt is simple to organise and demands little investment. Every fisherman can dry his own fish. Large quantities could be hung for drying during a short period. Farther south it is warmer and wetter, and cod preserved without salt will often spoil, making them unsuitable for long-distance

transport and the international market. The capital threshold for starting stockfish production was low, but only in the northern part of Norway.

Herring were occasionally also dried without salt, but the result is less satisfactory in terms of both quality and longevity with such an oily species, and the product could not be sold with a profit on the international market (cf. Heath 1968, 62). Smoked herring was produced in limited quantities for local markets all over northern Europe. The advantage of smoking was that preservation could take place even in a humid and warm climate; the disadvantage was that the process demanded large smoking houses, firewood and people to monitor the process. The quantity that could be smoked at one time was limited by the size of the available smoking houses, and herring is often caught in large quantities during a short period. Large 'houses' two or three floors high for smoking herring are mentioned in the Bohuslän herring fisheries in the sixteenth century (Nedkvitne 1983, 350–1; 1988, 478; 2014, 44–5; Tomfohrde 1914, 47–8, 87–8). In practice herring was often salted in barrels for long-distance export. In Norway salt production was minimal and the quality of the resulting salt was low. Therefore salt had to be imported and brought to the fishing grounds by merchants who salted the herring immediately when fishermen brought the catch ashore. Herring production thus demanded a large degree of merchant involvement and investment and the threshold for commercialisation was high. Therefore, unlike cod, the export of herring from medieval Norway was very modest. It probably began (and then only on a small scale) from the end of the thirteenth century, later than the emergence of the stockfish trade. The real expansion in Norwegian herring production and export started later (Nedkvitne 1988, 470–7; 2014, 42–5, 534–5; see Chapter 2).

Why did long-distance stockfish trade begin around AD 1100?

As discussed by Nielssen in Chapter 4, the first evidence implying a long-range stockfish trade in Norwegian sources relates to AD 1103–1107, when the three reigning kings confirmed a tax on those fishing in Lofoten, which was the main centre for the production of dried fish (Bagge *et al.* 1973, 20–1). One of these kings, Øystein (AD 1103–1123), also built huts and a church for seasonal fishermen there and established harbours and landmarks that facilitated sailing between Lofoten and Bergen (Andersson and Gade 2000, 326, 346–7; Aðaldbjarnarson 1951, 255). Concurrently, the town of Bergen entered a period of strong growth (Hansen 2005; 2015; Helle 1982, 84–170). The sources from around AD 1100 do not say that the stockfish trade had become international, but that is the most likely explanation for the evidence.

If stockfish producers were to send their goods abroad they had to get things in return. The most important goods which Norway imported were grain and fine cloth (Nedkvitne 1983, 46; 2014, 60–1). The earliest evidence for this trade is from England. Orderic Vitalis has the following to say regarding the year AD 1095:

> When four great ships called *canardes* were on their way from Norway to England, Robert [the Earl of Northumbria] and his nephew Morel with their minions waylaid them and violently robbed the peaceful merchants of their goods. The merchants, spoiled of their property, went to the king in great distress and laid a complaint about their loss. (Chibnall 1973, 281)

The word *canardes* must be a Latin translation of the Norse *knarr*, which is a specialised large trading ship. Thus the merchants were probably Norwegian and we can surmise that the exchange of bulky goods between Norway and England had already begun. The trade presumably entailed the exchange of stockfish for grain.

The main port of export from England to Norway was King's Lynn. Its great period of growth started after *c.* AD 1095, when the bishop of Norwich founded a monastery and a parish church there (Carus-Wilson 1964; cf. Hutcheson 2006). The economic basis for this growth was in part extensive land reclamation in the area inland from King's Lynn, which produced a surplus of grain and wool for export. Based on later sources, we know the wool went to Flanders. From the 1170s, the written sources provide solid evidence that grain was a main export to Norway, but there can be little doubt that this had been so from the start of the commercial expansion (Nedkvitne 1983, 19; 2014, 31). The Norwegian and English evidence thus corresponds regarding when the commercial exchange involving stockfish started.

The second area with which western Norway had direct commercial relations was the Rhine estuary. The cloth industry in Flanders grew at the end of the eleventh century (Carus-Wilson 1987, 626), but the earliest evidence for Norwegian trade in the Rhine estuary is an uncertain source of AD 1122 from the town of Utrecht (HU, vol. 1, no. 8) and a certain reference *c.* AD 1150 from the town of Deventer (Kålund 1908, 148–50). From the 1180s, the written sources tell us that the main imports to Norway were woollen and linen cloth, metal wares and wine; stockfish was sent in the opposite direction. Much of the stockfish was sold at markets along the Rhine and in Westphalia (Nedkvitne 1983, 20–2; 2014, 33–6). Documentation is in general better for England in this period, but it is not unreasonable to think that stockfish export to the Rhine area began later than that to England.

Norwegians had sailed with merchandise to Hedeby (Haithabu in German) in the Baltic at least since Ohthere's time, *c.* AD 890 (Bately and Englert 2007). The market for their goods there was probably the German interior. As discussed in Chapter 4, however, there is no convincing evidence that stockfish was involved at this early date, beyond a role as ship provisions.

The Norwegian town of Konungahella, near today's Gothenburg, was attacked by northern Baltic pirates in AD 1134, and at that time there were nine *austfararskip* (cargo ships used between Norway and the Baltic) moored in the town's port; just before that event, 13 *byrðinga* (inshore craft) had left Konungahella for Bergen. The source of this information is Snorri Sturluson, and it should be reliable because his foster father was aboard one of these ships (Aðalbjarnarson 1951, 289–90). In AD 1143, Lübeck was founded. It was rebuilt after changing hands in 1159. Lübeck's second ruler, Duke Henry the Lion, invited Norwegian, Danish, Swedish and Russian merchants to visit the new town (Schmeidler 1937, 169; cf. HU, vol. 1, no. 33). This indicates that Norwegian merchants were known to trade in these waters. The shipping lane between Norway and the western Baltic was used by many before the German commercial expansion started in earnest after AD 1159.

At the end of the twelfth century there was still a lively traffic of Norwegian *byrðinga* along the coast from Bergen to the market at Scania. In 1196 there were so many Norwegians at Haløre (the Norse name for the Scanian market) that the insurgency of the warrior band called *baglar* could be organised there (Sephton 1899, 159), and in 1203 the pretender Erling Steinvegg planned to do the same (Magerøy 1988, 13–14). In 1207 warships on their way to Bergen met 30 *byrðinga* at the coast of Vestfold, and at the coast of Agder they met other *byrðinga* coming from Bergen (Magerøy 1988, 104). Danish annals report large-scale clashes between Norwegians and Danes at the Scanian market in 1237 (Jørgensen and Jørgensen 1920, 109, 131, 137, 140, 146, 193, 196).

When did grain products become an important commodity in this trade? German Hansa merchants from the Baltic appeared in Bergen selling rye, malt and beer in the AD 1240s (DN, vol. 5, nos 1 and 2). But Norwegian merchants may have made grain part of their exchanges with the Scanian market some decades earlier. While the action in *Egils saga* took place in the tenth century, the saga was written *c.* AD 1220, and the author evidently describes the Scanian market of his own time: 'This was at the time when the ships sailed from the market at *Haløre*. There were many ships from Norway, as usual.' The Norwegians bought 'malt, wheat and honey' (Nordal 1933; author's translation). The export of grain products from the western Baltic on Norwegian keels had by AD 1220 probably been going on for some time.

On the basis of the extant written source material from Norway and northern European markets, the theory which has the best empirical basis is that stockfish started to become a commodity in long-distance trade perhaps a decade or two before AD 1100. This trade may have started as an exchange with eastern England, but after a few decades it had spread to the Rhine estuary and by the end of the twelfth century also to the Baltic, for grain supplies.

The eleventh and the twelfth centuries were the first part of what historians have traditionally named the 'Commercial Revolution' of the Middle Ages (Lopez 1971) – the start of long-distance trade in goods for mass consumption. Regions and commodities were drawn into it at different times. Why did the stockfish trade become part of it just around AD 1100?

Long-range exchange presupposes more or less professional merchants who sailed between polities. Such merchants existed in northern Europe at least from the Viking Age onwards and perhaps even earlier (Skre 2008; 2015). We know they traded in slaves, furs, cloth, and weapons and other metalwork. The increase in the number of merchants involved in Norwegian and north European foreign trade around AD 1100 should be seen as a consequence of and not a cause of the commercial revolution, which included the stockfish trade.

In my previous work (e.g. Nedkvitne 1983; 1988), I defended explanations for the 'internationalisation' of the stockfish trade around AD 1100 which could be labelled demographic or ecological. The Norwegian coastal population increased in the period between the so-called Justinian plague of the sixth century and the Black Death pandemic of AD 1349. Peasants had to extract more food resources. In practice this meant resources that were more labour intensive and more complex to organise. Stockfish production seems to have increased slowly based on internal consumption in the 1000-year period between *c.* AD 100 and 1100, before becoming part of long-range trade around 1100. However, the extension of the stockfish trade from regional to international cannot be explained by developments in Norway alone. Producers had to acquire useful goods in return. Therefore complementary economic needs had to arise in other regions. These causes explain both increasing regional exchanges within Norway before AD 1100 and the increase in long-range trade from AD 1100 onwards (Nedkvitne 1983, 17, 377–83, 387; 1988, 17, 596–8; 2014, 28–9, 535–49).

Alternative theories in some way or other connect the origin and development of the export stockfish trade with the process of state formation. The most prominent originators of this view were Johan Schreiner (1935; 1941) and Kåre Lunden (1967; 1972). Today I am more inclined to accept elements of these theories than I was in the 1980s.

Up to the eleventh century, Norway was ruled by local chieftains who had the power and right to impose on their clients duties in the form of labour or goods. Many fishing huts along the coast before AD 1000 (e.g. Johannessen 1996, 43, 46–7; Narmo 1994, 27–45) no doubt belonged to chieftains who sent some of their male servants and tenants to fish in periods when there was little work to do on the farm. Some archaeologists and historians have gone further, seeing stockfish production before AD 1100 as part of a redistributive economic system. Chieftains distributed the results of fishing among their clients, and they sent other peasants into the mountains to hunt – distributing the animal products in the same manner (Johannessen 1996, 85, 94–5; Lunden 1972, 42–9; Nilsen 1997, 86; see Chapter 4). This theory is impossible to verify because archaeological material does not tell us who sent the men to fish. Nevertheless, we are probably on safe ground if we say that up to the eleventh century a large part of the land along the Norwegian coast was owned and cultivated by chieftains, and that many fishermen in the known seasonal huts were probably servants producing food for people living on their chieftain's farms. This is a practice known from the thirteenth century and later, when the historical sources are richer (Nedkvitne 1983, 306–8; 1988, 282–91; 2014, 435–8, 442–9).

Between AD 1000 and 1200 there was a massive transfer of land ownership from local chieftains to the state and the church. The chieftains grew fewer and poorer. Church and state exploited land and peasants differently. The church demanded tithes, the state demanded taxes and landowners demanded land rents partly paid in money. Coastal farmers produced stockfish and sold it in Bergen, in Trondheim or at the marketplace Vågan in Lofoten to obtain coins to pay their dues. Church and state needed money to buy foreign prestige textiles, bronze, lead, weapons and church equipment – and to pay foreign master builders and other experts. The church was an international organisation, and the state was part of an international system. Both had international economic needs. State and church from the decades around AD 1100 imposed economic obligations, which pressured the peasants to sell part of their production for money. At the same time the state, starting in the eleventh century, created military safety and judicial predictability along the coast, which was a necessity for long-distance trade in large quantities to function.

The conclusion of what has been said above is that Norwegian stockfish production went international immediately before AD 1100 as a result of two factors. First, demographic pressures in peasant society made it desirable for peasants to exchange fish for grain to satisfy their own needs for food. Second, the new ruling class of a centralising Norwegian state needed tax money to buy international goods and services, while also providing enhanced security for traders.

The state-formation pressure only started to work in the eleventh century. Conversely, the demographic pressure started slowly (around AD 600, after the Justinian plague). As a result there was a protracted history of intensifying local exchange before the start of a long-range export trade in stockfish.

The golden age of stockfish trade in the late Middle Ages

Moving forward in time, stockfish has never fetched as favourable prices as it did in the century after the Black Death. In the late Middle Ages, AD 1350–1550, the fish were transported from fishing villages in Lofoten and elsewhere to Bergen mainly by the fishermen themselves, but also by local Norwegian merchants. Prices in Bergen therefore represent the exchange rate between fishermen and export merchants.

The most important items the fishermen received in return were Baltic rye, English wheat and high-quality woollen cloth. It is problematic to produce a price series for cloth because of variations in quality and size. Rye and wheat, however, were standard items, making it possible to reconstruct long-term price series.

Prices were recorded in money, but because the silver contents of the relevant coins are known, it is possible to standardise values as weights of silver. On this basis I have tabulated silver prices of stockfish and grain between AD 1270 and AD 1700 (Table 5.1). Even silver, however, was a commodity the price of which fluctuated. Thus an exchange ratio has been calculated. How many kilograms of grain did buyers exchange for one kilogram of stockfish?

Relevant prices from Bergen before AD 1500 are few indeed. Those that exist have been included in Table 5.1 (after Nedkvitne 1988, 42; 2014, 506). They require a brief discussion. To begin with the stockfish prices, the figures for AD 1270–1350 are based on prices in the English customs accounts that were meant to correspond to what the merchant had paid in Bergen. One market price said to be normal from Norway confirms this equivalence. The data for AD 1351–1440 are also based on the English customs, cross-checked against two prices from the fishing districts in northern Norway. For AD 1441–1500, the values are derived from four prices in Bergen and from English customs accounts from the AD 1440s. Customs officials had at this time started to use fixed values instead of market prices. The AD 1501–1650 data are from the account books of the sheriff (lensherre) of Bergen and the archbishop of Nidaros. The figures from AD 1651–1700 are taken from several account books (as detailed in Nedkvitne 1983, 594–9; 1988, 42, 602–4; 2014, 707–14).

Turning to the cost of rye, the data for AD 1270–1350 are based on two identical normative price assessments, from Bergen and western Norway. Their accuracy is corroborated by one market price from AD 1341 regarding a sale from a Hansa merchant to a Norwegian. The figures for AD 1351–1440 are based on two estimated values and one market price for rye flour owned by Hansa merchants in Bergen. The AD 1441–1500 figures are based on one estimated value and one statement about what was at that time a normal market price. The data for AD 1501–1700 are based on the same sources as the stockfish prices from the same periods (Nedkvitne 1983, 601–4; 1988, 605–6; 2014, 716–19).

In previous work I also collected prices from ports known to have imported Norwegian stockfish and compared them with prices in Bergen. Prices were higher than in Bergen, but changes in the relative cost of stockfish and rye showed similar trends. Thus these data corroborate the representativeness of the Bergen prices (Nedkvitne 1983, 341, 343; 2014, 503, 714–15).

In the century before the Black Death, stockfish had a low price compared with rye flour and grain in general. After the Black Death the price ratio improved dramatically. Fishermen seem to have received three times as much flour for their fish. Around AD 1500 at the latest the price ratio started to deteriorate, and around AD 1600 it seems to have been worse than in the High Middle Ages (Table 5.1).

These changes reflect long-term trends in medieval food prices. In a pre-industrial society there typically will be a close connection between population size and standard of

Table 5.1. Price ratio of rye flour to stockfish in Bergen AD 1270–1700 (Nedkvitne 1988, 42).

Year	Price of 100 kg stockfish (in g of silver)	Price of 100 kg of rye flour (in g of silver)	Kg of rye flour per kg of stockfish
1270–1350	102	48	2.1
1351–1440	262	36	7.3
1441–1500	155	24	6.5
1501–1550	101	34	3.0
1550–1600	124	64	1.9
1601–1650	118	64	1.8
1651–1700	102	75	1.4

living. In the Middle Ages, 80–90% of northern Europeans worked in agriculture. Thus growing populations could only be sustained if new households reclaimed land of a lower quality and/or if larger farms were subdivided. Both processes caused the standard of living of the average peasant to deteriorate. Creating new jobs for more people is a slow process in a pre-industrial society. More available hands therefore meant lower wages. In periods of population decrease, the reverse process took place. The remaining peasants were concentrated on the best farms, and their standard of living improved.

Grain was the staple food in northern Europe, consumed in the form of bread, porridge and beer. This was so for both aristocrats and peasants. Nevertheless, all but the most destitute supplemented their diet with protein from fish and meat. Both were desirable parts of the medieval European diet, but fish calories were more expensive than grain calories. Thus the consumption of fish increased when living conditions improved. Historians of pre-industrial Europe have long ago pointed out that the consumption of meat and fish was very high in the late Middle Ages in all social groups, a pattern attributed to the lower population after the Black Death (Abel 1966, 73; Pounds 1994, 448–9; van Bath 1963, 123, 140).

Strong competitors AD 1400–1700

The favourable fish prices after the Black Death encouraged merchants to develop fisheries for long-distance trade all over northern Europe. The main alternative to dried cod was salted herring. The price ratio between stockfish and herring will explain why the profitability of Bergen stockfish was gradually undermined. The price ratio calculated in Table 5.2 is the silver price of 100 kg stockfish divided by the silver price of 100 kg salted herring.

The data in Table 5.2 suggest that salted herring was increasingly profitable compared with stockfish. However, it could be criticised for combining data from geographically disparate areas; England was an important market for Bergen fish from the twelfth to the fifteenth centuries, whereas Holland was a major importer from the fifteenth to the seventeenth centuries (Nedkvitne 1983; Wubs Mrozewicz 2008). To circumvent this problem, Table 5.3 compares price series for England (after Rogers 1866–1902; Nedkvitne 1983, 346, 590–3, 604–6; 2014, 719–21) and Holland (after Posthumus 1946; 1964; Nedkvitne 1983, 346, 587–9, 592–3; 2014, 721).

The trend from Table 5.2 is confirmed in Table 5.3. Stockfish prices fell compared with those of salted herring throughout the period. Nevertheless, stockfish was consistently more expensive per kg. This was probably due in part to the fact that 60% of the weight of a stockfish is edible, compared with only 50% of that of a salted herring. Palatability must also have played a role. Moreover, stockfish contains 1.8 times more calories per weight unit than does salted herring (Nedkvitne 1988, 38; 2014, 510; Schulerud 1945, 68). Based on this figure and the data in Tables 5.2 and 5.3, Table 5.4 estimates the relative price of 1000 calories of the two foodstuffs on European markets.

Stockfish calories were more expensive than calories of herring all through the period AD 1259– 1550, but the difference diminished over time, to the point where after AD 1550, the two seem to have been on the same level. Stockfish

Table 5.3. Price ratio of salted herring to stockfish on northern European markets AD 1375–1700, by weight (after Nedkvitne 1988, 38)

Year	England	Holland
1375–1440	4.4	
1441–1500	3.7	
1501–1550	2.5	2.1
1550–1600	2.0	1.4
1601–1650		1.2
1651–1700		1.2

Table 5.2. Price ratio of salted herring to stockfish on north European markets AD 1259–1600. Ports that served as hubs for direct stockfish trade from Norway are omitted to exclude their atypical prices

Year	Price of 100 kg stockfish (in g of silver)	Price of 100 kg salted herring (in g of silver)	Kg of herring per kg stockfish
1259–1350	455	67	6.8
1351–1440	613	152	4.0
1441–1500	376	99	3.8
1501–1550	218	90	2.4
1551–1600	240	141	1.7

Table 5.4. Price ratio of stockfish to salted herring on northern European markets AD 1259–1700, by caloric content

Year	Northern Europe	England	Holland
1259–1350	3.8		
1351–1440	2.2	2.5	
1441–1500	2.1	2.1	
1501–1550	1.3	1.4	1.2
1551–1600	1.0	1.1	0.8
1601–1651			0.7
1651–1700			0.7

calories may even have become cheaper in Holland. How should this decline in the value (and perhaps popularity) of stockfish be explained? The answer is that the characteristics of stockfish remained the same through time, whereas the quality of salted herring, its main competitor, improved considerably.

The technology of salting herring in barrels was known and practised in the Baltic Sea region as early as the twelfth century. In Scania, for example, Danish peasant fishermen brought the herring ashore and sold it to German merchants, who salted it in barrels. Great quantities of salt were needed, produced in the salt mines near Lüneburg. Salt and empty barrels were brought to Scania, and the salting had to take place immediately after the fresh herring were landed. The barrels were loaded on board larger trading ships and transported to Lübeck or other ports and from there on river boats or wagons to the German interior (Eriksson 1980, 26–39; Jahnke 2000; Kuske 1905, 234–44; Lechner 1935, 58–9).

In the North Sea region, salting herring in barrels was more complicated because the catches were taken at changing locations over a larger area. With the probable exception of the East Anglian fishery (e.g. Campbell 2002), North Sea merchants thus invested less in herring for as long as prices remained as low as they were before AD 1350 (Nedkvitne 1983, 350–3; 2014, 515–518; Thompson 1983, 31–5). Fishermen from the Netherlands and England brought the herring ashore and sold it to fishmongers or merchants. Much of this catch may have been salted 'light' or 'green' (keeping it edible for a few days) and marketed close to the landing ports (Heath 1968, 61; Michell 1977, 142–3, 147; but cf. Chapter 3). In the Netherlands some was also smoked for medium-term preservation and sent to the German interior (Kuske 1905, 261–4). Before AD 1370, dried cod from Bergen may thus have had few competitors for consumers in the North Sea region who wanted fish for long-term storage.

After the Black Death, however, merchants on both sides of the North Sea showed new interest in the herring trade. Sea salt from western France (produced by solar evaporation) was imported in increasing quantities from the 1360s (Bridbury 1973, 43–4, 67, 102). It is highly likely that it was destined for the herring fisheries (Bridbury 1973, xviii). From the 1370s onwards, English and Dutch merchants equipped large fishing vessels that caught herring with drift nets in the North Sea and salted the catch in bins or barrels on board. These ships normally needed 6–8 weeks to be fully loaded. Back in their home ports, the herring was packed and salted in new barrels. These fishing ships, called 'doggers' in England and *haringbuizen* in the Netherlands, had crews of 20–30 men (Degryse 1944, 97–8, 106; Nedkvitne 1983, 353–5; 2014, 521–3; Vogel 1915, 305–7).

The Dutch called their herring fishery in the North Sea *de groote visscherij* (the great fishery), and it grew

to considerable proportions. In AD 1562, a contemporary source put the number of herring busses at 700, whereas in AD 1620 it was estimated to be 2000 (Beaujon 1885, 66; Dardel 1941, 153; Degryse 1944, 34–5, 43, 48, 108; Michell 1977, 148; Vogel 1915, 304–6, 318). Alternatively, some revisionist historians estimate a lower number (perhaps 800 herring busses) when the fishery was at its peak in the 1630s (Kranenburg 1946, 38–9; Michell 1977, 148–9; Nedkvitne 1988, 46; Poulsen 2008, 46). Circumstances changed after the middle of the seventeenth century. Between AD 1652 and the end of the century, Holland was almost constantly at war with England or France. 'The great fishery' suffered from piracy, and Dutch herring production slowly declined (Beaujon 1885, 67; Michell 1977, 149; Vogel 1915, 318). The Dutch also caught cod on the North Sea banks (Bennema and Rijnsdorp 2015). In AD 1628, 180 ships participated. The cod was mostly salted in barrels, although some was brought ashore fresh in wells (Beaujon 1885, 149–58; Nedkvitne 1988, 33).

English doggers caught cod on Dogger bank and salted it like the Dutch did (Heath 1968, 57; Michell 1977, 164–6; Nedkvitne 1988, 34–5). By AD 1412 onwards they also expanded to the cod grounds around Iceland. In AD 1528, 149 English ships are said to have visited Icelandic waters. Most were fishing doggers, but a few merchant ships also traded with the Icelanders (Carus-Wilson 1933, 161; Thorsteinsson 1957, 184; see Chapters 7–8). In earlier centuries Icelanders had sent some stockfish to Bergen, from where it was transhipped to English and other European markets. More direct access to the English market encouraged Icelandic production. Some stockfish from Iceland joined salted herring and cod from the North Sea and Iceland in replacing Bergen stockfish on the English market. At the beginning of the 1400s, stockfish was transported from Bergen to England on c. 16 Hanseatic and English ships annually. At the end of the century the entire trade could be conducted by one English ship (Nedkvitne 1983, 70, 110–11, 124–5; 2014, 105, 156–7, 159, 161, 172–3).

In AD 1497, an English expedition reached Newfoundland. In the following years an increasing number of Spanish and Portuguese fishing vessels visited the Grand Banks. The English do not seem to have extended their fisheries from Iceland to Newfoundland until the early 1500s, but from the final decades of the sixteenth century the Grand Banks became more important than Iceland for the English cod fisheries (Candow 2009; Innis 1954, 11, 13, 31–4, 69–70, 115, 127; Michell 1977, 155–8, 160–1; Nedkvitne 1988, 35; 2014, 527–8). The Newfoundland fisheries introduced a new product on the English and north European markets. The fish was first salted, then dried on rocks. This 'salted stockfish' was in Norwegian called *klippfisk* ('rock-fish') (Innis 1954, 50–1, 71; Michell 1977, 157; Nedkvitne 1988, 35–6; Vollan 1956, 12–13).

In this period the value of salted herring increased relative to stockfish (Tables 5.2–5.4), presumably because the quality of the former improved. Dutch fishing vessels had to sell their catches in Dutch ports, and all major fishing ports had a controller who saw to it that the herring was not too small or too old (Beaujon 1885, 53–65, 154–5; Michell 1977, 148, 152–3). After AD 1519, Dutch regulations stipulated that only new barrels could be used and that they had to have an identity mark burned into them so that producers of faulty herring could be identified. The amount and quality of salt in each barrel was also prescribed. After AD 1584, only refined sea salt was permitted (Beaujon 1885, 9–11, 39; HU, vol. 10, no. 916). This comprehensive control system was institutionalised for all fishing ports in Holland and Zeeland in AD 1575 by the Dutch ruler William of Orange (Beaujon 1885, 53–65, 154–5; Michell 1977, 148, 152–3). All fish products fell in price from the second half of the fifteenth century, but stockfish fell more than most other fish because producers had no potential to compensate by improving its quality.

The 'golden age' of the Norwegian fish trade in the late Middle Ages called into existence new competitors. The outcome was not advantageous for stockfish exports from Bergen. They were displaced on the English market by salt fish produced in North Sea and Icelandic waters, and also by stockfish from Iceland. Moreover, the formidable Dutch production of salted herring and cod marginalised stockfish in the Netherlands. Some Dutch salted fish were even traded to inland western Germany (Beaujon 1885, 44, 47–8), where Bergen stockfish did retain a significant market between the fifteenth century and AD 1700 (Kuske 1905, 261, 264–73; Nedkvitne 1983, 152–9; 2014, 234–40). Here stockfish had certain competitive advantages. It could be stored for several years, whereas a barrel of herring kept best for only six months to a year (Nedkvitne 1983, 349, 513–14). Stockfish was also simpler and cheaper to transport. Fresh cod is reduced to 24% of its original weight when dried, and needs no heavy packing. That was important when fish were to be transported in wagons and river boats from the Rhine estuary and Lübeck to the interior.

A sustainable exploitation of resources?

In recent years there has been an increasing focus, in public debate and in the field of environmental history, on the antiquity of human impacts on marine resources. This raises the question of whether medieval fishing was on a scale sufficient to have reduced fish stocks beyond the range of natural variability. Nobody will dispute that the situation was different before and after the introduction of industrialised fishing, which took place between c. AD 1870 and AD 1970. Before that time species and populations living close to the shore and therefore particularly vulnerable to fishing could have been overexploited (Erlandson and Rick

2008, 10). Until recently, however, it was uncontroversial to assume that pre-industrial fishermen lacked the technology to overfish widely distributed species. This traditional understanding is now being questioned, particularly for large marine taxa. Perhaps overexploitation of many more species goes further back in time than is often thought (e.g. Roberts 2007).

Let us consider the historical evidence regarding cod in Norway. From AD 1303, extant customs accounts make it possible to quantify stockfish exports from Bergen to English and Hansa towns in certain years. From AD 1518 there are accounts from Bergen and from AD 1577 there are relevant Øresund toll accounts. On the basis of these sources it is possible to attempt an estimate of the total exports from Bergen (see below).

The English customs accounts from AD 1303–11 show annual stockfish imports from Norway of more than 1500 metric tonnes by Hansa and Norwegian merchants. If the English merchants had been included, the total would probably have reached 2000 tonnes. Equivalent evidence is not available from Germany and the Netherlands, but by extrapolation we can estimate that the total exports from Norway most probably reached 3000–4000 tonnes (Nedkvitne 1983, 47–8, 60; 2014, 70–1, 92–3).

Between AD 1365 and 1400, the period after the Black Death, relevant customs accounts are preserved from Boston and King's Lynn in England and from Lübeck in Germany. Together they give evidence of annual stockfish imports of c. 700 metric tonnes. Other continental ports are not documented, but they probably would have brought the total to somewhere between 1000 and 2000 tonnes (Nedkvitne 1983, 78, 112, 124, 163; 2014, 114, 158, 173, 245–6).

The first extant customs accounts from Bergen that quantify stockfish exports date from AD 1577/8. A comparison with the Øresund accounts makes it possible to correct for under-reporting. The total annual export seems to have been c. 2800 metric tonnes (Nedkvitne 1983, 180–6, 385; 2014, 263). Finally, for the five years between AD 1650 and 1654 we have a reliable report that the average annual stockfish export from Bergen was 6100 tonnes (Nedkvitne 1983, 186; 2014, 265).

Fish exports have always experienced short-term and long-term fluctuations caused by the weather, sea temperatures, the size of the cod drifts and the number of participants in the fishery. But in the 360 years between AD 1303 and 1654, in a normal year the export seems to have been between 1000 and 6000 metric tonnes. Stockfish is reduced to 24% of its original weight if it is well dried. The export volumes must therefore have been based on 5000–25,000 tonnes of raw cod. How does this compare with later periods?

The stockfish from Bergen was produced from cod from northeastern Arctic stocks, which today are caught both when spawning along the Norwegian coast and in the North Atlantic by trawlers. The cod around Iceland belongs to a different population. In several years between 1955 and 1975, the total annual catch from the northeastern Arctic stock was above 900,000 tonnes of raw cod. This volume was not sustainable. It represented overfishing, and in the following period catches declined. In the years 2003–11 the annual recommendations of the official commission of scientists for sustainable exploitation varied between 300,000 and 700,000 tonnes, and the actual catches varied between the same figures (Norwegian Ministry of Fisheries 2012–13). Because in the pre-industrial context northeastern Arctic cod was caught only along the Norwegian coast, we can compare these recent figures with the export figures for the period AD 1650–1654. The catch in the period of industrial cod fishing was 12–28 times larger, a difference so considerable that one can claim with certainty that medieval fishermen were nowhere near to overfishing the Norwegian cod stock. Limits on the scale of production and export must have resulted from other variables (cf. Chapters 7 and 8 regarding Iceland).

Similar observations apply to the herring fisheries of the North Sea. In AD 1639, when Dutch herring production was at its maximum, a contemporary observer estimated that one herring buss could produce *c.* 560 barrels a year. The traditional estimate of 2000 active herring busses in this period thus implies an annual production of 1.12 million barrels. The revisionist estimate of 800 busses gives a lower figure, of 450,000 barrels (Beaujon 1885, 67; Michell 1977, 149; Nedkvitne 1988, 33) (see above). In comparison, between AD 1947 and 1964, the total catches of herring from the North Sea corresponded to an average of *c.* 5–9 million barrels annually, increasing to almost 13 million barrels in 1965. That constituted overfishing, and in the following decade the catches fell sharply (Gislason 2006, 420). Today the scientifically recommended annual catch corresponds to more than 5 million barrels (Poulsen 2008, 80). The catches are given in metric tonnes of herring. One barrel of herring (114 l) contains 90 kg of herring (information from Norges Sildesalgslag, see Nekvitne 1988, 47, note 53). Today only part of the herring catches are salted in barrels, but for the sake of comparison I have converted the whole catch into barrelled herring. Before they are put into barrels, some of the herring's intestines are removed, and some of the humidity in fresh herring meat is drawn into the brine. A herring weighs more when fresh than when it has been salted in a barrel.

Norwegian herring exports were of Atlanto-Scandic herring that spawned along the Norwegian west coast. The largest known volume of exports before 1800 occurred in AD 1756. The total for the whole of Norway was then 200,000 barrels. The top year in the nineteenth century was AD 1872, when 1.2 million barrels were exported. In the following decades the fisheries were slowly industrialised. The largest volume of Norwegian exports in the twentieth century

occurred in AD 1956, with a total of 12 million barrels, all of it caught along the Norwegian coast (Nedkvitne 1988, 14–15, 33; Solhaug 1976, vol. 2, 743; cf. Thompson 1983, 29–30). This represented overfishing, and the catches fell. In recent years the scientifically recommended quota for the catches of Atlanto-Scandic herring of all nations has corresponded to more than 10 million barrels. The much lower volumes of northern Europe's great pre-industrial fisheries (that is, before AD 1800), which preceded many years of increasing catches, were comparatively sustainable.

In the pre-industrial period technology made it problematic to live from fishing alone in most of Norway. Fishermen could not move to distant locations where the catches were best, because they lacked motor boats and electronic communication. Moreover, their equipment was not sufficiently effective to land great quantities during the spawning migration. Thus most households along the Norwegian coast combined fishing with cattle raising and normally also with grain production. They cultivated and took care of the soil within the boundaries of their farms, and each peasant saw to it that the exploitation of it was sustainable. Sea-fish, in contrast, would have been regarded as a resource without limits – and from a medieval perspective this was correct.

Conclusions

The medieval Norwegian cod fisheries took place at the Lofoten Islands and other spawning grounds close to shore. They could thus be conducted with small boats and simple fishing gear. Cod could be preserved as stockfish without major investments or a complex infrastructure. The fisheries took place in winter, when there was little to do in agriculture and much available labour in peasant communities. It is partly for these reasons that stockfish production was the first part of the Norwegian economy to be drawn into long-distance trade. Lofoten cod was the first fishery of northern Europe to be drawn into the incipient commercial revolution at the end of the eleventh century. Yet catches in the Middle Ages made only small inroads into the vast stocks of Arctic cod. Even at its height, the medieval fishery had ecological potential for increased harvests. In this instance pre-industrial limits on expansion were technological and logistical.

References

Abel, W. 1966. *Agrarkrisen und Agrarkonjunktur: eine Geschichte der Land- und Ernährungswirtschaft Mitteleuropas seit dem Hohen Mittelalter*. Hamburg: Paul Parey.

Aðalbjarnarson, B. (ed.) 1951. *Snorri Sturluson Heimskringla*. Vol. 1. Reykjavík: Hið Islenzka Fornritafélag.

Andersson, T. M & K. E. Gade (ed.) 2000. *Morkinskinna: the earliest Icelandic chronicle of the Norwegian kings (1030–1157)*, London: Cornell University Press.

Bagge, S., S. Holstad Smedsdal & K. Helle (eds) 1973. *Norske middelalderdokumenter*. Oslo: Universitetsforlaget.

Bately, J. & A. Englert (eds) 2007. *Ohthere's voyages: a late 9th-century account of voyages along the coasts of Norway and Denmark and its cultural context*. Roskilde: Viking Ships Museum.

Bennema, F. P. & A. D. Rijnsdorp. 2015. Fish abundance, fisheries, fish trade and consumption in sixteenth-century Netherlands as described by Adriaen Coenen. *Fisheries Research* 161: 384–99.

Beaujon, A. 1885. *Overzicht der geschiedenis van der nederlandsche zeevisscherijen*. Leiden: E. J. Brill.

Bridbury, A. 1973. *England and the Salt Trade in the Later Middle Ages*. Westport: Greenwood.

Campbell, J. 2002. Domesday herrings, in C. Harper-Bill, C. Rawcliffe & R. G. Wilson (eds) *East Anglia's History: studies in honour of Norman Scarfe*: 5–17. Woodbridge: Boydell Press.

Candow, J. E. 2009. The organisation and conduct of European and domestic fisheries in northeast North America, 1502–1854, in D. J. Starkey, J. T. Thór & I. Heidbrink (eds) *A History of the North Atlantic Fisheries. Volume 1: from early times to the mid-nineteenth century*: 387–415. Bremen: H. M. Hauschild.

Carus-Wilson, E. M. 1933. The Iceland trade, in E. Power & M. M. Postan (eds) *English Trade in the 15th Century*: 155–82. London: Routledge.

Carus-Wilson, E. M. 1964. The medieval trade of the ports of the Wash. *Medieval Archaeology* 6–7: 182–201.

Carus-Wilson, E. M. 1987. The woollen industry, in M. M. Postan, E. Miller & C. Postan (eds) *The Cambridge Economic History of Europe From the Decline of the Roman Empire. Volume 2: trade and industry in the Middle Ages*: 613–90 (2nd edition). Cambridge: Cambridge University Press.

Chibnall, M. (ed.) 1973. *The Ecclesiastical History of Orderic Vitalis*. Vol. 4. Oxford: Clarendon Press.

Dardel, E. 1941. *La pêche harenguière en France: étude d'histoire économique et sociale*. Paris: Presses universitaires de France.

Degryse, R. 1944. *Vlaanderens haringbedrijf in de middeleeuwen* (De Seizoenen 49). Antwerp: De Nederlandsche Boekhandel.

DN Diplomatarium Norvegicum 1847–2011. *Diplomatarium Norvegicum*. Vol. 5. Oslo: Dokumentasjonsprosjektet. Available at: http://www.dokpro.uio.no/dipl_norv/diplom_felt.html.

Eriksson, H. S. 1980. *Skånemarkedet*. Højbjerg: Wormianum.

Erlandson, J. M. & T. C. Rick. 2008. Archaeology, marine ecology, and human impacts on marine environments, in T. C. Rick & J. M. Erlandson (eds) *Human Impacts on Ancient Marine Ecosystems: a global perspective*: 1–19. Berkeley: University of California Press.

Gislason, H. 2006. Havet som menneskets spisekammer, in T. Fenchel (ed.) *Naturen i Danmark. Volume 1: havet*: 397–431. Copenhagen: Gyldendal.

Hansen, G. 2005. *Bergen c. 800–c. 1170: the emergence of a town*. Bergen: Fagbokforlaget.

Hansen, G. 2015. Bergen AD 1020/30–1170: between plans and reality, in J. H. Barrett & S. J. Gibbon (eds) *Maritime Societies of the Viking and Medieval World*: 182–97. Leeds: Maney/ Society for Medieval Archaeology.

HU Hansisches Urkundenbuch 1876–1939, ed. K. Höhlbaum *et al*. Vols 1, 10. Halle, Munich and Leipzig: von Duncker & Humblot.

Heath, P. 1968. North Sea fishing in the fifteenth century: the Scarborough fleet. *Northern History* 3: 53–65.

Helle, K. 1982. *Bergen bys historie*. Bergen, Oslo & Tromsø: Universitetsforlaget.

Hutcheson, A. R. J. 2006. The origins of King's Lynn? Control of wealth on the Wash prior to the Norman Conquest. *Medieval Archaeology* 50: 71–104.

Innis, H. 1954. *The Cod Fisheries*. Toronto: University of Toronto Press.

Jahnke, C. 2000. *Das Silber des Meeres: Fang und Vertrieb von Ostseehering zwischen Norwegen und Italien (12.–16. Jahrhundert)*. (Quellen und Darstellungen zur Hansischen Geschichte, Neue Folge 49). Cologne, Weimar, Vienna: Böhlau.

Johannessen, L. 1996. Fiskevær og fiskebuer i vestnorsk jernalder: en analyse av strandtufter i Hordaland. Unpublished Master's dissertation, University of Bergen.

Jørgensen, E. & E. Jørgensen (ed.) 1920. *Annales Danici medii aevi*. Copenhagen.

Kålund, K. (ed.) 1908. *Alfrædi islenzk: Islandsk encyclopedisk litteratur. Volume 1*. Copenhagen: Samfund til Udgivelse af Gammel Nordisk Litteratur.

Kranenburg, H. A. H. 1946. De zeevischerij van Holland in den tijd der Republiek. Unpublished PhD dissertation, Nederlandsche Economische Hoogeschool, Amsterdam.

Kuske, B. 1905. Der Kölner Fischhandel vom 14.–17. Jahrhundert. *Westdeutsche Zeitschrift für Geschichte und Kunst* 24: 227–313.

Lechner, G. 1935. *Die hansischen Pfundzollisten des Jahres 1368*. (Quellen und Darstellungen zur hansischen Geschichte, Neue Folge 10). Lübeck: Verlag des hansischen Geschichtsvereins.

Lopez, R. S. 1971. *The Commercial Revolution of the Middle Ages*. Englewood Cliffs: Prentice-Hall.

Lunden, K. 1967. Hanseatane og norsk økonomi i seinmellomalderen. *Historisk Tidsskrift* 46: 97–129.

Lunden, K. 1972. *Økonomi og samfunn*. Oslo: Universitetsforlaget.

Magerøy, H. (ed.) 1988. *Bøglunga sögur*. Vol. 2. Oslo: Solum.

Michell, A. R. 1977. The European fisheries in early modern history, in E. E. Rich & C. H. Wilson (eds) *The Cambridge Economic History of Europe From the Decline of the Roman Empire. Volume 5: the economic organization of early modern Europe*: 134–84. Cambridge: Cambridge University Press.

Narmo, L. E. 1994. Budar- et fiskevær fra vikingetid. *Gammalt frå Fræna* 14: 27–45.

Nedkvitne, A. 1983. Utenrikshandelen fra det vestafjelske Norge 1100–1600. Unpublished PhD dissertation, University of Bergen.

Nedkvitne, A. 1988. *'Mens Bønderne seilte og Jægterne for': Nordnorsk og vestnorsk kystøkonomi 1500–1730*. Oslo: Universitetsforlaget.

Nedkvitne, A. 2014. *The German Hansa and Bergen*. (Quellen und Darstellungen zur Hansischen Geschichte, Neue Folge 70) [revised version of PhD from 1983]. Vienna and Cologne: Böhlau Verlag.

Nilsen, G. 1997. Jernaldernaust på Vestvågøy i Lofoten. Unpublished Master's dissertation, University of Tromsø.

Nordal, S. 1933. *Egils saga Skalla-Grímssonar*. Reykjavík: Hið Islenzka Fornritafélag.

Norwegian Ministry of Fisheries. 2012–13. Previously available at: www.regjeringen.no/nb/dep/fkd/dok/regpubl/stmeld/2012-2013

Posthumus, N. W. 1946. *Inquiry into the History of Prices in Holland*. Vols 1 and 2. Leiden: E. J. Brill.

Poulsen, B. 2008. *Dutch Herring, an Environmental History, c. 1600–1800*. Amsterdam: Aksant.

Pounds, N. J. G. 1994. *An Economic History of Medieval Europe* (2nd edition). London: Pearson Education.

Roberts, C. 2007. *The Unnatural History of the Sea*. London: Gaia.

Rogers, J. E. T. 1866–1902. *A History of Agriculture and Prices in England*. Vols 1–7. Oxford: Clarendon Press.

Schmeidler, B. (ed.) 1937. *Helmoldi Cronica Slavorum*. (Monumenta Germaniae Historica 32). Hannover: Hahnsche Buchhanglung

Schreiner, J. 1935. *Hanseatene og Norges nedgang*. Oslo: Steenske.

Schreiner, J. 1941. *Hanseatene og Norge i det 16. århundre*. Oslo: Det norske videnskabsakademi.

Schulerud, A. 1945. *Norske næringsmidler*. Oslo: Aschehoug.

Sephton, J. (ed.) 1899. *The Saga of King Sverri of Norway*. London: David Nutt.

Skre, D. 2008. Post-substantivist towns and trade: AD 600–1000, in D. Skre (ed.) *Means of Exchange: dealing with silver in the Viking Age*: 327–41. Aarhus: Aarhus University Press.

Skre, D. 2015. Post-substantivist production and trade: specialized sites for trade and craft production in Scandinavia AD c 700–1000, in J. H. Barrett & S. J. Gibbon (eds) *Maritime Societies of the Viking and Medieval World*: 156–70. Leeds: Maney/Society for Medieval Archaeology.

Solhaug, T. 1976. *De norske fiskeriers historie 1815–1880*. Bergen: Universitetsforlaget.

Thompson, P. 1983. *Living the Fishing*. London: Routledge.

Thorsteinsson, B. 1957. Island, in A. E. Christensen *et al.* (eds) *Det nordiske syn på forbindelsen mellem Hansestæderne og Norden: det Nordiske Historikermøde i Århus 1957*: 165–95. Århus: Århus Stiftsbogtrykkerie.

Tomfohrde, T. 1914. Die Heringsfischereiperiode an der Bohuslen-Küste von 1556–1589. *Archiv für Fischereigeschichte* 3: 1–192.

van Bath, S. 1963. *The Agrarian History of Western Europe* AD *500–1850*. London: Edward Arnold.

Vogel, W. 1915. Zur Grösse der europäischen Handelsflotten im 15., 16., und 17. Jahrhundert, in *Festschrift für Dietrich Schäfer zum siebzigsten Geburtstag dargerbracht von seinen Schülern*: 268–333. Jena: Gustav Fisher.

Vollan, O. 1956. *Den norske klippfiskhandels historie*. Førde: Øens Forlag.

Wubs-Mrozewicz, J. 2008. *Traders, ties and tensions: the interaction of Lübeckers, Overijsslers and Hollanders in late medieval Bergen*. Hilversum: Uitgeverij Verloren.

6

The Birth of Commercial Fisheries and the Trade of Stockfish in the Borgundfjord, Norway

Helge Sørheim

Introduction

The sea and marine resources have always been important along the coast of Norway. The use of seagoing vessels must date back to the first settlement of the islands and fjords. Finds of fish-hooks, sinkers and rock carvings tell us about the use of both salt and fresh water from this initial period of settlement. In this paper I differentiate between the exploitation of marine resources for personal consumption and commercial fishing for exchange. When commercial production arose, and especially once it became common, it radically changed people's work life and social organisation. This transformation is particularly evident from the establishment of new settlements on the outermost coast, in areas not usable for farming, as near as possible to the fishing grounds (cf. Chapter 4).

By the twelfth century, the medieval expansion of northern European long-distance trade, the 'Commercial Revolution', had impacted Norway (Chapters 4–5). Large-scale trade increased and new types of goods appeared. Peasant tradesmen and proprietor tradesmen were gradually replaced by professional urban merchants (Nedkvitne 1983, 16).

Fish and fish products were among the first important large-scale export products of Norway. The commercial production of stockfish from *c.* AD 1100, or perhaps a little before, provided an export product for a larger European market that dominated Norwegian long-range trade until the seventeenth century. Medieval herring (*Clupea harengus*) production in Norway, conversely, rarely exceeded the needs of local markets (Nedkvitne 1988, 471–2; Chapter 5).

The main spawning ground for the pelagic northeastern Arctic cod (*Gadus morhua*) stocks is Lofoten. The commercial fisheries there date to the early twelfth century at the latest (Bertelsen 1994, 120). Vågan was the most

important marketplace for stockfish from northern Norway from the AD 1170s onwards (Nedkvitne 1983, 189). However, part of the northeastern Arctic cod population spawns as far south as Møre (Eliassen 1983, 4), particularly in the 12 km long Borgundfjord (Figure 6.1). In its relatively shallow waters, varying between 50 m and 125 m, an extensive cod fishery can be undertaken from February until April. Moreover, Sunnmøre is one of the southernmost areas with climatic conditions suitable for producing stockfish of reasonable quality.

Borgundfjord, and its western continuations Hessafjord and Storfjord, adjoin the deep, *c.* 400 m rift in the continental shelf known as Breisunddjupet, located between Godøy and Hareidlandet (Figures 6.2 and 6.3). According to fishermen, cod enter the fjord system through the northern side of Breisunddjupet, and the underwater slope towards the rift has been recorded as a much-used fishing ground for centuries (e.g. Strøm 1766, 52–3).

The nearby fjords, Sulafjord and Storfjord, are deep and have strong currents, making them less well suited for traditional fishing methods. Farther inland, in the Ørskogvika bay, the potential for cod fishing is variable.

In the Borgundfjord, conversely, conditions are ideal (Figure 6.4). With depths around 100 m, the waters are suitable for fishing with both lines and nets (Myklebust 1971). The extensive fisheries were described by Hans Strøm (1762, 470–6) in his account of Sunnmøre in the mid-1700s, when the cod net was commonly used. In AD 1756, 499 *fioringfar* (fishing boats) and 2994 fishermen took part in the Borgundfjord fisheries. Visiting fishermen were not included in this survey, so the number of participants probably exceeded the 3000 who took part in the Lofoten fisheries in the AD 1660s (Dyrvik 1979, 34). However, the cod fisheries in Sunnmøre were more

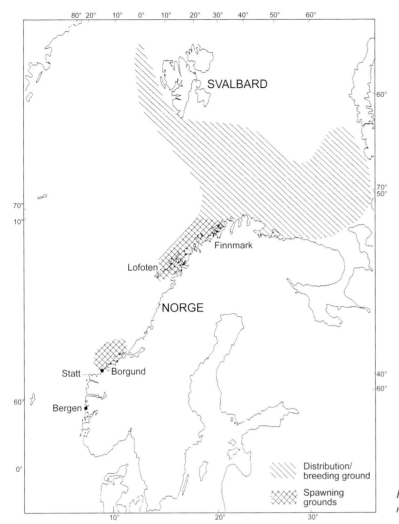

Figure 6.1. The distribution and spawning grounds of the northeastern Arctic cod stocks (after Eliassen 1983, 4).

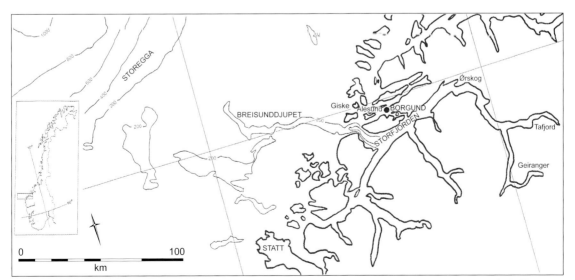

Figure 6.2. Sunnmøre, with its many islands and deep fjords. Note Breisunddjupet, the deep rift in the continental shelf (Drawing: Helge Sørheim).

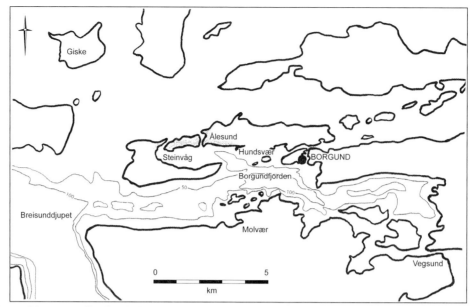

Figure 6.3. Borgundfjord with depths indicated. Note the deep Breisunddjupet (Drawing: Helge Sørheim).

unstable than those in Lofoten. At times they were almost at the level of the Lofoten fisheries, but in AD 1807, when the Lofoten fisheries were at a peak, Leopold von Buch noted that the fishing in Sunnmøre was poor (Solhaug 1976, 207–8).

By the end of the seventeenth century, fishermen in the Borgundfjord area had the most advanced cod-fishing equipment along the coast. Cod nets and long lines were first used in these protected waters. Documents from the early seventeenth century concerning the use and theft of long lines demonstrate that this equipment was already being employed by the late 1600s (Nedkvitne 1988, 438–40).

The question is whether commercial cod fishing in the Borgundfjord can be traced back to the early Middle Ages. One might also ask whether medieval fishermen remained in the shallow and protected Borgundfjord or, instead, travelled farther out to sea, to Storegga and the fishing banks, where the continental shelf drops into the Atlantic Ocean, some 70 km away. The evidence from excavations of medieval remains at Borgund throws light on these questions, by indicating the chronology of the settlement and by showing whether the fishing equipment used in the medieval period was of similar type as that used at the end of the AD 1600s.

Borgund

The medieval town of Borgund occupied a central location, not only as a point of intersection between the fjords and the coast, but also in relation to the richest

fishing grounds in western Norway, namely, that on the north side of Borgundfjord. It may have emerged as a central place during the transition between the Viking Age and the Middle Ages, both as a local market and as a fishing station for farmers who fished for their own consumption. Borgund is mentioned in Snorri Sturluson's *Heimskringla* in connection with events that allegedly should have taken place in the AD 1020s. The place name Hundsvær is also mentioned in this context, which must be the fishing village of that name that was situated on some small islands west of Borgund into the beginning of the twentieth century. Molvær, on the south side of the fjord, also has the suffix '-vær', which denotes a small fishing station or settlement.

The written sources regarding Borgund and the fisheries of Borgundfjord are sparse. My work (e.g. Sørheim 2004, 107–33) has thus mainly been based on the evidence of 20 seasons of excavation, conducted from the 1950s onwards. Even in this context, however, the methods of excavation and recording, together with poor conditions of preservation, make chronological and contextual interpretation of the portable artefacts difficult. It is mostly not possible to make chronological or spatial distinctions within the medieval material; therefore, finds from the entire medieval period of settlement, from the early eleventh century to *c.* AD 1500, and from all areas of the site must be treated more or less as a single provenience. The same caveat applies to the 24,593 fish bones recovered from Borgund, with the added complication that they were recovered exclusively by hand, without sieving the sediment through mesh. Thus the bones of small species, such as herring, are entirely absent

Figure 6.4. Borgundfjord viewed from the east. The fjord opens up at the Breisundet sound, at back left of the image. The medieval settlement of Borgund is located on the forested peninsula. The flat island of the manor of Giske is seen on the horizon at back right (Photo: Helge Sørheim).

and the remains of large fish, such as ling (*Molva molva*) (51% of the assemblage), are probably over-represented in comparison with those of cod (16%) and saithe (*Pollachius virens*) (32%) (Sunde 1972). Fortunately, however, the chronology and layout of the settlement's features – particularly its churches and cemetery – are known with greater precision.

The archaeological fieldwork (Figure 6.5) has established evidence of an urban settlement spread over an area of *c.* 45,000 m^2, of which approximately 5000 m^2 has been excavated. The medieval settlement is situated on a level area east of the present parish church, that was built on the site of its medieval predecessor, which was dedicated to St Peter. The settlement was orientated towards Borgundfjord to the south, where remains of large landing stages for boats can still be observed. To the east, the shallow and protected Klokkarsundet sound runs into the Katavågen bay (from Old Norse *kati*, meaning boat, a frequently used name for harbours in the district) (Figure 6.6). The sites of large boat-houses are visible, and rows of postholes are interpreted as parallel warehouses. Similar parallel rows of postholes have also been found farther north, in Katavågen. Measuring up to 30 m long, they stretch back from wooden piers at the beach (Figure 6.7). An alternative, albeit less likely, interpretation is that some of the postholes belonged to structures for drying fish. The

interiors of these postulated warehouses represent a surface area of approximately 600 m^2, indicating a significant storage capacity. Since large areas of the site have not yet been excavated, there could possibly be even more of them at Borgund. Traces of residential and commercial buildings were found behind the harbour area.

The three known stone churches, probably built in the twelfth century, demonstrate considerable economic strength and suggest a certain population size. They are unlikely to represent the earliest occupation of the site, however, as they are preceded by a cemetery of 300–400 graves that was in use by the early eleventh century (based on an English coin of Ethelred from AD 1004–9). The church eventually became a considerable economic power, owning, *c.* 25% of the land in the region by AD 1648 (Kvamme 1994, 272).

The aristocratic Giske lineage played an important role in the political and economic history of the region. Giske manor was located on a small island of the same name just outside Borgund. Its estate became one of the largest in medieval Norway, owning land all over Sunnmøre. It also had a tenement in Bergen. Because the harbour facilities at the island were poor, Borgund may have served as Giske's port. The rich fish resources in the Borgundfjord should also be considered in connection with the family's power and economic strength.

Figure 6.5. Overview of the medieval settlement of Borgund. The excavated area is located in the open fields in front of St Peter's church, the present parish church at Borgund. To the left rear, behind the quay, a boathouse of prehistoric type has been located. To the left, the remains of large landing stages for boats can be seen. To the front left, the Klokkarsundet sound runs into the Katavågen bay. The remains of long rows of parallel warehouses were found in both areas. A Christian graveyard from the early eleventh century has been located on the hill leading up to St Peter's. Another church (dedicated to St Matthew and/or to Christ) was located to the left of the current church, under the barn, and St Margaret's church was located on the outer promontory, against Klokkarsundet (Photo: R. Engvik).

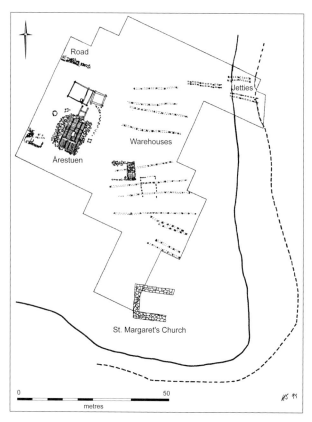

Figure 6.7. The eastern site, with traces of warehouses and piers (Drawing: Helge Sørheim).

Figure 6.6. Map of the medieval settlement area of Borgund (Drawing: Helge Sørheim).

Although fishing and the catching of sea mammals have long traditions as the backbone of Norway's coastal economy, they have, paradoxically, been only sparsely documented archaeologically. There are several reasons for

this. Many of the tools were made of organic materials that, with a few exceptions, have not survived. Another problem is that efficient fishing methods were developed at an early stage, resulting in few typological changes and concomitant difficulties with dating. Moreover, fishing equipment was used at sea and was often lost there. What the archaeologist finds is often in a secondary context. The large assemblage of fishing equipment from Borgund is thus unusual, although important comparative material from northern Norway has been analysed by Bjørn Ebba Helberg (1995) and a large corpus from the excavations in Bergen has been published by Ole Mikal Olsen (2004, 7–106).

One of the first tasks when working on an archaeological material such as fishing equipment is to interpret and explain the activities the artefacts reflect. We can resort to synchronous analogies based on contemporary material or to retrospective methods. To interpret these particular finds, we must use a retrospective method based on post-medieval fishing equipment, but we must do so with caution (Sørheim 1981, 174–5). We must be certain that a historical relationship exists between the archaeological material and the recent, documented material. Fortunately, fishing has been a traditional occupation. Historical examples of scepticism and resistance to new and more efficient fishing

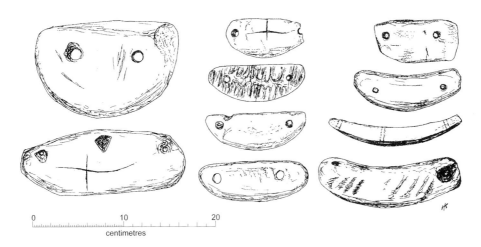

Figure 6.8. Sinkers for troll lines from Borgund (Drawing: Helge Sørheim).

methods abound, and these examples illustrate conservatism in relation to fishing methods. There was, for instance, great resistance to the introduction of long lines in the Borgund fisheries in the seventeenth century (Myklebust 1971, 18–21). This traditionalism, at least until the major innovations of the twentieth century, justifies the use of historical analogy in this study.

Few Norwegian researchers have, however, studied pre-industrial fishing tackle that is suitable for comparative studies of medieval fishing equipment. Norwegian folk museums, often focussed on romantic farming tradition, have until recently been less interested in fishing and coastal culture. Historical fishing equipment that would have helped in the interpretation of archaeological material has therefore only been preserved and documented to a limited degree. O. Nordgaard's (1908) *Træk fra fiskeriets utvikling i Norge* is an important exception employed in the present study. His book was written at a time when fishing equipment was still in use that can to some extent be compared with medieval equipment.

Fishing tackle from Borgund

Using this comparative method, the likely fishing gear identifiable from the Borgund material includes the hand line (Norwegian *djupagnsnøre* or *juksa*) and troll line (Norwegian *dorgesnøre*). These two types consist of a line, sinker, hook and snood (the cord between sinker and hook). The archaeological evidence relating to these elements is discussed below. The cod long line is not of relevance to the current discussion as it was not adopted until around 1600 (Myklebust 1971, 20). Similarly, it is doubtful whether cod nets were in use in the Middle Ages. As an aside, it is popularly believed, albeit mistakenly, that the inventor of the cod net, Claus Nielsen, lived in the Borgundfjord

area and that it was introduced there in 1685 (Strøm 1762, 448–9). Nets and closing nets were employed in the Middle Ages, but they were used for catching herring and salmon (*Salmo salar*) (Vollan 1960, 203–4). A few artefacts from Borgund are consistent with the use of nets, probably for herring in this instance.

Weights from Borgund

Weights comprise the largest group of fishing-related artefacts from Borgund. With few exceptions, they are all made of soapstone. Many were fashioned from broken pieces of soapstone pots.

When trolling, a hook and line are dragged through the water behind a boat. Consequently, it is important to prevent the gear from twisting. This is achieved by using an asymmetrical sinker. Weights for trolling (Norwegian *dorgesøkker*, Old Norse *dorg*), are frequently boat-shaped, with a straight back and a curved, hanging 'abdomen'. The line is attached at each end. Twelve of the trolling sinkers from Borgund have an almost straight or slightly convex back (Figure 6.8). Seven have a concave back, shaped like a banana. The shape causes the line to move slightly up and down in the water, but whether this was intentional or not is uncertain. Two of the weights were intentionally carved to be asymmetrical, while five others were made of fragments of soapstone vessels that predetermined this shape. According to Nordgaard (1908, 95) these sinkers are characteristic of western Norway. On the whole, the troll sinkers from Sunnmøre are heavier than those known from northern Norway.

The hand line is commonly known as the deep bait line in Norway (Norwegian *djupagnsnøre*, Old Norse *djúpshöfn*). In northern Norway it has the alternate name *juksa* (Aasen 1918, 108). The shape and weight of each sinker is adapted

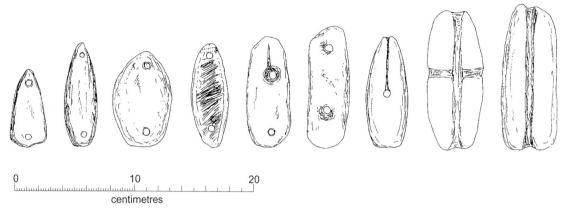

Figure 6.9. Sinkers for hand lines from Borgund (Drawing: Helge Sørheim).

to the desired depth of fishing and the anticipated currents. There is not the same need to prevent the line from twisting, so the sinkers are symmetrical (Figure 6.9). Different types of hand line weights have been found. The simplest is a stone with a latitudinal or longitudinal groove for the line. One sinker of this kind from Borgund weighed 740 g. An oval, flat stone with a latitudinal groove around the edge was heavier, weighing 909 g. Another type is represented by three long sinkers or fragments of sinkers with grooves stretching from the edge to a hole in the middle, similar to Nordgaard's (1908, 92) deep bait sinker from northern Norway. This type was known historically as a *jarstein* or *jarnsten* (ironstone) (Nordgaard 1908, 90). It was probably used exclusively with hand lines for cod and other deep-sea fish. The known sinkers from northern Norway are heavy, weighing 300–2103 g (Helberg 1995, 178). Only one example, resembling a large northern *jarstein* of Helberg's (1995, 117) type III, has been found in Borgund. Weighing 1299 g, it is one of the heaviest sinkers from the site. Nordgaard compares this type of weight to examples from Shetland, where the main line is attached over the head and the snood is tied around the stone (preventing the snood and hook from intertwining with the line). At a later stage, an iron wire was used in order to avoid this problem. A comparable heavy 'wired' sinker, using baleen as wire, comes from Uksnøy, Haram in Sunnmøre (Sunnmøre Museum 12285). It is an oblong stone of the northern *jarstein* type (Helberg's type III) and was used until 1931 for fishing halibut (*Hippoglossus hippoglossus*) (Figure 6.10).

The most abundant type of sinker involves long and straight stones with a symmetric axis and holes in both ends. The 15 complete examples from Borgund vary in weight from 105 g to 1380 g. Disregarding three exceptionally heavy sinkers (828–1380 g) their average weight is 222 g, much less than has been recorded for north Norwegian examples. Excluding the three heavy outliers, the weight of all the hand line sinkers is such that they were probably

Figure 6.10. Hand line of the type used for fishing halibut until the 1930s. Uksnøy, Haram, Møre and Romsdal (Photo: Helge Sørheim).

used for fishing in the shallow Borgundfjord rather than in deeper waters farther out to sea.

Hooks from Borgund

Although the shape of fish-hooks has varied greatly through time according to the limitations set by raw material – whether wood, shell, bone, stone or metal – their design has remained the same. Even so, within a single hook tradition there are also variations adapted to different species, fish sizes and fishing methods. A total of 19 complete or fragmentary iron hooks have been found at Borgund (Figure 6.11). All have crosswise eyes for attaching the line or snood. All but two have curved shanks. All have an open end, and most are barbed. The hooks from Borgund could have served for both hand lines and troll lines. They are a rarely known type in the material from northern Norway, where hooks with straight shanks and sheets for attaching the snood were common. The hooks from Borgund

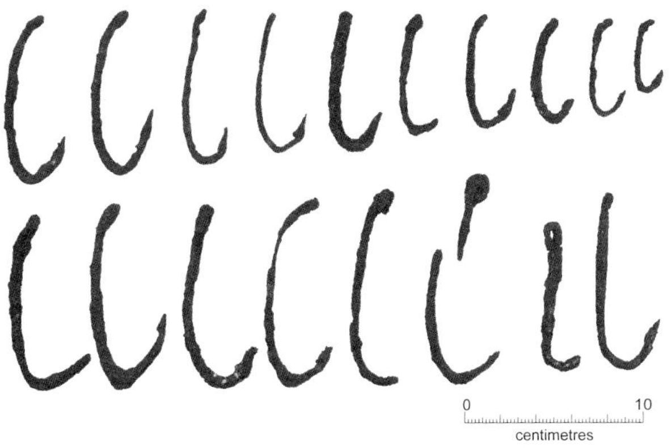

Figure 6.11. Fish-hooks from Borgund (Photo: Helge Sørheim).

Figure 6.12. Line runner made of bone from Borgund (Photo: Historisk museum, University of Bergen).

also appear to be slightly larger on average (7.9 cm) than the common hooks from northern Norway (6.8 cm). The hooks from Borgund are also higher and wider than those from both northern Norway and Bergen. It is unclear why the fishermen of western and northern Norway preferred different types of hooks – perhaps they reflect regional tradition, but the smaller size of the hooks from Bergen can also be explained by the fact that they were used mainly for fishing stationary populations of fish in the hinterland of the town. In contrast, unlike Sunnmøre and northern Norway, Bergen was a transhipment centre for dried fish rather than a focus of commercial fishing.

Line runner from Borgund

The only line runner found in Borgund is fork-shaped and designed for attachment to the gunwale by a wooden peg (Figure 6.12). It was made of bone. There is no trace of the rolling line runner, which has a rotating cylinder for pulling the line. Helberg (1995) found that this latter type was adopted at the same time as the heavy *jarstein*, from the eleventh–thirteenth centuries. It is one of the medieval technological changes in northern Norway that cannot be seen in the Borgund material.

Net equipment from Borgund

As noted above, nets and closing nets were mainly used for fishing salmon and herring in Norway (Vollan 1960, 327). A few net floats of different types found at Borgund indicate activity of this kind (Figure 6.13). According to Nordgaard's (1908) classification, only one barrel-shaped object with a longitudinal hole, weighing 233 g, can be identified as a probable net weight. The remaining 129 weights, also made of soapstone, could have served as either loom or net weights (Figure 6.14). Most are flat and pear-shaped, but their forms and finishes vary – ranging from re-used vessel fragments to neatly polished objects. None were found in a clear functional context at Borgund, but I interpret most of them as loom weights based on analogy with Norwegian grave and settlement finds.

Two wooden needles may have been used for creating or repairing nets, while a flat wooden peg with one convex side has been interpreted as equipment for knotting the mesh (Norwegian *re* or *kjølve*) (Herteig 1957, 447) (Figure 6.15). If this interpretation is correct, it would give a mesh of 2.5 cm measured from knot to knot, the same as a modern herring net. The former director of the excavations, Asbjørn E. Herteig (1957, 431), has claimed that fishing with nets must have been a widely used technique in Borgund. At 12 in total, however, the number of finds that indicate the use of nets is modest. As noted above, there is no indication that nets were used for fishing cod as early as the Middle Ages.

Discussion and conclusions

The location of Borgund and the fishing tackle found there indicate that activity in and around the settlement was, to a large extent, based on marine activities and maritime communications. The extensive remains of warehouses from the thirteenth and fourteenth centuries indicate that it was an important centre for the storage of staples. The large churchyard, extant by the eleventh century, and the minimum of three twelfth-century stone churches also reflect a dense population and the economic strength of the Borgund area in Norway's early Middle Ages. The powerful Giske chieftains and the church may both have played important roles in the activities of the town.

The main reasons for Borgund's development as more than a local centre and a fishing station have been the production and trade of stockfish. For much of the life of the town, this stockfish was taken to Bergen for further export. Based on the chronology of the Borgund cemetery and

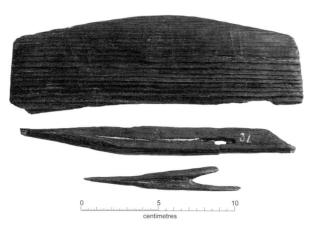

Figure 6.13. Net floats made of wood and pine bark from Borgund (Drawing: Helge Sørheim).

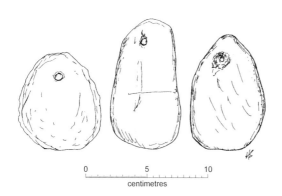

Figure 6.14. Weights made of soapstone from Borgund – either loom weights or net sinkers (Drawing: Helge Sørheim).

Figure 6.15. Top: wooden peg used for knotting the mesh in nets. Bottom: netting needles (Photo: S. Skare, Bergen Museum).

churches, however, it is not impossible that direct contact abroad took place as early as the eleventh century, prior to the consolidation of Bergen as Norway's primary port for long-range trade. Be that as it may, the development of Borgund must subsequently have been related to the growth of Bergen as the most important Norwegian commercial centre and, more widely, to the rapid demographic and economic expansion of northern Europe in the twelfth

century. It should also be taken into account that Borgund and Sunnmøre were much closer to European markets than was Lofoten. Commercial fisheries and large-scale stockfish export may actually have started here.

Based on analogy with more recent practice, I suggest that the troll line equipment found in Borgund was probably used mainly for fishing saithe and coastal cod intended for domestic consumption. Conversely, the hand line equipment was probably used for commercial fishing of migratory cod in the winter and spring. The light hand line sinkers bear witness to fishing in shallow waters, most likely in the Borgundfjord, rather than, or in addition to, deeper waters, such as Storegga (contra Nedkvitne 1983, 365; 2014, 532; Sulebust 1981, 273). The ling represented by fish bones at the site could have been caught all year round in the deep fjords of the area – and used for both local consumption and the production of stockfish for trade.

In northern Norway, the introduction of heavy deep-sea sinkers (juksa) and line runners with moving rollers occurred in the eleventh to twelfth centuries. Helberg (1995, 192, 218) links these technological changes to the commercialisation and professionalisation of stockfish production. The absence of corresponding equipment at Borgund may relate to the different circumstances there. The distance between the farms and the fishing grounds was shorter in Sunnmøre than farther north. The fishing season was also shorter. Given these characteristics, the social organisation of the fisheries also differed from that in Nordmøre and northern Norway. There was not the same need for fishing stations – fiskevær – because the town of Borgund and the many farms on the nearby islands could provide lodgings for seasonal fishermen.

Towards the end of the Middle Ages, Borgund declined and was deserted, like other similar small towns, such as Vågan in Lofoten and Veøy in Romsdal. The Black Death epidemic of AD 1349–50 and several later plague epidemics must have been one major factor; overall, 67% of the farms in this district were deserted. However, most of these were inland, suggesting that survivors moved to vacant farms near the fishing grounds (Sulebust 1994, 158–61). It may have been equally significant that Norwegian trade was reorganised in the late Middle Ages. A decree from 6 July

Figure 6.16. Cod fishing in the Borgundfjord c. 1900. Except for the klipfisk stack on the rocks in front and the type of sails, this could be an image from the Middle Ages (Photo: Aalesunds Museum).

1294 prohibiting foreign merchants from sailing farther north than Bergen (RN, vol. 2, no. 752) was lifted in the mid-fourteenth century. The fishermen began to take stockfish directly to Bergen, particularly to merchants of the German Hansa based at Bryggen in Bergen (Nedkvitne 2014, 277–332), instead of delivering it to Borgund. A document issued by Olav Håkonsson in AD 1384 reacted to the new situation and instructed people to maintain old traditions. People from Finnmark and Hålogland should sail with their goods to Vågan; people from Namdal, Nordmøre and Trøndelag to Trondheim; people from Romsdal to Veøy; and people from Sunnmøre to Borgund, 'as they had always done' (RN, vol. 7, no. 1191). This instruction seems to have failed and Borgund eventually became redundant as a staple port. Fluctuations in the fisheries are also a factor that should be considered. We know that in later periods migratory cod were absent from the Borgund area for shorter or longer periods. Such fluctuations may have occurred in the Middle Ages as well.

The commercial fisheries provided new possibilities for living on the outermost islands close to the fishing grounds, but without any possibility for traditional farming. An extreme example, slightly afield from Borgund, is the fishing community of Grip, located on some small rocks far out in the ocean, outside Kristiansund at Nordmøre.

Its fifteenth-century stave church lacks soil in which to bury the dead. Borgund's role as a staple for stockfish *en route* to Bergen had ended by AD 1500. Although fisheries in the Borgundfjord continued, the little town was deserted (Figure 6.16).

Acknowledgements

This paper draws on my 2004 article on Borgund and the Borgundfjord fisheries (Sørheim 2004).

References

Aasen, I. 1918. *Norsk ordbog*. Kristiania: Cammermeyer.

Bertelsen, R. 1994. Helgelendingene og Vågan i Lofoten, in B. Berglund (ed.) *Helgelands Historie*. Vol. 2: 113–32. Mosjøen: Helgelands Historielag.

Dyrvik, S. 1979. *Norsk økonomisk historie 1500–1970*. Vol. 1. Bergen: Universitetsforlaget.

Eliassen, J.-E. 1983. Fisk og fiskerier i Nord-Norge. *Ottar* 6: 3–7.

Helberg, B. H. 1995. *Fiskeriteknologi som uttrykk for sosial tilhørighet: en studie av nordnorsk fiske i perioden 400–1700 e.Kr.* (Stensilserie B 38). Tromsø: Institutt for Samfunnsvitenskap, Arkeologiseksjonen, Universitetet i Tromsø.

Herteig, A. E. 1957. Kaupangen på Borgund, in *Borgund og Giske*. Vol. 1: 421–72. Bergen: Borgund og Giske Bygdeboknemd.

Kvamme, S. 1994. Eigedomstilhøva på Sunnmøre: Giskegodset og anna gods, in S. U. Larsen & J. Sulebust (eds) *I balansepunktet, Sunnmøres eldste historie, ca. 800–1660*: 266–91. Ålesund: Sunnmørsposten.

Myklebust, B. 1971. *Borgundfjordfisket*. (Fiskeridirektoratets Skrifter, Serie Fiskeri 5(3)). Bergen: Fiskeridirektoratet.

Nedkvitne, A. 1983. Utenrikshandelen fra det vestafjeldske Norge 1100–1600. Unpublished PhD dissertation. University of Bergen.

Nedkvitne, A. 1988. *Mens Bønderne seilte og Jægterne for»: nordnorsk og vestnorsk kystøkonomi 1500–1730*. Bergen: Universitetsforlaget.

Nedkvitne, A. 2014. *The German Hansa and Bergen*. (Quellen und Darstellungen zur Hansischen Geschichte, Neue Folge 70) [revised version of PhD from 1983]. Vienna and Cologne: Böhlau Verlag.

Nordgaard, O. 1908. *Træk fra fiskeriets utvikling i Norge*. (Det kgl. Norske Videnskabers Selskabs Skrifter 1). Trondheim: DKNVS.

Olsen, O. M. 2004. Medieval fishing tackle from Bergen, in I. Øye (ed.) *Medieval Fishing Tackle from Bergen and Borgund*: 11–106. (Bryggen Papers Main Series 5). Bergen: Fagboklaget.

RN Regesta Norvegica 1978–2010. *Regesta Norvegica*. Vols 2 and 7. Oslo: Dokumentasjonsprosjektet. Available at: http://www.dokpro.uio.no/dipl_norv/regesta_felt.html.

Solhaug, T. 1976. *De norske fiskeriers historie*. Vol. 1. Bergen: Universitetsforlaget.

Sørheim, H. 1981. Synkrone og diakrone analogier i arkeologisk teoridannelse. *Universitetets oldsaksamling Årbok* 80–1: 169–77.

Sørheim, H. 2004. Borgund and the Borgundfjord fisheries, in I. Øye (ed.) *Medieval Fishing Tackle from Bergen and Borgund*: 107–31. (Bryggen Papers Main Series 5). Bergen: Fagboklaget.

Strøm, H. 1762. *Physisk og oekonomisk beskrivelse over fogderiet Søndmør, beliggende i Bergen stift i Norge. Første part: Sorøe*. (Facsimile edition 1957). Oslo: Børsums Forlag.

Strøm, H. 1766. *Physisk og oekonomisk beskrivelse over fogderiet Søndmør, beliggende i Bergen stift i Norge. Anden part: Sorøe*. (Facsimile edition 1957). Oslo: Børsums Forlag.

Sulebust, J. 1981. Borgund og næringsforhold i Borgund sogn i seinmellomalderen og nyere tid, in L. I. Hansen (ed.) *Seinmiddelalder i norske bygder. Utvalgte emner fra Det nordiske ødegårdsprosjekts norske punktundersøkelser*: 262–322. Oslo/Bergen/Tromsø: Universitetsforlaget.

Sulebust, J. 1994. Bosetning og befolkning fra yngre jernalder til 1665, in S. U. Larsen & J. Sulebust (eds) *I balansepunktet, Sunnmøres eldste historie, ca. 800–1660*: 150–66. Oslo: Sunnmørsposten.

Sunde, K. 1972. Unpublished letter to Senior Curator Asbjørn E. Herteig. Archive, Department of Archaeology, Medieval section, University of Bergen.

Vollan, O. 1960. *Garn og garnfiskeri*. (Kulturhistorisk Leksikon for Nordisk Middelalder 5). Malmö: Allhems.

Commercial Fishing and the Political Economy of Medieval Iceland

Orri Vésteinsson

Introduction

Fishing is the principal industry of the modern state of Iceland, generating nearly 50% of the state's export revenue (Sigurdardóttir *et al.* 2012, 15). Advances in fishing technology, beginning with decked vessels in the nineteenth century but primarily taking off with motorised vessels and trawlers in the first years of the twentieth century, laid the foundations for the building of this modern state, transforming one of the poorest economies in Europe into one of the most affluent in a matter of decades (Ólafsson 1993, 208).

In this light it is interesting that Icelandic historians have had a very ambivalent attitude towards the history of the fishing industry. For most of the twentieth century, leading scholars considered the roots of commercial fishing in the Middle Ages as a negative development. Even the more recent modernisation of the fishing industry in the nineteenth and twentieth centuries, with its obvious economic benefits for the Icelandic economy, cannot be said to have become a popular topic for research, despite the fact that this development is normally seen in a positive light (Þór 1996; see also Guðmundsson 1944–6; Óskarsson 1991; Starkey *et al.* 2009; Þór 2002–3; Þórðarson 1930; Valdimarsson 1997). Despite valiant efforts by such historians as Björn Þorsteinsson (1950; 1970; 1976; Þorsteinsson and Grímsdóttir 1989; 1990) and Helgi Þorláksson (2001; 2003a; 2003b; 2004), the commercial fishing of the Middle Ages has remained an issue of marginal concern in the research community. The aim of this paper is to review the historical discourse on the origins of commercial fishing in Iceland in the Middle Ages and to propose an alternative hypothesis to those in current circulation.

Historiography of commercial fishing in medieval Iceland

By the middle of the twentieth century there existed a well-formulated view which saw the political union of Norway and Iceland in AD 1262 in terms of loss of independence for the Icelanders. According to this view, Iceland had been independent up to that time, with its own indigenous political structure established in the tenth century. The Icelanders had originally been in control of their external trade, but this had been taken over by Norwegians from the twelfth century onwards. This takeover was seen to have contributed significantly to the perceived loss of independence in the late thirteenth century and to have resulted in the Icelandic economy becoming exposed to the undesirable influences of foreign markets. Evidence suggesting a marked increase in the export of dried fish from Iceland to Norway from the AD 1330s onwards was seen as a direct result of the loss of independence and the loss of control over trade with foreign countries, and was considered a negative development. In the words of Þorkell Jóhannesson, Iceland's most prominent economic historian in the first half of the twentieth century:

This economic change is a famous example of a whole society with deep-rooted economic traditions becoming suddenly upset by the influences of foreign entrepreneurs, for whom it was profitable to direct the work force into new channels. Hitherto the Icelanders had been an industrious agricultural people, as can still be seen in the ruins of magnificent buildings. From that time on the fishing industry grew steadily, and the economic fortunes of the country became more and more dependent upon the fishing and trade with foreign merchants. (Jóhannesson 1965, 25; see also pp. 48–9; author's translation)

Þorkell's view was that as long as the trade in the North Atlantic was in the hands of the Norwegians, the undesirable effects of market forces were not felt to any great extent in the remote farming idyll of Iceland. However, once the much more aggressive Hansa had gained control over the trade in Norway, and in Bergen in particular, in the closing years of the thirteenth century, Iceland became directly exposed to European market forces. His idea was that as the Hansa expanded their markets for dried fish in mainland Europe, so the pressure increased on merchants sailing to Iceland to secure this fish. They, in turn, forced the Icelanders to emphasise fishing instead of the production of homespun wool, which previously had been the principal export commodity of the Icelanders. As the fourteenth century wore on, the dependency of the Icelanders upon fishing exports grew. After AD 1400, when English vessels began to sail directly to Icelandic waters – with the Germans later following suit – fishing became the driving force of the Icelandic economy. The possession of coastal lands became a major asset. Fishing stations mushroomed, and the great magnates of the country moved their seats of power from centres of farming regions to hitherto marginal areas in the west, where the best fishing grounds were to be found.

To Þorkell Jóhannesson this was not necessarily a bad thing in the short run. He pointed out that at no other time during its union with the Norwegian, and later also Danish, kingdom had Iceland been so independent of Scandinavian royal control as in the fifteenth century. The Icelandic magnates were in direct contact with England and English power, and for a time in the fifteenth century these contacts were much more important for Icelandic politics than any connections with Scandinavia. This appealed to Þorkell and his generation of historians (Þorsteinsson 1970 is the classic study on the fifteenth century). To them this period represented a much needed respite in the otherwise steady eroding of Icelandic indigenous political power. Although the fifteenth century was one of the most turbulent in Icelandic history, these nationalistic historians were happy that Icelandic magnates were seen to be asserting their control over the country again, even if this was in fact primarily evidenced by a marked increase in political violence and economic disparities. However, they were not happy with the end result as they saw it. That was the steady deterioration of native autonomy from the sixteenth century onwards, with major setbacks associated with the Reformation and the introduction of a trade monopoly in AD 1602. These changes culminated in the most depressed period of Icelandic history, during the seventeenth and eighteenth centuries, when Iceland was reduced to colonial status. It had become a fish, wool and mutton factory for the merchants of Copenhagen and the Danish army (Gunnarsson 1983; Þorláksson 2003b, 306–13; Þorsteinsson 1964, 3–52; 1985, 131–3). This erosion of indigenous power was seen as a result of the native elite becoming ensnared in the intrigues of foreign courts

and dependent on foreign merchants. Because the foreign merchants were primarily interested in obtaining fish, they contributed to the weakening of the traditional, land-based economy. This, in turn, undermined the political clout of the landowning class, the feebleness of which allowed Denmark to exploit Iceland and keep it in a state of underdevelopment throughout the early modern period.

This view of Icelandic history – necessarily abridged and simplified here – does not stand up to scrutiny. In the past 30 years or so a re-evaluation has taken place of some parts of this view, although a complete rethinking has not been accomplished and the nationalistic formulation continues to affect the debate (for modified views, see Karlsson 2000). A key development has been a shift in the identification of culprits – from the Danes to the local landowning class. A central question is why, despite an ocean full of fish, the Icelandic fishing industry did not develop beyond the part-time, low-tech and low-yield subsidiary to farming activities it was until the nineteenth century. The contrast with the Newfoundland fisheries in the seventeenth century – a highly organised and profitable industry firmly integrated in the emerging capitalistic world system (Pope 2004) – could not be more striking. Gunnarsson (1983) has argued that it was the Icelandic landowners who kept the lid on, actively working against any modernising efforts by merchants or the Danish administration. As early as the fifteenth century negative attitudes towards fishing are evident in legislation prohibiting or setting up barriers to any further development of the industry. The growth of fishing stations was being curbed already at this date. People were not allowed to settle there permanently, and foreign merchants were not allowed to over-winter in Iceland, thus effectively preventing the formation of fishing villages and permanent trading points (Þorsteinsson 1970, 248–58). In the seventeenth century, attempts were even made to ban fishing nets, in a bid to keep the indigenous fisheries on such a low technological level that they had no chance of developing into an industry capable of sustaining a part of the populace independent of the farming economy (Gunnarsson 1987, 252–4). Gunnarsson's arguments suggest that the Icelandic elite was in fact remarkably resilient in resisting change and foreign interference, but that, at the same time, it was conservative and undynamic in the extreme, averse to risk-taking of any sort. While it controlled and relied on the fishing industry, it also feared and resisted any further development of it.

Also reflecting a lack of dynamism and a desire to insulate itself against the vagaries of foreign market forces is the success of the Icelandic elite in keeping prices stable. Prices remained unchanged in internal trade for centuries, and in the seventeenth and eighteenth centuries Icelandic leaders tried everything they could to make the Danish king ensure that Danish merchants – who by then had a monopoly over the Icelandic trade – only traded according to a fixed price list (Aðils 1919, 358, 403–6).

These are the symptoms of a society that valued stability more than opportunity. They paint a very different picture from the nationalistic view, which took it for granted that Icelandic society was dynamic, that it was responsive to foreign market forces and that it responded readily to new economic opportunities. The difference may to some extent lie in the fact that the nationalistic view is based on a study of medieval society, while more recent views are based largely on the study of the early modern evidence. It is entirely possible that a change occurred, say, in the course of the fifteenth and sixteenth centuries, whereby the Icelandic elite lost some of its confidence and became more conservative and inward-looking than it had been in the Middle Ages. If this were true, we would need to come up with some explanation about how and why such a change could have occurred. Even if we were able to explain this change, a number of important objections to the nationalistic view would remain, in particular those that reject notions of market forces as principal agents of change in medieval society.

The nature of trade in medieval Iceland and the rise of commercial fishing

Medieval historians have long been uneasy about the assumption, inherent in the nationalistic view, that the Icelanders were heavily dependent on trade and that trade conditions – who controlled the trade, how prices fluctuated and what goods were in demand abroad – were of vital importance to the Icelandic economy. In Iceland, Helgi Þorláksson has led the revision. In his main work on this subject, a study of the trade in homespun wool and prices in medieval Iceland (Þorláksson 1991), he demonstrated that the demand of foreign markets had no influence on the production or prices of homespun wool in thirteenth-century Iceland. He showed that reciprocity, and to some extent redistribution, were the principal modes of exchange in medieval Iceland and that external trade was such a marginal part of the economy of even the richest households – those that did buy prestige goods from abroad on a regular basis – that they did not see a reason to produce more than was needed to exchange for the items that were required. Production was not aimed at making a profit as it would in a market economy, but only at acquiring those goods that were deemed necessary in a reciprocal and redistributive economy. This also explains why the Icelanders were in favour of fixed prices. They were not interested in taking advantage of fluctuations in foreign markets. They were interested in maintaining stability in their trade with foreign merchants. That exchange was to a considerable extent based on personal contacts and prearranged agreements, so that the risk on both sides was minimal (Þorláksson 1991, 507–15).

This is not the sort of market economy that would be open to sudden pressures from foreign markets – an economy that would start producing large quantities of dried fish just because there was demand for it on a Bergen quayside. It is much more like what is known of early-modern Icelandic society, with its fixed prices and fear of unpredictable elements – such as too much fishing and trading – which could upset the social order. These aspects of Icelandic pre-industrial society were fairly stable. Society had been just as unresponsive to, and suspicious of, market forces in the High Middle Ages as it was in early modern times. How then did commercial fishing come about and what sort of function could it have in this kind of society?

Helgi Þorláksson (1991, 440–77; see also Karlsson 2000, 47) has tried to answer this question by suggesting that Christian fasting rules led to increased fish consumption within Iceland, and that as dried fish became a more regular item of exchange inside Iceland it became just as natural an export item as homespun wool had been earlier. The increase in dried-fish export that is evident from the AD 1330s onwards (see below) does then not necessarily signal the onset of commercial fishing. It only means that dried fish was available for export from Iceland. The context of these fisheries is likely to have been just as un-commercial as the production of home-spun cloth had been previously. The fish may have been caught primarily for home consumption; only a very limited portion of the catch may have been intended for exchange with foreign merchants. To continue this line of thought, one might suggest that this was precisely the reason why English and later German fishing vessels began to sail directly to Icelandic waters to catch the fish themselves in the fifteenth century. If Icelanders did not catch enough fish to meet demand on the European markets, it became necessary for the English and Germans to make the perilous journey themselves in order to acquire this precious commodity.

As we shall see, the evidence does not support the idea of an increase in fish consumption in Iceland as a result of the introduction of Christianity. Without that sort of evidence, we are left predominately with some sort of market influence as the most likely reason behind the increase in fish exports in the fourteenth century. It remains to consider what this evidence consists of.

Evidence for fishing and fish consumption in medieval Iceland

A Norwegian source from around AD 1340 explains that great amounts of good-quality merchandise in the form of fish and fish oil was then being imported from Iceland but that previously only home-spun cloth had come from there (DI, vol. 2, no. 729). An annalistic reference in an Icelandic source explains that in the year AD 1323 no dried fish was exported from Gásir, the main trade centre in northern Iceland (Grímsdóttir 1998, 362), but from the wording it is clear that the author considered fish a regular export item from that port at the time of writing, towards the end of the fourteenth century. From these two sources, scholars have dated the beginning of fish exports from Iceland to Norway to the AD 1330s or thereabouts (Karlsson 2000, 107; 2009, 180–2; Magerøy 1993, 148–9). There are plenty of references to dried-fish exports from Iceland to Norway after AD 1340 (Þorsteinsson and Grímsdóttir 1989, 177), but it is the development up to that date that requires a closer look.

The earliest unequivocal reference to fish exports from Iceland is a decree from AD 1294 that bans the export of dried fish from Iceland in times of famine (Jónsson 2004, 313). Icelandic fish is also mentioned in the cargo of a Norwegian merchant in Lynn, England, in AD 1307 (DN, vol. 19, no. 515). This suggests that by the closing years of the thirteenth century exports of fish had already begun. The decree from AD 1294, repeated in a resolution from AD 1319 (DI, vol. 2, no. 498), has resonance in a description of the Icelandic economy from around AD 1350 that says the populace is nourished by livestock and marine fish:

> The fish gives such powerful nutrition that destitute people become well off. The agricultural areas are dependent on this gift because dried sea-fish is bought and distributed all over the countryside. (Sigurðsson and Vigfússon 1856–78, vol. 2, 179)

These sources have usually been taken as evidence for a change in living conditions around AD 1300. The interpretation that this was really a change seems to be based on a reading of the saga literature, the large corpus of stories created in the thirteenth and fourteenth centuries dealing with events that were supposed to have taken place in the ninth to eleventh centuries. Scholars have taken it for granted that the society described in the sagas was almost entirely land-based and that marine resources were only utilised in the first decades after the settlement, while sufficient livestock was being raised. After that, fish played only a very limited role in the economy of ordinary households up until the onset of commercial fishing in the fourteenth century (e.g. J. Jóhannesson 1958, 139–41; Þ. Jóhannesson 1965, 48–56; Þorsteinsson 1978, 276–7).

This assumption is simply wrong. There are plenty of references in both the narrative literature and early charters to fishing and marine hunting, and it is quite clear from such descriptions that people went to considerable trouble to obtain fish, either by travelling long distances in order to catch them or by engaging in exchange (e.g. DI, vol. 1, nos. 422–3; vol. 2, nos 66, 426, 441–5, 468–9, 469, 617–18; vol. 3, no. 235; Halldórsson 1959, 258–9; Jóhannesson *et al.* 1946, vol. 1, 14–19; vol. 2, 131–2; Jónsson 1936, 26, 139; Kristjánsson 1956, 258–9; Nordal and Jónsson 1938, 156; Sveinsson 1954, 30–5, 46; Sveinsson and Þórðarson 1935, 18–19, 145–8; Þórólfsson and Jónsson 1943, 134–42, 147–9, 169; Vilmundarson and Vilhjálmsson 1991, 459). Many of the relevant charters are preserved in fourteenth-century or later collections, but in all these cases there are grounds to

believe that the originals dated to the thirteenth century. These references do not give any indication of the reliance of the economy as a whole on marine resources, nor can they be seen to describe the period before the twelfth century, but they clearly do indicate that fishing, particularly along the northern coast, was practised on a significant scale, and that it was integrated into the local economies.

The historical evidence is supported by archaeological data, which suggest that although there were differences in the relative importance of fish in people's diet from one region to the next, it was everywhere a significant element from the beginning of settlement in Iceland in the late ninth century (Perdikaris and McGovern 2009). It is particularly interesting that significant marine-fish remains have been uncovered at the very early inland site of Granastaðir, in Eyjafjörður, and at a number of sites in the Mývatn region, including Hofstaðir and Sveigakot. All are more than 40 km from the coast but despite the distance, the inhabitants of these ninth–tenth-century farms obtained marine fish and other marine species, including birds and molluscs, to complement their diet. This does not seem to have been because of a need to subsidise an incomplete source of nutrition. All the Mývatn sites also had access to trout (*Salmo trutta*) and char (*Salvelinus alpinus*) (both freshwater species), and to a lesser extent salmon (*Salmo salar*) (which lives part of its lifecycle in fresh water), and fish found in fresh water make up large portions of the bone assemblages there. It seems, rather, that the inhabitants had a preference for marine fish (McGovern *et al.* 2006; 2007; Perdikaris and McGovern 2009). It is also worth pointing out that at all of these inland sites the marine-fish bones come from contexts that are pre-Christian, showing that significant marine fishing predates the introduction of fasting rules. Only one early fishing station has been excavated and dated in Iceland. The site of Akurvík, in northwest Iceland, is a small station originally established in the early thirteenth century; it was subsequently abandoned and then reoccupied in the fifteenth century (Amundsen *et al.* 2005). Although this site could be used to suggest that commercial fishing began somewhat earlier than previously believed, that is, in the early thirteenth century, it presently stands alone – although fieldwork is underway at another fishing station, the site of Siglunes, in northern Iceland (Lárusdóttir *et al.* 2011). Although it is clearly a specialised fishing station, Akurvík does not necessarily indicate foreign exports of fish. It is, instead, consistent with descriptions in *Þorgils saga og Hafliða* – written in the thirteenth century about events taking place in the first quarter of the twelfth century – about hunting and fishing in precisely this area, owned by chieftains from the farming region of Húnaþing (Jóhannesson *et al.* 1946, vol. 1, 14–16). Moreover, the archaeological record does not support any sort of increase in the consumption of fish in Iceland after the conversion to Christianity. In fact fish consumption seems to have stayed fairly stable from the beginning of settlement through the

High Middle Ages. A significant increase in fish consumption only appears in the fifteenth century or later (Amorosi 1996, 417–18; Amorosi *et al.* 1994). If there was no increase in the consumption of fish in the thirteenth century, and if fishing had, in fact, been an important part of the economy from the beginning of settlement in Iceland, what – if anything – did happen around AD 1300?

The case for fish not having been a significant export item before the late thirteenth century really rests on only two sources (Finsen 1870, 192–5, 246–8). These are a price list preserved in the law code *Grágás* and a price list from Árnesþing, both predating the writing of the extant manuscripts in the mid-thirteenth century (see Þorláksson 1991, 98–101). The list from Árnesþing mentions fish, but it does not mention it clearly as an export item. The *Grágás* list, which is more concerned with foreign exchange, makes no mention of fish. More circumstantial, but also significant, is the fact that fish is nowhere mentioned as an export item before AD 1294, whereas there is good evidence for earlier exports of home-spun cloth, other wool products and a variety of other items, such as falcons and fox skins (Karlsson 2009, 244–6). Moreover, it has been pointed out that in the first half of the thirteenth century all the Icelandic chieftains had their estates in farming regions and that the greatest of these were invariably situated in the centres of such regions, usually at a considerable distance from the coast. By the AD 1270s and 1280s, however, important chieftains began to take up residence on coastal estates, which in many cases later came to dominate important fishing stations (Þorsteinsson 1980, 231–2). It can also be pointed out that the majority of chieftain families that survived the civil wars of the mid-thirteenth century and made it into the service of the Norwegian administration had estates in coastal areas in the western part of the country, where fishing stations later appeared (see Þorsteinsson and Líndal 1978, 79). Around AD 1300, several new parish churches were established in the west, and this has been interpreted as evidence for population growth in fishing regions at this time (Þorsteinsson 1980, 211–12). The western quarter was also the only part of the country that saw a significant increase in the number of tax-paying farmers between AD 1100 and 1300, and this again indicates population growth in fishing areas (Víkingur 1970, 188).

This evidence – while circumstantial – certainly suggests that in the late thirteenth century the fisheries became more politically important than they had been previously. The evidence is such that it does not preclude that the political significance of the trade only developed as a result of it having already been in place. Such a proposition would, however, require a modification of Helgi Þorláksson's convincing view of trade in medieval Iceland as firmly controlled by the ruling class. Indeed, this interpretation does need careful consideration in light of recent research in both northern Norway and Orkney suggesting that commercial fishing began considerably earlier than previously believed.

For Lofoten, there is good evidence that marketable size ranges of cod (*Gadus morhua*) were beginning to be prioritised in catches in the twelfth century (Perdikaris 1999; see Chapters 4–6), and in Orkney a case has been made for specialised fish exports in the eleventh to twelfth centuries (Barrett 1997; 2012). It is beginning to look as if the Icelanders were decidedly tardy in their involvement with commercial fishing.

An alternative hypothesis

While it is clear that Icelanders were significant consumers of marine fish from the beginning of settlement, there is no indication of any increase in fishing that might be related to a commercialisation of the industry until the late thirteenth century. The change was clearly being felt in Norwegian markets by at least the AD 1340s and possibly earlier. To that extent the old view stands up to scrutiny. There seems to have been a marked increase in fish exports from Iceland around AD 1300. The explanations for this change that have been put forward by earlier scholars are, however, either incorrect or at least not sufficient for a full appreciation of this development.

Based largely on Helgi Þorláksson's characterisation, it seems that Icelandic medieval society was not one that would have been particularly susceptible to outside market forces. Its reliance on foreign imports was minimal, and it was in general averse to any sort of exchange that involved risk-taking – or, rather, it tended to view risk in social rather than economic terms. It is therefore unlikely that the increase in fish exports in the late thirteenth century was a direct result of increased demand for dried fish in Norwegian markets. To put it differently, it is unlikely that such an increase in demand would have had any effect if there were not new conditions within the Icelandic economy to allow such a change. It is these conditions that we have to examine. Two aspects need particular consideration. First, we need to appreciate the role of fish in the Icelandic economy before it began to be exported in large quantities. Second, we need a better understanding of the importance of foreign imports for Icelandic society. What was being imported in exchange for the home-spun cloth or dried fish, and what was it used for?

In order to answer these questions we need to consider who controlled surplus production – that is, the production of the goods that were exchanged for foreign imports – and to what ends. The available evidence, both historical and archaeological, suggests that, with one exception, in the ninth–thirteenth centuries there was no regular import of materials or goods that can be considered as everyday objects or essentials for the running of a household. The only exception is whetstones, which the economy could not have been without in the long run, but which were used in small quantities. While whetstones came from a variety of sources during the Viking Age, from the eleventh century onwards stones from Eidsborg in Norway dominate the collections, suggesting a change towards a more controlled import of this commodity (Juel 2011). Less essential imports included baking plates and steatite vessels. Baking plates, imported from Norway, are not found until around AD 1100; they are known primarily in trading site contexts from the late Middle Ages (Gísladóttir and Snæsdóttir 2011). Steatite, conversely, is primarily found on Viking Age sites, becoming rare later in the Middle Ages. A recent study suggests that the stock of steatite objects found in Iceland is consistent with the view that they were mostly imported by the original settlers and only rarely replenished (Forster 2004). This is in no way compelling evidence of a general market for imported necessities. Instead, the available evidence, both archaeological and historical, suggests irregular trade connections where prestige objects were the main items of exchange. Weapons, jewellery, fine materials for clothing and timber for the construction of churches or feasting halls were the sort of goods the Icelandic aristocracy needed, but in fact the import of such items was so irregular that people often had to go to Norway themselves in order to secure a shipload of timber or a proper sword that would not bend at the first blow (Þorsteinsson 1953, 134). There was, even in the twelfth century, no regular import of construction timber, the material Icelanders would have been in most need of, as is evidenced by the many accounts of Icelandic chieftains undertaking journeys to Norway themselves to obtain timber for their churches.

Foreign trade was therefore only to a limited extent a source of the prestige needed by the chieftains and influential landowners to maintain or increase their status. It was too irregular for them to be able to depend on it, and in fact it seems that their prestige was often best maintained by going abroad themselves and doing the exchange there. This lack of reliance on trade is mirrored in the economic and political system of High Medieval Iceland. It was a landscape dominated by large estates, in a sea of smaller farmsteads that usually were bound by some ties of dependency to a nearby estate. Although each farm was largely self-sufficient, and the basic economic pattern therefore fairly similar from one farm to the next, there was some variation in the subsistence base. In particular, the extent to which wild species were exploited could vary greatly according to location. It is likely that coastal farms were all to some extent specialised in fishing – although it is not clear to what extent such farms exchanged their own fish farther inland (as barter or dues) or gave, or were forced to provide, access to the sea. In either case it is clear that there was a general cultural preference for variation in the food consumed. This emphasis on variety does not seem to vary with site status – at least not in Mývatnssveit, which is the only region where comparative data exists. The high-status site Hofstaðir has significant amounts of marine fish in much the same proportions as the nearby lower-status

sites of Sveigakot and Hrísheimar. The only difference in the archaeozoological assemblages seems to be that at Hofstaðir there is a higher proportion of prime beef. The beef has been interpreted as the remains of feasts held in the great hall at Hofstaðir (Lucas 2009, 188–95, 252, 404–7; Lucas and McGovern 2008).

Hosting splendid feasts was one of the principal methods of maintaining power and authority in Viking Age and High Medieval Iceland (Monclair 2003, 192–4). The sagas of Icelanders describe a society where the chieftains had constantly to be in personal contact with their followers, and where institutional ties of dependency did not begin to develop until after the middle of the twelfth century (Vésteinsson 2000, 16–17, 58). Before, a chieftain had very few ways in which he could make his power felt without actually being physically present whenever he wanted to bend someone's will. It was therefore imperative for the chieftains to maintain good relations with their following, and they did this mainly through peace-making, litigation on behalf of their clients, exchanging gifts and feasting. Along with assemblies, feasts were institutionalised and regular events. An average chieftain would probably have been expected to hold feasts for his followers several times a year. The implication is that the more food and beer, ale or mead the chieftain could offer, the more secure he could be of his following. The chieftain therefore needed to have at his command the means to obtain all this food. The considerable trouble people went to in order to grow grain in medieval Iceland should be seen in this light (Simpson *et al.* 2002). It was primarily for making ale to be consumed at the feasts (as in Jóhannesson *et al.* 1946, vol. 1, 24–7; vol. 2, 129). If it was beef that was primarily required for food at the feasts, this was presumably raised on both the chieftain's own estates and on the farms of his dependants. Marine fish consumption was clearly ubiquitous, and it may also have been tied into the chieftainly economy in complex ways. It may well be that such chieftains as Hafliði Másson – whom we hear of in the early twelfth century sending one of his men to the farm next to Akurvík (mentioned above) to do fishing and hunting – needed to procure fish to feed the cowboys who raised the beef for their feasts. If so, a complete reversal had taken place by the late Middle Ages, when the episcopal sees had an insatiable demand for butter to feed the fishermen on their boats (Grímsdóttir 2006, 126). But this reversal was in the kind of produce only. The system was essentially the same, where one type of foodstuff allowed the production of another that was more politically significant.

The political system changed in the course of the twelfth and thirteenth centuries. In this period there was a marked consolidation of power in Iceland. By the AD 1220s, all power was in the hands of five families, who spent the next 40 years in bloody civil wars over which of them should control the country (Karlsson 2000, 72–82).

These families established more permanent and absolute powers than previous generations of chieftains had ever managed. They had small armies and came to be largely independent of the following of large groups of clients. They introduced taxation and oaths of fealty and became knights at the Norwegian court. In short, they overcame the barriers to institutional power that had kept their ancestors in check. This new sort of magnate will have had less need for socialising with neighbouring farmers and other clients. Their power no longer rested on courting the acceptance of neighbours, but instead became dependent on recognition by, and support from, the Norwegian court and its king. In those changed circumstances the magnates needed to stress the visible differences between them and the less powerful landlords and local chieftains, rather than maintain an image of themselves as *primes inter pares*.

As well as becoming less dependent on hosting communal feasts, it also seems that this new type of magnate may have started, in the late thirteenth century, to detach themselves from the everyday running of their estates. It was no longer politically expedient for them to be seen to be closely involved. The landholding system began to change, whereby more and more emphasis was placed on obtaining rents in a currency that could be turned into merchandise exchangeable in foreign markets. The new magnates needed the means to finance long stays in Norway and to procure those goods which could set them apart from other, lesser landowners. We see during the late Middle Ages a reduction in some of the more costly and difficult economic activities. Iron extraction from bogs ceased. So did salt-making and grain cultivation. Pigs also disappeared. I suggest that, rather than indicating a general decline in the economic fortunes of the Icelanders, this indicates a reorganisation of the major estates. They were no longer kept at maximum size and complexity, but began to be simplified and homogenised. Each farm would produce a surplus of the goods that were most easily exchanged abroad. That this was dried fish in the late thirteenth century was probably just a coincidence. There may, in fact, have been demand for dried fish on the Norwegian markets for a very long time. It was only when the Icelandic magnates, those who actually controlled the production in Iceland, had reason to exploit this demand to their advantage that it began to be met from Iceland. It is also quite possible – and in fact likely – that fish was exported from Iceland earlier than the late thirteenth century as some sort of response to market demand, but that this export did not become significant in economic or political terms until the magnates began to exploit it to their advantage.

Paradoxically, although the Icelandic magnates affected a change in economic organisation in order to obtain more fish to sell in Norway, by the fifteenth century they had also set their minds against any further development of this industry. They had become much more powerful than their eleventh-century ancestors, but they seem to have had an ambivalent attitude towards the fishing industry – as if they had unleashed a beast they feared might be uncontrollable. Alternatively, it might be argued that the changes in the political system that began in the thirteenth century were not so fundamental. Although the Icelandic magnates did start to require more exchangeable revenues than before, and although this led to commercial fishing, they did not become powerful enough – or dynamic enough – to develop this industry to their further advantage. They were, in fact, remarkably successful at keeping a lid on the fisheries – so much so that their class did not have to accept defeat until the beginning of the twentieth century.

Conclusions

The hypothesis presented here suggests that commercial fishing started in Iceland in the late thirteenth century because of changes in the Icelandic political structure. Chieftains who had formerly based their power on the support of clients became dependent on the Norwegian court as the source of their authority. Thus a political economy that had previously been based on feasting and the obtaining of prestige objects through gift exchange now began to revolve around the procurement of currency to finance a particular lifestyle, including trips and stays abroad.

Many of the basic premises of this hypothesis are uncertain, and I have not dealt with many of its implications. In particular, it has important ramifications for our understanding of the development of the landholding system in late medieval and early modern times. Nor does this hypothesis take account of the possible influence of English and, later, German fishing fleets in Icelandic waters from shortly after AD 1400. Ongoing excavations of the fishing station at Gufuskálar in western Iceland, plausibly related to the presence of English fish merchants, promise to throw light on this issue (Pálsdóttir and Sveinbjarnarson 2011), but at present it is unclear to what extent the activities of these foreign merchants were integrated with the Icelandic economy. Did the English and German fishing fleets mostly catch and process the fish themselves, with limited involvement of the Icelanders? Or did the new foreign contacts intensify a development already underway, opening up new markets and channelling new ideas – and thus possibly transforming Iceland's political economy?

Trade in pre-industrial societies is a more complex phenomenon than we often care to remember. There has in recent years been a growing interest in the onset of commercial fishing in the North Atlantic during the Middle Ages. The Icelandic example reminds us that a full-fledged market economy need not accompany the emergence of a regular trade in dried fish or other commodities. Many different degrees of market dependency are evident. North Atlantic societies did not necessarily respond to market pressures at all, and when they did, they did so in different ways.

Acknowledgements

This paper develops ideas first aired in an unpublished keynote lecture at the NABO '97, Summit of the Sea conference, St John's, Newfoundland, 7 September 1997. I am grateful to Oddur Freyr Þorsteinsson for assistance with the references and indebted to Helgi Þorláksson both for inspiration and for useful critique.

References

Aðils, J. J. 1919. *Einokunarverzlun Dana á Íslandi 1602–1787.* Reykjavík: Verzlunarráð Íslands.

Amorosi, T. 1996. Icelandic zooarchaeology: new data applied to issues of historical ecology, paleoeconomy and global change. Unpublished PhD dissertation, City University of New York.

Amorosi, T., P. C. Buckland, K. Magnusson & T. H. McGovern. 1994. An archaeozoological examination of the midden at Nesstofa, Reykjavik, Iceland, in R. Luff & P. Rowley-Conwy (eds) *Whither Environmental Archaeology?*: 69–79. Oxford: Oxbow Books.

Amundsen, C. P., S. Perdikaris, T. H. McGovern, Y. Krivogorskaya, M. Brown, K. Smiarowski, S. Storm, S. Modugno, M. Frik & M. Koczela. 2005. Fishing booths and fishing strategies in medieval Iceland: an archaeofauna from Akurvík, north-west Iceland. *Environmental Archaeology* 10: 221–43.

Barrett, J. H. 1997. Fish trade in Norse Orkney and Caithness: a zooarchaeological approach. *Antiquity* 71: 616–38.

Barrett, J. H. 2012. Being an islander, in J. H. Barrett (ed.) *Being an Islander: production and identity at Quoygrew, Orkney, ad 900–1600*: 275–92. (McDonald Institute Monograph). Cambridge: McDonald Institute for Archaeological Research.

DI Diplomatarium Islandicum eða Íslenzkt Fornbréfasafn 1853–1976. Vols 1–3. Copenhagen and Reykjavík: Hið Íslenzka Bókmenntafélag.

DN Diplomatarium Norvegicum: Oldbreve til kundskab om Norges indre og ydre forhold, sprog, slägter, säder, lovgivning og rettergang i middelalderen 1847–1976. Vol. 19. Christiania and Oslo: Malling. Dokumentasjonsprosjektet. Available at: http://www.dokpro.uio.no/dipl_norv/diplom_felt.html

Finsen, V. (ed.) 1870. *Grágás: elzta lögbók Íslendinga. Volume 2.* Copenhagen: Nordiske Literatur-Samfund.

Forster, A. 2004. The soapstone trade in the North Atlantic: preliminary research of Viking and Norse period soapstone imports in Iceland, in G. Guðmundsson (ed.) *Current Issues in Nordic Archaeology*: 17–22. (Proceedings of the 21st conference of Nordic Archaeologists 6–9 September 2001, Akureyri, Iceland). Reykjavík: Félag Íslenskra Fornleifafræðinga.

Gísladóttir, G. A. & M. Snæsdóttir. 2011. Steinar fyrir brauð: Norsk eldhústíska á Íslandi, in O. Vésteinsson, G. Lucas, K. Þórsdóttir & R. Gylfadóttir (eds) *Upp á yfirborðið: Nýjar rannsóknir í íslenskri fornleifafræði*: 51–68. Reykjavík: Fornleifastofnun Íslands.

Grímsdóttir, G. Á. (ed.) 1998. *Biskupasögur 3* (Íslenzk fornrit 17). Reykjavík: Hið Íslenzka Fornritafélag.

Grímsdóttir, G. Á. 2006. Biskupsstóll í Skálholti, in G. Kristjánsson & Ó. Guðmundsson (eds) *Saga biskupsstólanna: Skálholt 950 ára – 2006 – Hólar 900 ára*: 21–245. Hólar: Bókaútgáfan Hólar.

Guðmundsson, G. 1944–6. *Skútuöldin.* Reykjavík: Bókaútgáfa Guðjóns Ó. Guðjónssonar.

Gunnarsson, G. 1983. *Monopoly Trade and Economic Stagnation: studies in the foreign trade of Iceland 1602–1787.* Lund: Ekomomisk-historiska Föreningen i Lund.

Gunnarsson, G. 1987. *Upp er boðið Ísaland: einokunarverslun og íslenskt samfélag 1602–1787.* Reykjavík: Örn og Örlygur.

Halldórsson, J. (ed.) 1959. *Kjalnesinga Saga.* (Íslenzk fornrit 14). Reykjavík: Hið Íslenzka Fornritafélag.

Jóhannesson, J. 1958. *Íslendinga saga. Volume 2: fyrirlestrar og ritgerðir um tímabilið 1262–1550.* Reykjavík: Almenna Bókafélagið.

Jóhannesson, J., M. Finnbogason & K. Eldjárn (eds) 1946. *Sturlunga saga.* Reykjavík: Sturlunguútgáfan.

Jóhannesson, Þ. 1965. *Lýðir og landshagir.* Vol. 1. Reykjavík: Almenna Bókafélagið.

Jónsson, G. (ed.) 1936. *Grettis Saga.* (Íslenzk fornrit 7). Reykjavík: Hið Íslenzka Fornritafélag.

Jónsson, M. (ed.) 2004. *Jónsbók: lögbók Íslendinga.* Reykjavík: Háskólaútgáfan.

Juel, S. 2011. The Icelandic whetstone material: an overview of recent research. *Archaeologia Islandica* 9: 65–76.

Karlsson, G. 2000. *The History of Iceland.* Minneapolis: University of Minnesota Press.

Karlsson, G. 2009. *Lífsbjörg íslendinga frá 10. öld til 16. aldar.* (Handbók í íslenskri miðaldasögu 1). Reykjavík: Háskólaútgáfan.

Kristjánsson, J. (ed.) 1956. *Eyfirðinga sogur.* (Íslenzk fornrit 9). Reykjavík: Hið Íslenzka Fornritafélag.

Lárusdóttir, B., H. Roberts, S. Þorgeirsdóttir, R. Harrison & M. Sigurgeirsson. 2011. *Siglunes: archaeological investigations in 2011.* Reykjavík: Fornleifastofnun Íslands.

Lucas, G. 2009. *Hofstaðir: excavations of a Viking Age feasting hall in north-eastern Iceland.* (Institute of Archaeology Monograph Series 1). Reykjavík: Fornleifastofnun Íslands.

Lucas, G. & T. H. McGovern. 2008. Bloody slaughter: ritual decapitation and display at the Viking settlement of Hofstaðir, Iceland. *European Journal of Archaeology* 10(1): 7–30.

Magerøy, H. 1993. *Soga om austmenn: Nordmenn som siglde til Island og Grønland i mellomalderen.* Oslo: Norske Samlaget.

McGovern, T. H., S. P. Perdikaris, Á. Einarsson & J. Sidel. 2006. Coastal connections, local fishing, and sustainable egg harvesting: patterns of Viking Age inland wild resource use in Mývatn District, northern Iceland. *Environmental Archaeology* 11(1): 187–206.

McGovern, T. H., O. Vésteinsson, A. Friðriksson, M. Church, I. Lawson, I. A. Simpson, Á. Einarsson, A. Dugmore, G. Cook, S. Perdikaris, K. J. Edwards, A. M. Thomson, W. P. Adderley, A. Newton, G. Lucas, R. Edvardsson, O. Aldred & E. Dunbar. 2007. Landscapes of settlement in northern Iceland: historical ecology of human impact and climate fluctuation on the millennial scale. *American Anthropologist* 109, 27–51.

Monclair, H. 2003. *Lederskapsideologi på island i det trettende århundret: en analyse av gavegivning, gjestebud og lederfremtoning i islandsk sagamateriale.* Oslo: Universitetet i Oslo.

Nordal, S. & G. Jónsson (eds) 1938. *Borgfirðinga sogur* (Íslenzk fornrit 3). Reykjavík: Hið Íslenzka Fornritafélag.

Ólafsson, S. 1993. Innreið nútímaþjóðfélags á Íslandi, in M. Snædal & T. Sigurðardóttir (eds) *Frændafundur*: 195–217. (Fyrirlestrar frá íslensk-færeyskri ráðstefnu í Reykjavík 20.–21. ágúst 1992). Reykjavík: Háskólaútgáfan.

Óskarsson, Þ. 1991. *Íslensk togaraútgerð 1945–1970*. Reykjavík: Menningarsjóður.

Pálsdóttir, L. B. & Ó. Sveinbjarnarson. 2011. *Under the glacier: 2011 archaeological investigations on the fishing station at Gufuskálar, Snæfellsnes*. Reykjavík: Fornleifastofnun Íslands.

Perdikaris, S. 1999. From chiefly provisioning to commercial fishery: long term economic change in Arctic Norway. *World Archaeology* 30(3): 388–402.

Perdikaris, S. & T. H. McGovern. 2009. Viking Age economics and the origins of commercial cod fisheries in the North Atlantic, in L. Sicking & D. Abreu-Ferreira (eds) *Beyond the Catch: fisheries of the North Atlantic, the North Sea and the Baltic, 900–1850*: 61–90. Leiden: Brill.

Pope, P. E. 2004. *Fish into Wine: the Newfoundland plantation in the seventeenth century*. Chapel Hill: University of North Carolina Press.

Sigurdardóttir, R., K. Áslaug Vignisdóttir, J. Hansen, J. Thórdarson, R. Jónsdóttir & H. Gudmundsdóttir. 2012. *Economy of Iceland*. Reykjavík: Central Bank of Iceland.

Sigurðsson, J. & G. Vigfússon (eds) 1856–78. *Biskupa sögur gefnar út af Hinu Íslenzka bókmentafélagi*. Copenhagen: Möller.

Simpson, I. A., P. Adderley, G. Guðmundsson, M. Hallsdóttir, M. Á. Sigurgeirsson & M. Snæsdóttir. 2002. Soil limitations to agrarian production in premodern Iceland. *Human ecology* 30(4): 423–43.

Starkey, D. J., J. Þ. Þór & I. Heidbrink (eds) 2009. *A History of the North Atlantic Fisheries. Volume 1: from early times to the mid-nineteenth century*. (Deutsche Maritime Studien/German Maritime Studies 6). Bremen: Hauschild.

Sveinsson, E. Ó. 1954. *Brennu-Njáls Saga*. (Íslenzk fornrit 12). Reykjavík: Hið Íslenzka Fornritafélag.

Sveinsson, E. Ó. & M. Þórðarson (ed.) 1935. *Eyrbyggja Saga*. (Islenzk fornrit 4). Reykjavik: Hið Íslenzka Fornritafélag.

Þór, J. Þ. 1996. Icelandic fishing history research: a survey, in P. Holm, D. J. Starkey & J. Þ. Þór (eds) *The North Atlantic Fisheries 1100–1976: national perspectives on a common resource*: 13–26. (Studia Atlantica 1). Reykjavík: North Atlantic Fisheries History Association.

Þór, J. Þ. 2002–3. *Saga sjávarútvegs á Íslandi*. Akureyri: Hólar.

Þórðarson, M. 1930. *Síldarsaga Íslands*. Copenhagen: Kaupmannhöfn.

Þorláksson, H. 1991. *Vaðmál og verðlag: vaðmál í utanlandsviðskiptum og búskap Íslendinga á 13. og 14. öld*. Reykjavík: Háskóli Íslands.

Þorláksson, H. 2001. Íslensk skreið skákar norskri, in G. Gíslason (ed.) *Líndæla*: 265–93. Reykjavík: Hið Íslenzka Bókmenntafélag.

Þorláksson, H. 2003a. Fiskur og höfðingjar á Vestfjörðum: atvinnuvegir og höfðingjar á Vestfjörðum fyrir 1500. *Ársrit Sögufélags Ísfirðinga* 43: 67–83.

Þorláksson, H. 2003b. Frá kirkjuvaldi til ríkisvalds, in S. Líndal (ed.) *Saga Íslands*. Vol. 6: 3–436. Reykjavík: Hið Íslenzka Bókmenntafélag.

Þorláksson, H. 2004. Undir einveldi, in S. Líndal (ed.) *Saga Íslands*. Vol. 7: 3–211. Reykjavík: Hið Íslenzka Bókmenntafélag.

Þórólfsson, B. & G. Jónsson (eds) 1943. *Vestfirðinga sogur*. (Íslenzk fornrit 6). Reykjavík: Hið Íslenzka Fornritafélag.

Þorsteinsson, B. 1950. Íslandsverzlun Englendinga á fyrri hluta 16. aldar. *Skírnir* 124: 83–112.

Þorsteinsson, B. 1953. *Íslenzka þjóðveldið*. Reykjavík: Heimskringla.

Þorsteinsson, B. 1964. Þættir úr verzlunarsögu: nokkur atriði úr norskri verzlunarsögu fyrir 1350. *Saga* 4: 3–52.

Þorsteinsson, B. 1970. *Enska öldin í sögu Íslendinga*. Reykjavík: Mál og Menning.

Þorsteinsson, B. 1976. *Tíu þorskastríð 1415–1976*. Reykjavík: Sögufélag.

Þorsteinsson, B. 1978. *Íslensk miðaldasaga*. Reykjavík: Sögufélag.

Þorsteinsson, B. 1980. *Íslensk miðaldasaga*. 2nd edition. Reykjavík: Sögufélag.

Þorsteinsson, B. 1985. Hvers vegna var ekkert atvinnuskipt þéttbýli á Íslandi á miðöldum? *Landnám Ingólfs* 2: 131–3.

Þorsteinsson, B. & G. Á. Grímsdóttir. 1989. Norska öldin, in S. Líndal (ed.) *Saga Íslands*. Vol. 4: 61–261. Reykjavík: Hið Íslenzka Bókmenntafélag.

Þorsteinsson, B. & G. Á. Grímsdóttir. 1990. Enska öldin, in S. Líndal (ed.) *Saga Íslands*. Vol. 5: 3–216. Reykjavík: Hið Íslenzka Bókmenntafélag.

Þorsteinsson, B. & S. Líndal. 1978. Lögfesting konungsvalds, in S. Líndal (ed.) *Saga Íslands*. Vol. 3: 19–111. Reykjavík: Hið Íslenzka Bókmenntafélag.

Valdimarsson, V. U. 1997. *Saltfiskur í sögu þjóðar: saga íslenskrar saltfiskframleiðslu og -verslunar frá 18. öld til okkar daga*. Reykjavík: Hið Íslenzka Bókmenntafélag.

Vésteinsson, O. 2000. *The Christiainization of Iceland: priests, power, and social change 1000–1300*. Oxford: Oxford University Press.

Víkingur, S. 1970. *Getið í eyður sögunnar*. Reykjavík: Kvöldvökuútgáfan.

Vilmundarson, Þ. & B. Vilhjálmsson (ed.) 1991. *Harðar Saga*. (Íslenzk fornrit 13). Reykjavík: Hið Íslenzka Fornritafélag.

The Character of Commercial Fishing in Icelandic Waters in the Fifteenth Century

Mark Gardiner

Introduction

Commercial fishing by its very nature was intimately connected with trade. Yet the subjects of fishing and trade in the Middle Ages have generally been studied separately. Without the ability to distribute and sell fresh fish very rapidly or to preserve fish for later sale and consumption, large-scale fishing was impossible. Some work on the connections between trade and fishing has been undertaken, and the broad outlines of their relationship are beginning to emerge. Barrett *et al.* (2004, 631) have argued that the advent of fishing on a substantial scale and the development of the fish trade around the end of the first millennium AD coincided with an increase in commerce in western Europe more generally. Kowaleski (2003) has identified a second stage in the commercialisation of fishing, which took place at the end of the fourteenth and during the course of the fifteenth century, during which the scale of production was substantially increased. The present paper examines further the concept of second-stage commercialisation by looking at fishing in Icelandic waters. It begins with a review of commerce between Scandinavia and northern Europe and sets the development of fishing within the context of the struggle for power between four parties: the English fishers and traders; the Icelanders; the Danish crown, as overlords of Iceland; and, later in the fifteenth century, Hanseatic merchants. The second issue considered here is the practical operation of fishing and trade. The paper concludes with an examination of the strategies adopted by all the parties as a means of understanding the rationality of their actions.

It would be possible to take a simple view of trade and fishing and argue that a larger market led to more extensive fishing; that the two were simply an example of demand stimulating supply. At the heart of this analysis is the view that fishing and trade must instead be studied as

the result of social processes as well as economic ones. This approach has much in common with the culturalist position adopted by such anthropologists as Gudeman (2001). It regards trading systems as embedded within, or contingent upon, the societies in which they operate. Economies operate according to a rationale defined by their society, a view termed 'situated reason' (Gudeman 2001, 38–42). Behaviour is reasoned and structured according to the perspectives and value-systems of the participants. However, the problems become more complex when we begin to consider trade between cultures. This necessitated an imperfect compromise between the different economic and social values. The parties suppress or set aside their differences while they both benefit through the exchange of goods. Trade, however, carries a cost in terms of its social and economic side-effects. The practices and regulations which surround the exchange seek to limit those undesired consequences, though inevitably they cannot entirely contain them.

The historical context

Commercial contacts between the North Atlantic and northwestern Europe date to at least the Viking Age and continued through the eleventh century. Trade between Grimsby, in England, and Norway in the reign of the English king William I is implied by the escape of a hostage on a boat sailing between the two places (Arnold 1885, 202). The Icelandic *Orkneyinga saga* claims that merchants from Norway and Orkney came to Grimsby early in the twelfth century (Guðmundsson 1965, 130–8). Norwegian merchants were certainly established in London by AD 1130, when they are mentioned in regulations for foreign merchants, but it is possible that these even date to the eleventh century

(Bateson 1902, 499; Brooke and Keir 1975, 267). Such links have also been demonstrated by the quantity of medieval pottery from southeastern and eastern England found in Bergen, in Norway. The pottery suggests that there was intensified contact between England and Norway from the later twelfth century, although the source within England shifted in the thirteenth century (e.g. Blackmore and Vince 1989, 112).

The strength of trade between Britain and Norway, and the mutual benefit which was perceived by both sides, led in AD 1223 to the first trade agreement concluded by either country (CPR 1216–25, 384). The spirit of this agreement was reflected the following year in an exemption for Norwegian vessels from a general arrest of ships (Hardy 1833, vol. 1, 606–7). The main imports to England were fish and wood, which were supplied in return for grain and cloth. The level of Norwegian trade seems to have been largely maintained until the end of the thirteenth century, but the diplomatic relationship gradually cooled in the second half of the century as Norway moved closer to Scotland in its interests (Helle 1967, 12–13, 16–17, 32–3; Lloyd 1982, 125–6).

Norwegian control over foreign trade with the North Atlantic was strengthened in AD 1294 when merchants were forbidden to sail beyond Bergen without licence. In the early fourteenth century it was further decreed that they could not sail to the Norwegian colonies, including Iceland, the Faroe Islands and Shetland (DN, vol. 5, no. 23; Keyser & Munch 1849, no. 53). This is unlikely to have caused many problems for English merchants at first, since most of the trade was conducted in Norwegian vessels and there was little, if any, trade between England and the North Atlantic islands. Trade between Norway and England was substantial in the first decade of the fourteenth century. There were 75 ships at King's Lynn in the three trading seasons from February 1303 to November 1305, and smaller numbers at Ravenser at the head of the Humber and at Hull, all of which were key ports visited by Norwegians. In addition, there were an estimated 20 ships carrying fish operated by Hansa merchants (Nedkvitne 1983, 579). The numbers began to fall in the second decade of the fourteenth century with allegations of ill-treatment of merchants. It was said in AD 1312 that four hundred English were detained in Norway, though no doubt the numbers were much exaggerated (Lloyd 1982, 125–6, 151–2). By the second quarter of the fourteenth century the trading activity by Norwegians was substantially curtailed by competition with Hanseatic merchants, who largely took over the export of fish. The Norwegians maintained links with the North Atlantic islands where the Hansa merchants were not allowed to trade, but these were limited in scope. Commerce there also declined in the second half of the fourteenth century (Gelsinger 1981, 185–6; Thorsteinsson 1957, 168).

The Hansa created a firm foothold in Bergen, the main trading port in Norway, and had established a *Kontor* there in AD 1366 (Gade 1951, 55; Nedkvitne 2014, 336). Relations between the Hansa and the Norwegian government deteriorated, and the German merchants withdrew in AD 1368 upon the outbreak of war, which allowed English merchants briefly to establish themselves in Bergen. When peace was re-established in AD 1370, Hansa merchants returned and attacked the English (DN, vol. 19, no. 591). Nevertheless, trade by English merchants continued, and it has been argued by Schreiner (1941, 93) that it expanded during the AD 1390s, with shipments of grain and malt being traded for stockfish (DN, vol. 19, nos 618–20, 624, 628–9, 631; Tuck 1972, 76–8).

It is rather more difficult to trace fishing by foreign vessels than it is to trace trade, since the former was less closely regulated. Fishing vessels in foreign waters were rarely subject to the payment of dues, unless they sold their catch abroad. Nevertheless, it is apparent that English fishing vessels were venturing out into the North Atlantic by the end of the fourteenth century. Boats from Norfolk were working in Scottish waters, and in AD 1383 fishermen from Cromer and Blakeney complained that their doggers (fishing vessels) had been arrested for use in the navy and asked that they should be allowed to leave to fish off the coasts of Denmark and Norway (Riley 1863–4, vol. 2, 246; TNA, SC 8/102/5100; for doggers, see Marcus 1954). It was said that in AD 1406 one hundred fishermen from the same ports and elsewhere in Norfolk were working off the Norwegian coast when they were attacked and driven into 'Wynford', or Fensfjord, near Bergen (DN, vol. 19, no. 707; CPR 1408–13, 384). The accounts of the early voyages to Iceland recorded in the *Nýi Annáll* suggest that one of the first boats to arrive, in AD 1412, was a fishing vessel which appeared at Dyrhólar, at the southern-most tip of the island. The following year a trading vessel appeared with a licence from the Danish king, and in the same year 30 or more doggers also made the voyage to Iceland (Þorsteinsson 1922, 18). However, this may not have been the start of trade there, since it had been reported that there were foreign, possibly English, merchants in Vestmannaeyjar (Westman Islands) in AD 1396/97 (Þorsteinsson 1922, 8–9).

Trading and fishing activity in the North Atlantic expanded considerably during the early fifteenth century. It had become one of the major fishing areas for boats from the east coast of England within a few years of the appearance of that fishing vessel in Iceland in AD 1412. In AD 1417–18, 11 ships went from Scarborough alone (Heath 1968, 63). At about the same time German merchants had begun to sail to Shetland, presumably to purchase dried fish and perhaps coarse cloth. The earliest recorded trader there is mentioned in AD 1415, and the following year the Hansetag forbade voyages to Orkney, Shetland and the Faroe Islands, which perhaps suggests that such journeys were already taking place to the other islands as well (Friedland 1973, 68).

There were substantial profits to be made from fishing in the North Atlantic. Scarborough fishermen going to Iceland in the period had average gross incomes of £85 (Heath 1968, 57). Similar voyages were made up the Norwegian coast as far north as Finnmark, well beyond Bergen, to which foreign vessels were supposed to be restricted (DN, vol. 1, no. 670; Strachey 1767, 4, 79).

Fishing and trade grew so rapidly that in AD 1414 the Danish king sent a letter to Iceland forbidding trade with foreigners. The following year King Eric also wrote to the king of England, complaining about damage done in Iceland and to the fisheries around it. The English response was merely to confirm that fishermen should continue to take fish to the staple at Bergen for the payment of custom, and this was proclaimed in towns on the East Coast, including Newcastle, Scarborough, Hull, Boston and Lynn and the fishing settlements of Whitby, Grimsby, Dersingham, Blakeney, Burnham and Cromer (DN, vol. 10, nos 733, 735–7; Þorsteinsson 1922, 20). This proclamation was largely ignored. The emphasis of English activity in Iceland at this time seems to have been on fishing rather than on trade, although a list of exchange rates for (stock)fish and goods at Vestmannaeyjar suggests that there was some commerce too (DN, vol. 10, nos 742, 753).

The fishermen and traders continued to circumvent the restrictions placed upon them by the Danish king until AD 1429, when voyages to both Finnmark and Iceland were forbidden in the English parliament and merchants were restricted to the staple town of Bergen (Strachey 1767, 4, 347). However, this statute was little more effective than the earlier proclamation. Within a few years, the English crown, which had no great enthusiasm to enforce it, began to issue licences to trade, and fishing also continued unabated (Childs 1995, 18, n. 22 indicates the numbers of licences issued).

The murder in AD 1467 of the Danish governor (*hirðstóri*), Björn Þorleifsson, by English sailors in Rif triggered a new crisis in the relations between England and Denmark (Þorsteinsson 1970, 209–12). Though the Danish king lacked leverage in Iceland, it was possible to take action against English ships trading in the Baltic, and in June 1468 seven ships were seized in the Øresund. The event led to war between England and the Hansa, which was only concluded by the Treaty of Utrecht in 1474 (Strachey 1767, 6, 65). Agreement between the English and Danish crowns was not reached until AD 1490, when English boats were again allowed to trade with Iceland and to fish, provided they purchased licences at seven-year intervals. However, when the treaty was presented to the Icelandic *Alþing*, it struck out the clauses concerning fishing rights and limited it to only those vessels that were also trading. The concern, expressed some years later, was that the larger English vessels operated farther out at sea were catching fish and so preventing them from coming inshore where they might be caught by the Icelanders themselves (Baasch 1889, 58). A decree forbidding

foreigners from over-wintering in Iceland was reiterated, and, if they were forced to stay in the country out of necessity, they were required to sell goods during the winter at the same price as in the summer (DI, vol. 7, no. 617; see also TNA, HCA 13/93, f. 253).

Ships from the German Hansa began journeying to Iceland in increasing numbers in the 1470s, and this inevitably brought them into conflict with the English vessels. In AD 1474, there was a fight between merchants from Hull and Bristol on the one hand and German traders on the other (DI, vol. 7, no. 66). The continuing lawlessness in Iceland led Richard III, in AD 1484, to direct ships from Norfolk and Suffolk to assemble in the Humber and go in convoy with those from Hull (Gairdner 1861–3, vol. 2, 287). Conflict persisted throughout the latter part of the fifteenth century and into the sixteenth century between the ships of the Hansa and those of England, even though hostilities had been formally concluded (Carus-Wilson 1954, 140–1; Seaver 1996, 205).

Fishing and trade

The businesses of fishing and trade were rarely separate activities for English merchants voyaging to Iceland. Routes pioneered by trading vessels seeking to obtain stockfish, caught by Icelanders and preserved by wind-drying, were followed by fishing vessels seeking to catch fish themselves and preserve it by salting, and *vice versa*. Once these routes were opened up, fishing and trading often took place on the same voyage. The evidence for this is abundant in the sixteenth century, but it is almost certain that the same was true in the fifteenth. The *Christopher* of Hull bought 4½ lasts and 60 stockfish (5460 fish), train (fish) oil and some cloth while fishing in Icelandic waters in AD 1430 (Childs 2000, 22). Three further examples drawn from the sixteenth century serve to illustrate this in greater detail. In a deposition, Richard Cutbert of Southolt (Suffolk) gave details of a voyage to Iceland *c.* AD 1535. He had chartered a boat, the *Anne*, from Snodland (Kent) to go to Iceland, but its departure was delayed. Eventually it returned with 6000 salted ling (*Molva molva*) and 4695 salted cod (*Gadus morhua*) worth about £450, and 27 lasts of stockfish (equivalent to 35,400 fish), 12 'wode' of coarse cloth and two hogshead of train oil, worth about £250 in total (TNA, SP 1/99, ff. 170–8). The salted ling and cod were probably caught by the *Anne*. Salt was not produced in Iceland, so fish caught by native fishermen there were wind-dried in the winter and spring, and exported as stockfish. The *Anne* thus carried back a mixed cargo obtained by fishing and trade.

A second example illustrates the opportunistic nature of trade in the North Atlantic and how it did not always run along the neat lines sometimes imagined. The *Jesus* of Tenby (in Wales) set out in AD 1564 to sail to Newfoundland to go fishing, but it was so late in the year that during the course of the voyage the ship turned back to Scotland and went to Orkney, where it met with two Flemish busses (sea-going fishing vessels). It hailed these and bought from them 144 barrels of herring (*Clupea harengus*). Subsequently, it encountered a ship from Hamburg that may have been sailing back from Iceland or Shetland, since it was loaded with stockfish. It bought part of that cargo too. Then the *Jesus* came into Papa Stour in Shetland, where it met another ship and bought from it salted ling and cod, which that vessel had evidently caught itself. Finally, the *Jesus* spent a couple of days fishing before returning to Tenby in the third week of September (TNA, HCA 13/15, f. 11).

This may not have been a typical voyage; equally, it was not wholly unusual. It reflects the pragmatic approach to the gathering of fish by English, or in this case, Welsh, vessels. Like many English ships, the *Jesus* was not just fishing, nor just trading, but sought to fill its hold with whatever fish, or indeed other cargo, might be obtained, wherever it might be obtained.

The third example is the *James* of Dunwich (in Suffolk), which went to Iceland in AD 1545 and appears to have taken a quite different approach. There survives for this voyage a list of laded goods. It is in some ways the most illuminating source of all, since from it we can infer the intentions of the venturers who equipped the ship. Most of the materials listed are victuals for feeding the ship's company and fishing gear, including hooks and lines for catching and salt for preserving the fish. The employees sent to Iceland included a merchant, Geoffrey Smythe, and the goods for sale are recorded separately from the company's victuals. These included food (butter, meal, wine), fabric, clothing and footwear (linen, broad cloth, shirts, shoes, boots) and utilitarian items (kettles, horseshoes, whetstones, wax) (Cooper 1939; Webb 1962, 82–3). Smythe would have been let off with the goods, presumably shortly after reaching Iceland, perhaps at one of the major ports in southwestern Iceland – for example, Vestmanaeyjar. There he would have set up a booth to store the merchandise and the stockfish taken in return. Towards the end of the season, in July or August, the *James* would have picked up Smythe and the stockfish that had been obtained in exchange before the ship set sail back to England.

Merchant ships required no adaption to undertake fishing. The fish were caught from small boats, skiffs or dogger boats, which were launched from the ships. The skiffs set long lines with hooks, often far from the coast (on such boats, see Webb 1962, 78, 81; TNA, SP 1/99, f. 170; on methods of fishing, see Jones 2000, 109). The mother ships might trade at small settlements when they came to land to take on fresh water, but this can only ever have been a minor activity. The greater part of trade was conducted by merchants who set up booths, or 'caves' as they are termed in contemporary English sources, close to the shore in ports.

The advantage of a mixed strategy combining fishing and commerce was that it ensured a return for the merchants, even in years when there was a paucity of stockfish to be obtained in Iceland. There was considerable competition among merchants for supplies of stockfish, and ships which arrived late in the season might find that others had already purchased all that was available (Childs 1995, 26). It was said that the violent events which led to the deaths of 40 English sailors in Grindavík in AD 1532 were precipitated by competition over supplies of stockfish. The English sailor John Bray bought stockfish which were to have been had by Hamburg merchants and then taunted the Germans to come and obtain them by force (H. Þorsteinsson 1922, 1, 92–3; B. Þorsteinsson 1957–61, 82).

Some smaller vessels engaged only in fishing. They may have lacked the expertise to trade or the capital to purchase goods to exchange for stockfish. The *Margaret Bonaventure* of Dunwich, a craft of 55 tonnes, was sent to fish between Scotland and Iceland in the 1560s equipped only with victuals for the crew and fishing gear (Williams 1988, 93). Though quite small, this vessel was close to the mean size of ships from the east coast and London voyaging to Iceland in AD 1533. As we might expect, ships from the major ports, such as Lynn, the Orwell (Ipswich) and London, tended to be amongst the largest. The 22 ships recorded under the heading of Dunwich, and presumably including the nearby ports of Southwold and Walberswick, were notably smaller, with a mean size of 37 tonnes (TNA, SP 1/80, ff. 60–78). It was evidently similar vessels – 22 fishing boats and 13 barks – which were drawn up on the beach at Walberswick in AD 1451 and were engaged in fishing and/or trading with Iceland, with the Faroe Islands and in the North Sea (Oppenheim 1907, 211).

The character of trade and fishing can be investigated further by examining English customs accounts. For this purpose, those of Hull have been chosen, because accounts of other ports have some limitations (as detailed by Childs 1995, 21). Hull was one of the major ports involved in Icelandic trade and fishing in the fifteenth century. Its commerce was as large or larger than that of Bristol and greater than that of London. The destination of the ships is not recorded, but it can be concluded with some certainty that it was Iceland from both the goods exported and the return cargo, which was formed exclusively of stockfish (Childs 1995, 20–1).

Table 8.1 shows the Hull ships that, according to the surviving custom accounts, made the most frequent voyages to Iceland and the number of stockfish which they bought on their return. The figures have been calculated on the basis that a 'hundred' of stockfish was often measured by a 'long hundred' of 120 (for salted fish the figure was 124 to the 'hundred'). A last was a 'thousand' stockfish, counted in a similar manner and so equal to 1200 fish (Hall

Table 8.1. Estimated number of stockfish imported to Hull from Iceland in selected ships (based on data in Childs 1994), mid– late fifteenth century. See text for method of estimating.

	1453	1460[1]	1461	1462[2]	1463[3]	1465	1468[4]	1471	1472	1473[5]
Mary of Hull	36,540	61,200	94,200					36,000		
Trinity of Hull			81,600		60,000	44,940	84,000			55,200
Anthony of Hull				193,080		66,840				
Peter of Hull					9600	7200				
Anne of Hull							143,580	37,260		
Total for all ships	100,500	73,200	175,800	207,490	73,200	133,380	227,580	73,260	25,200	94,440

Notes:

1. Account closes 12 August 1460; 2. Account opens 16 August 1462; 3. Account runs 6 July–26 August 1463; 4. Account runs 18 July–29 September 1468; 5. Account runs 6 August–29 September 1473

and Nicholas 1929, 17, 29; British Library, Lansdowne MS 21, f. 137).

There is notable variation in the figures, with, for example, the *Anthony* of Hull bringing back three times as many stockfish in AD 1462 as in 1465 and the *Anne* carrying a cargo in AD 1468 which was apparently four times as large as that in AD 1471. The general pattern of trade with Iceland from Hull suggests that there was a significant change after AD 1468, when it declined considerably. There are problems in refining this overall picture because the customs accounts in some years open too late or close too early to allow us to be certain that they include all ships trading with Iceland. Moreover, the volume of trade must have depended on the number of ships making the voyage, which was always rather small. Nevertheless, it is very likely that the much smaller numbers of stockfish imported in AD 1471 reflect the crisis following the murder of Björn Þorleifsson in AD 1467. The impact of this was not felt in AD 1468, but in June that year the English were banned from sailing to Iceland and war broke out with the Hansa. The persistence of trade, albeit at a lower level, suggests that, as in the past, Danish decrees were not fully effective so far from the seat of power. Nevertheless, they clearly did have an impact on the volume of commerce.

If we set these particular events to one side, we might suggest three different reasons for the varying quantities of stockfish brought back in the same ship. First, the cargo unloaded at Hull and subject to customs dues there might not have represented the full contents of the ship. It may have sailed on elsewhere with a hold part full of stockfish. Second, the return cargoes may have been affected by the catches of fish by Icelanders and the competition to obtain stockfish from them. In short, the size of cargo was supply-led. Third, the cargo of stockfish may have been demand-led. Merchants may have decided not to send large volumes of goods to Iceland and consequently the volume of fish obtained in return was smaller.

The first of these possible explanations can be rejected. For a few voyages we have customs accounts recording not only the individual merchants importing stockfish, but also the goods which were sent out for exchange. A total of 24 merchants laded goods on to the *Mary* of Hull in AD 1461 and are recorded as importing stockfish; two more export, but do not import; and seven import stockfish and have no recorded exports. The two exporters both have locative surnames – John Richemond and John Hebden – and it is possible that they appear amongst the unaccounted importers. Similarly, the *Trinity* of Hull, which sailed the same year, had 16 merchants who exported and brought back goods to Hull in return. There were only two exporters without imports which we can recognise, and 6 unaccounted importers. Clearly, in most cases the merchants laded and off-loaded goods at Hull.

It is more difficult to distinguish between the other alternatives. Did ships leave Hull with a cargo of goods to exchange for stockfish but found too few fish offered in exchange? In theory this hypothesis could be explored by looking at the match between the goods exported and the fish imported. If we find that the two are poorly correlated, then it would be likely that this was because the merchants in Iceland were unsuccessful in obtaining supplies of stockfish. Unfortunately, this is not easy to do. We can examine three voyages for which we have customs records for both the exported goods and the stockfish obtained in return. Table 8.2 sums the most common goods taken on each voyage. This excludes items, such as kettles and belts, swords and silk, which were freighted only occasionally. Proper comparison requires that we know the relative values of each of these goods in Iceland. We do have a list of values in terms of stockfish proclaimed in Vestmanaeyjar in AD 1420, but it is difficult to translate its information into the units of measure used by English merchants (DI, vol. 4, no. 337). Furthermore, we need to be aware that the fish listed in the AD 1420 table may be *vættir* fish, that is 'fish' used as a unit of currency rather

Table 8.2. Quantities of main commodities exported to Iceland and estimated number of stockfish imported to Hull (based on data in Childs 1994). See text for method of estimating.

	Mary *of* Hull 1460	Trinity *of* Hull 1461	Mary *of* Hull 1461
Exports			
Standard cloths	52	65	70
Canvas (ells)	130	216	92
Osmond (barrels)[1]	11	29	44
Beer (barrels)[1]	210	262	117
Meal (barrels)[1]	199 or 331	207 or 323	161 or 261
Malt (quarters)	0	38	39
Honey (barrels)	14	4	4
Wax (lbs)	0	50	244
Horseshoes	200	0	500
Imports			
Stockfish (n)	61,200	81,600	94,200

Note:

1. The figures use barrels as the measure where appropriate, though some goods were measured in both lasts and barrels. The figures have been calculated by taking 1 last of beer as equal to 12 barrels and 1 ship last of osmund (iron) as equivalent to 13 barrels (Hall and Nicholas 1929, 18, 23). Zupko (1977, 136) indicates that 1 last of meal was also equal to 12 barrels. A barrel contained a volume of meal equal to 32 gallons according to statutes of the late fourteenth and fifteenth centuries, and in the seventeenth century a last was said to contain 640 gallons of grain (Hall and Nicholas 1929, 21, 30, 49). This might suggest alternatively that 1 last was equivalent to 20 barrels. Both the upper and the lower figures have been given above.

than actual stockfish (Gunnarsson 1983, 18–19; Hastrup 1990, 139–41).

Given these uncertainties, we have inadequate evidence to decide between the second and third hypotheses. We cannot tell whether the variation in the size of imported cargoes of stockfish in the same ship in different years was a consequence of the availability of stockfish in Iceland or the volume of goods exported from Hull for exchange. Whichever was the case, it clearly had an impact on the profitability of the voyage. The hire of the ship, the wages of the crew and the supply of victuals were fixed, regardless of the volume of goods obtained in exchange. If the variation in imports was due to the availability of stockfish in Iceland, then this was more serious for merchants who had invested in goods for export and were unable to exchange them. There was a clear commercial logic in fishing from merchant boats, since it provided an alternative source of fish for import, though we cannot know whether this was what actually happened on the Hull ships examined here.

The emphasis in the discussion so far has been on fishing and trade by English merchants. However, much of the fishing was undertaken by Icelanders. The Hanseatic ships did not fish at all, and their cargoes were entirely made up of dried fish purchased from Icelandic fishers. We need to turn now to Icelandic fishing practices, but we must start off by acknowledging that the relevant historical evidence regarding the fifteenth century is very limited. We may be able to back-project traditional fishing practices, but we need to be cautious to avoid running the risk of creating an ethnographic present in which we imagine the medieval past to be no different from more recent practices (Krivogorskaya *et al.* 2005, 45). Nevertheless, it seems probable that in the fifteenth century fishing took place in the winter or spring, when there was otherwise little call on labour, so it did not compete with work on the farm (Kristjánsson 1982, 485–6). Farmers from inland areas may have moved to temporary settlements or outstations (*útver*) on the coast, where they occupied huts (*verbúðir*), as they did in later centuries. Fish were prepared and dried at the fishing stations (Amundsen *et al.* 2005). The fish could either have been brought back to the home farm at the end of the fishing season, or been kept at the fishing stations and taken by sea to the trading sites.

We can provide some historical substance to these suppositions. There survives an account book for AD 1558 drawn up by the Bremen merchant Claus Monnickhusen, which lists fishers who owed him for goods that had been supplied in advance (Hofmeister 2001). Monnickhusen had a trading site at Kumbaravogur, on the north coast of Snæfellsnes facing Breiðafjörður, which was an important fishing area (see Gardiner and Mehler 2007, 415–18 for details). It was one of a number of trading sites in the area. Others were to be found on the south side of Snæfellsnes at Buðir and Arnarstapi and on the north side at Rif, Grundarfjörður and Nesvogur (Figure 8.1). The account book lists the names of the individuals and the farms from which they came, allowing us to identify the location of indebted fisher–farmers within the hinterland of Kumbaravogur. We have no way of knowing whether these represented all the fisher–farmers who regularly came to trade at that site, but it is sufficient to provide an understanding of the Icelandic practice of fishing.

The distribution shows that most fisher–farmers selling stockfish at Kumbaravogur came from the north coast of Snæfellsnes, from the coasts of Fellströnd and Skarðsströnd, and from islands in Breiðafjörður. Almost all of these fishers were from farms that were situated on or very close to the sea, and they could have operated from their homes. It is likely that they would have brought stockfish to Kumbaravogur by boat. Indeed, there would have been no alternative for those fishers based on islands in Breiðafjörður. A smaller number came from the east end of

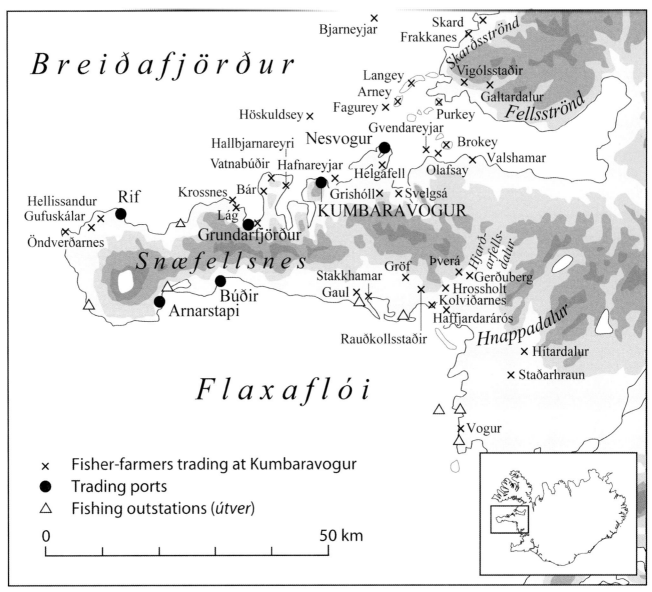

Figure 8.1. The trading hinterland of Kumbaravogur (Drawing: Libby Mulqueeny after Hofmeister 2001, 31, with outstations (útver) from Kristjánsson 1982).

the south coast of Snæfellsnes and even from inland farms in Hnappadalur. Presumably, those fishers based to the west of Gaul and Stakkhamar were trading at Arnarstapi and Búðir, and therefore had no dealings with Monnickhusen at Kumbaravogur. Again, most of the fishers operated from farmsteads close to the sea. Only two farmsteads were very far inland: those at Staðarhraun and nearby Hítardalur. Even these were little more than 10 km from the coast. It is much less likely that those from the south of Snæfellsnes and Hítardalur transported fish to Kumbaravogur by sea, since the journey around the west end of the peninsula is very long. It is probable that they brought fish by land over the pass through Hjarðarfellsdalur or over one of the passes to the east, climbing up from Hnappadalur.

In this area of Iceland no farm was very distant from the sea, since all the better land is situated towards the periphery. Most fishers would have been able to operate from home and many from bases situated on their own land (*heimræði*). Only a few outstations are know from more recent times, and some of these may have been used in the sixteenth century for farms, particularly those in Hítardalur, which were farther from the coast. The broad bays of Flaxaflói and Breiðafjörður were good fishing grounds, and the evidence from Monnickhusen's account book suggests that many farmsteads practised fishing. It suggests that in the mid-sixteenth century there were few constraints in getting access to the sea or selling dried fish to foreign merchants. Much the same was probably the case fifty years earlier, at the end of the fifteenth century. Kumbaravogur and Grundarfjörður had both been used as trading sites since before AD 1523, and there were other similar sites in the vicinity, including Rif ('Gamelwick'), which had been used since at least the late fifteenth century (DI, vol. 16, nos 234, 268, 285).

The evidence from England, Germany and Iceland has allowed us to construct a reasonable picture of fishing and trading patterns in the fifteenth century, which can now be summarised. Substantial numbers of ships sailed northwards each year from England to Icelandic waters. In AD 1528, as many as 149 ships made the voyage, though this was perhaps after some decades of growth in fishing; in the fifteenth century the numbers may have been rather smaller (Jones 2000, 105–6; Williams 1988, 95). English activity included both fishing and trading, often from the same vessels, though some smaller ships may have engaged only in fishing. Hansa activity, by contrast, was solely limited to trading: there is no evidence for fishing by Hanseatic vessels. Trade was carried out at a series of established centres around the coast of Iceland, particularly in the southwest, west and northwest, the areas with the best fishing grounds (Gardiner and Mehler 2007, 389–95). The trading sites allowed dried fish prepared by numerous fisher–farmers to be bulked for transport back to England and continental Europe. Fishing by Icelanders remained a small-scale enterprise in spite of the demand from foreign merchants.

The strategies of fishing and trade

Trade between peoples of different cultures involves careful negotiation to establish both the practices by which exchange is performed and agreement on systems of value. Once agreement has been reached on these matters, it needs to be continually renegotiated, since trade affects both socially and economically all those that it touches. By the end of the fifteenth century in Iceland, there were three parties at least to the trade – the Icelanders, the English and the Hansa merchants. The Danish crown was a fourth party; though not engaged in trade, it was seeking to regain power and control over activity in an area in which it had little influence. It is useful to examine each of the parties in turn to look at their strategies as they had evolved by AD 1500 in terms of their 'situated reason' (see above) and consider the impact these had on fishing.

Fundamental to the Icelandic conception of trade was the view that fishing should always be an adjunct of farming and not a separate activity in itself (see Chapter 7). The farm (bú) was treated as the unit of production and consumption. During the fifteenth century, when there were labour shortages due to decline in population following the outbreak of plague in AD 1404–5, a series of measures were enacted to tie labourers to work on farmsteads. The búalög, or household law, which is known from a manuscript dating to the mid-fifteenth century, regulated the duties and rights of farm servants (Hastrup 1990, 54–7). Cottagers (búðsetumenn) who did not own enough cattle to provide a living were not allowed to make a boat or to employ others at sea (Hastrup 1990, 109). The result was that fishing did not develop as an industry in itself, but was always an appendage which was subordinate to farming. The aim of fishing by Icelanders was not to accumulate capital, which, in any event, was difficult in a coinless society, but to provide a source of food and through exchange to acquire imported food and other goods. Fishers did not seek to build larger ships for this purpose or to expand their enterprise. It was necessary only to satisfy the needs of the farming household.

The absence of urban development in Iceland was a corollary of the centrality of the farm. The existence of towns depends upon trade and craft production, but in Iceland the latter could only take place within the context of the farm. Trade was accepted as a necessary part of Icelandic life, but traders were not. The sagas portray a society in the thirteenth century in which commerce was deeply alien. Instead, gift-giving, both willing and enforced, was more common. 'There was little time spent bargaining over price, the hasty abandonment of which marked the rejection of the mercantilist mode' (Miller 1986, 46). There is little evidence that attitudes had changed by the fifteenth century, when prices for trade in fish were still being set (DI, vol. 4, no. 337). The traders themselves were tolerated as a necessary evil, but they were kept at arm's length and largely excluded from Icelandic society. One of the principles was that foreigners should not be allowed to settle in the country, and this was achieved by preventing over-wintering. A decree against the year-round presence of foreigners was first recorded in Jónsbók in AD 1281, when it was aimed at Norwegians. It was repeated in AD 1431 (DI, vol. 4, no. 506), this time with the English in mind, and again in the Piningsdómur of AD 1490. The combined result of an emphasis on the farm and measures against the permanent settlement of traders was that urban centres failed to develop, even in major places of trade, such as Vestmannaeyjar and Hafnarfjörður.

The strategies of the English traders have already been discussed in part. The investment in chartering and outfitting a vessel for Iceland was considerable, and the risks were also very substantial. The James of Dunwich is one of the few Iceland-bound ships for which we have complete records for the expenses of outfitting. In AD 1545, the costs of equipment and victuals were £151, goods for sale were £37 and wages for the crew were £99, a total of £287 (Cooper 1939). The costs were not always as great as this. The expense of chartering the Christopher of Southwold in the 1530s with its equipments, stores and salt for preserving the catch was £120. It was reckoned by deponents in a subsequent trial that the two venturers would have made a profit of £20 each, equivalent to a return on capital of 33%. This figure is comparable with Newfoundland voyages by

French vessels in the mid-sixteenth century, which produced a return of 27.5% (Innis 1954, 21; TNA, SP 1/99, ff. 170–8). The investment and profit had to be set against the risk of the ships foundering, being captured by pirates off the Scottish coast or being attacked by Hansa merchants (Williams 1988, 90). One means to reduce the investment was to involve a larger number of parties in the costs of equipping a ship. This was clearly the approach taken to the ships leaving Hull in the fifteenth century, and some merchants chose to diversify their risks further by investing in goods dispatched on more than one ship. An example is Nicholas Stubbs, who in AD 1461 sent cloth, barley meal, honey and iron on the *Mary* of Hull and cloth, beer, belts, rye meal, knives and iron on the *Trinity* of Hull. Both ships were bound for Iceland and returned with cargoes of stockfish obtained in exchange (Childs 1984, 32, 33, 35, 37).

The approach of Hansa merchants to the Icelandic trade has not been considered so far. Their interests were well served by the establishment of the staple at Bergen, where there was a permanent settlement of merchants and a supply of stockfish from northern Norway. The Bergen *Kontor* allowed German merchants to dominate trade in the North Atlantic (Gade 1951, 92; Nedkvitne 2014, 277–332). The nature of trade in dispersed ports in Iceland made the establishment of something approaching a trade monopoly more difficult. This may explain the ambivalent attitude of the Hansa towns to direct trade with Iceland. Lübeck and Bergen were opposed to such a move and wished that trade continue to be channelled through Bergen (DI, vol. 6, no. 363). However, within a few years ships were regularly travelling from the German ports to Iceland. The Hansa merchants did not adopt the diversified approach to obtaining fish of their English counterparts and were dependent on acquiring sufficient stockfish. However, they also developed other exports that had been partially or wholly neglected by the English, including coarse cloth (*vaðmal*), hides, eider-down and sulphur (Marcus 1980, 154). There was sharp competition for the supply of stockfish, but it was more difficult to establish in Iceland the near-monopoly that had applied in Bergen. Many ports had developed in the southwest of Iceland, at Straum, Vatnleysa, Hafnarfjörður, Keflavík, Básendar and Grindavík, and it was from these that the English ships were progressively harassed or driven out in the last years of the fifteenth and early sixteenth centuries, though they continued to trade in the Vestmannaeyjar (Þorsteinsson 1970, 246–8; Historical Manuscripts Commission 1883, 13, 70).

The final party which had influence on fishing and trading in Iceland was the Danish crown. Up to AD 1449, successive kings of Denmark struggled to maintain the Bergen staple, through which fish should be traded. Throughout the fifteenth century, they worked to limit the influence of the English in Iceland and to restrain Hanseatic power in the North Atlantic. Lacking a naval presence and any military

capability in Iceland severely limited what they might hope to achieve. The only significant measure which might be adopted to force the English crown to negotiate was to seize ships in the Øresund, a tactic used twice, in AD 1447 and in 1468 (Þorsteinsson 1957–61, 70–1). But though this reduced English activity in Iceland, it did not prevent it. By the end of the fifteenth century, Danish policy had shifted to taxing both merchant and fishing vessels (Webb 1962, 84). Whether Danish policy extended to playing off the English against the Hansa merchants is unclear, though it was perhaps hardly necessary, since the relationship between the two groups remained very poor. The regulation of Hansa merchants was a good deal easier for Danish officials in Iceland than was the management of English ships, because the English mixture of fishing and trade was difficult to regulate and tax. Trade might take place when vessels came inshore to shelter and take on firewood or water, or it might be carried out in a more systematic way. In order to prevent this more systematic trade, fishing vessels departing for Iceland in AD 1491 were ordered to take no more victuals than were required for the voyage (Davis 2004, vol. 2, no. 824). The aim was to ensure that all exported goods paid customs dues in England and to prevent opportunistic trading in Iceland, which might otherwise not be taxed.

Conclusions

Rather than examining Icelandic fishing in terms of politics, economics and practical operation (e.g. Carus-Wilson 1954; Jones 2000; Þorsteinsson 1957–61; 1970), this study has sought to examine the strategies of the participants and to consider their different aims and cultural values. It has been argued above that the Icelandic enterprise was part of what Kowaleski (2003) identified (though she does not use the term) as second-stage commercialisation. This was a step-change in the level of investment and the scale of fishing that took place around the beginning of the fifteenth century. It was marked by longer-distance voyages in search of fish, the adoption of improved methods of fish preserving and the use of larger vessels (Kowaleski 2003, 220–7; Unger 1978, 345–9). Like the first stage of commercialisation of fishing, which Barrett *et al.* (2004) have identified taking place around AD 1000, this phenomenon seems to have affected not only Britain, but also much of northwestern Europe at about the same time (see comments by Unger 1978, 348–9, 353–6).

It is important to emphasise that the adoption of capital-intensive fishing did not drive out small-scale family or group fishing enterprises, which continued to work local waters (see, for example, Sweetinburgh 2006, 96). In Iceland, the second-stage fishing vessels worked by English sailors operated concurrently (though usually farther out to sea) with the smaller, 'first-stage' vessels of the Icelanders. The Icelandic fishers were farm-based, and their enterprise required less capital investment, particularly when it was operating from a home base. Their activity was complementary to English fishing and trading practices; it did not act in rivalry, and it allowed the English to diversify risk. The two enterprises met at trading sites where the stockfish could be bulked or amassed by English and Hanseatic traders for shipping back to their home ports.

Three key points have been advanced in this paper. The first is that for the English the businesses of fishing and trade in the North Atlantic were inseparable. This does not mean that every vessel did both, but rather that it is wrong to assume that most ships were engaged exclusively in either fishing or trade. This mixed approach stands in sharp contrast to that of the Hanseatic merchants, who were solely traders. They do not seem to have undertaken any fishing themselves. The second point is that the participants in fishing operations were working according to different strategies, which become comprehensible once we understand the cultural context in which their decisions were made. The final point is that the study of the fishing by the English and an understanding of the trading operations by the Hansa has helped to clarify the emerging concept of second-stage commercialisation of the fishing industry.

Acknowledgements

This paper is part of a study of fishing and trading in the North Atlantic that is a joint project with Dr Natascha Mehler of the University of Vienna, and I am indebted to her for valued comments and assistance in many ways. Anne Drewery and Christopher Whittick kindly gave me their notes on some documents at The National Archives. I am also grateful to Maryanne Kowaleski for sending me copies of her papers.

Archival references

British Library (London), Lansdowne MSS
TNA The National Archives (Kew, London):
SC 8, Ancient Petitions
SP 1, State Papers, Henry VIII: General Series
HCA 13, High Court of Admiralty, Instance and Prize Courts: Examinations and Answers

References

Amundsen, C., S. Perdikaris, T. H. McGovern, Y. Krivogorskaya, M. Brown, K. Smiarowski, S. Storm, S. Modugno, M. Frik & M. Koczela. 2005. Fishing booths and fishing strategies in medieval Iceland: an archaeofauna from the site of Akurvík, north-west Iceland. *Environmental Archaeology* 10: 141–98.

Arnold, T. (ed.) 1885. *Symeon Monachi opera omnia. Volume 2: historia regum.* (Rolls Series). London: Longman.

Baasch, E. 1889. *Die Islandfahrt der Deutschen, namentlich der Hamburger, vom 15. bis 17. Jahrhundert.* (Forschungen zur hamburgischen Handelsgeschichte 1). Hamburg: Herold Buchhandlung.

Barrett, J. H., A. M. Locker & C. M. Roberts. 2004. 'Dark Age Economics' revisited: the English fish bone evidence AD 600–1600. *Antiquity* 78: 618–36.

Bateson, M. 1902. A London municipal collection of the reign of John. *English Historical Review* 17: 480–511.

Blackmore, L. & A. Vince. 1989. *Medieval Pottery from Southeast England Found in the Bryggen Excavations, 1955–68. Volume 2: the Bryggen pottery.* (Bryggen Papers Supplementary Series 5). Bergen: University of Bergen.

Brooke, C. N. L. & K. Keir. 1975. *London, 800–1216: the shaping of a city.* London: Secker and Warburg.

CPR Calendar of Patent Rolls 1216–25. 1901. London: HMSO.

CPR Calendar of Patent Rolls 1408–13. 1909. London: HMSO.

Carus-Wilson, E. M. 1954. *Medieval merchant venturers: collected studies.* London: Methuen.

Childs, W. R. 1984. *The customs accounts of Hull, 1453–1490.* (Yorkshire Archaeological Society Record Series 144). Leeds: Yorkshire Archaeological Society.

Childs, W. R. 1995. England's Icelandic trade in the fifteenth century: the role of the port of Hull, in P. Holm, O. U. Janzen, O. Uwe & J. Thor (ed.) *Northern Seas Yearbook, 1995*: 11–31. (Fiskeri- og søfartsmuseets studieserie 5). Esbjerg: Fiskeri- og Søfartsmuseet.

Childs, W. R. 2000. Fishing and fisheries in the Middle Ages: the eastern fisheries, in D. J. Starkey, C. Reid & N. Ashcroft (eds) *England's Sea Fisheries: the commercial sea fisheries of England and Wales since 1300*: 19–23. London: Chatham.

Cooper, E. R. 1939. The Dunwich Iceland ships. *Mariner's Mirror* 25: 170–7.

Davis, N. 2004. *Paston letters and papers of the fifteenth century. Volume 2* (Early English Text Society Supplementary Series 21). Oxford: Clarendon Press.

DN Diplomatarium Norvegicum 1847–2011. *Diplomatarium Norvegicum.* Oslo: Dokumentasjonsprosjektet. http://www.dokpro.uio.no/dipl_norv/diplom_felt.html.

Friedland, K. 1973. Der Hansische Shetlandhandel, in K. Friedland (ed.) *Stadt and Land in der Geschichte des Ostseeraums*: 66–79. Lübeck: Verlag Max Schmidt-Romhild.

Gade, J. A. 1951. *The Hanseatic Control of Norwegian Commerce During the Late Middle Ages.* Leiden: Brill.

Gairdner, J. (ed.) 1861–3. *Letters and Papers of Richard III and Henry VII.* (Rolls Series). London: Longman.

Gardiner, M. F. & N. Mehler. 2007. English and Hanseatic trading and fishing sites in medieval Iceland: report on initial fieldwork. *Germania* 85: 385–427.

Gelsinger, B. E. 1981. *Icelandic Enterprise: commerce and economy in the Middle Ages.* Columbia: University of South Carolina Press.

Gudeman, S. 2001. *The Anthropology of Economy: community, market and culture.* Oxford: Blackwell.

Guðmundsson, F. 1965. *Orkneyinga saga.* Reykjavík: Hið Islenzka Fornritafélag.

Gunnarsson, G. 1983. *Monopoly Trade and Economic Stagnation: studies in the foreign trade of Iceland 1602–1787.* Lund: Ekonomisk-historiska Föreningen i Lund.

Hall, H. & F. J. Nicholas (eds) 1929. *Select Tracts and Table Books Relating to English Weights and Measures (1100–1742).* (Camden Miscellany 15, Camden Society 3rd Series 41). London: Royal Historical Society.

Hardy, T. D. (ed.) 1833. *Rotuli litterarum clausarum in turri Londinensi asservati*. Vol. 1. London: Record Commissioners.

Hastrup, K. 1990. *Nature and Policy in Iceland 1400–1800: an anthropological analysis of history and mentality*. Oxford: Oxford University Press.

Heath, P. 1968. North Sea fishing in the fifteenth century: the Scarborough fleet. *Northern History* 3: 53–69.

Helle, K. 1967. Trade and shipping between Norway and England in the reign of Håkon Håkonsson (1217– 63). *Sjøfartshistorisk Årbok* 1967: 7–34.

Historical Manuscripts Commission. 1883. *Calendar of Manuscripts of the Marquis of Salisbury Preserved at Hatfield House*. London: Historical Manuscripts Commission.

Hofmeister, A. E. 2001. Das Schuldbuch eines Bremer Islandfahrers aus dem Jahre 1558. *Bremisches Jahrbuch* 80: 20–50.

Innis, H. A. 1954. *The Cod Fisheries: the history of an international economy*. (Revised edition). Toronto: University of Toronto Press.

Jones, E. T. 2000. England's Icelandic fishery in the early modern period, in D. J. Starkey, C. Reid & N. Ashcroft (eds) *England's Sea Fisheries: the commercial fisheries of England and Wales since 1300*: 105–10. London: Chatham.

Keyser, R. & P. A. Munch (eds) 1849. *Norges gamle love indtil 1387*. Vol. 3. Christiania: Gröndahl.

Kowaleski, M. 2003. The commercialization of the sea fisheries in medieval England and Wales. *International Journal of Maritime History* 15(2), 177–232.

Kristjánsson, L. 1982. *Íslenzkir Sjávarhættir*. Vol. 2. Rekyjavík: Bókaútgáfa Memmingarsjóds.

Krivogorskaya, Y., S. Perdikaris & T. H. McGovern. 2005. Fish bones and fishermen: the potential of zooarchaeology in the Westfjords. *Archaeologia Islandica* 4: 31–50.

Lloyd, T. H. 1982. *Alien Merchants in England in the High Middle Ages*. Brighton: Harvester Press.

Marcus, G. J. 1954. The English dogger. *Mariner's Mirror* 40: 294–6.

Marcus, G. J. 1980. *The Conquest of the North Atlantic*. Woodbridge: Boydell.

Miller, W. I. 1986. Gift, sale, payment, raid: case studies in the negotiation and classification of exchange in medieval Iceland. *Speculum* 61: 18–50.

Nedkvitne, A. 1983. Der Stukturwandel im nordeuropäischen Seehandel vom 12.–14. Jahrhundert: seine Bedeutung für die norwegischen Seehandelsstädte, in G. P. Fehring (ed.) *Seehandelszentren des nördlichen Europa: Der Strukturwandel vom 12. zum 13. Jahrhundert*. (Beiträge des Ostsee-Kolloquiums, Lübeck 1981): 261–9. (Lübecker Schriften zur Archäologie und Kulturgeschichte 7). Bonn: Dr Rudolf Habelt.

Nedkvitne, A. 2014. *The German Hansa and Bergen*. (Quellen und Darstellungen zur Hansischen Geschichte, Neue Folge 70). Vienna and Cologne: Böhlau.

Oppenheim, M. 1907. Maritime history, in W. Page (ed.) *The Victoria History of the County of Suffolk*. Vol. 2: 199–246. London: Archibald Constable.

Riley, H. T. (ed.) 1863–4. *Chronica monasterii S. Albani: Thomae Walsingham, quondam monachi S. Albani, historia anglicana*. (Rolls Series). London: Longman.

Schreiner, J. (ed.) 1941. *Hanseatene og Norge i det 16. århundre*. Oslo: J. Dybwad.

Seaver, K. 1996. *The Frozen Echo: Greenland and the exploration of North America*. Stanford: Stanford University Press.

Strachey, L. (ed.) 1767. *Rotuli parliamentorum: ut et petitiones, et placita in Parliamento*. London: Record Commissioners.

Sweetinburgh, S. 2006. Strategies of inheritance among Kentish fishing communities in the later Middle Ages. *History of the Family* 11: 93–105.

Thorsteinsson, B. 1957. Island, in G. Authén, A. E. Christensen, E. Lönnsroth, V. Niitemaa & B. Thorsteinsson (eds) *Hansestæderne og Norden*: 165–95 (Det Nordiske historikermøde i Århus 7–9. August 1957). Århus: Universitetsforlaget i Århus.

Þorsteinsson, B. 1957–61. Henry VIII and Iceland. *Saga-Books of the Viking Society* 15: 67–101.

Þorsteinsson, B. 1970. *Enska öldin í sögu Íslendinga*. Reykjavík: Mál og Menning.

Þorsteinsson, H. (ed.) 1922. *Annales Islandici Posteriorum Saeculorum: Annáler 1400–1800. Volume 1*. Reykjavík: Félagsprentsmiðan.

Tuck, A. 1972. Some evidence for Anglo–Scandinavian relations at the end of the fourteenth century. *Mediaeval Scandinavia* 5: 75–88.

Unger, R. W. 1978. The Netherlands herring fishery in the late Middle Ages: the false legend of Willem Beukels of Biervliet. *Viator* 9: 335–56.

Webb, J. 1962. *The Great Tooley of Ipswich: portrait of an early Tudor merchant*. Ipswich: Suffolk Record Society.

Williams, N. J. 1988. *The Maritime Trade of East Anglian Ports 1550–1590*. Oxford: Clarendon Press.

Zupko, R. E. 1977. *British Weights and Measures: a history from antiquity to the seventeenth century*. Madison: University of Wisconsin Press.

9

Marine Fisheries and Society in Medieval Ireland

Colin Breen

Introduction

From about AD 1000, there is evidence for an expansion and intensification of fishing activity around the coasts of the island of Ireland. The collection of shellfish, near-shore fishing and the exploitation of beached cetaceans had been an integral part of the economy of previous centuries (Edwards 1990), but from the turn of the first and second millennia, and particularly from the middle of the fourteenth century, there were increases in the nature and extent of these activities. The drivers behind this intensification were multi-faceted, probably including changing environmental conditions, technological advances in boats and shipping, an influx of people and influences from the Scandinavian regions and a general rise in demand for marine produce across society. Much of this increase in demand came from rising population levels in expanding coastal towns and settlements. Our evidence for these changes comes from a range of historical sources, from the physical traces of past fishing activity and from bone assemblages recovered from excavations. This chapter offers a historical overview of the development of fishing through the medieval period and also provides a context for the fish-bone evidence associated with these activities to be discussed in Chapter 19.

The fishery resource

The seas around Ireland contained an abundance of marine resources throughout the Middle Ages. Extensive fish stocks, in particular, proved especially attractive to merchants and fishermen from abroad. A range of factors ensured high catch yields, including the bathymetry and seabed type off the coast, which were conducive to the spawning of various species. Fish are susceptible to changes in oceanographic conditions. In Irish coastal waters, a slight rise in water temperature, for example, can lead to an increase in stocks of some species (e.g. hake, [*Merluccius merluccius*]) and a decrease in others (e.g. cod [*Gadus morhua*]) (Simpson *et al.* 2011, 1567). A significant intensification of the fishing industry in medieval Ireland, during the middle of the fourteenth century in particular, could thus relate in part to changing environmental conditions. A number of fishing grounds came to prominence, and these became the focus of the industry over the following centuries. The grounds of the southern parts of the Irish Sea and the fisheries off the southwest coast were the most profitable and accessible. The Irish Sea supported a herring and cod fishery, with prawns being widely caught between Carlingford and Dublin at a later period. The southwestern fishery was a mixed one dominated by herring and white fish (i.e. demersal fish), with hake dominating during the spring months. Off the north and northwest coasts, haddock (*Melanogrammus aeglefinus*), cod and flatfish (Pleuronectiformes) were the most commonly caught fish, while large oyster grounds were also present off Malin Head and in parts of the Mayo coast. A salmon (*Salmo salar*) fishery was also present along the north and west coasts. Larger mammals that had become beached were utilised, while other cetaceans were occasionally hunted. In the eleventh century, the Spanish geographer Al-'Udhri (AD 1002–85) recorded that the Irish hunted young whale calves with harpoons from October–January yearly (James 1978). Following capture, the whales were brought ashore before being processed and salted. Adomnán's seventh-century *Life of St Columba* also refers to the hunting of seals (Murray *et al.* 2004; Sharpe 1995, 144). A number of other writers in turn commented on the wealth of resources around the coast. In the twelfth century, Giraldus Cambrensis recorded that 'sea-fishes are found in considerable abundance on all the coasts' of the island (Wright 1905, 25). Edmund Spenser in 1596 wrote of the 'plentiful' fishing on the coast (Renwick 1934), while

Fynes Moryson stated in the early part of the seventeenth century that 'no country [is] more abundant with fish, as well as sea-fish in the frequent harbours and upon all the coasts' (Kew 1998, 114).

An emergent fishing industry

The archaeological record shows that there was an increase in marine fish consumption following the establishment of a series of Hiberno-Norse settlements along the coast at such places as Dublin, Cork, Wexford and Waterford from the tenth century AD onwards (Figure 9.1; see also Chapter 19). While fish had been consumed throughout

earlier centuries, the extent of fishing and probably fish consumption expanded under the influence of the Vikings. This is marked by an increase in the number of small finds at such places as Dublin, where weights, floats and sinkers are all indicative of marine fishing.

Bones of marine species – including hake, cod, ling (*Molva molva*), plaice (*Pleuronectes platessa*) and herring – have been recovered from excavations at both Dublin and Wexford (Geraghty 1996, 55; MacCarthy 1998, 61), reflecting the role marine fish played both in the diet of the inhabitants of the towns and, potentially, as an export commodity. These species are also indicative of the expansion of the offshore fishing industry. The technological advances in boat building

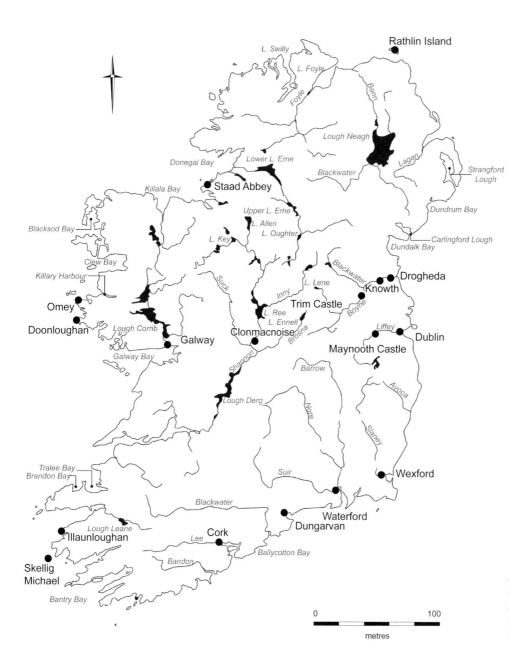

Figure 9.1. Map of Ireland showing the principal coastal locations mentioned in the text and the inland riverine systems (Drawing: Colin Breen).

that the Vikings brought with them enabled a rapid expansion of this form of activity. While it is clear that a small amount of deep-water fishing had been undertaken by the Irish prior to Norse settlement, it was only with the arrival of these new boat forms that these undertakings became more expansive and ambitious. The twelfth-century text *Acallam na Senórach* makes reference to deep-sea fishing off Dursey and Bere islands as an example of Irish fishing, but this was limited in nature and largely associated with subsistence fishing. From the limited available evidence, it appears that many of the vessels used by the Irish prior to the tenth century consisted of light, skin-covered craft that were not overly suited to offshore activity. The advent of large, wooden-built ocean-going craft and dedicated fishing boats greatly added to the technological development of the industry. The arrival of fishers from the northern Scandinavian regions led to commercial-level fishing and really introduced fish as a commodity.

With the arrival of the Anglo-Normans in Ireland in AD 1169, the fishing industry was further consolidated and brought formally within the remit of the Crown. Ports, waterfront facilities and mechanisms for operating the industry were formalised, and entrepreneurial activity associated with fishing expanded rapidly. Specific examples of this process include the development of quays at Drogheda in the thirteenth century and the construction of a stone-built waterfront at Dublin that superseded a series of earlier wooden fences and quays along the River Liffey. These developments paralleled changes in ship technology as vessels got bigger to accommodate expanding cargo sizes. In particular, vessels with larger, rounded hulls were developed to facilitate increased cargo sizes. The formalisation and expansion of these developments can be seen in the export of large amounts of fish from Ireland to Bristol and Chester, first documented in the late fourteenth and fifteenth centuries respectively (Childs and O'Neill 1993; Mac Laughlin 2010). The industry and its catches were formalised in line with centralised control of both government and economic practice. Prior to the introduction of these new controls, fishing had essentially expanded in a *laissez-faire* manner, with individual merchants and groups leading change. New forms of bureaucratic control changed this and led to an increased commodification of the resource. Exports were initially dominated by the English market, but rapidly grew to incorporate the broader western European region. Shellfish were also a common dietary commodity at this date, but they do not appear to have been harvested commercially for export. Instead they were consumed in considerable quantities on domestic sites along the coast. Excavations at Clough Castle in County Down, for example, have shown that the occupants consumed edible cockles, great scallops, periwinkles, mussels and oysters (Waterman 1954, 154). Mussels made up 55% of the molluscs recovered from the earlier thirteenth-century

levels, while oysters constituted nearly 60% of the shellfish assemblage from the later thirteenth- and fourteenth-century levels. Elsewhere in the county, at Rathmullan, the twelfth- and early thirteenth-century levels produced a large amount of molluscs, dominated by mussels (Lynn 1981–2, 159). Later thirteenth- and fourteenth-century levels produced even more substantial amounts of shells, especially of limpets and whelks. Shellfish, by and large, were confined to the home market until expanding European demand in the post-medieval period.

Across the eastern and southern coasts of the island a formalised system of ports now emerged, with each section of coastline having a lead or head port supported by a series of lesser or member ports and smaller fisher settlements. Drogheda, Dublin, Wexford and Cork emerged as the major fishery ports, but a myriad of smaller settlements involved in the fishing industry dotted the coastline (Figure 9.2). One such site is recorded in the *Cloyne Pipe Roll, c.* AD 1365: 10 fishermen are listed as holding cottages at Ballycotton in East Cork (MacCotter and Nicholls 1996). Geoffrey Cod held a cottage there for 12d yearly, eight others held theirs for 8d yearly, while John O Kynel held one for 3d yearly. The Pipe Roll adds that:

> all these are fishermen must serve the lord with fish, and what is worth 12d the lord can have for 8d. And in the season when they take langes (ling), the lord can have one for 2d, a mulewell (cod) for 1.5d, three haddocks for 1d. And the lord must not take more than is needed for himself and his house. And similarly, the said fishermen must make the lord's hay, and turn and gather (it), and weed the lord's corn, and make the lord's turf, or each of them gives the lord 3d, and for every cow of theirs 3d yearly, and six sheep are counted as a cow. (MacCotter and Nicholls 1996, 15)

This form of arrangement was common to most sites. It is clear that these settlements were not solely concerned with fishing, but also engaged with farming and other subsistence practices. Other sites of this type included Ballyhack in Waterford Harbour, which was listed in AD 1541 as containing nine tenements and eight fishermen's cottages (Went 1946, 48). The inhabitants were later licensed to export fish to France.

While these settlements were engaged in fishing from boats in the inshore waters, other forms of fishing practice also existed. Archaeological evidence is uncovering an emerging picture of a dynamic medieval near-shore fishery, which is exemplified by the discovery of clusters of fish traps at various locations around the coast. While weirs and traps were used and recorded in the riverine systems across the country, traps built in estuarine and coastal intertidal areas are increasingly common discoveries at this time. These medieval fishing engines consisted of wooden fences or stone walls constructed across the shoreline and were designed to trap fish in nets or baskets during flooding or receding tides. These traps were especially effective in shallow tidal areas, where fish follow the tidal movement and often feed off the nutrients introduced into the marine environment from river sources. There was no single design associated with these features; rather, their construction was adapted to the local morphology and environmental conditions associated with each location they were placed at. A cluster of traps has been identified at Greyabbey Bay in Strangford Lough on Down's eastern coast. The lough would have contained a range of fish species, including salmon, sea trout (*Salmo trutta*), plaice, flounder (*Platichthys flesus*), mackerel (*Scomber scombrus*), cod, grey mullet (Mugilidae) and skate (*Dipturus batis*) with large numbers of eels (*Anguilla anguilla*). Here, evidence for early medieval fish traps has been found at Chapel Island. A large wooden trap produced a calibrated radiocarbon date of AD 711–889 at two sigma (UB-3034, 1213+30 BP) (McErlean *et al.* 2002, 468). Three further V-shaped traps in the bay itself have produced dates ranging from the eleventh–late thirteenth centuries. Stone-built traps of varying morphologies have also been located in the bay, some of which can clearly be

seen to be stratigraphically later than the wooden examples (Figure 9.3). This was evidently a large inter-tidal fishing complex throughout the medieval period. A number of the traps were likely to have been associated with the Cistercian community that was established in the bay in AD 1193, because ecclesiastical communities of this kind were heavily engaged in landscape management and the development of their estates. The utilisation of their fishery resources would have been an integral part of this process. Their involvement at this scale is also reflective of the role religious obligations had in medieval diet, with red meat prohibited on many holy days and other occasions. Meat was also less accessible to many social categories of past societies.

Other fisheries and individual fish traps are known along the Down and Antrim coasts. A large curvilinear stone trap at Swinley Bay, Carnalea, and two stone-built traps on the foreshore in Dundrum Bay at Newcastle have been recorded, while other traps have been noted on the southern shores of Belfast Lough. The taxation of Pope Nicholas lists a church, valued at 2 marks, in the townland of Rossglass on the shores of north Dundrum Bay in AD 1306. This church was mentioned in a document dated 4 March 1305, when Walter de la Hay, the King's Escheator, gives an account of rents paid into the Irish Exchequer. Included is an account of '£3 16s 6d of the rents of farms, of a mill, of prizes, services, fisheries, and of the perquisities of the court of Rossglasse for the foresaid term'. The remains of a stone fish trap measuring more than 50 m in length, are visible on the foreshore below the church (Figure 9.4). It is hook-shaped, with a lead arm of roughly built stone wall leading to the low-water mark. It thus works on the ebb tide. Common features of many of these traps are their positioning close to freshwater streams that flowed into the sea and their proximity to ecclesiastical sites.

By the fifteenth century, fish had emerged as Ireland's main export commodity, with herring, in particular, being the

Figure 9.2. Late sixteenth-century Pacata Hibernia *map of the port and city of Cork.*

Figure 9.3. Aerial photograph of the large stone fish trap at Ogibly Island in Strangford Lough (Photo: Northern Ireland Environment Agency).

dominant species. The majority of vessels involved in these fisheries came from Britain and western Europe, but Irish vessels were also involved. For example, at certain periods between AD 1315 and AD 1465, nearly 5% of all fishing boats landing cargo at Exeter were Irish (Kowaleski 2000). At Bristol, in AD 1437/8, 28% of all fish ships were Irish, while the fish trade accounted for 70–90% of all Bristol's imports from Ireland in the second half of the fifteenth century (Kowaleski 2000, 434–5). Despite these substantial figures relating to Irish fishing, English boats initially dominated the fishery. Later, at the close of the fifteenth century, large numbers of European boats arrived. The push factors behind this influx of vessels to Irish waters were complex. Following the ravages of the fourteenth century, including the Black Death, the population again began to expand, and there was an increased demand for fish across southern England and parts of western Europe. It is possible that yields in some English grounds, especially on the east coast, were declining due to a degree of overfishing, and/ or that they were proving insufficient to supply increasing

demand. Many vessels moved into different grounds, where stocks were more diverse and plentiful. This movement also corresponded with the gradual westwards expansion of European fisheries in general, ultimately leading to the exploitation of the Newfoundland fisheries in the late fifteenth and sixteenth centuries (Candow 2009; Kowaleski 2000; Pope 2004).

The extent to which foreign vessels worked Irish waters began to raise serious concerns in the English administration about the loss of revenue this entailed through customs and sales. The Crown was also concerned about the extent to which these fleets supported the Gaelic lordships along the western seaboard that lay beyond the jurisdiction of either Dublin or London. A statute from AD 1465 refers to 'divers vessels of other lands … going to fish among the King's Irish enemies … whereby the said enemies are much advanced and strengthened, as well in victuals, harness and armour … also great tributes of money are given by every said vessels to the said enemies' (Barry 1914, 353). A taxation was subsequently introduced on all vessels involved in the herring fishery in the 1470s (Barry 1914, 389). But the problems remained, and in AD 1515 complaints were submitted that Breton fishermen were overfishing and exporting so many salmon, herring, ling and hake that there would be little left for English vessels (O'Neill 1987, 32). Large ventures appear to have been commonplace, as the treasurer of war in Ireland wrote in AD 1535 about a fleet of 600 English vessels fishing herring at Carlingford (Cal Carew MSS 1515–74, 85). The Iberian influence continued to increase, leading Lord Deputy St Leger to comment in AD 1543 that 'from Limerick to Cork the Spanish and Bretons have the trade as well as the fishing there' and also supplied the inhabitants with salt (Longfield 1929, 58). By AD 1572, it was reported that more than 600 Spanish vessels were fishing off the Irish coast.

Fisheries and Gaelic Ireland

For much of the medieval period, especially from the fourteenth century onwards, many parts of Ireland lay outside the control of the English administration in Dublin or London. These areas were mostly concentrated along the western and northern coasts and consisted of individual Gaelic lordships. These autonomous entities maintained diverse economies centred on agricultural activity, but fishing or fishing-related activities played a central part in their livelihoods. It is important to point out that few, if any, of these groups had extensive fishing fleets themselves. Instead, they put a loose infrastructure in place to facilitate fishing activity that was mostly undertaken by fleets drawn from the western seaboard of mainland Europe. This included the siting of their principal places of residence – mostly tower houses with associated settlement clusters – at harbours,

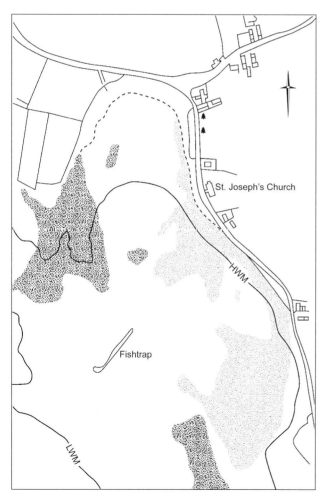

Figure 9.4. The location of the Rossglass fish trap on the foreshore below the site of a medieval church (Drawing: Colin Breen).

adjacent to anchorages or in view of the primary fishing grounds. The visiting fleets not only paid these groups taxes and levies to fish in their waters, but also would have been dependant on the inhabitants ashore to supply them with drinking water, foodstuffs, various supplies, repair facilities and indeed security. Enforcement of the lord's control was varied. Charles Smith in his *History of Cork* wrote of a Spanish fishing vessel seized by an English ship near Dursey Island in AD 1531 (Smith 1750, 36). Dermot O'Sullivan commandeered a small fleet of vessels and brought them to Berehaven, where he hanged the English Captain and released the others. At a less aggressive level, control was probably primarily enforced through the denial of much-needed supplies, such as fresh water, if the levies were not complied with. The cultural and economic interplay between the Gaelic inhabitants and visiting fishermen is evidenced in the material culture and traditions of the Gaelic groupings, who, rather than living isolated on the margins or periphery of European society, were an integral part of its economic and social structures.

A number of clan groups in particular rose to prominence due to their involvement with the industry. Most became involved intensively with fishing from the fifteenth century onwards, following the large-scale arrival of European fishing fleets in Irish waters. Already by AD 1449/50, the Crown attempted to limit commercial fishing off Baltimore in the territory of the O'Driscolls in west Cork (Kowaleski 2000). By the sixteenth century, the O'Donnells of Donegal were known as 'Lords of the Fish' (O'Neill 1987, 47). In AD 1561, the O'Donnells were referred to as exchanging 'fish always with foreign merchants for wine' (Cal Carew MSS 1515–74, 308), indicating that they were engaged in fishing themselves. On parts of the Galway and Mayo coasts the Uí Fhlaithbheartaigh and the Uí Mhaille extracted dues from visiting fishing fleets and traders (Naessens 2007, 220). Elsewhere, John Corbine, while passing through the territory of McCarthy, Earl of Glencarr, in the southwest of the country in AD 1569, remarked that 'every year two hundred sail fish there and carry away 2000 beef, hides and tallow and that no rent or customs are paid to the Queen's coffers' (Cal SP Ire 1509–73, 405). In AD 1572, it was reported that more than 600 Spanish ships and barks were fishing off Ireland, as well as others concentrated on the west Cork and Kerry coasts (Cal Carew MSS 1515–74, 422–3). English administrators complained in AD 1579 that every year 50 English ships came to Mayo, where 'they had been compelled to pay a great tribute to the O'Malleys' (Chambers 1998, 18). At the same time, MacFineen Duff of Ardee in Kenmare Bay received £300 a year from Spaniards wanting to fish there. By AD 1580, O'Sullivan Beare in west Cork was engaged in extracting heavy duties from visiting fishing vessels. His relative Domhnall O'Sullivan Bere was to write to Philip II of Spain in 1605, while he was in exile, 'that each year at least five hundred fishing boats came to his ports' (O'Neill 1987, 135). O'Sullivan was most likely extracting similar levies to those of the neighbouring O'Driscolls, who charged every ship or boat that fished between Fastnet and Toe Head a barrel of flour, a barrel of salt, a hogshead of beer and a dish of fish three times a week, as well as a sum of 19s 2d paid directly to the lord (Butler 1925, 167; Went 1949, 17–24).

An archaeological research theme that emerges from the extent of this activity is the nature of the interaction between fishers and the Gaelic communities. The repeated visits over many years of hundreds, if not thousands, of fishermen to the western seaboard must have led to the development of a complex series of relationships structured on a social and economic level. Groups of merchants and fishermen settled at large ports, such as Galway, while numerous other small coastal settlements developed. The small settlement cluster at Ballynacallagh on Dursey Island is interesting in this context (Breen 2005). It consisted of a group of seven houses positioned close to the landing place on the island and adjacent to the fishing grounds off Bantry Bay. Excavation has shown that they were roofed using Spanish tiles, while the material culture recovered from one of the houses consisted solely of Spanish ceramics similar to Cotrel or Seville coarse wares. It is likely then that this was a seasonal fishing settlement occupied by Spanish fishers in the late sixteenth century (Figure 9.5).

Foreshore fisheries were also a key feature of the western seaboard. Aidan O'Sullivan (2001) has discovered and recorded a large number of fish traps on the Deel and Fergus estuaries in south county Clare. There was a long sequence of trap construction on these mudflats from the fifth century onwards, with an intensification of activity

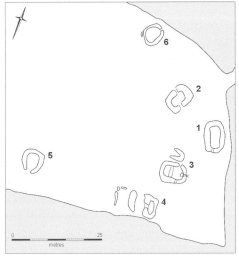

Figure 9.5. Plan of the settlement cluster at Ballynacallagh, Dursey Island (Drawing: Colin Breen).

occurring following the establishment of Norman manors at Askeaton and Bunratty and their involvement with the communities of the neighbouring Gaelic territories of the MacNamaras and the O'Briens. A number of V-shaped wooden traps on the Deel estuary have a date range from the eleventh to the fourteenth centuries. Similarly, at Bunratty two small ebb-tide traps have been dated to between the eleventh and thirteenth centuries. More recently, at Boarland Rock on the Fergus estuary, a number of large traps have been dated to the early to mid-fifteenth century. These post-and-wattle fence traps were in operation over a period of years. They may have been built and used by the monks of the Augustinian Abbey on Canon Island, *c.* 3 km farther to the south, or by the occupants of the fifteenth-century tower houses on the banks of the estuary.

Late medieval decline

By the end of the sixteenth century, the fishing industry had reached a hiatus and entered a period of relative stagnation. With the onset of the Nine Years' War in the 1590s and the subsequent collapse of Gaelic power in the early part of the seventeenth century, together with the extensive presence of English settlers and military forces across the country, many of the foreign fleets withdrew from Irish waters. New coastal fortifications, new settlements and expanded naval patrols all combined to essentially force the foreign fleets out and develop an effective monopoly for English and Scottish vessels in Irish waters. This withdrawal may also have corresponded with a natural, and perhaps anthropogenically influenced, decline in stocks, during which a number of fishing towns experienced significant deterioration. Ardglass on the County Down coast is a pertinent example: having experienced a boom during the fifteenth and sixteenth centuries, it was largely abandoned by the seventeenth century. It was not to experience a reversal of its fortunes until the eighteenth century, when fishing rebounded. The major ports around the coast adapted quickly to the decline, integrating into the increasingly globalised world and the new economic circumstances of the post-medieval period. New commodities and structures of trade were introduced, and Ireland's role in the European fishing markets decreased.

References

Barry, H. F. 1914. *Statute Rolls of the Parliament in Ireland First to the Twelfth years of King Edward IV*. Dublin: HMSO.

Breen, C. 2005. *The Gaelic Lordship of the O'Sullivan Beare: a landscape cultural history*. Dublin: Four Courts Press.

Butler, W. F. T. 1925. *Gleanings from Irish History*. London: Longman, Green.

Cal Carew MSS Calendar of the Carew manuscripts. 1867–8. London: Longmans.

Cal SP Ire Calendar of State Papers Relating to Ireland, 1509–1670. 1860–77. Vols 1–3. London: Longman

Candow, J. E. 2009. Migrants and residents: the interplay between European and domestic fisheries in northeast North America, 1502–1854, in D. J. Starkey, J. T. Thór & I. Heidbrink (eds) *A History of the North Atlantic Fisheries. Volume 1: from early times to the mid-nineteenth century*: 416–52. Bremen: H. M. Hauschild.

Chambers, A. 1998. *Granuaile: the life and times of Grace O' Malley c. 1530–1603*. Dublin: Wolfhound Press.

Childs, W. & T. O'Neill. 1993. Overseas trade, in A. Cosgrove (ed.) *A New History of Ireland. Volume 2: medieval Ireland 1169–1534*: 492–524. Oxford: Clarendon Press.

Edwards, N. 1990. *The Archaeology of Early Medieval Ireland*. London: Batsford.

Geraghty, S. 1996. *Viking Dublin: botanical evidence from Fishamble Street*. Dublin: Royal Irish Academy.

James, D. 1978. *Two medieval Arabic accounts of Ireland. Journal of the Royal Society of Antiquaries of Ireland* 108: 5–9.

Kew, G. 1998. *The Irish Sections of Fynes Moryson's Unpublished Itinerary*. Dublin: Irish Manuscripts Commission.

Kowaleski, M. 2000. The expansion of the south-western fisheries in late medieval England. *Economic History Review* 53: 429–54.

Longfield, A. K. 1929. *Anglo–Irish Trade in the Sixteenth Century*. London: Routledge.

Lynn, C. J. 1981–2. The excavation of Rathmullan, a raised rath and motte in County Down. *Ulster Journal of Archaeology* 44/45: 65–171.

McCarthy, M. 1998. Archaeozoological studies and early medieval Munster, in M. A. Monk & J. Sheehan (eds) *Early Medieval Munster*: 59–64. Cork: Cork University Press.

MacCotter, P. & K. W. Nicholls (eds) 1996. *The Cloyne Pipe Roll*. Cork: Cloyne Literary and Historical Society.

Mac Laughlin, J. 2010. *Troubled Waters: a social and cultural history of Ireland's sea fisheries*. Dublin: Four Courts Press.

McErlean, T., R. McConkey & W. Forsythe. 2002. *Strangford Lough: an archaeological survey of the maritime cultural landscape*. Belfast: Blackstaff Press.

Murray, E., F. McCormick & G. Plunkett. 2004. The food economies of Atlantic island monasteries: the documentary and archaeo-environmental evidence. *Environmental Archaeology* 9: 179–88.

Naessens, P. 2007. Gaelic lords of the sea: the coastal tower houses of south Connemara, in L. Doran & J. Lyttleton (eds) *Lordship in Medieval Ireland*: 217–35. Dublin: Four Courts Press.

O'Neill, T. 1987. *Merchants and Mariners in Medieval Ireland*. Dublin: Irish Academic Press.

O'Sullivan, A. 2001. *Foragers, Farmers and Fishers in a Coastal Landscape*. Dublin: Royal Irish Academy.

Pope, P. E. 2004. *Fish into Wine: the Newfoundland plantation in the seventeenth century*. Chapel Hill: University of North Carolina Press.

Renwick, W. L. (ed.) 1934. *A View of the Present State of Ireland by Edmund Spenser*. London: E. Partridge.

Sharpe, R. (ed.). 1995. *Life of St Columba*. London: Penguin.

Simpson, S. D., S. Jennings, M. P. Johnson, J. L. Blanchard, P. J. Schön, D. W. Sims & M. J. Genner. 2011. Continental shelf–wide response of a fish assemblage to rapid warming of the sea. *Current Biology* 21: 1565–70.

Smith, C. 1750. *History of Cork*. Dublin: Guy.

Waterman, D. M. 1954. Excavations at Clough Castle. *Ulster Journal of Archaeology* 27: 103–63.

Went, A. E. J. 1946. The Irish hake fishery, 1504–1824. *Journal of the Cork Historical & Archaeological Society* 51: 41–51.

Went, A. E. 1949. Foreign fishing fleets along the Irish coasts. *Journal of the Cork Historical & Archaeological Society* 54: 17–24.

Wright, T. 1905. *The Historical Works of Giraldus Cambrensis*. Toronto: Dent.

The Decline in the Consumption of Stored Cod and Herring in Post-medieval and Early Industrialised England: A Change in Food Culture

Alison Locker

This chapter looks beyond the peak period of stored-fish consumption in medieval England and seeks to explain the pattern of decline in the popularity of dried and salted cod (*Gadus morhua*) and salted herring (*Clupea harengus*) from the beginning of the sixteenth to the end of the nineteenth centuries. It begins by placing these products in their historical and culinary contexts. Influential changes will then be examined by century in order to assess contemporaneous social and technological factors.

Prior to the secularising effects of the Reformation (AD 1534), the Christian church calendar decreed abstinence from meat for all persons during nearly half the year (Wilson 1973, 31), including two significant fast periods, Advent and Lent, as well as fixed weekdays. Stored cod and herring were important commodities in this pre-industrial society, where contemporary transportation methods restricted the inland distribution of fresh marine fish to a limited, wealthy sector of society. The revolution in fishing from fleeting to steam trawling and the advent of railway transport and preservation by icing and canning, which made fresh fish more widely accessible, lay far ahead in the nineteenth century. Yet well before then, heavy cures for stored cod and herring found an increasingly polarised market.

Prior to the Reformation, the church dispensed food alms and helped care for the poor and sick as part of its ethos of Christian duty. Some wealthy secular households also courted heavenly dispensations through contributions to hospitals, almshouses and poor relief, which often took the form of food, including stored fish. After the Reformation, the state became increasingly responsible for the provisioning of nationalised welfare and such institutions as hospitals, prisons and workhouses, as the concept of

support through taxation evolved. The establishment of a standing navy by Henry VIII necessitated a standardised approach to victualling, though a fixed daily ration for both sailors and soldiers had long been in place. Daily food allowances were costed and implemented in these new institutions, some more quickly than others, as will be shown. Although stored fish had long been issued on three days a week in military rations, this food did not find favour in other post-Reformation contexts despite, or because of, having been a staple in the calendar of the Roman Catholic Church.

The simplest curing methods are 'hard', meaning based on drying, salting and smoking. Air drying is the most basic method for storing animal protein, in which longevity is dependent on a low fat content and low air humidity during drying. Cod is ideal in this respect because a fresh fillet has only 0.3% fat (Cutting 1962, 164). When air dried to a 15% moisture content, stockfish has a long storage life, potentially seven years (Wubs-Mrozewicz 2009, 188). The fish had to be rehydrated prior to being consumed. The stockfish hammer was used to break up the fibres before soaking, and this became an automated process with the development of 'stockfish mills' in southern Germany in the early sixteenth century, where, as a result, harder Icelandic stockfish became preferred over those from Bergen. However, despite the large number of watermills in England, there is, as yet, no evidence to support a similar development, an area which may repay further documentary study (Wubs-Mrozewicz 2009, 197). Salted and dried cod, with a moisture content of 40% (Davidson 1999, 688), requires, in addition to reconstitution by soaking, several changes of water to reduce the salt content. Stockfish could be transported in bundles, while salt cod was packed in barrels. Both were suitable for slow transport and long storage.

Herring are much oilier fish, averaging 15% fat content. They soon become rancid and are, therefore, unsuitable for long-term storage by drying alone. Combinations of drying, salting and smoking were practised, producing the salted 'whites' and hard salted and smoked 'reds'. Being high in calories and fatty, herring provides a feeling of satisfaction. It was the fish equivalent of bacon, adding flavour to such staples as bread and potatoes.

Stored cod and herring shared similar processes of preservation, but in the kitchen both stockfish and salt cod required more preparation. Salt cod could, for example, be simmered with vegetables such as parsnips in half milk and half water and served with a yellow egg sauce (Hartley 1979, 240). Stored cod was typically more expensive than herring and more suited to an equipped kitchen, where the cook had time and facilities to create a complete dish. Herring was more of a convenience food, at least in post-medieval contexts (cf. Chapter 2). Although herring was also preserved with a high level of salt, cooking could be more elementary, by rinsing and heating. 'Reds' were also edible raw from October to December (Davidson 1979, 453). Today we have lost the taste for such salty foods, which survive only for a niche market. Salt cod, for example, is still produced in the north of Scotland at Aberdeen and in Orkney (Mason and Brown 1999, 74). 'Red' herring are still cured at Lowestoft and Yarmouth, but today only a small number are dry salted and cold smoked for the full four to six weeks. A lighter cure is more popular – brined fish are cold smoked for three to seven days. 'White' herring (salted and barrelled) are no longer produced, and even the milder nineteenth-century cures (bloaters and kippers) have a very small market share in fish sales today.

The original highly salted cod and herring belong to a long tradition of strong-tasting foods preserved with honey, sugar, salt and/or vinegar, produced at a time when such flavours were common. They fall into Capatti's (1999, 494) 'industrial' category, being made on a large scale with repeatable quality – despite pre-dating the late eighteenth-century beginnings of England's Industrial Revolution. These fish retained a market sector at a period when other 'medieval' elements of the diet were in decline, but their long past was also their weakness. In post-medieval England they proved to be an unhappy reminder of abstinence in an increasingly secular society. Flandrin (1999, 349) views the Reformation as forming an end to the medieval regime, where diversification of cuisine on national lines began. In England national cuisine was to be characterised by simple dishes where meat (particularly beef), as the main ingredient, was the focus. This plain cooking is thought to have been fostered in the country by gentry who spent time at their prestigious estates as well as their London houses. Elaborate

French cuisine with complex sauces also influenced high society, but neither favoured the inclusion of salt fish, which became increasingly relegated to an informal, 'at home' dish.

The sixteenth century

At the beginning of the sixteenth century, stored fish was still very much part of the church calendar. The clergy still observed long periods of self-denial, exemplified by the monks of St Swithun's Priory in Winchester, Hampshire. The Compotus Rolls for AD 1515 include stockfish, 'dri milwelle', 'milewelle', 'drylynge' and 'haburden' – all forms of dried cod (some with salt, some without) (Kitchin 1892). They were regularly eaten on fast days together with 'red' and 'white' herring. Preserved fish was eaten more often than fresh fish, although the latter was more frequent on the abbot's table. Winchester received a lot of preserved fish from Southampton. The port books for AD 1527–8 show much herring and two instances of Newfoundland cod (Stevens and Olding 1985). This is an early record for the import of Newfoundland cod to England, the earliest known being from Bristol in 1502 (Pope 2004, 15).

Even among the elite, stored fish was still commonly eaten. In AD 1548 Sir William Petrie (minister to Henry VIII and secretary to Elizabeth I) laid in stockfish, haberdine (a type of salted cod), ling (*Molva molva*) and herring (*Clupea harengus*) sourced from London, Stourbridge Fair (Cambridge) and the ports of Kings Lynn and Harwich for his house at Ingatestone, Essex (Emmison 1961, 139). Stocks were to be ample for emergencies, and account books for AD 1551–9 show that salt fish was still eaten on Fridays and Saturdays. Fresh marine fish was also bought, and the estate had fishponds ensuring a supply of freshwater fish, as did many other noble houses (Emmison 1961, 139, 308). However, fish from estate ponds do not usually show up in household accounting, only in records of pond stocking and maintenance. The accounts of the Willoughbys of Middleton, Warwickshire, show that salted white-fleshed fish (ling, cod and pollock [*Pollachius pollachius*]) was eaten throughout the year on fast days and that herring (both 'red' and 'white') was consumed in Lent. The amount of salted white fish purchased seems to have remained the same throughout the sixteenth century, though that of herring declined (Dawson 2008a, 24). A similar trend has also been detected at Westminster Abbey (Harvey 1993, 49) and for three noble houses of the thirteenth, fifteenth and sixteenth centuries (Woolgar 1999, 113). Dawson (2008b, 90) has suggested that since 70–80% of all fish purchased for Middleton was preserved, when the Willoughbys were away most of their servants ate only stored fish in Lent. The decline in the amount of herring consumed by the elite represents a change in status. Having been ubiquitous in the medieval period (see Chapter 3), it later became regarded as inferior to stored white-fleshed fish. Dawson (2008a,

25) has suggested that any shortfall may have promoted an increased consumption of dairy foods.

After the dissolution of the monasteries the loss of church alms, which fed many poor, led to a compulsory poor-tax for London and five other major towns. This tax, which was used for poor relief, was extended nationally by AD 1572 (Higginbotham 2008, 11). Some of the poor were housed in 'poorhouses'. Houses of correction were also opened for the able-bodied who refused to work. Meat rations at Bury in Suffolk in AD 1588 were a quarter pound of meat a day or, on fish days, one good (fresh?) herring, or two 'white' or 'red' herring (Higginbotham 2008, 13). This is one of the few early references to stored fish in such institutions, where stored cod seems not to have been served. Royal hospitals, such as St Bartholomew's and St Thomas' in London (both founded in the twelfth century), were gradually reopened after the Reformation and funded by parish collections. Unfortunately there is no evidence for their dietaries at this time.

The Tudor soldier's ration allowance was six pence a day. A victualler's contract of AD 1575 lists two pounds of beef or mutton on four meat days and a quarter of a stored cod or ling, or seven to eight 'red' or 'white' herring on fish days – bulked up with bread, a little butter or cheese and beer. This generous allowance applied to soldiers on the borders and may have also supported their families (Fussell 1949, 34). In AD 1586, Chester merchants purchased 200,000 Newfoundland cod at 20 shillings per 100 for the army in Ireland (Cutting 1956, 132). However, the target market for English-caught dried and salted Newfoundland cod was primarily Catholic Europe, especially Spain, where the English style of split, lightly salted and dried cure was popular (Janzen 2013, 5). The home demand for stored cod at this time was mostly met by the well-established Icelandic fishery, mainly serviced by east coast ports.

At sea, cod bones from the wreck of the *Mary Rose* (Hamilton-Dyer 1995; Hutchinson *et al.* 2015), which sank fully laden in AD 1545, represent part of the ship's stores. Stored fish was part of the seaman's ration three days a week. This was usually cod, variously described as stockfish, haberdine and North Sea cod. Whether the stockfish was always the authentic air-dried northern cod is debatable, as the term was sometimes used for dried salt cod. The light weight of stockfish would have been advantageous on board ship. Salted and pickled herring were also used in rations, but these were heavier. Dried or salted fish would have been eaten as a stew with bread or biscuit, the latter being up to 70% of the ration in the Middle Ages (Rodger 1997, 136). By the end of Henry VIII's reign, the supply of victuals was regularised, with the first 'General Surveyor of the Victuals for the Seas' appointed in AD 1550. This appointment recognised the importance of victualling for a standing fleet, though quantity and quality continued to be a problem. On the three fish days at sea in AD 1570, half a stockfish or four

herring were standard daily issue (Hattendorf *et al.* 1993, 102). At this time English warships could only store victuals for three or four months, a major weakness according to Rodger (1997, 236), who considers that the single factor allowing eighteenth-century ships to make longer voyages was a 'steady unspectacular improvement over 300 years in the quality of victualling'. One feature of this improvement was the gradual exclusion of stored fish from the ration.

The continuing regular consumption of stored fish at a domestic level was further supported by the introduction of secular fish days three times a week, plus Ember days and Lent, in AD 1563. Non-observance without an exempting license could lead to a fine or three months in prison (Spencer 2004, 119). The purpose of continuing fish days was to safeguard the fishing industry, which was itself a source of naval recruitment. Protectionism, banning the import of cod and ling, further encouraged home fisheries (Spencer 2004, 114).

The evidence for stored fish suggests that little had changed in the sixteenth century, but in other areas the culinary landscape was evolving. Thirsk (2007, 9), in a study of food from AD 1500 to 1760, considers that prior to the agricultural and industrial revolutions English food was less monotonous than often described. Imported fruits and spices were commonplace, as were a wide variety of seasonal vegetables, varying regionally and socially. The impact of printing and the circulation of books on food, cuisine and social etiquette, both at home and abroad, spread new ideas and food fashions. However, in food preservation, technology remained static. It continued to be based around drying, salting and smoking.

The seventeenth century

Grain shortages occurred in the latter half of the sixteenth century, and in the early seventeenth century serious harvest failures affected supplies of bread and beer, both staples of the poor. Higher up the social scale rising prices had less impact, and Thirsk (2007) highlights the interest in fruit and vegetables from the beginning of the century. Imported varieties were increasingly used as signs of wealth in place of ostentatious displays of meat and fish. There was a growing interest in different diets as travel brought new food experiences and more diversity to English high tables, which filtered down society as merchant and professional middle classes sought to reflect their status (Thirsk 2007, 66).

In the early years of the century, a medieval style prevailed, shown by fish-day purchases for the Star Chamber, the Privy Council of James VI and I in AD 1612. Mostly fresh fish was bought, including freshwater species, but some stored fish (e.g. 'old linges') was also purchased (Drummond and Wilbrahim 1994, 109). Later changes are reflected in the cookery book of La Varenne (chef to the Marquis d'Uxelles), translated into English in AD 1653, which became very popular and influential in England. A showcase for advances in French cooking (with Italian influence), it retained few medieval influences and included New World ingredients. There are many recipes for fish, but hardly any for stored fish. The few examples include Newfoundland cod desalted and served with parsley sauce, smoked herring soaked in milk and roasted, and salt herring desalted and roasted with mustard and peas (Scully 2006, 279). La Varenne also refers to salt salmon (*Salmo salar*) and salt hake (*Merluccius merluccius*), but with few recipes. It would seem that these traditional cures had become less popular among the elite.

The inclusion of Newfoundland cod in such an influential book highlights the importance of this fishery at this time, one in which French fishermen played an important role. It was also a valuable commodity for English merchant venturers, who sent goods out to Newfoundland, where they bought fish to sell in Spain, Portugal and Italy. There, they bought wine and other goods to bring back to England – a three-way trade. The letters of John Paige, a London merchant, in the mid-seventeenth century are much concerned with the price of stored cod in Newfoundland and their sale in Bilbao (Steckley 1984).

Among the poor the place of stored fish seems to have been stable in the early part of the century. Herring was cheap. Samuel Hartlib's dietary of AD 1640 for London's poor children in residential care included herring (or another fish) four times a week (Thirsk 2007, 121). As herring catches are seasonal, they must have been salted for part of the year. By AD 1696, however, at St Cross Almshouses in Winchester, Hampshire, 9 d was paid in lieu of the herring formerly served to the poor during Lent in the Hundred Mennes Hall. By this date it may be that only the religious brothers still ate 'greenfish' and herring (Hopewell 1995, 94).

The Poor Relief Act of AD 1601 was the foundation of future poor laws, initially assisting with housing, food and work. Only paupers who refused work were placed in 'houses of correction' (Higginbotham 2008, 14). The early workhouses were non-residential places of work, often in textiles. Consequently there is little evidence for the food supplied. By the beginning of the eighteenth century, conversely, they were compulsorily residential. In prisons, the food, if supplied at all, was bread, often of poor quality and in short supply. All other food had to be bought. For hospitals there is more information. The new diet sheet of AD 1687 for St Bartholomew's in London was based on beef, mutton and broth with bread, pottage, butter, cheese and beer. There is no mention of fish (Drummond and Wilbrahim 1994, 104). Stored fish was considered by the medical establishment at this time to be suitable only for the hardy, of strong nature, employed in trade, and given to much exercise (Drummond and Wilbrahim 1994, 122). The growing numbers of cattle brought to town for slaughter would have increased the availability of fresh meat, including the cheaper cuts suitable for meeting budgets to

feed the poor, sick and/or incarcerated (Drummond and Wilbrahim 1994, 105).

At sea initially no change was evident. Three weekly fish days continued, and dried fish was issued with butter, cheese, biscuit and a bottle of beer (Clowes 1996, 24). Although the allowance rate had risen by around 40% in AD 1622, rises in wheat prices and corruption in the victualling service meant the food was worse than in the preceding century – being of poor quality and short measures. British seamen expected, and were issued, far more meat in rations than those of other European nations. Tuscan galleys in the late sixteenth century carried some salt meat, but also salt dried cod, barrels of sardines and herring, and salt tuna (Hémardinquer 1970, 85). Dutch sailors travelling to Brazil in AD 1648 had equal weights of stockfish, oil, butter and cheese per man (Morineau 1970, 107). Armitage (1985) refers to observations by Nicolaas de Graaff on voyages of Dutch East Indiamen between AD 1639 and AD 1687 that stockfish was served five days a week, with meat and bacon used on the remaining days. The Spanish Armada de le Guardia usually carried dried and salted cod for their fish days, which, by the early seventeenth century, numbered nine per month (Phillips 1986,164, 241).

Any alterations to the British rations were fiercely resisted. Sailors were notoriously conservative, but some changes were forced upon them. In AD 1667, the meat ration was restored to 8 lb (*c*. 3.6 kg) per week, but for voyages south of 39° latitude it was replaced by flour, suet and raisins for pudding, while rice and oil replaced fish and butter. Higher temperatures affected salted meat and fish adversely, but this 'lighter' diet was not popular.

After his appointment to the Navy Board in AD 1677, the diarist Samuel Pepys attempted to establish some control on quality and quantity. On fish days, the allowance per man was to be the eighth part of a 24 inch (61 cm) North Sea cod, or a sixth of a 22 inch (56 cm) haberdine, or a pound of 'poor John', salted hake (Tanner 1920, 61). However, without a strict system of inspection, in reality little changed.

There was already some concern that high quantities of salted provisions were the cause of poor of health and infections, including scurvy, which plagued long voyages. It was noted that other nations ate more rice, oatmeal, biscuits, figs and olive oil, but that if that was unacceptable to British sailors, at least they should eat less salted flesh (Hope 1990, 185). Nevertheless, the only significant changes in victualling at this time were those implemented on longer, more southerly voyages.

There were some advances in preservation. Pickling patents registered in AD 1691 suggest innovation. Meat was soaked in brine, dried, rubbed with salt and hung in a warm place to dry again. Thirsk (2007, 145) suggests this was preferable to normal salting as it was quicker and the brine could be re-used. Other experimentation involved roasted meats sealed with butter and raw meat stored in alcohol

(Drummond and Wilbrahim 1994, 117). The use of stored cod and herring was now only important in the traditional victuals of the navy at sea in cooler climes, with fresh fish and meat issued while in harbour. The army excluded fish earlier. By 1670, a soldiers' field allowance was based on meat, bread and cheese (Drummond and Wilbrahim 1994, 103).

Domestically, herring was increasingly a cheap food for the poor or a breakfast dish, and stored cod was becoming a niche product. Strong, salty tastes were relegated to the past as palates became accustomed to more variety, more fresh foods and ever sweeter dishes, supported by evidence of the rise of imported sugar in customs records (Thirsk 2007, 94).

The eighteenth century

Against a background of increasing enclosures, which further impoverished small farmers dependent on common land, there was an increase in poverty both in the countryside and among the growing urban populations following a rural exodus. Residential workhouses were established across the country, and the first social surveys recorded the living conditions of the poor. The Workhouse Act of AD 1723 allowed parishes to build workhouses, and by AD 1776 there were nearly 2000, usually with 25–30 inmates each (Briggs 1994, 193).

The residential workhouse was not yet the grim institution epitomised by Dickens in *Oliver Twist* in the mid-nineteenth century. Many felt better fed there than they had been at home. Before the harsh reform of the *Poor Law* in AD 1834, workhouses were parish-based, and conditions, including diet, varied in quality. Meat seems to have been largely fresh, using cheap cuts, such as ox cheek and sheep's head, as at Bury-King, Essex, in AD 1724 (Higginbotham 2008, 28), where meat was served every day, but no fish. For institutions operating on a restricted budget and a rotating weekly menu, stored cod and herring, in particular, could have been economical. It is significant that none was incorporated into the workhouse dietaries. In AD 1702, the Quaker Workhouse, Clerkenwell, London, included roasted and boiled meat three times weekly. The rest of the week was vegetarian, but there was flexibility to include mackerel, herring and salt fish (Higginbotham 2008, 17) – a rare exception. This was not a punitive house, and the standard of food was relatively high. Several other eighteenth-century workhouse diet sheets are cited by Fussell (1949). All served meat three times a week, but no fish.

The evidence for hospital food is similar. The Foundling Hospital for orphans in AD 1747 and Christ's Hospital in AD 1704 and 1782 (both in London) regularly served meat, including pickled pork at the latter, but no fish (Drummond and Wilbrahim 1994, 227). The diet for Winchester Hospital in the AD 1740s included three meat days, with no evidence for fish (Carpenter Turner 1986, 15). The serving of meat three times a week was an echo of medieval fast days,

continued for economy and conserving meat supplies, but the remaining days were vegetarian.

There was still no obligation to supplement the bread ration in prisons. Other food had to be bought. In AD 1729, a parliamentary committee reported that 300 inmates had died of starvation in three months in the Marshalsea debtors' prison in London, where Dickens' own father had been kept (it was the setting for his novel *Little Dorrit*). By AD 1776 the need for prison reform, including the introduction of a more balanced diet, had been highlighted by John Howard and William Smith. The latter advocated 'codfish' as the only flesh that should be served, while Howard suggested that boiled beef and broth could be served on a Sunday dependent on good behaviour (Drummond and Wilbrahim 1994, 230). Naval prisoners on the Woolwich hulks were issued very poor victuals, including rotten meat with their biscuit and broth (Drummond and Wilbrahim 1994, 231).

In AD 1797, Eden noted that the poor by the coast ate some fish, especially in Yarmouth and Norfolk, and that pilchards (*Sardina pilchardus*) were much used in Cornwall (Rule 1986, 46). The availability of salt, however, determined whether there was a fish supply beyond the season. There was great variation between years in the quantities of 'red' and 'white' herring brought to the port of London. 'Reds' ranged from 9000 in AD 1751 to more than 2,000,000 in AD 1777, and 'whites' also showed great variation (Minchinton 1995, 39). 'Reds' were only produced in England, principally at Yarmouth and Lowestoft, on the east coast. The maximum of 2,000,000 in AD 1777 can be compared with the 50,000,000 cited by Mayhew for Billingsgate around AD 1850 (Beeton 1982, 172), reflecting the increase in demand from London's growing population of poor.

Salt fish was included in *The Art of Cookery* by Hannah Glasse, published in AD 1747 as a guide for women running a household. At a time when, increasingly, more books were written by women for women, economy was as important as cooking skills, and the emphasis was on English cooking. Advice is given on choosing dried ling and pickled and red herring (Glasse 1995, 163). The desalting of 'old ling' (considered the best), 'water cod' and 'scotch haddocks' (*Melanogrammus aeglefinus*) is discussed as a prelude to serving with a milk and butter sauce, boiled eggs, parsnips and bacon (Glasse 1995, 91). Salt fish pie and the pickling of herring and mackerel (*Scomber scombrus*) are also described (Glasse 1995, 114, 116).

Stockfish and salt fish still seem to have been considered suitable for an occasional dish at the highest level. Stockfish pie was served to George II on 3 February 1740, along with squabs, pullets and mutton (Lehmann 2003, 359). Salt fish was eaten 'at home' by the Duke of Newcastle at Claremont, Esher, on 28 February 1761 (Lehmann 2003, 362). However, evidence that salt fish was not considered prestigious is found in the AD 1758–1802 diaries of the

Reverend Woodforde in Norfolk, who, when entertaining local farmers, his social inferiors, usually served boiled mutton, salt fish and rabbit (Lehmann 2003, 374).

At sea some change was made in the conservative world of naval victualling. By the first quarter of the eighteenth century, stockfish had been dropped from standard rations. There were also some improvements in the administration of victualling, if none in the technology of preservation. By AD 1734, the fish of the three traditional 'fish days' had been replaced by peas, oatmeal, butter and cheese. There was some flexibility on 'foreign' voyages, for example rice could replace oatmeal (Baugh 1965, 376). Stockfish was recorded as scarce and expensive, and showed a high percentage of spoilage compared with other victuals. Between AD 1750 and AD 1757, nearly 8% of stockfish was condemned, comparing poorly with all other goods, all less than 1% each (Rodger 1988, 85). By that time stockfish was already reduced to a small percentage of animal protein. Of all the beef, pork and stockfish issued, only 1.5% was stockfish (Rodger 1988, 83). Herring seems to have been excluded earlier, as no records appear in navy stores.

The concern that salted provisions caused scurvy is exemplified by Anson's circumnavigation in the AD 1740s, when the crew suffered greatly from this disease. Having stopped at the island of Juan Fernandez off the coast of Chile in AD 1741, one boat carrying two Newfoundland fishermen salted 'cod' to add to their stores. However, fears of scurvy meant little use was made of this salted fish (Walter and Robins 1974, 133). Given the location, they could not have caught cod; it was probably *Polyprion oxygenois* (wreckfish), known colloquially as *bacalao de Juan Fernandez* (Ceron Carrasco pers. comm.). *Bacalao* is the Spanish word for cod. This anecdote captures a time when salt cod was still remembered as traditional victuals, but considered undesirable because of an emerging view that it caused scurvy. In AD 1684, Dampier also stopped off at Juan Fernandez and noted a 'rockfish' or 'grooper' resembling cod and called 'baccalao' by the Spanish, which was plentiful off the coasts of Peru and Chile (Dampier 1968, 70).

In summary, stored fish was removed from naval rations during the eighteenth century, having performed increasingly poorly as voyages grew longer. On land, these vestiges of medieval diet and Christian fasting faded from English tables as revolutions in agriculture and industry transformed society. The price of stockfish may also have been influential in its demise. A comparison of French costs by calorie near the end of the eighteenth century showed that dried cod was often more expensive than meat, especially inland and in times of war (Mollat 1987, 163). Price fluctuations may be another reason why stored cod never found a place in the kitchens of workhouses, but salt herring did find a niche as food for the new urban poor of nineteenth-century London and many growing towns.

The nineteenth century

In the nineteenth century, major changes took place in the fishing industry that both supplied and fuelled the demand for fresh marine fish. These included the use of trawlers, icing and live fish wells at sea and of transportation by rail and canning on land. The shift from traditional methods to milder cures for herring (e.g. bloaters and kippers), which use less salt and have a shorter shelf life, also reflected changing tastes.

Increasing urbanisation enlarged the target market for fresh marine fish, which required swift delivery from landing to table. In AD 1801, a fifth of the population were town dwellers, rising to half by mid-century and four-fifths by AD 1911 (Burnett 1989, 3). Consumer expectations for fresh fish were realised in urban areas and in towns served by fast fishing vessels and the railway network. Conversely, inland rural villages, only served by poor roads, felt little benefit.

Victorian ideas on cookery and housekeeping are encapsulated in *Mrs Beeton's Book of Household Management*, published around AD 1859. Many different species of fish are included, with details regarding their season, price, preparation and cooking. The references to stored fish are 'dried haddock' for breakfast and a recipe for salt cod at 6 d a pound, described as an 'Ash Wednesday dish' (Beeton 1982, 119) – continuing the historic connection with fasting. 'Red' and 'white' herring (the latter may be fresh in this case) are also included, with the 'white' comparatively cheap at 1 d each (Beeton 1982, 135). Two sauces are recommended for salt fish. 'Dutch sauce' was composed of flour, butter, vinegar, egg yolk and lemon, cooled and eaten as a salad with flaked cod. 'Egg sauce' was of chopped boiled eggs, butter and lemon (Beeton 1982, 197). Mrs Beeton recognised that salt fish needed a robust sauce, but these recipes, harking back to a pre-Reformation cuisine, are of little significance in the large section on fish.

Mrs Beeton lived in London and was familiar with the supply of fish from Billingsgate Fish Market. She cites Mayhew's mid-century data (Beeton 1982, 172) for 186,805 metric tonnes of fresh fish supplied annually. More than 70% was herring, followed by plaice (*Pleuronectes platessa*), sole (*Solea solea*) and mackerel. Cod, listed as 'live', represented only 1% of the supply, with similarly low weights of salmon and turbot (*Scophthalmus maximus*) – though by price these were the most valuable of the wet fish. Only 21,650 metric tons of stored fish was listed, reflecting the low demand. 'Red' herring was first by weight (29%), then haddock (23%), bloaters (22%), dried salt cod (17%), barrelled cod (9%) and sprat (*Sprattus sprattus*) (0.02%).

British fishermen still involved in catching cod for salting and drying locally were generally based in the northeast of England, Scotland, Shetland and Orkney. In the nineteenth century, they found little demand in England and exported most of it. Nevertheless, some of the Aberdeenshire ports still sent salt cod to the London market in AD 1843. Any cod that did not survive live transport in wells on the boats of the Dogger Bank line fishery were salted on board, which also supplied the dwindling London market (Cutting 1956, 167).

The need to move fresh fish quickly to the consumer meant that 'low-grade' white-fleshed fish, or any fish past optimum freshness, were sold on quickly and cheaply through costermongers. These often ended up as 'fryers', sold on the streets as 'fast food' – in the sense of speed rather than abstinence. Plaice and sole were most popular, but whiting (*Merlangius merlangus*), haddock and flounder (*Platichthys flesus*) were also cut into portions, fried, and sold with a slice of bread (chips are a more recent accompaniment). Herring, both salt and fresh, were also cheap, but were not used as fryers. Mayhew (1985), who interviewed many of London's poor and recorded their struggle to make a living in the AD 1840s, cites a hatter who lived mostly on coffee and bread. Fresh or salt herring (1–3 depending on price) were a treat and near the price of half a pound (about 225 g) of sausages (at 2 d), the only meat he could ever afford (Thompson and Yeo 1971, 545).

The abolition of the salt tax in 1825 (Cutting 1956, 103) was some stimulus to the failing traditional cures. Mayhew's figures show 'red herring' as the major component of stored fish at Billingsgate (Beeton 1982, 172), but from the AD 1840s, milder cures such as bloaters and kippers became more popular. These found their way inland to rural villages, such as Candleford in Oxfordshire, where bloaters were sold for a penny each, but they were still beyond the pocket of most villagers, who relied on bread and tea (Thompson 1978, 119).

Welfare relief in the form of workhouses was to become harsher. Escalating costs led to the *New Poor Law Act* of AD 1832. A uniform new workhouse test was applied, and outdoor relief was phased out. The workhouse was to be seen as the last option for the destitute. Husbands, wives and children were all separated, and food costs were cut where possible. Some appalling examples are described by Engels (1987, 283). One of the most infamous was Andover in Hampshire (Higginbotham 2008, 61). Violent protest led to some reform of the act (Higginbotham 2008, 50), and six model dietaries were published in AD 1835. As economy was crucial, salt and 'red' herring would have seemed ideal, cheap and storable. There is no evidence for salt fish, or indeed any fish, being included on a regular basis in England, though in Scotland, where fish consumption was higher, white fish was allowed (Higginbotham 2008, 123). Later, during the AD 1880s, when fish was cheaper than meat, 'fish dinners for the poor' were introduced in England, using fresh cod. They were not popular, however, as the inmates considered beef more filling. The Edmonton workhouse (in London) served cod with butter and anchovy, which was a saving on meat, but this practice was discontinued as it was perceived as an extravagance, on the grounds that some ratepayers could not afford such a meal (Priestland 1972, 30).

Prison food remained a scandal: mainly bread, some thin gruel and occasionally a little meat or cheese. By AD 1842, the home secretary decreed that there should be three meals a day provided, of which two should be hot (Drummond and Wilbrahim 1994, 368). These were based on bread, potatoes and meat, soup, gruel and cocoa, but no fish.

In the armed forces rations, stored fish were a distant memory. The Royal Hampshire Regiment in Jamaica in AD 1834 was issued fresh beef, salt pork, bread and tea (Atkinson 1950). General field rations by AD 1884 were bread, biscuit, fresh or salted meat, coffee, tea, salt and sugar. Live animals accompanied troops as meat on the hoof, together with portable bread ovens, reflecting a new emphasis on fresh food. Only specific groups (e.g. the Chinese Corps in AD 1860) included salt fish as part of their ration (Wolseley 1886).

At sea, in AD 1811, meat (beef or pork) was now served daily with cheese, butter, suet, sugar, bread, flour and beer as naval rations (Drummond and Wilbrahim 1994, 465). Merchant seamen described to Mayhew (Thompson and Yeo 1971, 366) the old, condemned stores of rotten beef, pork and weevil-infested biscuit issued to both seamen and poor immigrants. They considered themselves worse off than naval sailors, but regarded the American merchant navy best for wages, food and work conditions. On board the English East Indiamen of the nineteenth century, live animals were carried to supply fresh meat at the high table, with salt beef and pork of varying grades being reserved for sailors and steerage passengers. Armitage (2002) found a few cod bones from the wreck of the *Earl of Abergavenny* in Weymouth Bay (which sank in AD 1805), which, he suggested, based on the presence of cleithra (see Chapter 1; Chapter 18), could be from salt cod or stockfish, but documentary evidence for ships stores at that time only refers to salmon, herring and 'fish'.

A new form of preservation, canning, had been in development since the AD 1800s. An Englishman, Durand, had further developed the Frenchman Appert's technique for sealed glass jars to work with tin containers. He patented it in AD 1810. From the AD 1820s, canned meat began to be used in ships stores. There were initial problems with quality, but it was cheap (half the price of fresh meat) and convenient. Fray Bentos corned beef from Argentina was soon popular among the working classes (Spencer 2004, 283). By the early AD 1900s, several kinds of canned fish were available, mainly salmon, 'sardines' (canned in France since the AD 1850s) and herring. Soldiers' rations in the First World War included a variety of cans, but no fish.

During this century of unprecedented change in agriculture, industry and population, the fishing industry also evolved – delivering fresh marine fish quickly, in bulk and more cheaply, transported on land by rail and kept fresh on ice. Hard cured fish became increasingly marginalised as an occasional traditional dish, such as stockfish pie, or as cheap food for the poor (e.g. 'red' herring). The development of canning and milder cures for herring effectively superseded the storage advantages of the old cures.

Discussion

The low consumption of fish of all types in England has intrigued recent researchers from a variety of fields. The strong taste of dried and salted cod and herring was a casualty of the early modern period, before that of salted and pickled meat. The combination of a hard, salty cure and fish seem to have ensured an early demise, preceding new preservation techniques. Various theories have been suggested to explain the more recent English dislike for fish. These usually reach back into past associations with fast days and austerity and, within living memory, the Second World War. Franklin (1997, 252) views 'oily and other preserved fish' as firmly identified with poverty. He also describes preserved fish as monotonous, unappetising, difficult to prepare and thus an unpopular staple (Franklin 1997, 245). As shown in this chapter, however, the motives and drivers influencing their consumption were more complex.

The decline of stored-fish consumption in England accelerated since at least the eighteenth century, when the historic association of stockfish and salt fish with naval and, even earlier, army stores was broken. The rise of the welfare state did not revive the use of stored fish in dietaries based on economy, a phenomenon on which the fluctuating price of stored cod may have some bearing. The decline was already established before the advent of fleeting and icing for fast delivery of fresh fish, before canning, and before milder cures. Urban poverty supported demand for stored herring which, on price and strong taste, shared the same relish factor as fatty bacon. Mayhew's data show that, in the mid-nineteenth century, herring constituted 70% of the fresh fish delivered to Billingsgate and 'red' herring made up the main share of stored fish.

The English retained little enthusiasm for already declining stored cod dishes once canning and freezing became common. Cod was still being salted on the northeast coast of England and in parts of Scotland (e.g. Aberdeen) in the nineteenth century, but mostly for export. There was little affection for this part of culinary history, whereas other nations still celebrate their long association with salt fish as part of a Catholic past with such dishes such *stocaficada*, a pungent Niçois stockfish stew (Medecin 2002, 104); *bacalau a la Vizcaina*, a Basque dish; or the Portuguese *Bacalhau dorado* (Davidson 1979, 267, 278). Modern Caribbeans still recall their past with 'salt fish and ackee', a legacy of salt cod, often of low quality, sent to feed plantation workers, who never saw a fresh or whole cod.

Mrs Beeton's inclusion of salt cod as an Ash Wednesday dish is perhaps a key to its ultimate demise. She both influenced and reflected householders in Victorian England, and her book, surprisingly, remained a reference work up to the AD 1930s (Spencer 2004, 306). Her opinion reflected the then perceived status of salt cod as for very occasional use on a day of traditional self-denial in preparation for Lent. Herring and smoked haddock were marginalised as breakfast or supper dishes, unsuitable for the main meal of the day. The Victorian poor increasingly relied on tea, bread and bacon, or convenience cooked food bought on the street. In Britain, a nation of meat eaters, stored fish was largely lost from cultural memory, restricted to the northernmost areas and islands. There, the slower pace of modernisation ensured the use of traditional products continued for longer. Even in these contexts, however, the taste for salted preserved food was easily lost. In one revealing anecdote, an inhabitant of Lewis, in the Western Isles of Scotland, spent a year in Glasgow in the AD 1960s before returning to the island as the first there to afford a freezer (in preference to a television), having no wish to return to salted provisions (Murray 2008, 62).

References

Armitage, P. 1985. Mammalian, bird and fish bones from the 'Amsterdam', St Leonards, East Sussex: 2nd interim report. Report prepared for Save the Amsterdam Foundation, Brighton.

Armitage, P. 2002. Study of the animal bones, in E. Cumming (ed.) *The Earl of Abergavenny: historical research and wreck excavation*. CD ROM MIBEC Enterprises. ISDN 0-954 2104-0-9.

Atkinson, C. T. 1950. *Royal Hampshire Regiment*. Glasgow: Maclehouse.

Baugh, D. A. 1965. *British Naval Administration in the Age of Walpole*. Princeton: Princeton University Press.

Beeton, I. 1982. *The Book of Household Management*. (Facsimile Edition). London: Chancellor Press.

Briggs, A. 1994. *A Social History of England*. London: Weidenfeld and Nicolson.

Burnett, J. 1989. *Plenty and Want*. London: Routledge.

Capatti, A. 1999. The taste for canned and preserved food, in J.-L. Flandrin & M. Montanari (eds) *Food: a culinary history from antiquity to the present*: 492–9. New York: Columbia University Press.

Carpenter Turner, B. 1986. *A History of the Royal Hampshire County Hospital*. Stroud: Phillimore.

Clowes, W. L. 1996. *The Royal Navy: a history from earliest times to 1900*. Vol. 2. London: Chatham.

Cutting, C. L. 1956. *Fish Saving: a history of fish processing from ancient to modern times*. New York: Philosophical Library.

Cutting, J. L. 1962. The influence of drying, salting and smoking on the nutritive value of fish, in E. Heen & R. Kreuzer (eds) *Fish in Nutrition*: 161–79. London: Fishing News Books.

Dampier, W. 1968. *A New Voyage Round the World*. New York: Dover Publications.

Davidson, A. 1979. *North Atlantic Seafood*. London: Penguin.

Davidson, A. 1999. *Oxford Companion to Food*. Oxford: Oxford University Press

Dawson, M. 2008a. Changing tastes in sixteenth century England: evidence from the household accounts of the Willoughby family, in S. Baker, M. Allen, S. Middle & K. Poole (eds) *Food and Drink in Archaeology*. Vol. 1: 20–8 (University of Nottingham Post-graduate Conference 2007). Blackawton: Prospect Books.

Dawson, M. 2008b. The food year in the Willoughby household at Wollaton and Middleton in the sixteenth century. *Petits Propos Culinaire* 85: 77–96.

Drummond, J. C. & A. Wilbrahim. 1994. *An Englishman's Food*. London: Pimlico.

Emmison, F. G. 1961. *Tudor Secretary: Sir William Petrie at court and home*. London: Longmans.

Engels, F. 1987. *The Condition of the Working Class in England*. London: Penguin Classics.

Flandrin, J.-L. 1999. The early modern period, in J.-L. Flandrin & M. Montanari (eds) *Food: a culinary history from antiquity to the present*: 349–73. New York: Columbia University Press.

Franklin, A. 1997. An unpopular food? The distaste for fish and the decline of fish consumption in Britain. *Food and Foodways* 7(4): 227–64.

Fussell, G. E. 1949. *The English Rural Labourer*. London: Batchworth Press.

Glasse, H. 1995. *First Catch your Hare: the art of cookery made plain and easy*. Blackawton: Prospect Books.

Hamilton-Dyer, S. 1995. Fish in Tudor naval diet – with reference to the *Mary Rose*. *Archaeofauna* 4: 27–32.

Hartley, D. 1979. *Food in England*. London: Macdonald and Jane.

Harvey, B. 1993. *Living and Dying in England 1100–1540*. Oxford: Clarendon Press.

Hattendorf, J. B., R. J. B. Knight, A. W. H. Pearsall, N. A. M. Rodger & G. Till (eds) 1993. *British Naval Documents 1204–1960*. (Navy Records Society 131). London: Navy Records Society.

Hémardinquer, J.-J. 1970. Sur les galères de Toscane au XVI siècle, in J.-J. Hémardinquer (ed.) *Pour une histoire de l'alimentation*: 85–92. (Cahiers des annales 28). Paris: Librairie Armand Colin.

Higginbotham, P. 2008. *The Workhouse Cookbook*. Stroud: History Press.

Hope, R. 1990. *A New History of British Shipping*. London: John Murray.

Hopewell, P. 1995. *St Cross: England's oldest almshouse*. Chichester: Phillimore.

Hutchinson, W. F., M. Culling, D. C. Orton, B. Hänfling, L. Lawson Handley, S. Hamilton-Dyer, T. C. O'Connell, M. P. Richards. & J. H. Barrett. 2015. The globalization of naval provisioning: ancient DNA and stable isotope analyses of stored cod from the wreck of the *Mary Rose*, AD 1545. *Royal Society Open Science* 2: 150199.

Janzen, O. 2013. The logic of English saltcod: an historiographical revision. *Northern Mariner/Le marin du nord* 23(2): 1–12.

Kitchin, G. W. 1892. *Compotus Rolls of the Obedientaries of St. Swithun's Priory, Winchester*. Winchester: Warren & Son.

Lehmann, G. 2003. *The British Housewife: cookery books, cooking and society in eighteenth-century Britain*. Totnes: Prospect Books.

Mason, L. with C. Brown. 1999. *Traditional Foods of Britain*. Totnes: Prospect Books.

Mayhew, H. 1985. *London Labour and the London Poor*. London: Penguin Classics.

Medecin, J. 2002. *Cuisine Niçoise: recipes from a Mediterranean kitchen*. London: Grub Street.

Minchinton, W. 1995. London fisheries in the eighteenth century, in P. Holm, O. Jansen & J. Thór (ed.) *Northern seas yearbook 1995* (Fiskeri-og Sofartmuseets Studieserie 5): 33–50. Esbjerg: Association for the History of the Northern Seas.

Mollat, M. 1987. *Histoire des pêches maritimes en France*. Toulouse: Bibliothèque Historique Privat.

Morineau, M. 1970. En Hollande au XVII siécle, in J.-J. Hémardinquer (ed.) *Pour une histoire de l'alimentation*: 107–14. (Cahiers des annales 28). Paris: Librairie Armand Colin.

Murray, D. 2008. *The Guga Hunters*. Edinburgh: Birlinn.

Phillips, C. R. 1986. *Six Galleons for the King of Spain*. Baltimore: Johns Hopkins University Press.

Pope, P. 2004. *Fish into Wine: the Newfoundland plantation in the seventeenth century*. Chapel Hill: University of North Carolina Press.

Priestland, G. 1972. *Frying Tonight: the saga of fish and chips*. London: Gentry Books.

Rodger, N. A. M. 1988. *The Wooden World: an anatomy of the Georgian navy*. London: Fontana.

Rodger, N. A. M. 1997. *The Safeguard of the Sea: a naval history of Britain. Volume 1: 600–1649*. London: HarperCollins.

Rule, J. 1986. *The Labouring Classes in Early Industrial England: 1750–1850*. London: Longman.

Scully, T. (trans. & ed.) 2006. *La Varenne's Cookery*. Totnes: Prospect Books.

Spencer, C. 2004. *British Food: an extraordinary thousand years of history*. London: Grub Street.

Steckley, G. F. (ed.) 1984. The letters of John Paige, London merchant, 1648–58. *London Record Society* 21. Available at: British History Online. http://www.british-history.ac.uk/london-record-soc/vol21

Stevens, K. F. & T. E. Olding. 1985. *The Brokerage Books of Southampton 1477–8 and 1527–8*. Southampton: Southampton University Press.

Tanner, J. R. 1920. *Samuel Pepys and the Royal Navy*. Cambridge: Cambridge University Press.

Thirsk, J. 2007. *Food in Early Modern England: phases, fads and fashions 1500–1760*. London: Hambledon Continuum.

Thompson, E. P. & E. Yeo. 1971. *The Unknown Mayhew*. London: Penguin Classics.

Thompson, F. 1978. *Lark Rise to Candleford*. London: Penguin.

Walter, R. & Robins, B. 1974. *A Voyage Round the World in the Years MDCCXL, I, II, III and IV by George Anson*. Oxford: Oxford University Press.

Wilson, C. A. 1973. *Food and Drink in Britain*. London: Constable.

Wolseley, G. J. 1886. *The Soldier's Pocket-book for Field Service*. London: Macmillan.

Woolgar, C. M. 1999. *The Great Household in Late Medieval England*. New Haven: Yale University Press.

Wubs-Mrozewicz, J. 2009. Fish, stock and barrel. Changes in the stockfish trade in northern Europe, c. 1360–1560, in L. Sicking & D. Abreu-Ferreira (eds) *Beyond the Catch*: 187–208. (Northern World 41). Leiden: Brill.

Part II

Perspectives from Zooarchaeology and Stable Isotope Analysis

Fishing and Fish Trade During the Viking Age and Middle Ages in the Eastern and Western Baltic Sea Regions

Lembi Lõugas

Introduction

Information about fishing in Viking Age (eighth–eleventh centuries AD) and medieval (twelfth–sixteenth centuries AD) times in the Baltic Sea region has been obtained from finds of preserved fish remains. The Swedish Viking Age is particularly well represented. One of the best known Viking Age assemblages is that from the town of Birka, in eastern Sweden. Its rich fish-bone material indicates the importance of fishing, including some sea fishing, and possibly of fish trade. Conversely, the bones from a Viking Age settlement site on Saaremaa Island in Estonia suggest an exclusive focus on local freshwater fishing. In the Middle Ages, fishing continued to be an important local activity in Estonia, but long-distance fish trade also emerged. In this chapter the development of fishing and fish trade in Estonia is considered against the background of other countries around the Baltic at the end of the first millennium and in the first half of the second millennium AD.

The Viking Age

Some researchers argue that long-distance trade, and thus activities along the Baltic coast, developed in the seventh or eighth centuries AD and accelerated during the subsequent Viking Age (e.g. Callmer 2007). It is clear that marine fishing did become more common during these centuries, as evidenced by increasing numbers of archaeological fish-bone remains of herring from settlements around the Baltic Sea (Figures 11.1 and 11.2). This fact alone, however, does not prove that they were the subject of long-distance trade. Moreover, many areas engaged in very limited marine fishing at this time.

Birka, located on a small island in Lake Mälaren, is considered one of the main Viking Age trading centres

in eastern Sweden. Excavations led by Björn Ambrosiani from 1990 to 1995 recovered six metric tonnes of mammal bone (see Wigh 2001). Fish bones were very abundant. However, marine fish made up only a tiny proportion of 15,826 identified fish specimens. The majority of the fish remains were freshwater species, such as pikeperch (*Sander lucioperca*), pike (*Esox lucius*) and bream (*Abramis brama*) (Figure 11.2; Lõugas 2001; 2008). Of the marine species, herring (*Clupea harengus*) was most plentiful in terms of number of identified specimens (NISP). Cod (*Gadus morhua*) bones were very rare, and flounder (*Pleuronectes flesus*) bones were absent (Lõugas 2001; 2008). For reasons that will be discussed further below, the presence of herring is

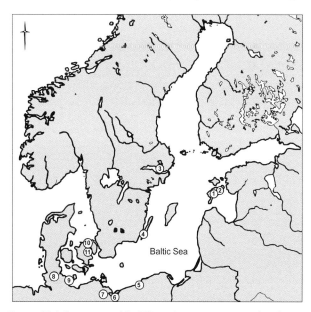

Figure 11.1. Locations of the Viking Age sites mentioned in the text, as numbered in Figure 11.2 (Drawing: Lembi Lõugas).

interesting. It remains to be determined whether it represents trade from southern or western Baltic regions or, instead, eastern Baltic stocks of the subspecies Baltic herring (*Clupea harengus membras*) obtained through local fishing.

Tornimäe is another Viking Age site of the Baltic region that has yielded a large fish-bone assemblage. At this harbour site on Saaremaa Island in western Estonia (Mägi 2005), fish bones were collected by dry sieving on 5 mm mesh. Analysis of the fish bones is still ongoing, but preliminary results (based on *c.* 1000 identified specimens) show that it is an assemblage of freshwater fish. Perch (*Perca fluviatilis*) are most abundant, followed by ide (*Leuciscus idus*), pike and various cyprinids (Cyprinidae). Few bones were from pikeperch. No herring, cod or flounder have been identified so far (Lõugas 2008).

Pöide stronghold, in the southeastern part of Saaremaa Island, tells a similar story (Lõugas 1991; Lõugas and Mägi-Lõugas 1994). It was occupied during the eighth–ninth centuries – immediately prior to, and very early in, the Viking Age. Although the assemblage is very small and was recovered by hand collecting rather than sieving, it may, nevertheless, be significant that all the fish species identified are of freshwater origin. Perch predominate, with 18 bone fragments and 148 scales (Enghoff 1999; Lõugas

1997; 1999; 2001). There were also nine bones of pike and two of roach (*Rutilus rutilus*). The assemblage may be so small because the stronghold was used only as a refuge, in the case of emergency.

The Middle Ages

Medieval (AD 1227–1558) Estonia is characterised by influences from Sweden, Denmark and Germany to the west and from Russia to the east. These influences concerned most spheres of life, including the use of aquatic resources. Medieval fishing and fish trade in the eastern Baltic region can be explored based on ten fish-bone assemblages from medieval towns in Estonia (Tallinn, Pärnu, Tartu and Viljandi) and Latvia (Valmiera) (Figures 11.3–11.4).

Number 4 Rahukohtu Street in Tallinn is situated on Toompea Hill, in the 'upper town'. It was excavated between 2000 and 2002 by Peeter Talvar of Tael Ltd. One of the excavated plots represents a thirteenth-century residence of Danish conquerors in Toompea Castle. It produced a tiny but interesting assemblage of ten hand-collected fish bones. In addition to bones of herring, flounder and pike, there were four cod vertebrae from the layer dated to the end of the thirteenth century or the fourteenth century. For reasons to

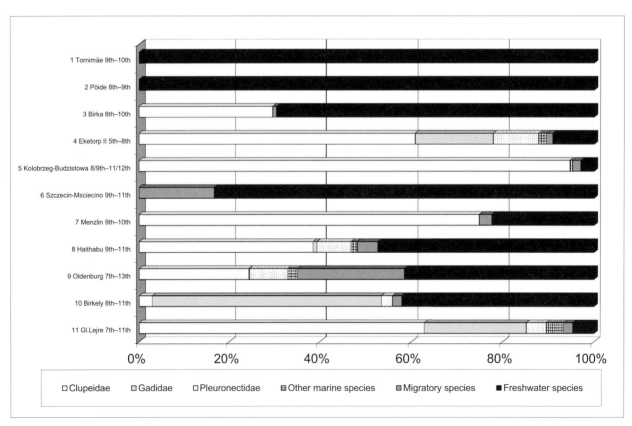

Figure 11.2. Relative frequency of different fish groups in the Viking Age bone assemblages by NISP, including dates in centuries AD (based on Enghoff 1999 and references therein; Lõugas 2001; 2008).

be discussed below, the latter may represent imports from beyond the Baltic Sea region.

Number 10 Sauna Street in Tallinn is situated between Town Hall Square, which was a market place even before the town hall was built in the fourteenth century, and the eastern

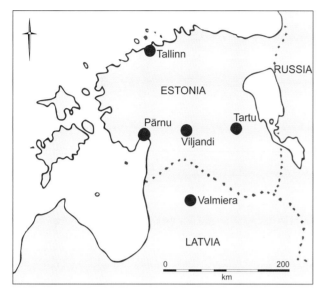

Figure 11.3. Locations of the medieval towns mentioned in the text (Drawing: Lembi Lõugas).

curtain-wall of the 'lower town'. In 1998 and 1999, Tael Ltd. conducted archaeological rescue excavations in a yard and cellar on the site, resulting in the excavation of 520 m² of the yard and 100 m² of the cellar of the house (V. Sokolovski pers. comm.). The site yielded abundant finds, with very good preservation of organic materials, such as leather. The fish-bone assemblage was small (NISP = 164) and recovered without sieving, but nevertheless included 11 taxa: pike (NISP = 44), perch (NISP = 37), cod (NISP = 26), bream (NISP = 19), flounder (NISP = 13), herring (NISP = 6), pikeperch (NISP = 4), turbot (*Scophthalmus maximus*) (NISP = 2), whitefish (*Coregonus*) (NISP = 1), ide (NISP = 1) and fourhorn sculpin (*Myoxocephalus quadricornis*) (NISP = 1) (Lõugas 2001; unpublished data). Eight cod vertebrae from a layer with finds characteristic of the thirteenth century or the early fourteenth century may represent fish imports from the west (see below).

Number 10 Viru Street, also in Tallinn's 'lower town', was excavated in 1994 by Agu EMS Ltd. (Talvar 1995). The archaeozoological material was hand-collected. Only 32 fish bones were identified (Lõugas 2001), mainly from layers dated to the second half of the thirteenth century (Talvar 1995). These include both freshwater and marine fishes: pike (NISP = 13), cod (NISP = 12), perch (NISP = 1), flounder (NISP = 1), bream (NISP = 1) and four bones of other cyprinids. Three of the cod bones (two vertebrae, one cleithrum) imply the presence of imported dried fish based on isotope analysis (see below).

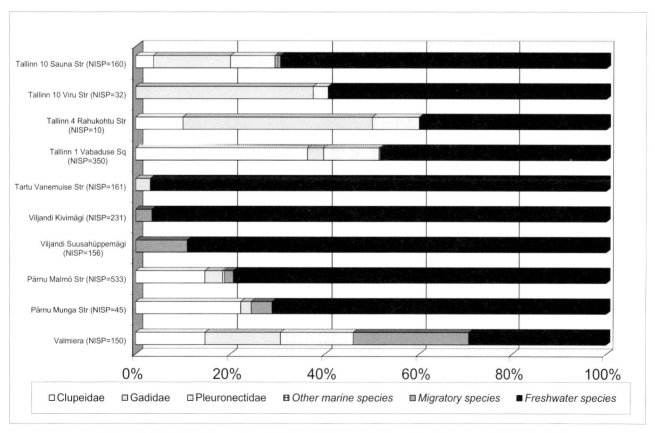

Figure 11.4. Relative frequencies of different fish groups in the medieval bone assemblages by NISP.

Number 1 Vabaduse Square is situated just outside the southern part of Tallinn's 'old town'. Large-scale excavations around eighteenth-century bastions in 2008 and 2009 produced some *in situ* medieval material (Kadakas *et al.* 2010). Fish bones were recovered from a thirteenth–fourteenth-century layer (excavation area I, sector J). Approximately 2 m^3 of soil was sieved using 5 mm mesh and *c.* 100 litres from the <5 mm fraction was then water sieved using 1.5 mm mesh. Altogether, 350 bones or bone fragments were identified (not considering scales). The identification of 127 herring bones illuminates the possible significance of a small-sized taxon that is underrepresented among the hand-collected material from Tallinn discussed thus far. Other marine fish are represented by 12 cod (be they local or imported) and 41 flounder bones. Freshwater species include pike (NISP = 30), perch (NISP = 75) and cyprinids (NISP = 64), including 33 roach, 4 bream, 1 ide and 1 dace (*Leuciscus leuciscus*). In addition, one vertebra of whitefish was recorded.

The coastal town of Pärnu has also produced fish-bone assemblages of medieval date. Two are included in this paper. Excavations at number 2 Munga Street (Ülle Tamla pers. comm.) yielded 45 hand-collected fish bones from a fourteenth-century layer. A larger collection, of 547 identified specimens, was recovered by a combination of hand collecting and sieving of fourteenth–sixteenth-century layers at 15 Malmö Street (Enghoff 1999; Lõugas 2001; 2008; Aivar Kriiska pers. comm.; Aldur Vunk pers. comm.). Both assemblages include marine, migratory and freshwater fish, all of which could have been caught near Pärnu. Perch predominates in the material (NISP = 268), followed by herring (NISP = 88) and pikeperch (NISP = 76). There were also 21 cod bones, but the small size of these specimens and the presence of head bones suggest that they represent Baltic fish rather than imports. Of the 99 bones of cyprinids, 80 come from roach, 5 from bream, 5 from vimba bream (*Vimba vimba*) and 1 from dace. Other species from medieval contexts in Pärnu are trout (*Salmo trutta*) (NISP = 5), eel (*Anguilla anguilla*) (NISP = 2), flounder (NISP = 2), whitefish (NISP = 5) and sturgeon (*Acipenser*) (NISP = 1).

In the inland town of Tartu, a 140 m long and 1–4 m wide trench was excavated in Vanemuise Street in 1995 (Aun 1996). These were rescue excavations carried out by the Institute of History, Tallinn, in the context of canalisation. Only hand collecting was conducted. Vanemuise Street is between Toome Hill and the Emajõgi River, a 'lower town' area in medieval times. A medieval layer was present along 100 m of the excavated section, from which 161 fish bones have been analysed (Lõugas 2001). Almost all are of freshwater species common in local water bodies. The five identified cod bones, however, likely all come from traded fish, because stable isotope analysis of two cod vertebrae and one cod cleithrum suggests an ultimate origin from beyond the Baltic Sea (Orton *et al.* 2011; see below).

Other taxa identified among the material from Vanemuise Street are as follows: perch (NISP = 76), pike (NISP = 43), bream (NISP = 32), unidentified cyprinids (NISP = 29), pikeperch (NISP = 4), ide (NISP = 2), burbot (*Lota lota*) (NISP = 2), roach (NISP = 1), vimba bream (NISP = 1), tench (*Tinca tinca*) (NISP = 1) and wels catfish (*Silurus glanis*) (NISP = 1).

Several excavations outside the medieval inland town of Viljandi, on the high bank of the lake of the same name, also produced fish bones (Haak 2001; Lõugas 2001; Valk 2000; 2001). All bones were hand-collected. The Kivimägi site produced 231 identified fish bones (59 pike, 72 perch, 8 whitefish, 30 bream, 18 roach, 18 ide, 1 vimba bream, 1 tench and 36 unidentified cyprinids) dating to between the tenth–eleventh centuries and AD 1223. The high-status Suusahüppemägi site yielded 156 identified fish bones (30 pike, 68 perch, 3 pikeperch, 15 whitefish, 2 trout, 2 bream, 3 roach, 3 ide, 1 dace, 28 unidentified cyprinids and 1 burbot) dating between the middle of the twelfth century and the beginning of the thirteenth century.

Moving to inland Latvia, excavations in Valmiera 'old town' in 2006 resulted in a small assemblage of 150 identified fish bones from the medieval period (Lõugas 2007). Despite its inland location, the migratory and/or marine species twaite shad (*Alosa fallax*) (NISP = 35), cod (NISP = 24), flounder (NISP = 23) and herring (NISP = 22) dominated the material. Twaite shad is today quite rare in the Baltic, but at the beginning of the twentieth century its population in the Gulf of Riga was large enough to allow for a good catch (Veldre 2003). It is a migratory fish that is distributed in coastal waters and makes spawning migrations into rivers. Valmiera is situated on the bank of the Gauja River, which flows into the Gulf of Riga, so this species could have been caught close to the settlement during the spawning migration or transported up river from the sea, as was probably the case with the other marine fish. Freshwater fish, such as perch (NISP = 19), bream (NISP = 10) and dace (NISP = 4), and migratory fish, such as trout (NISP = 2), are also represented in the assemblage. Surprisingly, however, the pike, which is a very common fish in the eastern Baltic (including the coastal waters of that sea) is not well represented (NISP = 1).

Overview

Fishing in the eastern Baltic Sea region during the first millennium AD seems to have targeted freshwater species and (to a lesser degree) local marine fish that could be caught in inshore waters (even in the case of islands, such as Saaremaa). This situation differed from that in earlier millennia – specifically the late Neolithic – when cod fishing on a large scale was carried out in the eastern Baltic Sea (e.g. Lõugas 1999). It appears that after the transition of the Baltic from the salty Littorina Sea stage to the brackish

Limnea Sea stage (*c.* 2500–1800 BC) human interest in the seashore and marine fishing diminished. This change is less clear in western Baltic contexts such as Denmark, where the salinity of the water did not change as much as in the east. In the context of this environmental change and subsequent social and economic developments, marine fishing ceased to be relevant to the coastal communities of Estonia, Latvia, Lithuania and Poland during the centuries from *c.* 500 BC to *c.* AD 1000 (Enghoff 1999; Lõugas 1997; 1999; Makowiecki 1998; 1999; Chapter 12). Agriculture, hunting and fishing in lakes and rivers using weirs, traps, lines and spears became the preferred options for food provision.

In Estonian coastal areas the economy was predominantly based on agricultural activity during the first millennium AD. Fish bones indicate local fishing, but their absolute numbers and relative abundance *vis-à-vis* the bones of livestock are quite small (Lõugas 2001). In some cases, however, this small number is likely the result of excavation techniques rather than the absence of bones, because bones were often collected only by hand, rather than through more fine-grained recovery methods such as sieving.

Marine fisheries for herring, cod and flounder in the eastern Baltic Sea developed once again during medieval times. This conclusion is based on finds from Tallinn and Pärnu and (in the case of herring) a number of sites dating to the eleventh to thirteenth centuries (or earlier) around the Baltic Sea (cf. Enghoff 1999). The capacity of ships increased at the same time and trade became more intensive (Lõugas 2001).

Evidence of fish trade?

The Viking Age assemblages considered in this paper are mostly characteristic of local fishing. The trade centre at Birka is the only possible exception. It was strategically a good place for the trade in furs and handicrafts, but also for fishing and perhaps the trade in fish. The connection between Lake Mälaren and the Baltic Sea (which has since been severed by isostatic rebound) probably favoured a diversity of fish fauna around the small island of Björkö, on which the settlement was built. Not only freshwater species but also fish tolerant of brackish water could live there. Most of the fish identified in the Birka material could thus have been caught in Lake Mälaren. Theoretically, even a form of the Baltic herring could live in such a lightly brackish, lake-like body of water. If it did live in Lake Mälaren, then we have no firm proof of fish trade at Birka. However, estimation of the body size of the herring shows that some individuals were 25–30 cm long (Lõugas 2001), which is rather large for the eastern population of Baltic herring. Therefore there is a possibility that the herring remains represent salted imports from more distant marine waters (e.g. the southern or western Baltic) (Lõugas 2001).

Into the Middle Ages, most of the fish recovered from the sites considered above were probably also caught in local waters. Nevertheless, a few remains of marine fish from inland towns provide unambiguous evidence of fish trade on at least a regional scale. The cod bones from medieval Tartu are a clear example, as are the cod, (Baltic) herring, flounder and (perhaps) twaite shad bones from Valmiera. The five herring and 11 smelt (*Osmerus eperlanus*) bones from fourteenth-century Viljandi tell a similar story (T. Paaver pers. comm.). Inge Enghoff (1999) noted similar examples from Denmark, Sweden, Germany and Poland. Moreover, bones of haddock (*Melanogrammus aeglefinus*) and ling (*Molva molva*) found at a few sites in eastern Sweden, at Eketorp and Uppsala, are likely to represent imports to the region on biogeographical grounds (Enghoff 1999).

Enghoff (1999; Chapter 13) has also called attention to the existence of large cod (more than 1 m in estimated total length) in Viking Age and medieval assemblages. In recent decades cod from the Baltic population rarely exceed *c.* 75 cm in length and are typically less than *c.* 60 cm long (e.g. Limburg *et al.* 2008). A large late Neolithic assemblage of cod bones from the Loona site, on Saaremaa Island, indicates that lengths of 40–60 cm should be expected for local catches (cf. Limburg *et al.* 2008; Lõugas 1997). Occasional larger individuals do occur in the Baltic, but they must be considered very rare. Thus bones from cod of greater than 1 m total length have a high probability of representing imports. Examples include medieval specimens from Tallinn and Tartu that represent fish with estimated total lengths of between 0.95 m and 1.25 m – including ten vertebrae from Viru Street, seven vertebrae from Sauna Street and at least eight vertebrae from Rahukohtu Street (Lõugas 2001). Such large cod, represented only by vertebrae, are interpreted as the remains of dried and/or salted fish, perhaps Scandinavian stockfish (Enghoff 1999; Jonsson 1986; Lie 1988; Chapter 18). The examples from Tallinn, and perhaps also those from Tartu, date to the thirteenth–fourteenth centuries and are thus probably connected with Danish rule in northern Estonia. Stable isotope analysis of cod bones from Tallinn and Tartu returned $\delta^{13}C$ and $\delta^{15}N$ signatures that differ from Baltic control samples, instead matching sources such as Arctic Norway, where stockfish were known to have been produced for commercial export in the Middle Ages (Orton *et al.* 2011).[1] The question still remains, who were the consumers of these Atlantic cod in Estonia – the local population or migrants and merchants from the west?

A trade in seafood is also evidenced by oyster finds from medieval Pärnu (from PäMu 14350: A 2501) and Tallinn (from Vabaduse Square). Moreover, written sources from the fourteenth and fifteenth centuries in the Tallinn city archives note fish imports in the town's market, including swordfish, 'flack visch' (stockfish), sturgeon from the Neva River and 'Tartu pike' (the latter being distinguished from other pike) (Põltsam-Jürjo 2013).

Note

1 "In 2015, after this article was written, an important discovery was made in old sea sediments at the southern beach of Tallinn Bay. A fourteenth-century shipwreck included a herring barrel and bunches of large cod vertebrae, presumably representing stockfish (Lõugas forthcoming)."

Acknowledgements

This paper was completed within the Medieval Origins of Commercial Sea Fishing project, supported by the Leverhulme Trust. I am very grateful to the project team. Osteological analysis of the fish bones was carried out within a project supported by the Estonian Science Foundation (grant 6899).

References

Aun, M. 1996. Археологические исследования на улице Ванемуйзе в Тарту. *Proceedings of the Estonian Academy of Sciences, Humanitarian and Social Sciences* 45(4): 451–5.

Callmer, J. 2007. Urbanisation in northern and eastern Europe, ca. AD 700–1100, in J. Henning (ed.) *Post-Roman Towns, Trade and Settlement in Europe and Byzantium. Volume 1: the heirs of the Roman West*: 233–70. (Millennium-Studien/Millennium Studies 5/1). Berlin: Walter de Gruyter.

Enghoff, I. B. 1999. Fishing in the Baltic region from the 5th century BC to the 16th century AD: evidence from fish bones. *Archaeofauna* 8: 41–85.

Haak, A. 2001. Archaeological investigations of the castle ruins, and at Pikk Street in Viljandi. *Archaeological Fieldwork in Estonia* 2000: 108–16.

Jonsson, L. 1986. Finska gäddor och Bergenfisk ett försök att belysa Uppsalas fiskimport under medeltid och yngre Vasatid, in N. Cnattingius & T. Neréus (eds) *Uppsala stads historia. Volume 7, Från Östra Aros till Uppsala: en samling uppsatser kring det medeltida Uppsala*: 122–39. Uppsala: Almqvist and Wiksell.

Kadakas, V., R. Nurk, G. Püüa, G. Toos, L. Lõugas, S. Hiie & K. Kihno. 2010. Rescue excavations in Tallinn Vabaduse Square and Ingermanland bastion 2008–2009. *Archeological Fieldwork in Estonia* 2009: 49–69.

Lie, R. W. 1988. Animal bones, in E. Schia (ed.) *De arkeologiske utgravninger i Gamlebyen, Oslo. Volume 5: 'Mindets tomt'–'Sondre Felt' animal bones, moss-, plant-, insect- and parasite remains*: 153–95. Oslo: Alvheim & Eide.

Limburg, K. E., Y. Walther, B. Hong, C. Olson & J. Storå. 2008. Prehistoric versus modern Baltic Sea cod fisheries: selectivity across the millennia. *Proceedings of the Royal Society B* 275(1652): 2659–66. doi:10.1098/rspb.2008.0711

Lõugas, L. 1997. *Postglacial Development of Vertebrate Fauna in Estonian Water Bodies: a palaeozoological study*. (Dissertationes Biologicae Universitatis Tartuensis 32). Tartu: Tartu University Press.

Lõugas, L. 1999. Postglacial development of fish and seal faunas in the eastern Baltic water systems, in N. Benecke (ed.) *The Holocene History of the European Vertebrate Fauna: modern aspects of research*: 185–200. Berlin: Marie Leidorf.

Lõugas, L. 2001. Development of fishery during the 1st and 2nd millennia AD in the Baltic region. *Journal of Estonian Archaeology* 5: 128–47.

Lõugas, L. 2007. Analyses of fish bones from Valmiera. Report on file, Institute of History, University of Tallinn, Estonia.

Lõugas, L. 2008. Fishing during the Viking Age in the eastern and western Baltic Sea, in P. Béarez, S. Grouard & B. Clavel (eds) *Archéologie du poisson, 30 ans d'archéo-ichtyologie au CNRS*: 27–33. Antibes: Éditions APDCA.

Lõugas, L. forthcoming. Long and short distance fish trade during the Middle Ages in the eastern Baltic region, in S. Gabriel & E. Reitz (eds) *Fishing through Time: Archaeoichthyology, biodiversity, ecology and human impact on aquatic environments*. Lisbon: Direcção Geral do Património Cultural.

Lõugas, V. 1991. Investigation of Pöide (Kahutsi) ancient monuments. *Proceedings of the Estonian Academy of Sciences, Humanitarian and Social Sciences* 40: 373–5.

Lõugas, V. & M. Mägi-Lõugas. 1994. Investigation of ancient monuments at Pöide 1991–1992. *Proceedings of the Estonian Academy of Sciences, Humanitarian and Social Sciences* 43: 27–33.

Mägi, M. 2005. Viking Age harbour site at Tornimäe, eastern Saaremaa. *Archaeological Fieldwork in Estonia* 2004: 65–75.

Makowiecki, D. 1998. *About the History of Fishing and its Development in Prehistoric and Medieval Times in Poland*: 113–27 (Beiträge zum Oder-Projekt 5). Berlin: Deutsches Archäologisches Institut.

Makowiecki, D. 1999. Some aspects of studies on the evolution of fish faunas and fishing in prehistoric and historic times in Poland, in N. Benecke (ed.) *The Holocene History of the European Vertebrate Fauna: modern aspects of research*: 171–84. Berlin: Marie Leidorf.

Orton, D. C., D. Makowiecki, T. de Roo, C. Johnstone, J. Harland, L. Jonsson, D. Heinrich, I. B. Enghoff, L. Lõugas, W. Van Neer, A. Ervynck, A. K. Hufthammer, C. Amundsen, A. K. G. Jones, A. Locker, S. Hamilton-Dyer, P. Pope, B. R. MacKenzie, M. Richards, T. C. O'Connell & J. H. Barrett. 2011. Stable isotope evidence for late medieval (14th–15th C) origins of the eastern Baltic cod (*Gadus morhua*) fishery. *PLoS ONE* 6: e27568. doi: 10.1371/journal.pone.0027568

Põltsam-Jürjo, I. 2013. *Pidusöögist näljahädani: söömine-joomine keskaja Tallinnas*. Tallinn: Hea Lugu OÜ.

Talvar, P. 1995. Arheoloogilised kaevamised Viru tn. 10 hoovi alal. Report on file, Tallinna Kultuuriväärtuste Amet.

Valk, H. 2000. Archaeological investigations in late prehistoric–early medieval Viljandi and in Pilistvere churchyard. *Archaeological Fieldwork in Estonia* 1999: 39–53.

Valk, H. 2001. Besieging constructions from 1223 in Viljandi. *Archaeological Fieldwork in Estonia* 2000: 65–79.

Veldre, I. 2003. Twaite shad, *Alosa fallax* (Lacépedè), in E. Ojaveer, E. Pihu & T. Saat (eds) *Fishes of Estonia*: 88–9. Tallinn: Estonian Academy Publishers.

Wigh, B. 2001. *Animal Husbandry in the Viking Age Town of Birka and its Hinterland: excavations in the black earth 1990–95*. (Birka Studies 7). Stockholm: Birka Project, Riksantikvarieämbetet.

Cod and Herring in Medieval Poland

Daniel Makowiecki, David C. Orton, and James H. Barrett

Introduction

The study of the emergence and elaboration of long-range trade in marine fish has become an important aspect of understanding the development of centralised states, market trade and urbanism in medieval Europe (e.g. Barrett *et al.* 2004; 2011; Enghoff 2000; Makowiecki 2008). The beginning of the trade in herring (*Clupea harengus*) and cod (*Gadus morhua*) in Poland provides a particularly good case study. This story combines a series of illuminating circumstances:

- naturally abundant herring and cod stocks in the Baltic Sea;
- a natural source of salt production at Kołobrzeg, which was exploited as early as the sixth–seventh centuries (Leciejewicz and Rębkowski 2007, 306);
- major rivers (the Oder and the Vistula) facilitating rapid transport inland;
- the existence of coastal emporia at sites such as Wolin and Truso by the beginning of the tenth century at the latest (Filipowiak 2004; Jagodziński 2010);
- the formation of a Polish state centred around inland Wielkopolska between the Oder and the Vistula in the mid-tenth century (it conquered Wolin in AD 967) (Buko 2007; Filipowiak 2004);
- the introduction of western food customs with, for example, the Teutonic Knights in the thirteenth to fourteenth centuries (Dembińska 1963; Pluskowski *et al.* 2011); and
- the development of later medieval (second half of the thirteenth century) towns, such as Gdańsk, Kołobrzeg, Elbląg and Szczecin, which became part of the Hansa and its well-documented trade in Baltic herring and Norwegian stockfish (dried cod) (Dollinger 1970; Nedkvitne 2014).

As elsewhere in Europe, however, there is a potential discrepancy between when fish trade is first clearly documented in written records and the (possibly earlier) date at which it began. To address this issue, it is necessary to study the bones of herring and cod recovered from archaeological excavations conducted in the emporia, elite centres, castles, rural settlements, monasteries and towns of medieval Poland. This paper does so using zooarchaeology, which allows us to identify the presence of herring and cod in sites of different date, location and function. It can also provide evidence regarding how the fish were prepared for storage (based on what skeletal elements are present or absent) and what sizes of fish are represented (with implications for where they were caught). In the case of cod, stable isotope analysis can then be applied to protein extracted from the bones to ascertain – within reasonable limits of certainty – their approximate region of catch (Barrett *et al.* 2008; Orton *et al.* 2011). Was it locally in the Baltic Sea, farther away in the North Sea, or even in Arctic Norway, where the dried cod known as stockfish were traditionally prepared?

In this chapter, we will discuss the emergence of sea fishing and the trade of marine fish from the eighth century until the end of the Middle Ages. The discussion divides naturally into two parts. The first part concerns the emergence of herring fishing and the development of herring trade with sites in the Polish interior between the ninth and eleventh centuries. It may be reasonable to argue that such inland transport was contemporary with the production of a surplus storable product that may also have been exported to other regions of northern Europe via the Baltic Sea. The second part of the paper considers finds of cod in Polish archaeological sites – which first appear in any significant quantity during the thirteenth century. Here we may witness the beginning of an import trade from the northwest, which subsequently inspired the development of a new fishery for cod from the Baltic Sea.

Herring

Herring fishing is recorded in Polish prehistory, at the late Neolithic site of Rzucewo, for example (Makowiecki 2003; Makowiecki and Van Neer 1996). Marine fishing may have subsequently become less important, although the number of fish bone assemblages between the Neolithic and the early Middle Ages is too small to confirm this. In the early Iron Age, Halstatt sites near Vistula Bay produced no marine fish (Filuk 1966a; 1966b; Makowiecki 1999; 2000a; 2003; 2012b). The fish-bone record then improves from the eighth century AD (Figure 12.1). At Kędrzyno, near Kołobrzeg, an assemblage of 550 fish bones dating from the second half of the eighth–second half of the ninth centuries includes two herring bones among dominant freshwater fish, with

salmon/trout (20 specimens) and a single specimen of sturgeon, a migratory species. This site is 12 km from the Baltic coast, and the bones presumably represent the use of a local resource rather than long-range trade. The earliest herring bones from Kołobrzeg-Budzistowo, a stronghold settlement which preceded the later town of Kołobrzeg, occur in a layer dating from the second half of the eighth–first half of the ninth century AD (Leciejewicz 2007a, 82; 2007b, 195). Moreover, herring bones dating to the first half of the eighth century have been found during excavation of the trading emporium at Truso (Makowiecki 2012b). Later, by the tenth century, herring bones appear in larger numbers in coastal sites, including Kołobrzeg-Budzistowo, the trading emporium of Wolin and the stronghold of Sopot (Table 12.1).

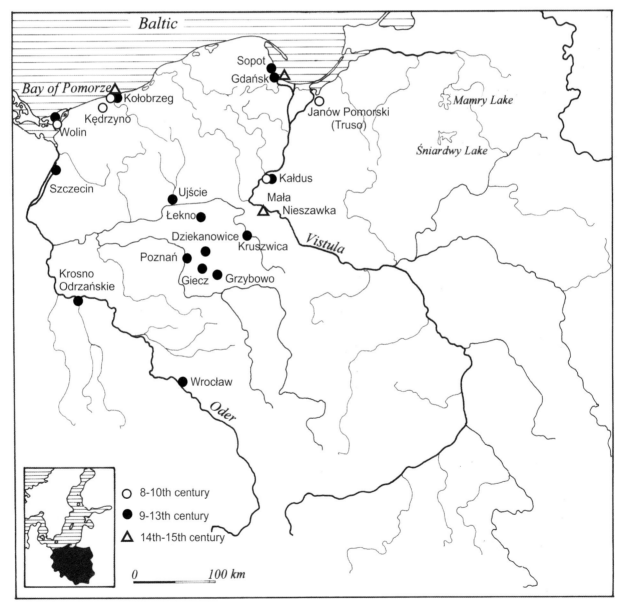

Figure 12.1. Locations of Polish sites with finds of medieval herring bones (Drawing: Daniel Makowiecki).

Table 12.1. Catalogue of Polish sites with herring bones (medieval only) and cod bones (medieval and post-medieval) (Makowiecka and Makowiecki 2006; Makowiecki 2003; 2010; 2012a; 2012b). The term agglomeration refers to urbanised complexes (strongholds and settlements) that developed in the ninth–twelfth/thirteenth centuries AD, prior to the foundation of towns with municipal rights.

Locality (site)	Herring	Cod	Pleuronectidae (flatfish)	Others	Approx. distance to Baltic Sea (km)	Total NISP fish remains	Type of site	Chronology (years or centuries AD)	Comments/reference
Kędzyrno 1	2				12	550	stronghold	2nd half 8th–2nd half 9th	
Wolin 1 (Town)	204 + Af=236 + Aa=2				0	3517	agglomeration	8th–10th	Af: *Alosa falax*, Aa: *Alosa alosa*
Kołobrzeg – Budzistowo 1	4				0	21	settlement on the Salt Island	2nd half 8th–1st half 9th	Leciejewicz (2007b)
Truso – Janów Pomorski 1	420		1		14*	4729	agglomeration	2nd half 8th–1st half 10th	*the present-day distance of Druzno Lake from Vistula Lagoon (Cieśliński 2004)
Sopot 1	2	1			0	25	stronghold	9th–10th	
Wolin 1 (Harbour)	10				0	3537	agglomeration	end 9th–13th	
Giecz 1	present				300		stronghold	9th–13th	no numbers known, 10 taxa identified
Poznań 3 – (Ostrów Tumski, Wieżowa St 2–4)	present				250		stronghold	9th–13th	no numbers known
Grzybowo 1	1				250	250	stronghold	10th–11th	sieved
Dziekanowice 22	2				220	30	settlement	10th–12th	
Gdańsk 1	19 Cl				0	501	stronghold	10th–13th	only identified as Clupeidae
Kruszwica 2–4	present				230		stronghold	10th–13th	
Wolin 1 (Town)	119 + Af=49 + Aa=1				0	6126	agglomeration	2nd half 10th–13th	Af: *Alosa falax*, Aa: *Alosa alosa*
Wrocław (Ostrów Tumski)	16				375	288	stronghold	10th–14th	
Szczecin (Podzamcze)	1				50	1193	suburb or agglomeration	2nd half 12th	
Poznań (Zagórze – Alumnat)	2				250	80	stronghold	2nd half 11th	
Ujście 1	present				180		stronghold	11th–13th	9 species identified
Krosno Odrzańskie 1C	1				215	57	settlement	11th–13th	

(Continued)

Table 12.1. Catalogue of Polish sites with herring bones (medieval only) and cod bones (medieval and post-medieval) (Makowiecka and Makowiecki 2006; Makowiecki 2003; 2010; 2012a; 2012b). The term agglomeration refers to urbanised complexes (strongholds and settlements) that developed in the ninth-twelfth/thirteenth centuries AD, prior to the foundation of towns with municipal rights. (Continued)

Locality (site)	Herring	Cod	Pleuronectidae (flatfish)	Others	Approx. distance to Baltic Sea (km)	Total NISP fish remains	Type of site	Chronology (years or centuries AD)	Comments/reference
Kałdus 3	18				120	237	stronghold	2nd half 11th	site is part of the stronghold complex
Kałdus 2	10				120	1531	settlement	2nd half 12th–1st half 13th	site is part of the stronghold complex
Łekno 3	117				240	949	Cistercian monastery	1153–14th	
Szczecin (Vegetable Market)	1 Af				50	979	agglomeration	beginning 12th	Af: *Alosa falax* (after Kłyszejko et al. 2004)
Szczecin (Vegetable Market)	1 Af				50	83	agglomeration	1175–1200	Af: *Alosa falax* (after Kłyszejko et al. 2004)
Kołobrzeg – Budzistowo 1	3235	3	16		0	3418	stronghold	9th–13th	2 cod bones to end of 11th–mid-12th centuries (Leciejewicz 2007b)
Gdańsk 1	present, Aa, Cl=342	10	42		0	18681	stronghold	9th–1308	Cl: Clupeidae, Aa: *Alosa alosa*; 10 bones of cod from 1 individual dated to 13th C
Elbląg (Bednarska St 23)		4			0	831	town, wooden house of fisherman	1250–1275	
Gdańsk (Olejarna St 2)		4			0	198	suburb	1255–1295	
Gdańsk (Olejarna St 2)		16	1		0	350	town	1295–1350	
Kołobrzeg (Ratuszowa St 9–13)	28		14		0	194	town	13th–14th	
Kołobrzeg (Armii Krajowej St 19)	8	17	4		0	95	town	13th–14th	
Kołobrzeg (Rynek)		10			0	17	town	13th–15th	
Stargard Szczeciński 11a		6 +2 Ga +2 Ma			60	17	Augustinian Eremites monastery	4th quarter 13th– beginning 15th	Ga: Gadidae; Ma: *Melanogrammus aeglefinus*

(Continued)

Table 12.1. Catalogue of Polish sites with herring bones (medieval only) and cod bones (medieval and post-medieval) (Makowiecka and Makowiecki 2006; Makowiecki 2003; 2010; 2012a; 2012b). The term agglomeration refers to urbanised complexes (strongholds and settlements) that developed in the ninth–twelfth/thirteenth centuries AD, prior to the foundation of towns with municipal rights. (Continued)

Locality (site)	Herring	Cod	Pleuronectidae (flatfish)	Others	Approx. distance to Baltic Sea (km)	Total NISP fish remains	Type of site	Chronology (years or centuries AD)	Comments/reference
Stargard Szczeciński 11a		4			60	10	Augustinian Eremites monastery	1299–1st quarter 14th	
Gdańsk (Olejarna St 2)		17	1		0	104	town	1350–1400	
Mała Nieszawka	16	982			160	1649	castle	14th–15th	
Gdańsk (Green Gate)	55 + Asp=3	73	27		0	677	town	14th–15th	Asp: *Alosa* sp.
Kołobrzeg (Gierczak St)	1	10	3		0	54	town	2nd half 14th	
Kołobrzeg (Narutowicza St 36)		2			0	2	town	14th–15th	
Gdańsk (Na Piaskach St)		1			0	21	town	14th–17th	
Chojnice (Stary Rynek)		1			110	2	town	14th–16th	abdominal vertebrae, total length = 50–60 cm
Gdańsk (Granary Island)		123	13		0	269	town	14th–18th/19th	
Gdańsk (Green Gate)	7	40	11	1 Pm	0	280	town	14th–17th	Pm: *Scophthalmus maximus*
Gdańsk (Olejarna St 2)		1			0	12	town	1400–1550?	
Gdańsk (Kładki St 24)		40	3	2 Pm	0	137	town	15th–17th/18th	Pm: *Scophthalmus maximus*
Gdańsk (Heweliusz Square)		23		1 Bb	0	157	town	15th–17th	Bb: *Belone belone*
Gdańsk (Długie Ogrody St)		12			0		town	16th–17th	
Poznań (Żydowska St 13/14)			3		250	34	town	16th–20th	
Poznań (Szyperska St)		2			250	5	town	17th–18th	
Gdańsk (Młyńska St)			1		0	5	town	17th	
Stawiec 44		7			180	15	complex of manor house	1663–1767	
Gdańsk (Młyńska St)	9		10		0	24	town	19th–20th	
Poznań (Gołębia St 2)		1			250	1	town	19th	

The earliest *inland* medieval herring find is a single bone from a tenth–eleventh-century layer at Grzybowo, a stronghold of the Piast dynasty in Wielkopolska, 250 km from the Baltic Sea. We cannot exclude earlier occurrence of the fish at inland sites, such as Grzybowo and Poznań, but the chronology of the relevant fish assemblages is very approximate (from the ninth–thirteenth centuries; see Table 12.1). Subsequently, in the eleventh century, herring is commonly found (albeit in small numbers) at inland sites. Examples include Wrocław (375 km from the Baltic), Poznań (250 km), Kruszwica (230 km) and Kałdus 2 and 3 (150 km) (see Table 12.1). Later still, in the twelfth–fourteenth centuries, herring occurs at the inland monastic site of Łekno 3.

The processing of herring for salting and trade is now well understood based on zooarchaeological finds of waste from medieval production centres (e.g. Enghoff 1996; Chapter 13) and consumer settlements (e.g. Van Neer and Ervynck 1996, 161). Unlike cod, this species is too oily to be preserved for long-term storage by drying alone. Smoking can prolong the usable life of herring for a short time, but for long-range trade they were salted and, by the twelfth–thirteenth centuries AD if not earlier, also packed in barrels filled with brine (cf. Chapters 2–3; Chapter 13). Some bones, such as the cleithrum, were removed while gutting the fish. Most other skeletal elements, including the skull bones, remained in the salted herring, which were exported packed in barrels. In theory, fish trade can therefore be inferred by analysing whether characteristic bones are present or absent from an archaeological collection – with the caveat that there were some spatial and chronological variations in processing methods (cf. Chapter 15).

The best Polish herring assemblage with which to assess butchery methods is a collection of 3235 bones from Kołobrzeg-Budzistowo dating to between the ninth and thirteenth centuries (Makowiecki 2000b). A second good example is a group of 420 herring bones from a single feature (no. 13) at Truso – part of an assemblage of thousands of fish remains from the site as a whole (Makowiecki 2012b). Table 12.2 compares the element distribution from the stronghold settlement and trading emporium with the remains from a medieval herring production site at Selsø-Vestby in Denmark (Enghoff 1996; see also Chapter 13). The absence of cleithra from the Polish assemblages may imply that they represent salted fish of the kind that would be appropriate for export in barrels. This pattern is unlikely to result from recovery bias because the Truso material was sieved with mesh not exceeding 2 mm and the distinctive Kołobrzeg assemblage was the result of very careful hand collecting. The alternative interpretation is that the cleithra are missing due to the taphonomic vagaries of fish-bone preservation. If the fish bones do indicate salted herring, they represent consumption, given that the finished product

Table 12.2. Skeletal element distribution of herring from Kołobrzeg (Makowiecki 2000b) and Truso (Makowiecki 2012b) in Poland and Selsø-Vestby in Denmark (Enghoff 1996).

Anatomical element	Kołobrzeg	Truso	Selsø-Vestby
neurocranium	0	1	0
vomer	1	0	0
frontal	54	3	0
posttemporal	5	0	0
supraoccipital	1	0	0
exoccipital	1	0	0
prootic	0	5	0
pterotic	0	1	0
basioccipital	5	0	0
basisphenoid	4	0	0
parasphenoid	6	0	0
quadrate	11	0	0
articular	171	9	0
dentary	338	4	0
maxilla	268	14	0
metapterygoid	24	2	0
interopercular	217		0
opercular	607	9	0
subopercular	286	7	0
preopercular	713	10	0
branchial region	0	1	0
infrapharyngeal	0	0	0
cleithrum	0	0	465
supracleithrum	0	0	51
abdominal/caudal vertebra	209	308	0
rib	0	0	0
scapula	0	0	79
pterygiophore	2	0	0
coracoid	0	0	219
ceratohyal	171	20	47
epihyal	38	1	34
hyomandibular	73	1	2
urohyal	30	1	134
other	0	23	0

is represented rather than production waste. Nevertheless, the very existence of salted herring makes long-range trade plausible. The feature in Truso was interpreted as a sunken part (cellar?) of a building. The associated finds – including coins, pendants, numerous glass beads and raw amber weighing a total of 0.5 kg – are indicators of trade activity. In the case of Kołobrzeg, it is believed with great probability that salt was produced on Solna (Salt) Island, known in the Middle Ages as Mons Salis Salzberg (Leciejewicz and Rębkowski 2004, 37). Thus numerous remains of this species found in the stronghold area can be considered as a result of local fishing and at the same time preservation with salt for inland transport. The recipients were the centres of the Piast monarchy, where remains of herring are documented (Figure 12.1). At Kołobrzeg, herring constitute 95% of the fish assemblage. This species makes up *c.* 13% of the fish remains from Truso.

Turning to historical evidence, the first records of herring trade in Poland were written in the twelfth and thirteenth centuries – with some reference to earlier, eleventh-century events (Łęga 1949; Rulewicz 1994, 361–2). The most important sources are Herbord's *Life of Otto* (written AD 1158–59) and the early twelfth-century chronicle of Gallus Anonymous. The latter includes a description of the siege of Kołobrzeg by the army of Bolesław Krzywousty of the Piast dynasty from Wielkopolska: 'salted and stinking fish suffices for our forebears, we come for fresh ones splashing in the ocean' (Rulewicz 1994, 362; see Grodecki 1982, 89, 91). This passage is illuminating insofar as it implies the trade of salted herring to the Polish interior. Its emphasis on access to fresh fish is less significant. The Piast dynasty had already conquered Wolin – which has archaeological evidence of early herring fishing, as noted above – in AD 967 (Filipowiak 2004, 68). Moreover, most herring must have been salted for storage, transport and/or trade rather than eaten fresh – as implied by the consumption of what may have been cured herring at Kołobrzeg-Budzistowo itself (see above).

The socio-economic context of the growth in herring trade to the Polish interior would appear to be the formation of the Polish state and the adoption of Christianity (Leciejewicz 1985; Makowiecki 2003, 122, 144, 195; 2008). Almost all of the early inland herring finds are from strongholds or settlements that are part of stronghold complexes in Wielkopolska (Giecz, Grzybowo, Dziekanowice, Poznań, Kruszwica and Ujście, plus Krosno Odrzańskie in Lubuska Land). The creation of state infrastructure presumably combined increased demand for surplus goods with improved opportunities for long-range exchange (be it market trade or substantivist redistribution, cf. Skre 2015). Moreover, the adoption of Christianity by Mieszko I in AD 966 probably led to the adoption of Christian fasting practices among at least some elements of the population.

The conquest of Wolin and then Kołobrzeg by the Piast dynasty is indicative of the relationship between Polish state formation and the development of the herring trade. Wolin was a major port of trade in the tenth century, with abundant evidence of fishing for herring and related herring family (Clupeidae) species (Table 12.1) (Chełkowski *et al.* 2001; Filipiak and Chełkowski 2000a,b; Makowiecki 2003).

Kołobrzeg was a stronghold settlement and community of fishermen at this time, rather than a trading emporium. Nevertheless, its salt springs were clearly critical to the development of long-range trade in herring. Given the relationship between fish trade and Christian fasting practices, it may not be a coincidence that Kołobrzeg was one of three bishoprics of the Polish state created in the year AD 1000 (Barford 2001, 264; Leciejewicz and Rębkowski 2004).

Despite these observations, it remains unclear whether the export of salted herring by sea to regions outside Poland preceded, was contemporary with, or post-dated the tenth-century development of trade with the Polish interior. If we apply Ockham's Razor, it would seem plausible that the developments were contemporary. However, Wolin was functioning as a trading emporium from at least the beginning of the tenth century, and it is possible that herring was one of its earliest exports. Following this model, its conquest by Mieszko I in AD 967 could be interpreted as an effort to gain control of a pre-existing coastal fish trade. If this hypothesis proves to be correct, it would revise the existing historical narrative regarding the chronology and location of the earliest Baltic export trade in salted herring (cf. Chapter 2).

Another issue for future consideration is the degree to which there was continuity in the herring trade between the tenth-century developments just discussed and the formal foundation of medieval towns: at Szczecin (in AD 1243) and Kołobrzeg (in AD 1255) by the Dukes of Pomerania; at Elbląg (in AD 1237) by the Teutonic Knights; and at Gdańsk, first (in AD 1257–63) by the Duke of Gdańsk Pomerania and later (in AD 1308/1343) by the Teutonic Knights (Cieślak and Biernat 1969, 31, 41; 1988, 29, 41). When these towns were founded, the Polish state lost political and cultural influence in the whole of Pomerania. In AD 1181, the dukes of West Pomerania came under the control of Frederick I Barbarossa of Germany. From AD 1231, they were under the political control of the Margravite of Brandenburg (Gąsiorowski 1981, 138). Gdańsk and its hinterland were ruled independently by local dukes until AD 1308, when they were first conquered by the Teutonic Knights (Gąsiorowski 1981, 150). These towns became integral centres of the emerging Hansa of merchants which promoted trade between the Baltic Sea, the North Sea and (via Norwegian intermediaries) the North Atlantic (Czaja 2005; Dollinger 1970; Nedkvitne 2014). Of course, two of the Hansa's important trade goods were herring traded westward and North Atlantic stockfish (dried cod) traded eastward. Thus it is now time to turn to cod.

Cod

Fishing for cod, like herring, is evidenced in late Neolithic Poland. This species then disappears from the Polish zooarchaeological record until the ninth to tenth centuries, when a single specimen occurs at the coastal stronghold

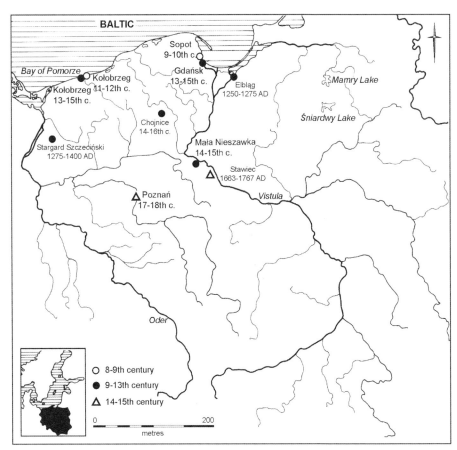

Figure 12.2. Locations of Polish sites with finds of medieval and post-medieval cod bones (Drawing: Daniel Makowiecki).

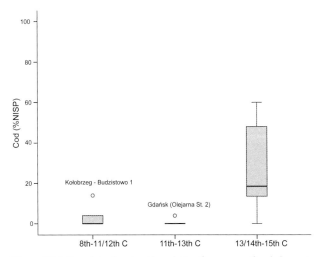

Figure 12.3. Boxplots showing the relative frequency of cod elements in Polish fish-bone assemblages of medieval date (eighth–eleventh/ twelfth centuries n = 5 sites; eleventh–thirteenth centuries n = 10 sites; thirteenth/fourteenth–fifteenth centuries n = 8 sites).

of Sopot (Makowiecki 2000a; 2003) (Figure 12.2). This, however, is an unusually early example. Another single find, from the coastal stronghold of Kołobrzeg-Budzistowo, is dated to the eleventh century. At this site an additional two cod bones occur in a layer of eleventh- to twelfth-century date (Leciejewicz 2007a, 71, 73; 2007b, 195). Cod first appear as

more than one or two specimens in layers of thirteenth-century date at the new town of Gdańsk (formally founded in AD 1257–63). Thereafter, cod bones are relatively common finds from both Gdańsk and the town of Kołobrzeg. They also occur at the towns of Elbląg and Chojnice, and at a monastic site in Stargard (near Szczecin). A large collection of cod bones dating to the fourteenth–fifteenth century was also found at Mała Nieszawka, a castle of the Teutonic Knights close to the Vistula, near Toruń. Although they are outside the main focus of this paper, later, post-medieval finds of cod are known from the town of Poznań and from a rural manor at Stawiec.

In almost all cases the number of cod bones recovered from Polish sites has been small. This is at least partly because the assemblages were recovered by hand collecting rather than sieving. From the thirteenth century onwards, however, cod bones constitute a meaningful proportion of the fish-bone collections from which they derive – despite the presence of bones of other large species, such as catfish (*Silurus glanis*), sturgeon (*Acipenser* sp.), pike (*Esox lucius*) and pike-perch (*Sander lucioperca*), which might be favoured by hand recovery (Figure 12.3). One may thus hypothesise that the recovered cod bones represent the biased residue of a larger fishery.

The small number of bones per site makes it difficult to interpret butchery patterns in terms of possible traded cod products (Table 12.3). In the North Atlantic region

Table 12.3. Skeletal element distribution of cod from Polish sites (*2 specimens of Gadidae (cod?) and 2 specimens of haddock (Melanogrammus aeglefinus)).

	Sopot - 1 (9th–10th)	Kołobrzeg – Budzistowo 1 (11th–mid-12th)	Gdańsk 1 (13th)	Elbląg – Bednarska St 23 (1250–1275)	Gdańsk – Olejarna St 2 (1255–1295)	Gdańsk – Olejarna St 2 (1295–1350)	Kołobrzeg – Ratuszowa 9–13 (13th–14th)	Kołobrzeg – Armii Krajowej St 19 (13th–14th)	Kołobrzeg – Rynek (13th–15th)	Stargard Szczeciński 11a (4th quarter 13th–beginning 15th)	Stargard Szczeciński 11a (1299–1st quarter 14th)	Gdańsk – Olejarna St 2 (1350–1400)	Mała Nieszawka (14th–15th)	Gdańsk – Green Gate (14th–15th)	Kołobrzeg – Gierczak (2nd half 14th)	Kołobrzeg - Narutowicza St 36 (14th–15th)	Gdańsk - Na Piaskach St (14th–17th)	Chojnice – Stary Rynek (14th–17th)	Gdańsk – Granary Island (14th–18th/19th)	Gdańsk – Green Gate (14th–17th)	Gdańsk – Olejarna St 2 (1400–1550?)	Gdańsk – Kładki 24 (15th–17th/18th)	Gdańsk – Plac Heweliusza (15th–17th)	Gdańsk – Długie Ogrody St (16th–17th)	Poznań – Szyperska St (17th–18th)	Stawiec (1663–1767)	Poznań – Gołębia St 2 (19th)
basioccipital				1									5	1							1						
ethmoid		1							1				11														
exoccipital																											
frontal				1									93						2	1	2						
lacrimal													38														
ethmoid																					1						
opisthotic														1													
parasphenoid					2								14						2			1					
posttemporal													2	1							1						
prefrontal													1								1						
prootic													1														
pterotic													1	1													
sphenotic													6														
supraoccipital																											
vomer													10														
angular																											
articular				1		1						1	99	1					10	3	1						
dentary	3			4							1	3	141		1				22	1	4	2	3				
ectopterygoid												1	31						1			2					
entopterygoid																					1						
maxilla				3		1							107	1					8	1	2		1				
metapterygoid															2												
palatine						1							20								1						
premaxilla				1							1		55	1					7	1							
quadrate						1							25	2							1						
ceratohyal						1						1	108	3			1		11	1	1		1				

(Continued)

*Table 12.3. Skeletal element distribution of cod from Polish sites (*2 specimens of Gadidae (cod?) and 2 specimens of haddock (*Melanogrammus aeglefinus*)). (Continued)*

	(9th–10th)	(11th–mid-12th)	(13th)	(1250–1275)	(1255–1295)	(1295–1350)	(13th–14th)	(13th–14th)	(13th–15th)	(4th quarter 13th–beginning 15th)	(1299–1st quarter 14th)	(1350–1400)	(14th–15th)	(14th–15th)	(2nd half 14th)	(14th–15th)	(14th–17th)	(14th–17th)	(14th–18th/19th)	(14th–17th)	(1400–1550?)	(15th–17th/18th)	(15th–17th)	(16th–17th)	(17th–18th)	(1663–1767)	(19th)
	Sopot - 1	Kołobrzeg – Budzistowo 1	Gdańsk 1	Elbląg – Bednarska St 23	Gdańsk – Olejarna St 2	Gdańsk – Olejarna St 2	Kołobrzeg – Ratuszowa 9–13	Kołobrzeg – Armii Krajowej St 19	Kołobrzeg – Rynek	Stargard Szczeciński 11a	Stargard Szczeciński 11a	Gdańsk - Olejarna St 2	Mała Nieszawka	Gdańsk – Green Gate	Kołobrzeg – Gierczak	Kołobrzeg - Narutowicza St 36	Gdańsk - Na Piaskach St	Chojnice – Stary Rynek	Gdańsk – Granary Island	Gdańsk – Green Gate	Gdańsk – Olejarna St 2	Gdańsk – Kładki 24	Gdańsk – Plac Heweliusza	Gdańsk – Długie Ogrody St	Poznań – Szyperska St	Stawiec	Poznań – Gołębia St 2
epihyal													20	2										1			
hyomandibular													16	1	1	1			6								
hypohyal													3														
interhyal																											
interopercular													13	1							1						
opercular												1	12						4								
preopercular					2									3	1									2	2		
subopercular					1		1					2	7	1					2			1					
symplectic													19	2													
urohyal													3														
branchial region														4													
branchiostegal ray							1	1					7						1	1							
ceratobranchial																			2								
cleithrum	1	1		1	3	2	2			4*			39	2	1	1			14	5		19		2	2	1	1
postcleithrum													2	2						1				1			
supracleithrum													4	2					2	1		1					
1st vertebra													4	1													
abdominal vert.		1		4	3	2	4	4				6	15	18	5			1	22	8		5		1	1		
abdominal/caudal vert.									3	6																	
caudal vert.							1	1				2	19	9					1	6		4		3		1	
rib							1	4			1	1	32	6					6	2				5	1		
lepidotrich																							4				
pterygiophore																											
otolith													4	1							1						
not identified		10																				5					

(including Norway, Iceland and the Northern Isles of Scotland), dried cod were sometimes made by removing the cranium but leaving some vertebrae and often also the cleithrum (an appendicular element found just posterior to the skull) in the exported product (Barrett 1997; Harland 2007; Hufthammer 2003; Krivogorskaya *et al.* 2005; see Chapter 1; Chapter 18). The early find from Sopot was a cleithrum, which could thus represent either an imported or a locally caught fish. The eleventh-century specimen from Kołobrzeg-Budzistowo was a cranial element, most consistent with a local catch. In the thirteenth century, at Gdańsk Olejarna Street 2, only vertebrae were found. This could imply the presence of imported stockfish, but the number involved (four specimens) is far too small to justify such a claim on the basis of this evidence alone (see below). Less precisely dated thirteenth- to fourteenth-century collections from Gdańsk and the town of Kołobrzeg include both vertebrae and cranial elements – consistent with either local fishing or a mixture of local and imported, preserved fish.

The large sample of well-preserved cod bones from Mała Nieszawka presents a different pattern (Makowiecki 2003). It includes mainly cranial and appendicular elements, with very few vertebrae. Some cleithra are cut in half, indicating that the cod were also purposefully decapitated. Several interpretations are possible. It is likely that this collection represents a stage in the butchery either of fresh cod shipped up the Vistula River, or of fish preserved in a way that differed from that employed in the North Atlantic – perhaps the eastern Baltic dried product known as *strekfusz* in historical records (Hoffmann 2009).

Here we interpret the Mała Nieszawka collection as the remains of a Baltic cod product, based on its similarity with a broadly contemporary fifteenth–sixteenth-century assemblage from Uppsala in Sweden (Jonsson 1986). In the Uppsala example, most vertebrae had been removed from both the cod and the pike found at the site. Processed pike are known to have been a product of the medieval Baltic region (Dembińska 1963; Hoffman 2009), and they could not have been an import from the eastern North Atlantic littoral, where this species did not naturally occur in the Middle Ages. Thus the similarly butchered cod at both Uppsala and Mała Nieszawka were probably also of Baltic origin (cf. Jonsson 1986). This need not imply that the Uppsala and Mała Nieszawka fish were from the same producer – there were few pike at the latter site – but simply that they reflect the same late medieval Baltic method of preparing fish for later consumption. It may be relevant that inventories relating to Mała Nieszawka (from AD 1411 to 1435) mention cod but not stockfish, whereas for other castles of the Teutonic Knights stockfish are explicitly listed (Ziesemer 1916; 1921).

Fish size estimates can also contribute to the interpretation of whether the Polish cod were of Baltic origin or imported

from the west. Fish total length (TL) can be estimated using bone measurements and regression equations when sample sizes are large and many examples of particular skeletal elements exist (e.g. Desse and Desse-Berset 1996). This is not the case for Poland, where size estimates must instead be made using small numbers of many different skeletal elements. Thus a semi-quantitative, ordinal scale was employed, by comparing the archaeological bones with reference specimens from fish of known TL.

The resulting size distributions for a selection of key settlements and chronological groups (Figure 12.4) facilitate three observations. First, many of the cod were smaller than *c.* 70 cm – entirely consistent with local Baltic catches on the basis of comparative size data from Neolithic assemblages (Makowiecki and Van Neer 1996; Olson and Walther 2007) and modern fisheries research (Bagge *et al.* 1994; Limburg *et al.* 2008). Second, a mode between *c.* 70 cm and *c.* 100 cm TL could conceivably represent either targeted catches of the largest cod from the Baltic or imports from the North Sea/North Atlantic (where cod grow to a greater size). Third, a small number of specimens from fish larger than 100 cm seem highly unlikely to be from the Baltic Sea based on their size. If the available TL data from medieval Polish sites are re-plotted by date (Figure 12.5), it is evident that these cod of greater than 100 cm TL (possibly imported) are more characteristic of the thirteenth to fourteenth centuries than of the later Middle Ages.

To investigate these issues further, a selection of archaeological cod bones was subjected to stable isotope ($\delta^{13}C$ and $\delta^{15}N$) analysis as part of the Medieval Origins of Commercial Sea Fishing project (Barrett *et al.* 2008; Orton *et al.* 2011). First, cranial bones from Mała Nieszawka, Gdańsk and Uppsala were used as 'control' specimens to determine the isotopic character of cod from the eastern Baltic Sea during the Middle Ages. Analogous control specimens were also analysed from the Kattegat/western Baltic, the southern North Sea (including the English Channel), the northeastern Atlantic (northern Scotland and Iceland), Arctic Norway, and the northwestern Atlantic (Newfoundland). Together, the results provide a guide to the isotopic signature of cod from each major region. Control samples from medieval archaeological contexts were used because isotopic signatures can change through time, particularly in regions subject to water pollution from such sources as mineral fertilisers (Barrett *et al.* 2008; 2011 and references therein). Having established these baseline data, 'target' specimens were then chosen from Gdańsk, Kołobrzeg, Elbląg, Mała Nieszawka and Stargard Szczeciński. These were all vertebrae and cleithra – bones that might have been imported to Poland in stockfish or similar dried products (cf. Barrett 1997).

For fish with estimated TL of 50–100 cm, Discriminant Function Analysis (DFA) of the control samples was used to assign target specimens to probable source regions.

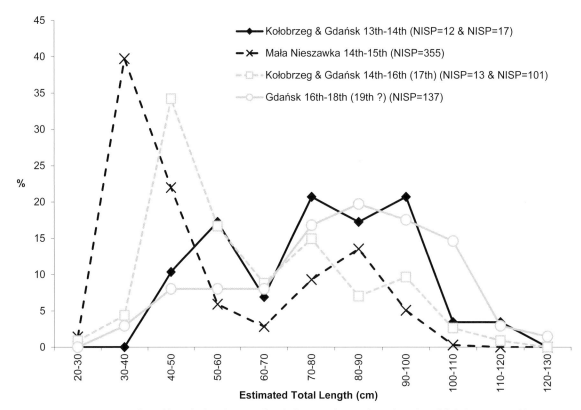

Figure 12.4. Estimated total length distribution of cod elements from selected medieval fish-bone assemblages.

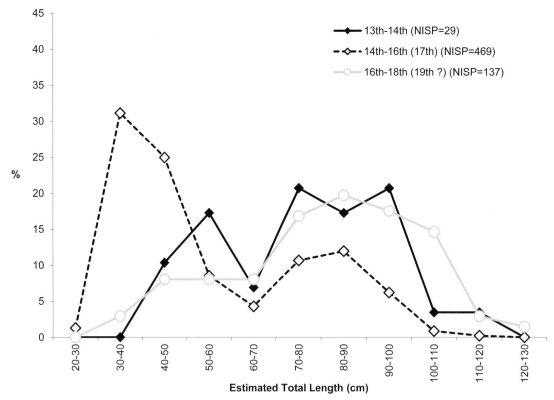

Figure 12.5. Estimated total length distribution of cod elements from two stages of the Middle Ages.

For larger fish there were only two control specimens from the eastern Baltic, preventing the use of DFA. These two specimens, however, do stand out very clearly from all the other control regions. As an alternative to formal DFA, control specimens from the possible source regions for imports (Arctic Norway, southern North Sea, northeastern Atlantic and northwestern Atlantic) were treated as a single group, and target specimens tentatively designated as either imported or local, according to their two-dimensional (Mahalanobis) distance from the mean $\delta^{13}C$ and $\delta^{15}N$ values for this group (i.e. the group centroid). Specimens falling within a 95% confidence zone around the centroid are assumed to be imports; those lying outside this envelope – and in the direction of the two large eastern Baltic controls – may have been locally caught (Orton *et al.* 2011).

The target specimens were divided into an earlier (thirteenth–fourteenth century) and a later (fourteenth–sixteenth century) group. This is necessarily an arbitrary division, and the fourteenth-century overlap is the result of a number of specimens dated to the fourteenth–fifteenth centuries being included in the later group. The target specimens from fish of 50–100 cm TL show a clear shift between the two time periods. In the earlier group, four out of five vertebrae are imported, three probably from Arctic Norway and the fourth possibly from the southern North Sea or the English Channel. The remaining vertebra and the only cleithrum analysed appear to be from the eastern Baltic. By the later period, however, the majority of vertebrae (three out of four) and all seven analysed cleithra are almost certainly local catches, with just one vertebral specimen possibly deriving from Arctic Norway. The data for larger fish (>100 cm TL) are consistent with this pattern. In the earlier period five out of six target specimens are probable imports, while one falls just outside the 95% confidence zone and is classified as indeterminate; it is, again, likely to have been imported. In the later date group there is one probable import – a vertebra – and one likely local specimen, a cleithrum.

It is thus conceivable that the first cod consumed in Poland's medieval towns were imported stockfish. Did these imports only slowly inspire the demand for a local cod fishery in the Baltic? The stable isotope data from the fourteenth–sixteenth centuries suggest that the emergence of a local, eastern Baltic, cod fishery was a later development, occurring after an initial period in which the majority of cod consumed were imported stockfish.

Historical records leave us in no doubt that dried cod (stockfish) from the North Atlantic region were traded into the towns of the southern Baltic region, probably from the thirteenth century onwards (Czaja 2004; Dembińska 1963; Molaug 2005). According to inventories of the Teutonic Knights, stockfish was consumed as early as AD 1379. On the other hand, there are also records from Szakarpawa of cod-fishing vessels belonging to the Teutonic Knights from as early as AD 1387, indicating that at least some local cod fishing was being carried out in the Bay of Gdańsk at this date (Ziesemer 1916, 54; cf. Nedkvitne 2014, 130). The archaeological evidence discussed above adds to these observations in several ways. Perhaps surprisingly, it would seem that cod was neither caught nor consumed on any scale in Poland before the thirteenth century. Perhaps there was no demand for it among the Slavs – an example of social rather than environmental causation (cf. Makowiecki 2003; Schmölcke 2004). Alternatively, it is conceivable that hydrographic conditions were poor for cod productivity before that date. The spawning success of Baltic populations of cod is known to be susceptible to fluctuations in temperature and salinity (Bagge *et al.* 1994; Cushing 1982; Enghoff *et al.* 2007). In recent times, however, these variables have influenced the northernmost limit of cod in the Baltic Sea, rather than the extent of its southern distribution.

Concluding remarks

Combining different types of data (archaeoichthyological, archaeological, isotopic and historical) has enabled us to infer a clear sequence of herring and cod fishing and consumption. First of all, herring fishing for trade preceded the exploitation of cod. In the western Slavonic and particularly the Polish context it was related to the creation of the Polish state in the tenth century by the Piast dynasty – and to the demographic and economic development of the new political structure. The Christian practice of fish consumption during fasting days may also have been an important determining factor. The consumption of cod, and its trade to the interior, was a second step. It is notable that this stage was associated with the emergence of the Hansa; the development of towns; a new stratification of society; and, particularly, the creation of burghers and merchants. Concurrently, Pomerania, formerly an integral part of the Polish state, became parts of other political entities culturally influenced by Germans and Scandinavians.

All of the historical events mentioned here probably played a role in the importation of stockfish and the adoption of cod consumption. It is worth emphasising that the application of isotopic analysis has corroborated the presence of stockfish on the tables of 'Pomeranian' citizens from the very beginning of cod consumption, as previously hypothesised by Makowiecki (2003) on the basis of macroscopic analysis. However, establishing the cultural and economic scale of early cod consumption remains at present a difficult matter due to a shortage of both conventional zooarchaeological data and stable isotope data.

Acknowledgements

The first author would like to thank the Polish colleagues and institutions who made available cod specimens for research. Among them are Romualda Uziembło (District Museum in Toruń), Henryk Paner (Archeological Museum in Gdańsk), Marcin Majewski (Museum in Stargard Szczeciński), Marian Rębkowski (Institute of Archaeology and Ethnology, Polish Academy of Sciences), Maria Kasprzycka, Marek Jagodziński, Mirosław Marcinkowski and Beata Jurkiewicz (Museum of Archaeology and History in Elbląg). The research was partly supported by 'The Medieval Origins of Commercial Sea Fishing' project funded by the Leverhulme Trust.

References

Bagge, O., F. Thurow, E. Steffensen & J. Bay. 1994. The Baltic cod. *Dana* 10: 2–28.

Barford, P. M. 2001. *The Early Slavs: culture and society in early medieval eastern Europe*. London: British Museum Press.

Barrett, J. H. 1997. Fish trade in Norse Orkney and Caithness: a zooarchaeological approach. *Antiquity* 71: 616–38.

Barrett, J. H., A. M. Locker & C. M. Roberts. 2004. 'Dark Age Economics' revisited: the English fish bone evidence AD 600–1600. *Antiquity* 78: 618–36.

Barrett, J. H., C. Johnstone, J. Harland, W. Van Neer, A. Ervynck, D. Makowiecki, D. Heinrich, A. K. Hufthammer, I. B. Enghoff, C. Amundsen, J. S. Christiansen, A. K. G. Jones, A. Locker, S. Hamilton-Dyer, L. Jonsson, L. Lõugas, C. Roberts & M. Richards. 2008. Detecting the medieval cod trade: a new method and first results. *Journal of Archaeological Science* 35(4): 850–61.

Barrett, J. H., D. Orton, C. Johnstone, J. Harland, W. Van Neer, A. Ervynck, C. Roberts, A. Locker, C. Amundsen, I. B. Enghoff, S. Hamilton-Dyer, D. Heinrich, A. K. Hufthammer, A. K. G. Jones, L. Jonsson, D. Makowiecki, P. Pope, T. C. O'Connell, T. de Roo & M. Richards. 2011. Interpreting the expansion of sea fishing in medieval Europe using stable isotope analysis of archaeological cod bones. *Journal of Archaeological Science* 38: 1516–24.

Buko, A. 2007. *The Archaeology of Early Medieval Poland: discoveries, hypotheses, interpretations*. Leiden: Brill.

Chełkowski, Z., J. Filipiak & B. Chełkowska. 2001. Studies on ichthyofauna from an archaeological excavation on Wolin-town (site 1, pit 6). *Acta Ichthyologica et Piscatoria* 31: 61–80.

Cieślak, E. & C. Biernat. 1969. *Dzieje Gdańska*. Gdańsk: Wydawnictwo Morskie.

Cieślak, E. & C. Biernat. 1988. *History of Gdańsk*. Gdańsk: Wydawnictwo Morskie.

Cieśliński, R. 2004. Application of arithmetic formulae in determining volume of sea waters inflow into Elbląska Bay – river Elbląg – lake Druzno hydrological system. *Acta Geophysica Polonica* 52: 521–39.

Cushing, D. H. 1982. *Climate and Fisheries*. London: Academic Press.

Czaja, R. 2004. Uczty rady miejskiej w średniowiecznym Elblągu, in R. Czaja, G. Nawrolska, M. Rębkowski & J. Tandecki (eds) *Archeologia et urbana*: 263–7. Elbląg: Muzeum w Elblągu.

Czaja, R. 2005. Hanza w XIII wieku nad Bałtykiem i jej rola w procesie urbanizacji, in L. Leciejewicz & M. Rębkowski (eds) *Civitas Cholbergiensis: transformacja kulturowa w strefie nadbałtyckiej w XIII wieku*, 30–40. Kołobrzeg: Wydawnictwo La Petit Café.

Dembińska, M. 1963. *Konsumpcja żywnościowa w Polsce średniowiecznej*. Wrocław: Zakład Narodowy Imienia Ossolińskich.

Desse, J. & N. Desse-Berset. 1996. On the boundaries of osteometry applied to fish. *Archaeofauna* 5: 171–9.

Dollinger, P. 1970. *The German Hansa*. London: Macmillan.

Enghoff, I. B. 1996. A medieval herring industry in Denmark and the importance of herring in eastern Denmark. *Archaeofauna* 5: 43–7.

Enghoff, I. B. 2000. Fishing in the southern North Sea region from the 1st to the 16th century ad: evidence from fish bones. *Archaeofauna* 9: 59–132.

Enghoff, I. B., B. R. MacKenzie & E. E. Nielsen. 2007. The Danish fish fauna during the warm Atlantic period (ca. 7000–3900 BC): forerunner of future changes? *Fisheries Research* 87: 167–80.

Filipiak, J. & Z. Chełkowski. 2000a. Osteological characteristics of fish remains from the early medieval town of Wolin (site 1, pit 6). *Piscaria* 27: 55–68.

Filipiak, J. & Z. Chełkowski. 2000b. Osteological characteristics of fish remains from early medieval sedimentary layers of the port in the town of Wolin. *Acta Ichthyologica et Piscatoria* 30: 135–50.

Filipowiak, W. 2004. Some aspects of the development of Wolin in the 8th–11th centuries in the light of new research, in P. Urbańczyk (ed.) *Polish Lands at the Turn of the First and the Second Millennia*: 47–74. Warsaw: Institute of Archaeology and Ethnology, Polish Academy of Sciences.

Filuk, J. 1966a. Szczątki ryb z grodziska kultury łużyckiej w Tolkmicku nad zalewem Wiślanym. *Wiadomości Archeologiczne* 32: 223–34.

Filuk, J. 1966b Pozostałości ryb z grodziska w miejscowości Łęcze, pow. Elbląg. *Wiadomości Archeologiczne* 32: 405–8.

Gąsiorowski, A. 1981. Rozbicie dzielnicowe i odrodzenie królestwa Polskiego (1138–1333), in J. Topolski (ed.) *Dzieje Polski*: 133–79. Warsaw: Państwowe Wydawnictwo Naukowe.

Grodecki, R. 1982. *Anonim tzw. Gall. Kronika Polska*. Wrocław: Ossolineum.

Harland, J. 2007. Status and space in the 'fish event horizon': initial results from Quoygrew and Earl's Bu, Viking Age and medieval sites in Orkney, Scotland, in H. Hüster Plogmann (ed.) *The Role of Fish in Ancient Time*: 63–8. (Proceedings of the 13th Meeting of the ICAZ Fish Remains Working Group). Rahden: Marie Leidorf.

Hoffmann, R. 2009. *Strekfusz*: a fish dish links Jagiellonian Cracow to distant waters, in P. Górecki & N. van Dusen (eds) *Central and Eastern Europe in the Middle Ages: a cultural history*: 116–24. London: I. B. Tauris.

Hufthammer, A. K. 2003. Med kjøtt og fisk på menyen, in O. Skevik (ed.) *Middelalder-gården i trøndelag*: 182–96. Stiklestad: Stiklestad Nasjonale Kultursenter.

Jagodziński, M. F. 2010. *Truso: między Weonodlandem a Witlandem*. Elbląg: Muzeum Archeologiczno-Historyczne w Elblągu.

Jonsson, L. 1986. Finska gäddor och Bergenfisk: ett försök att belysa Uppsalas fiskimport under medeltid och yngre Vasatid, in N. Cnattingius & T. Neréus (eds) *Uppsala stads historia. Volume 7, Från Östra Aros till Uppsala: en samling uppsatser kring det medeltida Uppsala*: 122–39. Uppsala: Almqvist & Wiksell.

Krivogorskaya, Y., S. Perdikaris & T. H. McGovern. 2005. Fish bones and fishermen: the potential of zooarchaeology in the Westfjords. *Archaeologica Islandica* 4: 31–50.

Leciejewicz, L. 1985. Z denara otrzymasz wóz świeżych śledzi, in S. K. Kuczyński & S. Suchodolski (eds) *Nummus et historia: pieniądz Europy średniowiecznej*: 103–9. Warsaw: Polskie Towarzystwo Archeologiczne i Numizmatyczne.

Leciejewicz, L. 2007a. Grodzisko w Budzistowie: badania w latach 1954–1958, in L. Leciejewicz & M. Rębkowski (eds) *Kołobrzeg: wczesne miasto nad Bałtykiem:* 31–110. Warsaw: Institute of Archaeology and Ethnology, Polish Academy of Sciences.

Leciejewicz, L. 2007b. Podstawy gospodarcze w świetle wykopalisk, in L. Leciejewicz & M. Rębkowski (ed.) *Kołobrzeg: wczesne miasto nad Bałtykiem*: 186–97. Warsaw: Institute of Archaeology and Ethnology, Polish Academy of Sciences.

Leciejewicz, L. & M. Rębkowski. 2004. Kołobrzeg: an early town on the Baltic coast, in P. Urbańczyk (ed.) *Polish Lands at the Turn of the First and the Second Millennia*: 33–46. Warsaw: Institute of Archaeology and Ethnology, Polish Academy of Sciences.

Leciejewicz, L. & M. Rębkowski. 2007. Uwagi końcowe: początki Kołobrzegu w świetle rozpoznania archeologicznego, in L. Leciejewicz & M Rębkowski (eds) *Kołobrzeg: wczesne miasto nad Bałtykiem*: 298–317. Warsaw: Institute of Archaeology and Ethnology, Polish Academy of Sciences.

Łęga, W. 1949. *Obraz gospodarczy Pomorza Gdańskiego w XII i XIII wieku*. Poznań: Instytut Zachodni.

Limburg, K. E., Y. Walther, B. Hong, C. Olson & J. Storå. 2008. Prehistoric versus modern Baltic Sea cod fisheries: selectivity across the millennia. *Proceedings of the Royal Society B* 275(1652): 2659–65. doi:10.1098/rspb.2008.0711

Makowiecki, D. 1999. Some aspects of studies on the evolution of fish faunas and fishing in prehistoric and historic times in Poland, in N. Benecke (ed.) *The Holocene Hhistory of the European Vertebrate Fauna: modern aspects of research*: 171–84. (Workshop, 6–9 April 1998 Berlin). Rahden: Marie Leidorf.

Makowiecki, D. 2000a. Catalogue of subfossil fish remains from Poland. *Archaeofauna* 9: 133–49.

Makowiecki, D. 2000b. Rybołówstwo i konsumpcja ryb w średniowiecznym Kołobrzegu, in L. Leciejewicz & M. Rębkowski (eds) *Salsa Cholbergiensis: Kołobrzeg w średniowieczu*: 223–32. Kołobrzeg: La Petit Café.

Makowiecki, D. 2003. *Historia ryb i rybołówstwa w holocenie na Niżu Polskim w świetle badań archeoichtiologicznych*. Poznań: Institute of Archaeology and Ethnology, Polish Academy of Sciences.

Makowiecki, D. 2008. Exploitation of early medieval aquatic environments in Poland and other Baltic Sea countries: an archaeozoological consideration, in *L'acqua nei secoli altomedievali*: 753–77. Spoleto: Fundazione Centro Italiano di Studi Sull'Alto Medioevo.

Makowiecki, D. 2010. *Wczesnośredniowieczna gospodarka zwierzętami i socjotopografia in Culmine na Pomorzu Nadwiślańskim*. (Mons Sancti Laurentii 6). Toruń: Instytut Archeologii, Uniwersytet Mikołaja Kopernika.

Makowiecki, D. 2012a. Badania archeozoologiczne, in K. Kwiatkowski (ed.) *Archeologia Stargardu. Volume 1: badania zachodniej części kwartału V*: 259–80. Stargard: Muzeum w Stargardzie.

Makowiecki, D. 2012b. Badania archeoichtiologiczne szczątków ze stanowiska Janów Pomorski 1, in M. Bogucki & B. Jurkiewicz (eds) *Janów Pomorski: wyniki ratowniczych badań archeologicznych w latach 2007–2008*: 302–23. Elbląg: Muzeum Archeologiczno-Historyczne.

Makowiecki, D. & W. Van Neer. 1996. Fish remains from the late Neolithic site of Rzucewo (Baltic coast, Poland). *Archaeofauna* 5: 111–19.

Makowiecka, M., & D. Makowiecki. 2006. Studia nad średniowieczną gospodarką zwierzętami w strefie środkowej Odry na podstawie analiz archeozoologicznych materiałów z Krosna Odrzańskiego, in M. Magda-Nawrocka, A. Nawojska & L. Szymczak (eds) *Archeologia w studiach nad najstarszymi dziejami Krosna Odrzańskiego*: 133–74. Krosno Odrzańskie: Gmina Krosno Odrzańskie.

Molaug, P. B. 2005. Medieval urbanisation of Norway, in L. Leciejewicz & M. Rębkowski (eds) *Civitas Cholbergiensis: transformacja kulturowa w strefie nadbałtyckiej w XIII wieku*: 115–25. Kołobrzeg: Wydawnictwo La Petit Café.

Nedkvitne, A. 2014. *The German Hansa and Bergen 1100–1600*. Köln: Böhlau.

Olson, C. & Y. Walther. 2007. Neolithic cod (*Gadus morhua*) and herring (*Clupea harengus*) fisheries in the Baltic Sea, in the light of fine-mesh sieving: a comparative study of subfossil fishbone from the late Stone Age sites of Ajvide, Gotland, Sweden and Jettböle, Åland, Finland. *Environmental Archaeology* 12: 175–85.

Orton, D. C., D. Makowiecki, T. de Roo, C. Johnstone, J. Harland, L. Jonsson, D. Heinrich, I. B. Enghoff, L. Lõugas, W. Van Neer, A. Ervynck, A. K. Hufthammer, C. Amundsen, A. K. G. Jones, A. Locker, S. Hamilton-Dyer, P. Pope, B. R. MacKenzie, M. Richards, T. C. O'Connell & J. H. Barrett. 2011. Stable isotope evidence for late medieval (14th–15th C) origins of the eastern Baltic cod (*Gadus morhua*) fishery. *PLoS ONE* 6: e27568.

Pluskowski, A., A. J. Boas & C. Gerrard. 2011. The ecology of crusading: investigating the environmental impact of holy war and colonisation at the frontiers of medieval Europe. *Medieval Archaeology* 55: 192–225.

Rulewicz, M. 1994. *Rybołówstwo Gdańska na tle ośrodków miejskich Pomorza od IX do XIII wieku*. Wrocław: Ossolineum.

Schmölcke, U. 2004. *Nutztierhaltung, Jagd und Fischfang: zur Nahrungsmittelwirtschaft des frühgeschichtlichen Handelsplatzes von Groß Strömkendorf, Landkreis Nordwestmecklenburg*. Lübstorf: Archäologisches Landesmuseum Mecklenburg-Vorpommern.

Škre, D. 2015. Post-substantivist production and trade: specialized sites for trade and craft production in Scandinavia AD c 700–1000, in J. H. Barrett & S. J. Gibbon (eds) *Maritime Societies of the Viking and Medieval World*: 156–70. Leeds: Maney/Society for Medieval Archaeology.

Van Neer, W. & A. Ervynck. 1996. Food rules and status: patterns of fish consumption in a monastic community (Ename, Belgium). *Archaeofauna* 5: 155–64.

Ziesemer, W (ed.). 1916. *Das Marienburger Ämterbuch (1375–1442)*. Danzig: A. W. Kafemann.

Ziesemar, W. 1921 *Das Grosse Ämterbuch des Deutschen Ordens*. Danzig: A. W. Kafemann.

Herring and Cod in Denmark

Inge Bødker Enghoff

Introduction

Long-term trends in herring (*Clupea harengus*) and cod (*Gadus morhua*) fishing in Denmark can be inferred from zooarchaeological evidence ranging from the Mesolithic to the modern era in date. Traditions of harvesting these species stretch into prehistory. Nevertheless, important chronological watersheds are evident based on the relative abundance of herring and cod bones in archaeological assemblages, the skeletal elements present and the sizes of fish targeted. Some of these changes – such as line fishing for large cod (in eastern Jutland and Zealand) during the Bronze Age and net fishing for herring in the Baltic Sea during the Iron Age – long pre-date the medieval focus of this book. Others – such as herring processing on a commercial scale in Zealand and line fishing for large cod in the North Sea – appear to be innovations of the Middle Ages. Changes in the absolute scale of fishing and fish consumption through time are also likely based on stable isotope analyses of human remains (e.g. Fischer *et al.* 2007; Tauber 1981a; Yoder 2010; 2012). However, obtaining a systematic isotope time-series from the Mesolithic to the end of the Middle Ages remains a challenge for the future.

Materials and methods

In this chapter fishing is described on the basis of bone assemblages from a number of selected settlements that were excavated using modern techniques, including sieving of sediment through a fine mesh. The mesh size varied between sites (Table 13.1) and, unfortunately, has not always been reported. Knowledge of the mesh size used is important because it allows the reader to assign appropriate significance to the distributions of fish species and sizes represented in each collection.

The geographical and temporal distribution of suitable fish-bone assemblages is uneven, but an attempt is made here to consider both peninsular Denmark (Jutland) and the Danish islands (Zealand etc.) (Figure 13.1). The assemblages discussed here are all interpreted as resulting from human activities, rather than being deposits of naturally deceased fish (cf. Enghoff 1987; Noe-Nygaard 1988). Dates are given as calibrated calendar years unless otherwise indicated.

The size of cod represented in the archaeological collections is important for interpreting harvesting methods. Thus chronological trends in cod fishing can be explored by considering variations in the total length (TL) of the caught cod. TL has been estimated using formulae based on measurements of premaxillae and dentaries (Cardell 1995), first and second vertebrae (Enghoff 1994), and third and fourth vertebrae.[1] Much less size variation is represented by the recovered herring bones, so TL estimates have not been used for this taxon.

Herring have been a target of Danish fishing since the Mesolithic, but never more so than in the Middle Ages. Figure 13.2 illustrates the representation of this taxon through time based on the total number of identified fish bones from a number of Danish sites. The variability is meaningful when recovery was by means of fine mesh sifting (see Table 13.1) and can in some cases be further explored based on the relative representation of different skeletal elements, which can illuminate butchery, preservation and trade.

Herring
Mesolithic to Bronze Age

The oldest known animal bone material from a Danish marine coastal site comes from the settlement Blak II, of the Mesolithic Kongemose culture (6800–5400 BC). It is an underwater site currently lying at 4 m depth off the islet of Blak in Roskilde Fjord (Zealand). Both settlement and midden areas have been identified, with radiocarbon determinations on material from the latter giving ages

Table 13.1. Relative frequency of herring and gadids in selected fish-bone assemblages from Danish sites. For the gadids, identification to species was done only on selected skeletal elements. The identified gadid species from each site are listed.

General cultural period/site	Cultural period/ chronology (years or centuries)*	Recovery method†/mesh size (mm)	Herring (% of NISP)‡	Gadidae total (% of NISP)‡	Gadid species§	Other dominant taxa‖	No. identified fish bones	Reference(s)
MESOLITHIC								
Blak II	M/KM	5	+	0	–	Flatfish	10,041	Enghoff 2009b
Stationsvej 19	KM	3.5	4	79	Cod!! pollack, whiting	Flatfish	4547	Enghoff 1994
Vænget Nord	KM	3.5	+	44	Cod!! pollack/ saithe, whiting	Flatfish	734	Enghoff 1994
Gøngehusvej 7	KM	3.5	1	56	Cod!! pollack! saithe	Flatfish	11,116	Enghoff 2011
Maglemosegårds Vænge	KM/EBK	3.5	+	23	Cod!! pollack, whiting	Flatfish	6324	Enghoff 1994
Nivå 10	KM	4 + 1.7	1	38	Cod!! pollack, whiting	Flatfish	8606	Enghoff 2011
Yderhede	EBK	1.4	0	1	Cod	Flatfish	528	Enghoff 1994
Maglemosegård	EBK	3.5	+	48	Cod!! pollack, saithe, whiting	Flatfish	12,547	Enghoff 1983; 1994
Henriksholm-Bøgebakken	EBK	3.5	+	51	Cod!! pollack, saithe, whiting	Flatfish	10,893	Enghoff 1994
Nederst	EBK	1.5	14	23	Cod!! saithe	Flatfish	6476	Enghoff 1994
Lystrup Enge	EBK	S	1	74	Cod!! pollack, whiting	Flatfish	9278	Enghoff 1994
Nivågård	EBK	H +S 1.5	7	30	Cod!! haddock, pollack	Flatfish	4966	Enghoff 2011
Vængesø III (total)	EBK	H + S	8	5	Cod!! pollack, saithe, whiting	Flatfish	6396	Enghoff 2011
Vængesø III (experiment)	EBK	1	23	35	Cod!! pollack, saithe, whiting	Flatfish	2255	Enghoff 2011
Magleholm	EBK/TRB	3.5	+	70	Cod		670	Enghoff 1994
Smakkerup Huse	EBK	4	+	70	Cod!! whiting		9332	Larsen 2005
Ertebølle	EBK	M	1	8	Cod! saithe! pollack/saithe!		9462	Enghoff 1987; 1994
Bjørnsholm	EBK	2–3	+	10	Cod!! saithe		11,490	Enghoff 1993; 1994

(Continued)

General cultural period/site	*Cultural period/ chronology (years or centuries)**	*Recovery method†/mesh size (mm)*	*Herring (% of NISP)‡*	*Gadidae total (% of NISP)‡*	*Gadid species§*	*Other dominant taxa‖*	*No. identified fish bones*	*Reference(s)*
Norsminde	EBK	H + S	3	29	Cod!! saithe	Flatfish	8921	Enghoff 1991; 1994
Grisby	EBK	3–4	+	87	Cod		1680	Enghoff 1994
Krabbesholm II	EBK	4 + 0.5	15#	1	Cod!! saithe		3220	Enghoff 2011
NEOLITHIC								
Bjørnsholm	TRB	2–3	0	19	Cod		252	Enghoff 1993; 1994
Kainsbakke	GBK	1	+	9	Cod!! whiting	Greater weever, eel	5416	Richter 1987
BRONZE AGE								
Højbjerggård	11th–8th BC	0.5	+	100	Cod!! haddock, saithe, whiting		12,475	Enghoff 2008a
Dybendahl	11th–8th BC	H + 0.25	0	92	Cod!! haddock, ling		780	Enghoff 2008b
Engdraget 9	9th–6th BC	H + 0.5	0	100	Cod!! haddock, saithe		1012	Enghoff 2008c
IRON AGE AND LATER								
Nørre Hedegård	600 BC–AD 70	H + 5	0	0		Flatfish	2398	Enghoff 2009a
Smedegård	4th BC–2nd AD	4	0	3	Cod!! haddock	Flatfish	518	Enghoff 1999
Seden syd	4th–5th AD	3–3.5	0	11	Cod	Flatfish	2207	Gotfredsen et al. 2009
Bulbrogård	6th AD	4	1	84	Cod		1754	Gotfredsen 2006
Sortemuld	6th–7th AD	3	96	4	Cod		13,935	Enghoff 1999
Fugledegård	AD 600–1000	4	+	73	Cod! haddock		2725	Gotfredsen 2006
Gammel Lejre	7th–11th AD	4	63	22	Cod!! haddock		1537	Enghoff 1999
Ribe	AD 705–790	3–4	0	4	Cod!! haddock	Flatfish	2108	Enghoff 2000; 2006
Ribe	AD 790–820	3–4	0	5	Cod!! haddock	Flatfish	492	Enghoff 2000; 2006
Selsø-Vestby	8th–9th AD	4	18	31	Cod!! saithe	Flatfish	3485	Enghoff 1996a,b; 1999
Birkely	8th–11th AD	4	3	51	Cod!! haddock		1106	Enghoff 1999

(Continued)

General cultural period/site	Cultural period/chronology (years or centuries)*	Recovery method†/mesh size (mm)	Herring (% of NISP)‡	Gadidae total (% of NISP)‡	Gadid species§	Other dominant taxa‖	No. identified fish bones	Reference(s)
Ribe	AD 820–850	3–4	0	83	Cod! haddock! ling		382	Enghoff 2000; 2006
Viborg Søndersø	10th AD	0.5	9	+	Cod		28,883**	Enghoff 2005a
Selsø-Vestby	10th–11th AD	4	42	20	Cod!! haddock	Flatfish	860	Enghoff 1996a,b; 1999
Tårnby	AD 950–13th	3	77	17	Cod!! whiting, ling		13,136	Enghoff 1999; 2005b
Møllebjerg	11th AD	3	95	5	Cod		3160	Enghoff 1999
Kobbegård	11th AD	3	67	27	Cod		1940	Enghoff 1999
Munkerup	11th AD	3	81	19	Cod		3705	Enghoff 1999
Ribe	12th–13th AD	3–4	0	97	Cod! haddock! ling		181	Enghoff 2000; 2006
Selsø-Vestby	13th AD	M	100	0	–		>>9770	Enghoff 1996a,b; 1999
Fyrbakken	AD 1290–1306	5	0	91	Cod!! haddock! ling		423	Enghoff 2002
Kastelsbakken	AD 1290–1306	5	51	37	Cod!! haddock! pollack, saithe, whiting, ling		3161	Enghoff 2002
Holbæk	13th–15th AD	H + 1.5 + 0.5	19	39	Cod!! haddock! saithe, whiting, ling	Flatfish	3843	Enghoff 1995; 1997
Tårnby	14th–15th AD	3	62	23	Cod!! haddock		2655	Enghoff 1999; 2005b
Tårnby	16th–19th AD	3	33	29	Cod!! haddock		2078	Enghoff 1999; 2005b

*M = Maglemose culture, KM = Kongemose culture, EBK = Ertebølle culture, TRB = Funnel Beaker culture, GBK = Pitted Ware culture.

†S = unspecified sieving, H = only hand collecting, M = all sediment studied microscopically

‡Non-zero percentages < 0.5 are indicated with '+'.

§Dominant species are indicated with '!', strongly dominant species with '!!'.

‖'Flatfish' here refers to the plaice/flounder/dab group.

#An additional 3.66% of the identified fish bones are from anchovy, which is another species of clupeid fishes.

**Number of fish bones from 957 L sediment. The material is composed of hand-collected bones, bones collected on 3 mm mesh, and bones collected on fine meshes down to 0.5 mm. Fine sieving was done only on subsamples, the results of which have been standardized to bones per 957 L.

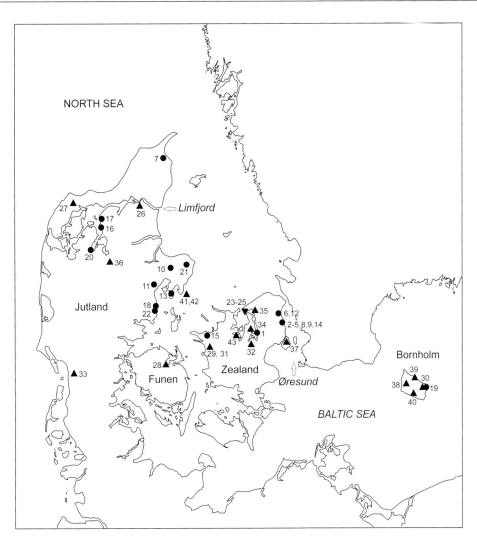

Figure 13.1. Map of Denmark showing the settlements mentioned in the text. ●: Stone Age (Mesolithic and Neolithic), ▼: Bronze Age, ▲: Iron Age, Viking Age and medieval. 1: Blak II, 2: Stationsvej 19, 3: Vænget Nord, 4: Gøngehusvej 7, 5: Maglemosegårds Vænge, 6: Nivå 10, 7: Yderhede, 8: Maglemosegård, 9: Henriksholm-Bøgebakken, 10: Nederst, 11: Lystrup Enge, 12: Nivågård, 13: Vængesø III, 14: Magleholm, 15: Smakkerup Huse, 16: Ertebølle, 17: Bjørnsholm, 18: Norsminde, 19: Grisby, 20: Krabbesholm II, 21: Kainsbakke, 22: Højbjerggård, 23: Dybendahl, 24: Engdraget 9, 25: Nr. Hedegård, 26: Smedegård, 27: Seden Syd, 28: Bulbrogård, 29: Sortemuld, 30: Fugledegård, 31: Gammel Lejre, 32: Ribe, 33: Selsø Vestby, 34: Birkely, 35: Viborg Søndersø, 36: Tårnby, 37: Møllebjerg, 38: Kobbegård, 39: Munkerup, 40: Fyrbakken, 41: Kastelsbakken, 42: Holbæk (Drawing: Knud Rosenlund).

around 6300 BC. Deposition is assumed to have ended with a transgression over the site around 6200–6100 BC. Excavation was by suction, with materials collected in nets with 5 mm apertures (Sørensen 1996). The smallest fish bones will thus generally have been missed. Despite this limitation in recovery method, six vertebrae of herring were recovered, constituting the earliest evidence of fishing for this species in Denmark (Table 13.2) (Enghoff 2009b; 2011).

Herring is ubiquitous on subsequent sites of the Kongemose culture, but in all cases accounts for only a small percentage of the total number of fish bones. Moreover, the relevant sites are all situated on the coast of Øresund, famous for its rich herring fishing during some later periods. The highest percentage of herring bones from a Kongemose

site (4%) is found at Stationsvej 19, situated on a small island where the ancient Vedbæk Fjord flowed into Øresund. The second highest percentage (1%) occurs at Nivå 10 (Jensen 2009), where the herring bones were found in three floor levels around the fireplace in a hut (Enghoff 2011).

Low-intensity herring fishing continued into the Mesolithic Ertebølle culture (5400–3900 BC). Again, a number of sites on the Øresund coast show continuity in this regard. At one, Nivågård, herring was relatively important (7% of the identified fish bones) (Enghoff 2011). In contrast, herring fishing seems to have been without importance at Grisby, a site on the Baltic island of Bornholm, which is also known for rich occurrences of herring in later times (see below) (Enghoff 1994).

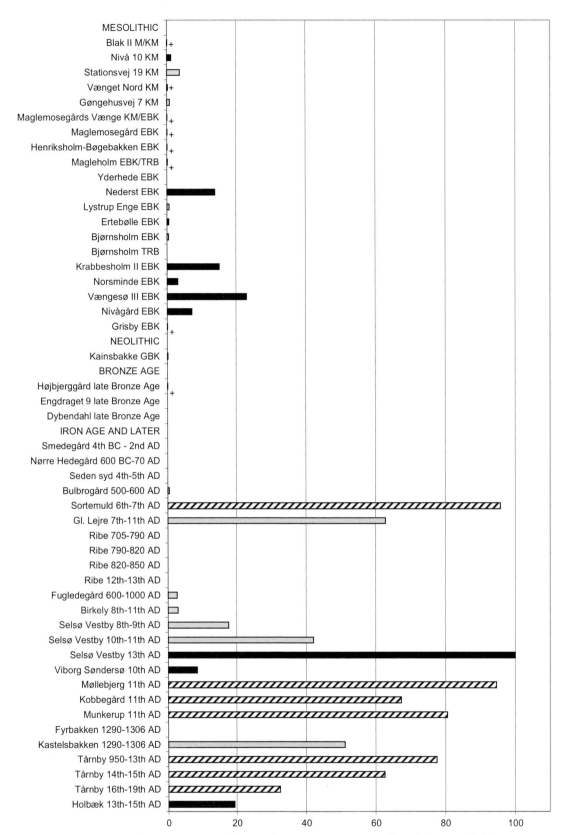

Figure 13.2. Relative frequency of herring (number of herring bones as a percentage of total identified fish bones) on Danish sites. M = Maglemose culture, K = Kongemose culture, EBK = Ertebølle culture, TRB = Funnel Beaker Culture, GBK = Pitted Ware culture. Black: sieved through 0.5–1.7 mm mesh; obliquely striped: sieved through 3 mm mesh; grey: sieved through 3.5–5 mm mesh.

a. 3 mm mesh, n = 1057

b. 3 mm + 2 mm meshes, n = 2021

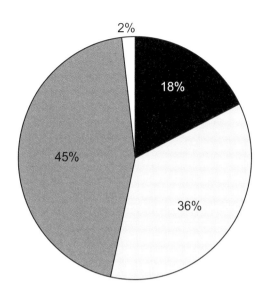

■ herring
□ gadids
▨ flatfish
□ others

c. 3 mm + 2 mm + 1 mm meshes, n = 2255

d.

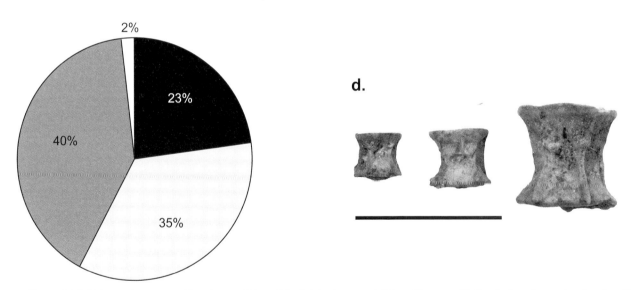

Figure 13.3. Relative frequency of herring, gadids and flatfish in the material from Vængesø III, Ertebølle culture, showing the effect of mesh size. Insert: herring vertebrae found on (from left) 1 mm, 2 mm and 3 mm mesh, respectively. Scale 1 cm (Photo: Geert Brovad).

Herring fishing was important on the east coast of Jutland. At Nederst and Vængesø III herring bones constitute 14% and 23% of the identified fish bones, respectively. At both sites, sieving through very fine meshes (1–1.5 mm) was employed (see Table 13.1).

At Vængesø III, a sieving experiment was conducted by the author in order to illustrate the effect of mesh size on the documentation of herring fishing. Sediment from two selected squares was sieved consecutively through 3 mm, 2 mm, and 1 mm mesh. The relative frequencies of herring

bones, gadid bones and bones from other fishes are quite different in the three samples (Figure 13.3) (Enghoff 2011). In the 3 mm fraction, herring bones constitute 4% of the identified fish bones. If the fraction retained by the 2 mm mesh is added, the frequency of herring increases to 18%, and adding the 1 mm fraction raises the frequency to 23%. The experiment clearly illustrates that if only 2 mm, 3 mm, or even coarser meshes are used, the presence of herring in the material may be documented, but the real importance of herring cannot be assessed. This experiment has led me to

Table 13.2. Fish bones from Blak II (c. 6300 BC, Kongemose culture), the oldest known fish-bone material from Denmark (identifications by the author).

	N	%
Eel	50	0.5
Herring	6	0.1
Cyprinids unspec.	2	<0.1
Bream	1	<0.1
Pike	31	0.3
Salmon/trout	84	0.8
Garpike	3	<0.1
Pikeperch	1	<0.1
Viviparous eelpout	1	<0.1
Turbot/brill	1	<0.1
Plaice/flounder/dab	9739	97.9
Plaice/flounder	2	<0.1
Flounder	26	0.2
Total identified	9947	
Unidentified fish	92	

conclude that, unfortunately, results of fish-bone analyses from only a minority of sites are reliable as far as the relative frequencies of fish species are concerned.

Enghoff (1994) analysed fish-bone assemblages from a large number of Danish Ertebølle sites and concluded that the main fishing during the period was done with stationary fish traps set in shallow water during the summer. This conclusion is supported by finds of remains of fish traps and weirs (e.g. Becker and Troels-Smith 1941; Petersen *et al.* 1979; Price and Gebauer 2005, 79–90). Nevertheless, finds of fish hooks and leister prongs, and possible finds of fishing nets, indicate that other methods were used as well (Andersen 1993; Enghoff 1994; Larsen 2005; Price and Gebauer 2005, 115–122).

There are several famous Stone Age kitchen middens along the Limfjord in northern Jutland. One of these, Krabbesholm II, is stratified, with layers from the Mesolithic Ertebølle culture directly overlaid by layers from the earliest phase of the Neolithic, namely, the early Funnel Beaker culture (Andersen 2005). Herring is one of the important species in both layers (Table 13.3; see also Figure 13.2) (Enghoff 2011).

Interestingly, at this site the herring is accompanied by another well-known and related species, the anchovy (*Engraulis encrasicolus*) (4%) (Table 13.3; Figure 13.4). It occurred in Danish waters during the Ertebølle culture, when the Atlantic warm period reached its maximum, with the mean temperature *c.* 3°C higher than today (Antonsson 2006; Brooks 2003; Enghoff 2011; Enghoff *et al.* 2007; Mörner 1980). In very recent years, when sea temperatures have risen to 0.5°C above 'normal', the anchovy again

Table 13.3. Comparison of fish bones from Krabbesholm II, Ertebølle culture (Mesolithic) and transitional Ertebølle–Funnel Beaker culture (Neolithic) phase (identifications by the author).

	Ertebølle culture		Ertebølle–Funnel Beaker culture transition	
	N	%	N	%
Eel	172	46.6	1085	40.1
Clupeids unspec.			1	<0.1
Herring	42	11.4	427	15.8
Anchovy	14	3.8	103	3.8
Cyprinids unspec.			79	2.9
Roach			1	<0.1
Salmonids unspec.	5	1.4	17	0.6
Whitefish			1	<0.1
Gadids unspec.	2	0.5	20	0.7
Cod/whiting			4	0.1
Cod			8	0.3
Saithe	1	0.3		
Garpike			87 (+7?)	3.5
Three-spined stickleback	120	32.5	792	29.3
Gurnard unspec.			2	0.1
Perch			3	0.1
Atlantic horse-mackerel			2	0.1
Viviparous eelpout	3	0.8	14	0.5
Sand-eel			4	0.1
Goby unspec.			10	0.4
Black goby			1?	<0.1
Atlantic mackerel	1	0.3	26	1.0
Plaice/flounder/dab	5	1.4	9	0.3
Flounder	4	1.1	3	0.1
Total identified	369		2705	

has entered Danish waters, and in such abundance that a commercial fishing has commenced (Enghoff *et al.* 2007).

The early Funnel Beaker culture of the Neolithic is represented by unmixed layers at a few sites, such as Magleholm, Norsminde and Bjørnsholm. Few fish bones have been found in these layers. The highest number (252) is from Bjørnsholm (Table 13.4). In the Mesolithic layers from Bjørnsholm, 40 herring bones were found, constituting less than 0.5% of the identified fish. There were no herring bones at all in the much smaller Neolithic sample (Enghoff 1993).

The much lower total numbers of fish bones in the Neolithic layers suggest that fishing was much less intensive. This agrees with results from δ13C analyses, which show that Mesolithic Danes subsisted mainly on marine food,

Table 13.4. Comparison of fish bones from Bjørnsholm, Ertebølle culture (Mesolithic) and Funnel Beaker culture (Neolithic) (data from Enghoff 1993).

	Ertebølle culture		Funnel Beaker culture	
	N	%	N	%
Smooth-hound	1	0	2	0.8
Spurdog	10	0.1	1	0.4
Common stingray			1	0.4
Eel	6460	56.1	154	61.1
Clupeids unspec.	12	0.1		
Herring	40	0.3		
Cyprinids unspec.	1511	13.1	8	3.2
Roach	111	1.0	5	2
Rudd	8	0.1	1	0.4
Tench	9	0.1		
Pike	26	0.2		
Salmon/trout	20	0.2		
Trout	1	<0.1		
Whitefish	2	<0.1		
Gadids unspec.	898	7.8	41	16.3
Cod	253	2.2	7	2.8
Saithe	8	0.1		
Garpike	41	0.4	4	1.6
Three-spined stickleback	754	6.5		
Gurnard unspec.	15	0.1	1	0.4
Grey gurnard	11	0.1		
Bullrout	8	0.1	2	0.8
European sea bass	11	0.1		
Perch	56	0.5		
Atlantic horse mackerel	25	0.2	1	0.4
Black sea bream	103	0.9	1	0.4
Viviparous eelpout	51	0.4	1	0.4
Sand-eel	2	0		
Greater weever	721	6.3	1	0.4
Gobiids unspec.	2	<0.1		
Atlantic mackerel	177	1.5	12	4.8
Flatfishes unspec.	2	<0.1		
Turbot	3	<0.1		
Plaice/flounder/dab	147	1.3	9	3.6
Flounder	17	0.1		
Total identified	11,490		252	

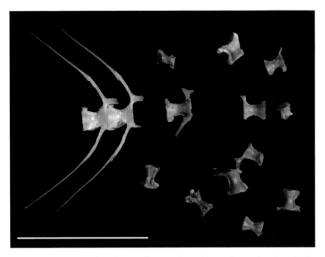

Figure 13.4. Subfossil vertebrae of anchovy from the Ertebølle culture (Krabbesholm II). Recent vertebrae are shown to the left for comparison. Scale 1 cm (Photo: Geert Brovad).

suggest that the types of fishing that were characteristic of the Ertebølle culture continued more or less unchanged into the early Neolithic. Moreover, numerous finds of fish weirs (Pedersen 1995; 1997a–c) suggest that the fishing method still involved stationary traps on a large scale.

There are only a few finds of fish bones from later in the Danish Neolithic, but quite a large sample is available from Kainsbakke, dating to the middle Neolithic Pitted Ware culture (Rasmussen 1991). Despite the use of fine sieving (1 mm mesh), only eight herring specimens (<1% of the identified fish bones) were found (Richter 1987; 1991). No herring bones were found on the late Neolithic (Single Grave culture) site of Kalvø, situated in Norsminde Fjord on the east coast of Jutland (Andersen 1983), where material was exclusively hand collected (Enghoff 2011).

Until very recently, there were few fish-bone samples from Danish Bronze Age sites, and none of these had been excavated with modern recovery techniques. Berntsson (2005) discussed Bronze Age fishing in Denmark based on this material and does not mention herring. However, during 2005–7, three late Bronze Age sites were excavated in North Zealand: Dybendahl, Engdraget 9 and Højbjerggård (Pantmann 2008; Søsted 2008). At these sites, sediment from several pits and structures was sieved through a series of meshes, down to 0.5 mm. The resulting fish-bone samples have been analysed by the author, and the results are published here. The assemblage from Højbjerggård is particularly large. The fish bones from all three sites are exceptionally well preserved, indicating that conditions for preservation were good at these sites. On this basis one may conclude that the chances of recovering herring bones are high. Nevertheless, only one single herring bone was found at Højbjerggård, and none at the two other sites (Table 13.5). Herring fishing therefore seems to have been unimportant. Most of the fish bones from these three Bronze Age sites are from large gadids (see below). This

whereas Neolithic Danes mostly lived on terrestrial food (Fischer *et al.* 2007; Price and Gebauer 2005, 137–58; Tauber 1981a; 1981b). Nevertheless, Tables 13.2 and 13.3

Table 13.5. Comparison of fish bones from three Bronze
Age sites, Højbjerggård, Dybendahl and Engdraget 9.
Percentages include probable identifications (identifications
by the author). Chronology is given in centuries.

| | Højbjerggård | | Dybendahl | | Engdraget 9 | |
| | 11th–8th BC | | 11th–8th BC | | 9th–6th BC | |
	N	%	N	%	N	%
Eel	11	0.1	3	0.4		
Herring	1	<0.1				
Cyprinids unspec.	1	<0.1				
Rudd	1	<0.1				
Pike	1	<0.1				
Gadids unspec.	5725	45.9	279 (+60?)	43.5		
Gadids unspec., not haddock	5058	40.5			698	69.0
Cod	1469 (+38?)	12.1	392	50.3	302	29.8
Whiting	1	<0.1				
Saithe/pollack	2	<0.1			2	0.2
Saithe	1 (+1?)	<0.1			1	0.1
Haddock	127 (+2?)	1.0	26	3.3	7 (+1?)	0.8
Ling			19	2.4		
Garpike			1	0.1	1	0.1
Goby (black?)	1	<0.1				
Turbot/brill	3	<0.1				
Turbot	1	<0.1				
Plaice/flounder/ dab	27	0.2				
Flounder	4	<0.1				
Total identified	12,475		780		1012	

implies that fishing was conducted with hooks and lines in
deep water, meaning a radical change compared with the Stone
Age methods discussed above.

The first and second millennia AD (Iron Age to modern)

There are a few fish-bone assemblages from Danish early Iron
Age sites, such as Nørre Hedegård (600 BC–AD 70) (Enghoff
2009a) and Smedegård (300 BC–AD 100) (Enghoff 1999),
but no herring bones were found. This may be because only
coarse mesh sieving was used for bone recovery during the
relevant excavations. The oldest Iron Age site with abundant
herring bones is Sorte Muld, on the Baltic island of Bornholm,
dating to the sixth–seventh centuries AD. Here, fine sieving
was applied because numerous small gold foil figures

(guldgubber) had been found (Enghoff 1999). No fewer than
13,332 herring bones were recovered, corresponding to 96%
of the identified fish bones. Since the mesh size employed was
3 mm, there were probably even more herring bones present
in the soil matrix that fell through the sieve.

Three fish-bone samples of the eleventh century AD from
Bornholm (Møllebjerg, Kobbegård and Munkerup) are also
strongly dominated by herring and confirm the continued
importance of this species in the Øresund and the Baltic
Sea (Enghoff 1999, bones identified by H. Høier). A 3 mm
mesh was used on these sites. As was the case at Sorte Muld,
probably far more herring bones would have been found if
a finer mesh had been used.

The many finds of herring bones from the Danish Iron
Age and later may be due to a change in emphasis, from
stationary traps to net fishing. This is in agreement with finds
of net sinkers on sites from the period (Enghoff 1999). From
this point onwards, herring bones abound in many samples.
Figure 13.2 clearly shows this conspicuous shift in the relative
importance of herring. The samples in which herring is
abundant are mainly from eastern Denmark (Enghoff 1996a;
1999). Not a single herring bone is present in the fish-bone
assemblage from the marketplace of Ribe, on the west
coast of Jutland (Enghoff 2006). This is true for all phases,
covering the period AD 705–1300 (Table 13.1; Figure 13.2).
The sieves employed had a relatively coarse mesh, 3–4 mm,
but 3 mm is usually sufficient for retaining the largest pieces
of herring bone (see above). Interestingly, however, herring
was imported to Viborg Søndersø, an inland site in Jutland
dating from the Viking Age, where it played a significant role
(Figure 13.2; Table 13.1) (Enghoff 2005a; 2007).

An increase in the exploitation of herring through time
is suggested by Selsø Vestby, on the island of Zealand. We
see an increase from 18% herring bones in the eighth–ninth
centuries to 42% in the tenth–eleventh centuries, culminating
in Denmark's oldest archaeologically attested commercial
herring industry in the Middle Ages (thirteenth–fourteenth
centuries) (Enghoff 1996a,b; 1999; note revision of original
dating). Figure 13.5 shows an unsorted sediment sample from
the medieval layers at Selsø Vestby, exclusively containing
bones from herring. Only a limited selection of skeletal
elements is represented in this sediment, namely, bones from
the hyoid arch, the gill arches and the shoulder girdle (Figure
13.6). The large amount of this material, combined with the
element representation, indicate that it represents refuse from
an industry where the gill region was removed before the
herring were further processed. The context was radiocarbon
dated to AD 1290–1380 at one sigma (lab number Ka-6987).

At Tårnby, on the Øresund coast, a number of consecutive
farms, together spanning the period from c. AD 950 to the
nineteenth century, have been excavated, allowing us to
follow the development of fishing through several centuries,
including the rise and decline of medieval herring fishing
(Table 13.1; Figure 13.2; Enghoff 2005b). The first phase

Figure 13.5. Medieval herring remains from Selsø Vestby. Scale 5 cm (Photo: Geert Brovad).

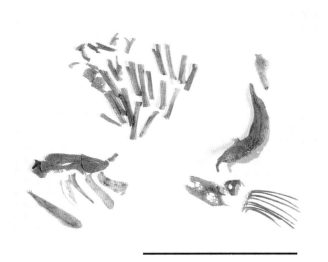

Figure 13.6. Remains of a medieval herring industry from Selsø Vestby. Exclusively bones from the hyoid arc, gill arches and shoulder girdle are represented. Scale 5 cm (Photo: Geert Brovad).

at Tårnby (*c.* AD 950 to thirteenth century), corresponds to the famous 'herring adventure' in Øresund, when it is said that the species was so abundant that it was hardly possible to row a boat, and herring could literally be caught by hand (Eriksson 1980; see Chapter 2). No less than 77% of the fish bones from this period are from herring. Herring later declined, constituting 62% of the fish bones in the fourteenth–fifteenth centuries and only 33% from the sixteenth century onwards. The decline of herring is accompanied by an increase in fishing for other species, such as eel (*Anguilla anguilla*), bullrout (*Myoxocephalus scorpius*) and perch (*Perca fluviatilis*) (Enghoff 2005b). A decline in commercial herring fishing over these centuries is documented in written sources, which show that it came to a halt in the AD 1600s (Enghoff 2005b; Frandsen and Jarrum 1992; Jarrum 1995; cf. Chapter 2).

Cod

Cod belongs to the Gadidae family, which also includes other important species, such as haddock (*Melanogrammus aeglefinus*), saithe (*Pollachius virens*), pollack (*Pollachius pollachius*), whiting (*Merlangius merlangus*) and ling (*Molva molva*). Many bones from this group, notably the vertebrae, are difficult or impossible to identify to the species level, although they can easily be recognised as 'gadids'. Furthermore, the number of bones recovered is often overwhelmingly large, and in such cases the analyst may choose only to identify selected skeletal elements to the species level and leave the rest as unspecified gadids. This general category therefore often appears as the most abundant taxon in data tables. In most cases, however, cod is by far the most abundant gadid species based on those bones which *can* be identified to species level.

The Mesolithic to Bronze Age

Cod fishing in Denmark can be documented as far back as the Mesolithic Kongemose culture (6800–5400 BC). Bones of gadids – mainly cod – strongly dominate the fish-bone material from several Kongemose sites. Stationsvej 19 at Vedbæk, on the Øresund coast, is an example, and gadids were also very important in all three floor levels in the Kongemose culture hut at Nivå 10 (Table 13.1; Figure 13.7; see also Jensen 2009).

Gadids, again mainly cod, continued to be important on sites from the Ertebølle culture – even though at this time the Atlantic warm period reached its maximum. Thus, on the Ertebølle site of Maglemosegård (Vedbæk), up to *c.* 40,000 gadid bones per square metre were found in midden layers (Figure 13.8; Aaris-Sørensen 1980). Sea temperatures are today rising again, and cod seem to be disappearing from Danish waters, but the subfossil evidence shows that the current decline cannot be due to temperature alone (Enghoff *et al.* 2007).

In contrast to herring (see above), cod and other gadids constitute a large fraction of the fishbone assemblages from

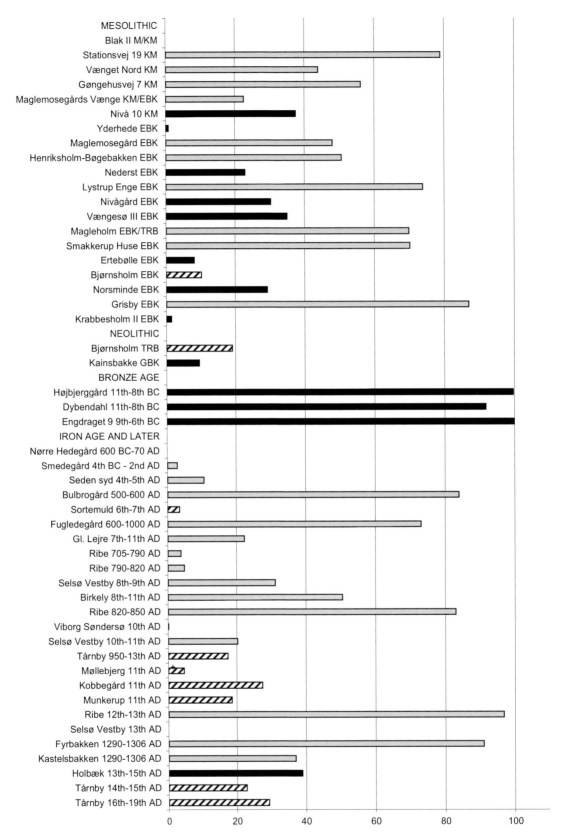

Figure 13.7. Relative frequency of gadids (number of gadid bones as a percentage of total identified fish bones) on Danish sites.
M = Maglemose culture, K = Kongemose culture, EBK = Ertebølle culture, TRB = Funnel Beaker Culture, GBK = Pitted Ware culture.
Black: sieved through 0.5–1.7 mm mesh; obliquely striped: sieved through 3 mm mesh; grey: sieved through 3.5–5 mm mesh.

Figure 13.8. Gadid bones from Maglemosegård, Ertebølle culture. This site yielded up to 40,000 bones per m². Scale 10 cm (Photo: Geert Brovad)

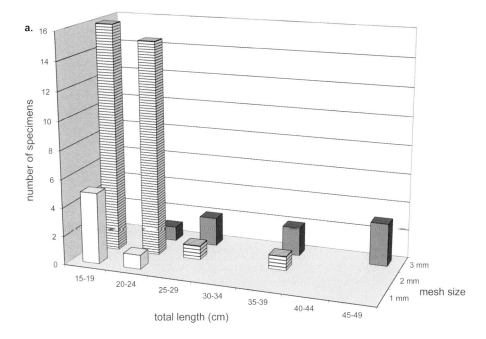

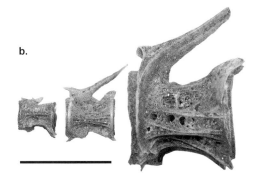

Figure 13.9. Size frequency diagram for cod (sieved material only) from Vængesø III, Ertebølle culture, showing the effect of mesh size. Insert: cod vertebrae found on 1 mm, 2 mm and 3 mm mesh, respectively. Scale 3mm (Photo: Geert Brovad).

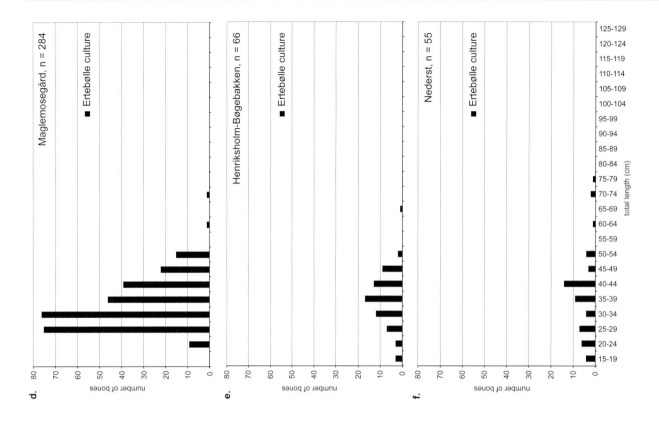

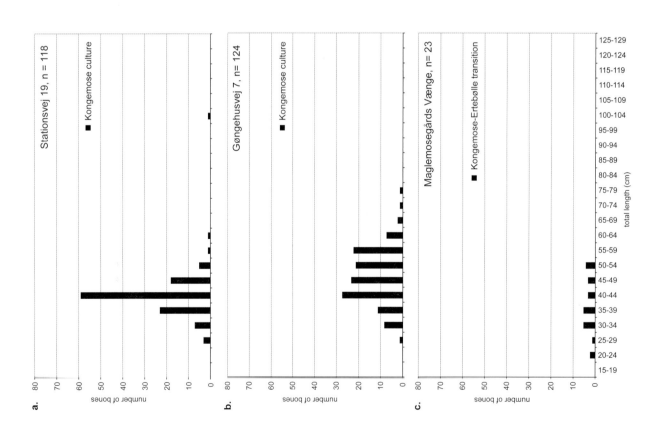

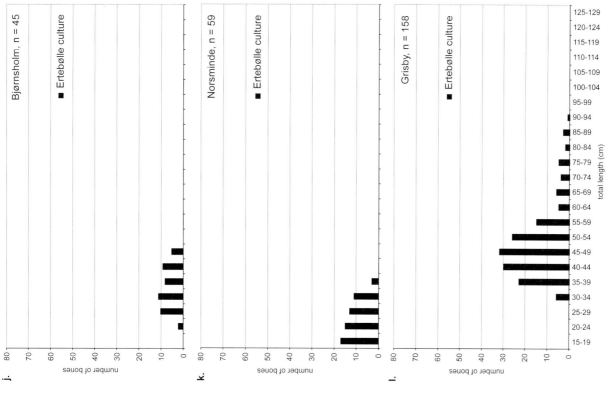

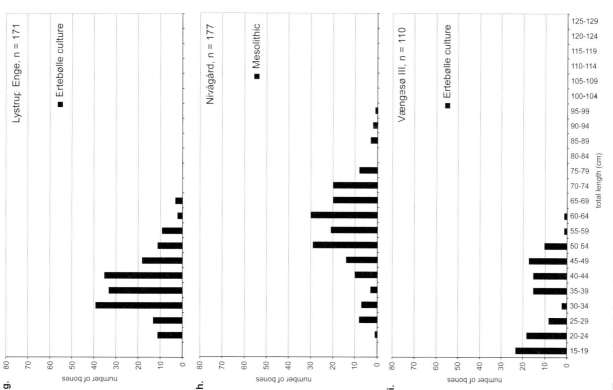

Figure 13.10. Selected size frequency diagrams for cod from the Danish Mesolithic period. a: Stationsvej 19, b: Gongehusvej 7, c: Maglemosegårds Vænge, d: Maglemosegård, e: Henriksholm-Bøgebakken, f: Nederst, g: Lystrup Enge, h: Nivågård, i: Vængesø III, j: Bjørnsholm, k: Norsminde, l: Grisby (after references in Table 13.1).

most periods in Denmark's past (Figure 13.7). Another difference between the two species is that the size of cod covers a much larger range.

We now have a good deal of sieved Mesolithic Kongemose and Ertebølle culture fish bone from Denmark, allowing us to conclude that the cod caught during this period were generally small. The vast majority were less than 50 cm in total length (TL) (Figures 13.9–13.10). The Ertebølle site Vængesø III provides an example. Cod vary between 15 and 49 cm in estimated TL. The columns of Figure 13.9 show the size distribution in material retained by 3 mm, 2 mm and 1 mm mesh. This case also makes it clear that fine sieving (e.g. 2 mm or less) is necessary to gain a realistic representation of the size distribution of cod and hence a correct interpretation of fishing methods.

Although there are very few finds of fish bone from the Danish Neolithic (3900–1700 BC), we have enough evidence to surmise that the character of cod fishing probably continued without change. This can be seen in rare stratified middens that contain continuous layers from both the Mesolithic Ertebølle culture and the Neolithic Funnel Beaker culture. Bjørnsholm (Limfjord, northern Jutland) is one such stratified kitchen midden (Andersen 1993; Enghoff 1993). The same species are present, largely in the same proportions, in both the Ertebølle and early Neolithic phases (Table 13.4).

In contradiction to the rarity of Neolithic assemblages of fish bone, several extensive systems of fish weirs and traps dating from the Neolithic have been found (Pedersen 1995; 1997a–c; Price and Gebauer 2005, 79–90). These finds support Enghoff's (1993; 1994) conclusion that the main fishing method used during the Mesolithic, involving stationary fish traps in shallow water during the summer, continued to dominate in the Neolithic.

Moving forward in time, cod bones constitute 9% of the identified fish bones in the large middle Neolithic (Pitted Ware culture) material from Kainsbakke (eastern Jutland). The TL of the cod varied between 36 cm and 53 cm. Here the dominant species were greater weever (*Trachinus draco*), followed by eel (Richter 1987).

A small late Neolithic (Single Grave culture) fish-bone assemblage has also been excavated at Kalvø (Andersen 1983). Gadids (64%), especially cod (45%), dominate the material, but interestingly there are also bones from haddock, a species which is rare on Danish Stone Age sites. The TL represented by the measured cod bones ranges from 42 to 81 cm, and the material also includes some unmeasured vertebrae other than numbers 1–4 that clearly came from individuals *c.* 1 m long (Enghoff 2011; see Figure 13.11). However, this size distribution, and the relative abundance of different species, will be biased because the fish bones from Kalvø were exclusively hand collected. For this reason the site is not included in Figure 13.7.

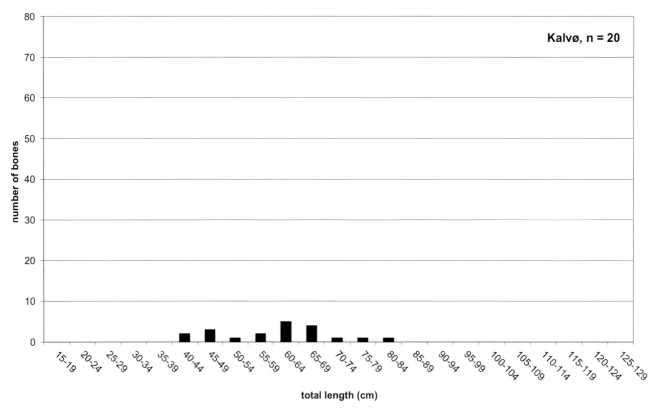

Figure 13.11. Size frequency diagram for cod from Kalvø, Single Grave culture, Late Neolithic period.

Until very recently, we knew little about Danish fishing in the Bronze Age, even though some bones from large gadids had been found on sites of this age (Berntsson 2005). Now we have analysed fish-bone assemblages from Højbjerggård, Engdraget 9 and Dybendahl, where fine sieving has been applied. Gadids, mainly cod, dominate strongly (Table 13.5; Figure 13.12). Size estimates of the cod from these sites show that, unlike in the Mesolithic, large cod were routinely caught (Figure 13.13). The majority of cod from the three sites were more than 50 cm long, often much longer. From two sites there are individuals with estimated TL of as much as 119 cm. The occurrence of the deep-water species ling at Dybendahl is noteworthy. From Højbjerggård and Dybendahl there are also smaller numbers of small cod, with TL as low as 30 cm and 20 cm, respectively, consistent with

some shallow-water fishing of the kind previously known from the Mesolithic and Neolithic.

The quantitative shift from an emphasis on small cod during the Mesolithic and Neolithic to an emphasis on large cod and such species as ling during the Bronze Age suggests a change to fishing mainly with hook and line in deep water. This is in agreement with findings of fishhooks on Bronze Age sites (Berntsson 2005). Furthermore, we know from decorations on razors and rock carvings that seagoing ships were available in this period (Berntsson 2005; Kaul 2004). Because we have few fish-bone assemblages from the Neolithic period in general, we cannot know whether the change in fishing methods may have begun already in the late Neolithic, as represented by the (admittedly unsieved) Kalvø material.

Figure 13.12. Bones of large individuals of cod from Højbjerggård, Bronze Age (Photo: Inge Enghoff).

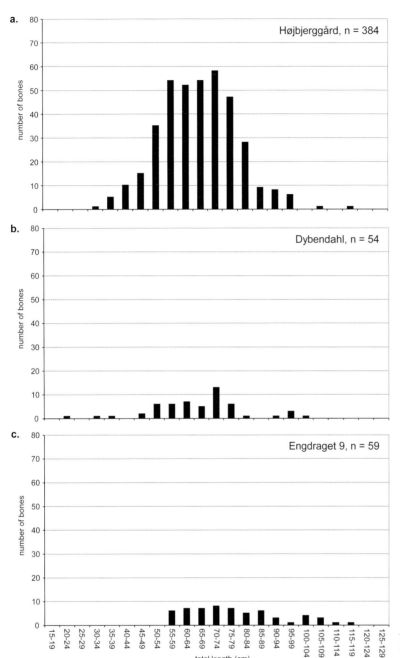

Figure 13.13. Size frequency diagrams for cod from three Bronze Age sites: a: Højbjerggård, b: Dybendahl, c: Engdraget 9.

The first and second millennia AD (Iron Age to modern)

On the banks of Lake Tissø on Zealand, one of northern Europe's largest and richest centres of power existed from the Migration Period (fifth–sixth centuries AD) to the end of the Viking Age (ninth–eleventh centuries AD). A large assemblage of fish bones was found on two sites here, Bulbrogård and Fugledegård (sixth–eleventh centuries AD). Cod measuring 35–88 cm TL were present, with the majority being 38–62 cm (Gotfredsen 2006).

Similarly, cod from the chieftain's farm of Gammel Lejre (seventh–eleventh centuries AD) were mainly 35–55 cm

long, but there were three much larger individuals, up to 109 cm long (Figure 13.14). On the roughly contemporaneous site of Birkely, on Lake Arresø in Zealand, most cod were 45–70 cm long, although the total size range was 35–119 cm (Figure 13.14).

Well-preserved fish-bone assemblages from the west coast of Jutland (Danish North Sea coast) are very rare, but one has been recovered from a Germanic Iron Age, Viking Age and medieval market place in Ribe (Figure 13.1). The material has been dated by dendrochronology to narrowly defined phases, and the bones are exceptionally well preserved.

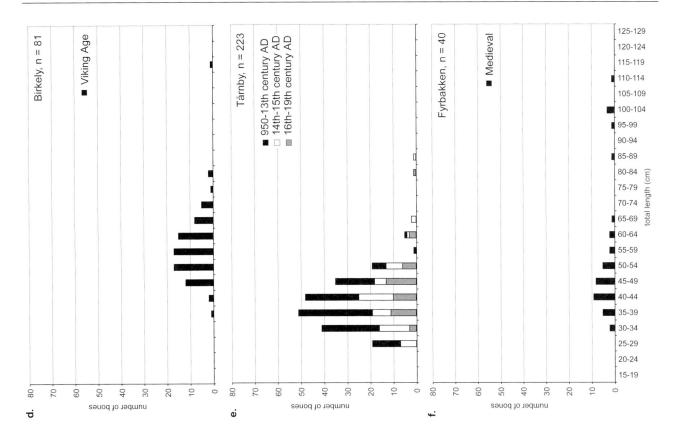

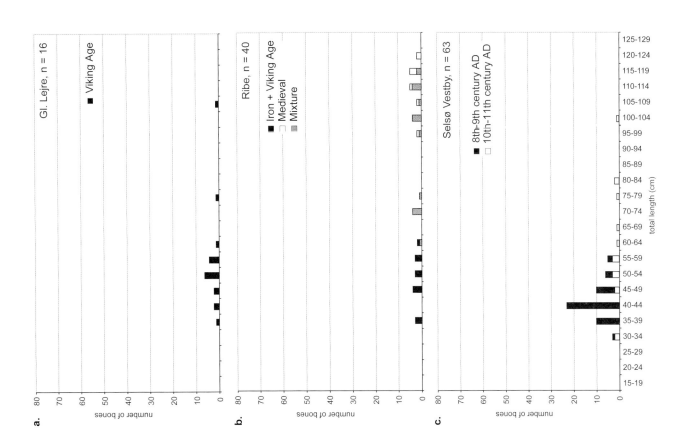

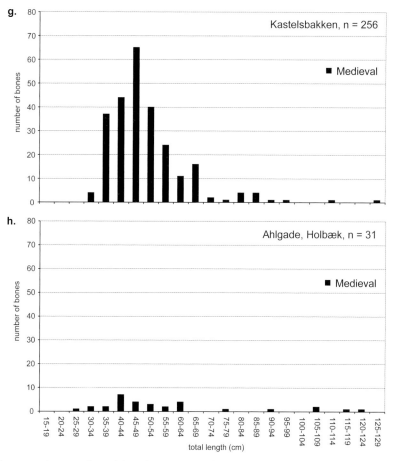

Figure 13.14. Selected size frequency diagrams for cod from the Iron Age, Viking Age and Middle Ages. a: Gammel Lejre, b: Ribe, c: Selsø Vestby, d: Birkely, e: Tårnby, f: Fyrbakken, g: Kastelsbakken, h: Ahlgade, Holbæk.

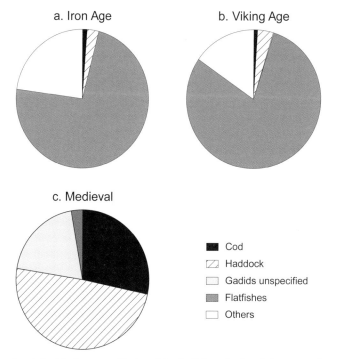

Figure 13.15. Relative frequency of cod, haddock, unspecified gadids, flatfish and other fish species from the marketplace at Ribe during three different periods: Iron Age, Viking Age, and Middle Ages.

Whereas the cod family, the gadids, constitute only about 4–5% of the fish bones from the Germanic Iron Age and Viking Age contexts, they account for 97% of the fish bones from the twelfth–thirteenth centuries (Table 13.1). Cod was exploited throughout all phases, but its relative the haddock was more important (Figure 13.15). The deep-water species ling was only caught during the medieval phase.

There was also a change in cod size from Iron Age and Viking Age Ribe to medieval Ribe (Figure 13.14). The Iron Age and Viking Age cod were of moderate size, 35–65 cm long, whereas in the Middle Ages they were very large, 95–125 cm (see also Enghoff 2006). There are no indications of production or import of stockfish: entire skeletons of cod are represented, and no cut marks were found on the bones.

At Selsø Vestby, where herring became increasingly important from the Iron Age through the Viking Age to the Middle Ages, the relative importance of cod decreased accordingly (Table 13.1). The size of the cod from Selsø Vestby also changed over time. During the eighth–ninth centuries AD they were 30–60 cm long, but during the tenth–eleventh centuries they were 30–105 cm long (Figure 13.14) (see also Enghoff 1996a; 1999).

Large cod are known from many medieval sites, but on some sites smaller cod predominate. This is for instance the case in the material from Kastelsbakken and Fyrbakken, on the small island of Hjelm (Figure 13.14). Although a few large cod are represented, the vast majority of the bones are from small individuals similar in size to those seen in the majority of Stone Age samples. This is understandable if interpreted in light of written documentation. Hjelm was the home base of the notorious Marsk Stig, an outlaw who took refuge on this small island. He and his followers may have had to be content with whatever they could catch close to the coast (Enghoff 2002).

The cod from Holbæk, a medieval town on the island of Zealand, included both small individuals and very large ones. Four-fifths of the measured bones are from individuals 30–60 cm long, but there are a few bones from large (90–130 cm TL) cod (Figure 13.14). The bones are probably partly from local fishing and partly from large imported individuals – be they from western Jutland (cf. Chapter 2), the North Atlantic (cf. Chapter 4) or elsewhere.

The Viking Age, medieval and post-medieval material from Tårnby, on the Øresund coast, reveals both the peak and the decline of Denmark's medieval 'herring adventure' (see above). Cod were also caught during all the centuries represented (Table 13.1), but unlike the medieval finds at sites like Ribe, they were consistently of small individuals (Figure 13.14). In this context they may represent a secondary catch obtained during herring fishing. A similar combination of herring and small cod has been seen at other Danish and Swedish sites, such as Møllebjerg, Kobbegård and Munkerup (Enghoff 1999).

Discussion

This paper describes fishing for herring and cod in Denmark based on zooarchaeological evidence, starting with the oldest known, that is, Mesolithic, coastal sites and ending with post-medieval assemblages from Tårnby on the Øresund. The bones of herring are present from the start of settlement on the coast. However, during the Stone Age, this species seems to have been of slight significance and during the Bronze Age herring bone finds are almost non-existent. In the Iron Age a sudden change to intensive herring fishing occurred, with the oldest finds of large amounts of herring bones dating back to the sixth–seventh centuries AD. From that time onwards there are numerous finds indicating intensive herring fishing. The sudden shift is interpreted as a result of changing fishing methods: from mainly stationary fish traps to large-scale net fishing. Herring were transported to at least one inland Danish site in the Viking Age, and we have evidence for the processing of this species for storage and probably long-range trade from the thirteenth–fourteenth-century layers at Selsø Vestby in Zealand. After the Middle Ages, the importance of this species declines in Danish zooarchaeological assemblages.

Cod fishing has always been important in Denmark, from the Mesolithic Kongemose culture onwards. The development of cod fishing can most easily be illustrated through the size of the landed cod at different times. Small cod were caught during the Mesolithic. Large cod are first seen in the late Neolithic material from Kalvø and become abundant in bone samples from the Bronze Age, probably reflecting a change in fishing method. It is likely that cod were mainly caught in stationary fish traps during the Mesolithic, early Neolithic (Funnel Beaker culture) and middle Neolithic (Pitted Ware culture). Starting in the late Neolithic at Kalvø and continuing during the late Bronze Age, we see a marked change in the fish bone assemblages, probably the result of a shift to hook-and-line fishing in deep water. This type of fishing was also conducted during the Iron Age, the Viking Age, and the Middle Ages. Smaller cod were, however, still being caught, probably at least in part as a secondary catch in connection with herring fishing. It is interesting to note, given that cod is currently predicted to disappear from Danish waters as sea temperatures rise from global warming, that there were cod in Danish waters during the post-glacial (Atlantic) warm period.

Note

1 3rd vertebra: TL (mm) = 84.527 × W(mm)$^{0.8181}$; 4th vertebra: TL (mm) = 84.365 × W(mm)$^{0.8349}$

Acknowledgements

A large number of Danish archaeologists excavated fish bones and placed the material at my disposal. Henrik Enghoff translated the manuscript into English and provided discussion and advice. Geert Brovad provided photographs and Knud Rosenlund produced the map of Denmark. Annica Cardell, Anne Birgitte Gotfredsen and Henrik Høier gave access to their unpublished results. My research project 'Regionality and Biotope Exploitation in Danish Ertebølle and Adjacent Periods' has been supported by the Danish Council for Independent Research/Humanities (formerly: the Danish Research Council for the Humanities) and the Farumgård foundation. Support has also been provided by the Heritage Agency of Denmark.

References

Aaris-Sørensen, K. 1980. Atlantic fish, reptile, and bird remains from the Mesolithic settlement at Vedbæk, North Zealand. *Videnskabelige Meddelelser Dansk Naturhistorisk Forening* 142: 139–49.

Andersen, S. H. 1983. Kalvø – A coastal site of the Single Grave culture. *Journal of Danish Archaeology* 2: 71–80.

Andersen, S. H. 1993. Bjørnsholm: a stratified køkkenmødding on the central Limfjord, North Jutland. *Journal of Danish Archaeology* 10: 59–96.

Andersen, S. H. 2005. Køkkenmøddingerne ved Krabbesholm: ny forskning i stenalderens kystbopladser. *Nationalmuseets Arbejdsmark* 2005: 151–71.

Antonsson, K. 2006. Holocene climate in central and southern Sweden. Unpublished PhD dissertation, Uppsala University.

Becker, C. J. & J. Troels-Smith. 1941. Fund af ruser fra Danmarks stenalder. *Aarbøger for Nordisk Oldkyndighed og Historie* 1941: 131–54.

Berntsson, A. 2005. *Två män i en båt – om människans relation till havet i bronsåldern.* Lund: University of Lund, Institute of Archaeology.

Brooks, S. J. 2003. Chironomid analysis to interpret and quantify Holocene climate change, in A. MacKay, R. Battarbee, J. Birks & F. Oldfield (eds) *Global Changes in the Holocene*: 328–41. London: Arnold.

Cardell, A. 1995. Fisk & fiske: osteologisk analys av fiskbensmaterialet från kökkenmöddingen Sandeplan. Unpublished Master's dissertation, Arkeologiska Institutionen vid Lunds Universitet.

Enghoff, I. B. 1983. Size distribution of cod (*Gadus morhua* L.) and whiting (*Merlangius merlangus* (L.)) (Pisces, Gadidae) from a Mesolithic settlement at Vedbæk, North Zealand, Denmark. *Videnskabelige Meddelelser fra Dansk Naturhistorisk Forening* 144: 83–7.

Enghoff, I. B. 1987. Freshwater fishing from a sea-coast settlement – the Ertebølle *locus classicus* revisited. *Journal of Danish Archaeology* 5: 62–76.

Enghoff, I. B. 1991. Fishing from the Stone Age settlement Norsminde. *Journal of Danish Archaeology* 8: 41–50.

Enghoff, I. B. 1993. Mesolithic eel-fishing at Bjørnsholm, Denmark, spiced with exotic species. *Journal of Danish Archaeology* 10: 105–18.

Enghoff, I. B. 1994. Fishing in Denmark during the Ertebølle period. *International Journal of Osteoarchaeology* 4: 65–96.

Enghoff, I. B. 1995. Fishing from medieval Holbæk – with notes to reversed *Platichthys*. *Offa* 51: 299–303.

Enghoff, I. B. 1996a. A medieval herring industry in Denmark and the importance of herring in eastern Denmark. *Archaeofauna* 5: 43–7.

Enghoff, I. B. 1996b. Denmark's first herring industry? *Maritime Archaeology Newsletter from Roskilde, Denmark* 6: 2–4.

Enghoff, I. B. 1997. Fiskeri fra Holbæk i Middelalderen. *Aarbøger for Nordisk Oldkyndighed og Historie* 1994–1995: 205–14.

Enghoff, I. B. 1999. Fishing in the Baltic region from the 5th century BC to the 16th century AD: evidence from fish bones. *Archaeofauna* 8: 41–85.

Enghoff, I. B. 2000. Fishing in the southern North Sea region from the 1st to the 16th century AD: evidence from fish bones. *Archaeofauna* 9: 59–132.

Enghoff, I. B. 2002. De fredløses måltider, in P. Asingh & N. Engberg (eds) *Marsk Stig og de fredløse på hjelm*: 221–43, 296–306, 311–17, 335–8. Ebeltoft, Højbjerg: Ebeltoft Museum, Jysk Arkæologisk Selskab.

Enghoff, I. B. 2005a. Dyreknogler fra Vikingetidens Viborg, in M. Iversen, D. E. Robinson, J. Hjermind & C. Christensen (eds) *Viborg Søndersø 1018–1030: arkæologi og naturvidenskab i et værkstedsområde fra vikingetid*: 239–68. Viborg: Stiftsmuseum, Jysk Arkæologisk Selskab.

Enghoff, I. B. 2005b. Fisk og fiskeri, in M. Svart Kristiansen (ed.) *Tårnby, gård og landsby gennem 1000 år*: 469–78. Højbjerg: Jysk Arkæologisk Selskab.

Enghoff, I. B. 2006. Fiskeknogler fra markedspladsen i Ribe, ASR 9 Posthuset, in C. Feveile (ed.) *Det ældste Ribe: udgravninger på nordsiden af Ribe å 1984–2000*: 155–66. Højbjerg: Jysk Arkæologisk Selskab.

Enghoff, I. B. 2007. Viking Age fishing in Denmark, with a particular focus on the freshwater site Viborg, methods of excavation and smelt fishing, in H. Hüster Plogmann (ed.) *The Role of Fish in Ancient Time*: 69–76. (Proceedings of the 13th meeting of the ICAZ Fish Remains Working Group, 4–9 October, Basel/Augst 2005). Rahden: Verlag Marie Leidorf.

Enghoff, I. B. 2008a. Rapport over bestemmelser af dyreknogle-materiale fra lokaliteten Højbjerggård. Report 54/2007/NFH A 2039 on file, Zoologisk Museum, Kopenhagen.

Enghoff, I. B. 2008b. Rapport over bestemmelser af dyreknogle-materiale fra lokaliteten Dybendahl. Report 14/2006/NFH A 2420 on file, Zoologisk Museum, Kopenhagen.

Enghoff, I. B. 2008c. Rapport over bestemmelser af dyreknogle-materiale fra lokaliteten Engdraget 9. Report 26/2008/NFH A 2605 on file, Zoologisk Museum, Kopenhagen.

Enghoff, I. B. 2009a. Fiskeri fra Nr. Hedegård, in M. Runge (ed.) *Nr. Hedegård: en nordjysk byhøj fra ældre jernalder*: 233–8. Højbjerg: Jysk Arkæologisk Selskabs Skrifter.

Enghoff, I. B. 2009b. Rapport Blak II. Report 33/1990/MFG 13/89 on file, Zoologisk Museum, Kopenhagen.

Enghoff, I. B. 2011. Regionality and biotope exploitation in Danish Ertebølle and adjoining periods. *Scientia Danica, Series B, Biologica* 1: 1–394.

Enghoff, I. B., B. R. MacKenzie & E. E. Nielsen. 2007. The Danish fish fauna during the warm Atlantic period (ca. 7000–3900 BC): forerunner of future changes? *Fisheries Research* 87: 167–80.

Eriksson, H. S. 1980. *Skånemarkedet.* Højbjerg: Wormianum.

Fischer, A., J. Olsen, M. Richards, J. Heinemeier, A. Sveinbjörnsdottir & P. Bennike. 2007. Coast–inland mobility and diet in the Danish Mesolithic and Neolithic: evidence from stable isotope values of humans and dogs. *Journal of Archaeological Science* 34: 2125–50.

Frandsen, S. & E. A. Jarrum. 1992. Sæsonfiskelejer, åresild og helårsfiskerlejer ved Sjællands nordkyst. *Gilleleje Museum Årbog* 29: 105–39.

Gotfredsen, A. B. 2006. Jagt og husdyrbrug i sen jernalder og vikingetid på stormandssædet ved Tissø. *Årets Gang Kalundborg og Omegns Museum* 2005: 28–34.

Gotfredsen, A. B., M. B. Henriksen, J. Kveiborg & K. G. Therkelsen. 2009. Fjordfiskere, strandjægere, håndværkere og handelsmænd i jernalderens Seden. *Fynske Minder* 2009: 73–105.

Jarrum, E. A. 1995. Torskefiskeri i Kronborg Len 1576–1600. *Gilleleje Museum* 32: 43–57.

Jensen, O. L. 2009. Dwellings and graves from the late Mesolithic site of Nivå 10, eastern Denmark, in S. MacCartan, R. Schulting, G. Warren & P. Woodman (eds) *Mesolithic Horizons*: 465–72. Oxford: Oxbow Books.

Kaul, F. 2004. *Bronzealderens religion: studier af den nordiske bronzealders ikonografi*. Copenhagen: Kongelige Nordiske Oldskriftselskab.

Larsen, C. S. 2005. Fish bones and shell, in T. D. Price & A. B. Gebauer (eds) *Smakkerup Huse: a late Mesolithic site in northwest Zealand, Denmark*: 103–13. Aarhus: Aarhus University Press.

Mörner, N.-A. 1980. A 10,700 years' paleotemperature record from Gotland and Pleistocene/Holocene boundary events in Sweden. *Boreas* 9: 283–7.

Noe-Nygaard, N. 1988. Taphonomy in archaeology with a special emphasis on man as a biasing factor. *Journal of Danish Archaeology* 6: 7–62.

Pantmann, P. 2008. Bronzealderfolket på Halsnæs. *Nomus [Medlemsblad for Nordsjællands Museumsforening]* 3: 13–18.

Pedersen, L. 1995. 7000 years of fishing: stationary fishing structures in the Mesolithic and afterwards, in A. Fischer (ed.) *Man and Sea in the Mesolithic: coastal settlement above and below present sea level: 75–86.* (Oxbow Monograph 53). Oxford: Oxbow Books.

Pedersen, L. 1997a. Settlement and subsistence in the late Mesolithic and early Neolithic, in L. Pedersen, A. Fischer & B. Aaby (eds) *The Danish Storebælt Since the Ice Age: man, sea and forest*: 109–15. Copenhagen: A/S Storbælt Fixed Link and Kalundborg Regional Museum.

Pedersen, L. 1997b. They put fences in the sea, in L. Pedersen, A. Fischer & B. Aaby (eds) *The Danish Storebælt Since the Ice Age: man, sea and forest*: 124–43. Copenhagen: A/S Storbælt Fixed Link and Kalundborg Regional Museum.

Pedersen, L. 1997c. Wooden eel weirs – a technology that lasted 7000 years, in L. Pedersen, A. Fischer & B. Aaby (eds) *The Danish Storebælt Since the Ice Age: man, sea and forest*: 144–6. Copenhagen: A/S Storbælt Fixed Link and Kalundborg Regional Museum.

Petersen, E. B., V. Alexandersen, P. V. Petersen & C. Christensen. 1979. Vedbækprojektet: ny og gammel forskning. *Søllerødbogen* 1979: 21–97.

Price, T. D. & A. B. Gebauer (eds). 2005. *Smakkerup Huse: a late Mesolithic coastal site in northwest Zealand, Denmark*. Aarhus: Aarhus University Press.

Rasmussen, L. W. 1991. Kainsbakke – en kystboplads fra yngre stenalder, in L. W. Rasmussen & J. Richter (eds) *Kainsbakke – en kystboplads fra yngre stenalder – aspects of the palaeoecology of Neolithic man*: 11–69. Grenaa: Djurslands Museum/Dansk Fiskerimuseum.

Richter, J. 1987. Evidence for a natural deposition of fish in the middle Neolithic site, Kainsbakke, East Jutland. *Journal of Danish Archaeology* 5: 116–24.

Richter, J. 1991. Kainsbakke: aspects of the palaeoecology of Neolithic man, in L. W. Rasmussen & J. Richter (ed.s *Kainsbakke – en kystboplads fra yngre stenalder – aspects of the palaeoecology of Neolithic man*: 71–127. Grenaa: Djurslands Museum/Dansk Fiskerimuseum.

Sørensen, S. A. 1996. *Kongemosekulturen i Sydskandinavien*. Jægerspris: Egnsmuseet Færgegården.

Søsted, K. 2008. Højbjerggård – en fiskerbygd fra yngre bronzealder. *Nomus* 1: 22–5.

Tauber, H. 1981a. δ^{13}C evidence for dietary habits of prehistoric man in Denmark. *Nature* 292: 332–3.

Tauber, H. 1981b. Kostvaner i forhistorisk tid – belyst ved C-13 målinger, in R. Egevang (ed.) *Det skabende menneske*, 112–26. Copenhagen: National Museum of Denmark.

Yoder, C. 2010. Diet in medieval Denmark: a regional and temporal comparison. *Journal of Archaeological Science* 37: 2224–36.

Yoder, C. 2012. Let them eat cake? Status-based differences in diet in medieval Denmark. *Journal of Archaeological Science* 39: 1183–93.

The Rise of Sea-Fish Consumption in Inland Flanders, Belgium

Wim Van Neer and Anton Ervynck

Introduction

When one wants to imagine how the medieval urban markets looked, the paintings of Joachim Beuckelaer and his contemporaries are often used as a reference. One of the most famous in this genre is the *Fish Market*, painted by Beuckelaer at Antwerpen in 1569 (Verbraeken 1986, 125) (Figure 14.1). The fish are depicted in realistic detail and are therefore easily identifiable (Gijzen 1986). Without doubt, all attention is focused upon a large basket in the lower right part of the painting, containing six large (and fresh) gadids, all representatives of the species *Gadus morhua*, the cod. Flatfish play a less prominent role (in the centre of the painting), while herring are hidden towards the back of the scene. The other species, except for a ray and some salmon cuts, represent the freshwater fauna. In general, the image feeds the idea that cod was the most important commodity on the fish market, and that it was rather common.

Of course, the informative value of Beuckelaer's *Fish Market* needs to be evaluated critically. First of all, the painting is not medieval, and it is not certain to what extent the scene depicted can be extrapolated into older periods. Moreover, art historical research proves that the market scene represents a clear symbolic framework (Kavaler 1986). The fish are there to tell a story, not as part of a realistic picture of a former commercial activity. Still, the social context of an artwork of this quality (meant to be displayed in the best rooms of the houses of the urban elite, and thus showing food items that represented the cuisine of the wealthy) indicates that, at the time of painting, cod was a 'prize beast' on the fish market. Nevertheless, the question remains: did this interpretation also hold in the High or late Middle Ages, and, more generally, how did the abundance and commercial appreciation of this fish and other marine species evolve on the markets? Ultimately, it is crucial to get new information regarding when and how sea-fish trade and consumption initially developed in inland Flanders. Of course, for this research theme to be developed, more is needed than the study of ancient paintings. The subject could be approached through the written sources, but the possibility remains that the onset of the trade goes back to times only scarcely (or even not) documented by texts. An archaeological approach is another option, albeit one with its own advantages and disadvantages. The archaeological record has inevitable geographical, chronological and social biases, as well as inherent problems of differential preservation, sampling and recovery of animal material from excavation sites.

In what follows, an overview is presented of zooarchaeological information on the development of the consumption of marine fish in inland Vlaanderen (Flanders, Belgium), particularly focusing on cod. It should be noticed that the term 'Flanders' does not refer to the medieval feudal entity with the same name, nor to the present-day political entity within Belgium. In this overview, the term is used to refer to sites in the basin of the River Schelde (Scheldt), as well as one major medieval town just outside of this hydrogeographic unit: Brugge (Bruges). As for chronology, the terms 'early', 'High' and 'late Middle Ages' will be used, here defined as covering the periods from the fifth to the ninth centuries, the tenth and eleventh centuries, and the twelfth to the fifteenth centuries AD, respectively. The post-medieval period thus begins in AD 1500. A general historical framework for the data presented can be found in the volumes of *Algemene geschiedenis der Nederlanden*, in particular, the chapters by Asaert (1980), Boelmans Kranenburg (1979; 1980) and Van Uytven (1979). More accessible to non-Dutch-speaking readers is *Medieval Flanders* by Nicholas (1992). The overview presented below builds on previously published reviews (Ervynck *et al.* 2004; Van Neer and Ervynck 2006; 2007), but it revises a number of interpretations put forward in those reviews. The location of the archaeological sites mentioned is depicted in Figure 14.2.

Figure 14.1. The Fish Market by Joachim Beuckelaer (Antwerpen, 1569) (©Wikimedia Commons, original in the National Gallery, London).

Flemish sea-fish consumption before the Middle Ages

In Flanders, the conditions at pre- and proto-historic inland sites are generally unfavourable for the preservation of bone or shell, mainly due to the presence of sandy or loamy soils (see Ervynck *et al.* 2008). Nevertheless, a small number of sites have yielded some animal remains. The scarce fish bones found represent only freshwater species, while marine molluscs are always absent (Dobney and Ervynck 2007; see also references in Ervynck *et al.* 2008). Although the small data set does not allow for any definitive interpretations, this lack of marine taxa possibly suggests that an inland trade of marine products did not exist before Roman times. Unfortunately, there is no information about economic activities at pre- and proto-historic coastal sites because the coastline from those times has been inundated due to the Holocene rise in sea level (see Baeteman and Denys 1997). It can be envisaged that pre- and proto-historic coastal settlements existed, and that they exercised some type of coastal or marine fishery, or gathering of molluscs, within the frame of a local subsistence economy, but this cannot be illustrated by archaeological finds.

During the Roman period, large marine fish from the North Sea generally remained absent at inland settlements, even at larger sites, such as the *vicus* of Tienen or the town of Tongeren (Ervynck *et al.* 2008 and references therein). Exceptions are the find of a single flatfish bone (Pleuronectidae sp.) from Nevele (Ervynck *et al.* 1997) and more numerous remains from the same taxonomic group excavated at two sites at Tournai (sites 'CV12' and 'Cloîtres': Brulet *et al.* 2004; Lentacker *et al.* in press; Pigière 2009) and Liberchies (Van Neer *et al.* 2009). Amongst the material from Tournai 'CV12' and Liberchies, plaice (*Pleuronectes platessa*) is certainly present, indicating that (some of) the flatfish consumed at these inland sites derived from coastal waters. The flatfish remains that could not be identified to species level are more difficult to use as evidence for trade, since the material could belong to flounder (*Platichthys flesus*), a species that lives in the brackish waters of estuaries but also migrates far into the fresh water of river basins. However, since these upstream flounders are always immature specimens, less than 3–4 years of age (Philippart and Vranken 1983) and measuring less than 20–25 cm standard length (SL, from the snout to the base of the tail), archaeological specimens with a reconstructed SL over 20–25 cm can be safely considered to be imports from a coastal or estuarine area. Such large

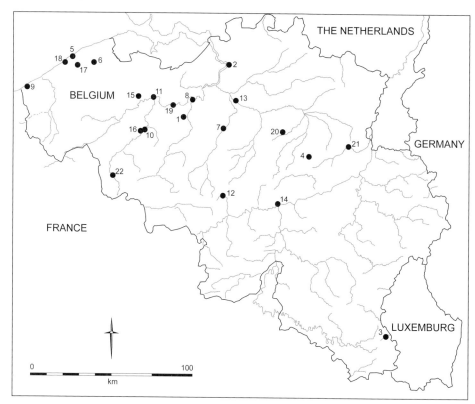

Figure 14.2. Location of the archaeological sites mentioned: 1. Aalst; 2. Antwerpen; 3. Arlon; 4. Braives; 5. Bredene; 6. Brugge; 7. Brussel; 8. Dendermonde; 9. De Panne; 10. Ename; 11. Gent; 12. Liberchies; 13. Mechelen; 14. Namur; 15. Nevele; 16. Oudenaarde; 17. Oudenburg; 18. Raversijde; 19. Schellebelle; 20. Tienen; 21. Tongeren; 22. Tournai (Drawing: Flemish Heritage).

specimens have been found at both Tournai 'Cloîtres' and Nevele. Finally, a single Roman site, 'Wijmeersen C' at Schellebelle, has yielded a small number of herring bones (Van Neer and Ervynck unpublished data). The material dates from the end of the first to the beginning of the third century (see Bogemans *et al.* 2008).

It should be mentioned that the exact catchment area of the marine fish consumed at the Roman inland sites remains unknown. Taking into account the road system that existed in Roman times, there are various possibilities: not only the North Sea coast or the estuarine areas of the rivers Scheldt, Meuse and Rhine, but also the more southern coasts of the Channel. Thus far, however, production sites have not yet been excavated. The dynamic nature of the estuarine area is responsible for the disappearance or present inaccessibility of most Roman sites. Moreover, for the Roman period, there is virtually no archaeological information from Flemish coastal sites. Like the prehistoric coast, the Roman coastline has also disappeared into the sea (Baeteman and Denys 1997, see also the map by Baeteman *et al.* in Thoen 1987, 104–5). However, some information is available from sites located near the present-day coast (on the former coastal plain). At De Panne a single fish skeletal element (of the ray family, Rajidae) was found (Thoen 1987, 67). Amongst the finds from Bredene, a skeletal element of a cod family fish (Gadidae) was recognised (Peters 1987). Consumption refuse from a dwelling platform erected in the intertidal zone, dating from the end of the first to the first half of the second century AD, demonstrates a rather

restricted exploitation of coastal waters (Demey *et al.* 2013). Consumption refuse from the late Roman *castellum* at Oudenburg, at the edge of the coastal plain, shows that some fishing took place in tidal inlets and possibly also in shallow coastal waters (Vanhoutte *et al.* 2009). In general, this consumption of marine fish at sites near the former coast must be regarded as part of the local subsistence economy rather than as the base of a flourishing inland trade.

Taking into account the large zooarchaeological data set for Roman inland sites, the scarcity of finds of large marine fish could imply that this food item was almost meaningless in economic terms. However, the few large flatfish and even the herring found may have been regarded as high-status, luxury food items, since the status of fish products varies through time (see Van Neer and Ervynck 2004). Within the Roman cuisine, fish was certainly important, a conclusion that is corroborated by the import of salted fish, *salsamenta*, from southern Europe into the newly conquered northern parts of the empire. The same is true for the varieties of fish sauce that followed the same trade route (Van Neer *et al.* 2010). Remarkably, zooarchaeological remains show that, from the second century AD onwards, local production of fish sauce, mostly using small herring and sprat (*Sprattus sprattus*), must have started in local estuaries along the North Sea coast, the Channel and possibly even farther south. This production most probably began as a competitive enterprise *vis-à-vis* southern imports, or as a reaction against a failing or insufficient supply of southern products (Van Neer and Ervynck 2004, Van Neer *et al.* 2010). A comparable local production of North Sea *salsamenta* has not yet been demonstrated by means of archaeological finds. Traces of North Sea fish sauce have been found in Tienen (Lentacker *et al.* 2004; Van Neer *et al.* 2005), Braives (Van Neer and Lentacker 1994) and Arlon (Van Neer *et al.* 2010). The production sites are unknown, but they should probably be sought in the vicinity of salt production sites (Van Neer and Ervynck 2006; 2007; Van Neer *et al.* 2010).

Finally, it must also be mentioned that, in contrast to large marine fish, marine molluscs were traded to inland sites, as proven by the finds of shell fragments of mussels (*Mytilus edulis*) and oysters (*Ostrea edulis*) – at Tongeren, for example (Vanderhoeven *et al.* 1992). The existence of this transport of marine molluscs makes it all the more remarkable that hardly any large fish were traded. Possibly, however, the indigenous culture in northern Gaul was characterised by a lack of interest in fish in general, possibly even as the result of ideology (Dobney and Ervynck 2007).

Medieval sea-fish consumption until *c.* AD 1000

The scarcity of zooarchaeological material dating back to the early Middle Ages (see Ervynck *et al.* 2008) makes it difficult to evaluate whether local fish-sauce production

persisted beyond the end of the western Roman Empire. The taxation of an unidentified kind of fish sauce mentioned in a northwestern European early medieval culinary text (Plouvier 1990; van Winter 1976) could suggest continuation of local production (a more likely option than renewal of Mediterranean imports). However, material evidence for this economic activity is still lacking, and it is not even certain whether the early medieval product described as 'fish sauce' is the same as the Roman product. In any case, production must have ceased before the High Middle Ages, since the product does not appear in culinary texts from that or younger periods – and there are no archaeological finds.

Whether the import of larger marine fish survived into early medieval times in the Scheldt basin is again difficult to evaluate due to the scarcity of early sites. A sixth–seventh-century context from Tournai ('Cloîtres' – TN2.1) contained seven flatfish bones, of which two could be identified as plaice. However, there is a strong possibility that residual Roman material is present within this assemblage. Residuality was certainly a characteristic of an older, fifth–sixth-century context from the same site containing a single (unidentifiable) flatfish bone (Pigière 2009). The continuation of the import of marine flatfish at Tournai can thus not be proven beyond doubt. Overall, the

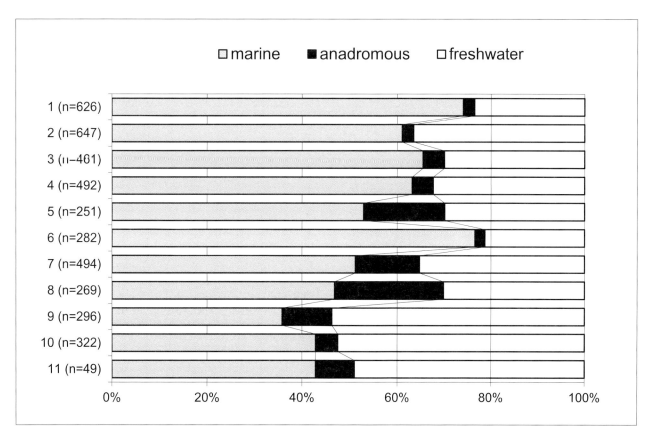

Figure 14.3. Frequencies of marine, anadromous and freshwater fish in the medieval urban refuse deposit of the portus *of Gent (sieved samples, >1 mm). The time span covered ranges from the middle of the tenth century (layer 11) to the end of the twelfth century (layer 1).*

early medieval contexts from this town are dominated by freshwater fish (Pigière 2009).

With the oldest zooarchaeological context from Mechelen being only vaguely dated (ninth–twelfth century) (Lettany 2003), all other evidence about fish consumption in the Scheldt basin before AD 1000 comes from excavations in Gent. The material represents three different habitation units that would later all fuse together when the town expanded considerably during the High Middle Ages (see Decavele 1989 for an overview of the history of this town). A first, small assemblage derives from the fill of a well that was part of a civilian settlement linked to the abbey of Sint-Baafs. The fill dates from the first half of the eighth century. The hand-collected assemblage of animal remains contained freshwater fish and marine molluscs, but no bones of sea-fish (Ervynck and Van Neer 1999). At another site, the habitation around the abbey of Sint-Pieters, animal remains have been found dating back to the ninth and tenth centuries (Van den Bremt and Vermeiren 2004). Amongst the hand-collected and sieved material (>1 mm), a small number of flatfish bones and a single herring bone were present. Within the flatfish group, the two bones out of 13 that could be identified to species were of flounder. Freshwater fish dominate the collection, representing more than 92%

of the finds. When only the sieved samples are taken into account, this becomes almost 100% (Van Neer and Ervynck unpublished data). The third site comprises a refuse deposit from the first independent urban settlement with a trade function, known as the *portus* of Gent (Laleman and Stoops 1996). The deposit has gradually accumulated over a long time span. The oldest, lowest part dates back to the middle of the tenth century, while the highest part stems from the end of the twelfth century, based on unpublished radiocarbon dates. The excavators differentiated eleven layers of equal depth. The sieved samples (>1 mm) show that already in the tenth century a significant proportion of fish consumption was based on marine species (>40%), a trend that becomes even more pronounced in the younger layers (>70% in the highest part of the deposit) (Figure 14.3). Herring and flatfish dominate the marine fish remains, with gadids only gaining some importance in the youngest layers (Figure 14.4) (Van Neer and Ervynck unpublished data). It should be noted that these observations differ from preliminary statements made earlier (Ervynck *et al.* 2004), which were based on only a partial study of the finds material.

The three sites from Gent thus appear to show clear differences in fish consumption, but some methodological considerations have to be taken into account. First, the

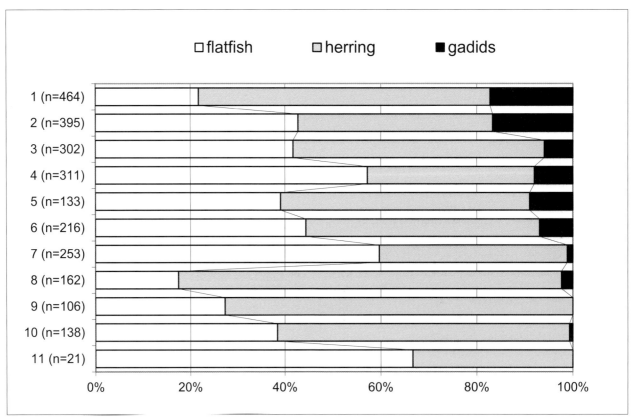

Figure 14.4. Frequencies of flatfish, herring and gadids in the medieval urban refuse deposit of the portus *of Gent (sieved samples, >1 mm). The time span covered ranges from the middle of the tenth century (layer 11) to the end of the twelfth century (layer 1).*

context from Sint-Baafs was not sieved, making it possible that, originally, herring and (small) flatfish bones were present and that these have not been recovered. However, the absence of large flatfish remains or of bones from any other large marine species, such as gadids, cannot be explained by recovery bias. Second, for the sites of Sint-Pieters and the *portus*, it remains uncertain which flatfish bones represent marine versus freshwater fishing. For Gent, this distinction is especially important because flounder does migrate that far upstream from the estuary (Poll 1947). Only a species identification of the pleuronectid bones could solve this, but an important proportion of the relevant finds remains unidentified due to the inherent lack of discriminating characteristics on the skeletal elements within this fish family. Recently, new criteria have been defined (Wouters *et al.* 2007), but this methodological breakthrough now necessitates that all collections studied previously be revisited. Since this has not yet been done, in what follows, the flatfish remains will mainly be treated as a group, although we assume that a significant proportion of the specimens was caught at sea or in the Scheldt estuary, thus representing traded items. In fact, except for the collection from Sint-Pieters, plaice has been recognised in all of the assemblages.

What can now be concluded regarding the material from Gent? The numerical importance of flatfish and herring in the oldest layer of the consumption refuse from the *portus* is striking, but it is better understandable if we accept the existence of a trade in marine fish before the middle of the tenth century. This, in turn, implies that the real origin of the trade must be sought in older periods, although the archaeological record cannot (yet) adequately document this. Possibly, there is some continuity between the import of plaice in Roman times and that at the end of the first millennium AD (corroborated by the early medieval finds from Tournai, provided these are not residual). If this putative continuity ultimately proves to be unsubstantiated, the origin of the import of marine flatfish remains obscure. The same is true for the origin of the herring trade. Evidence for a continuation of import from Roman to medieval times is lacking. Possibly, the herring trade was interrupted but started anew in the early Middle Ages. When exactly is impossible to say, but, as mentioned above, the development most probably must be situated earlier than the contexts from Gent suggest. This is corroborated by finds outside of the Scheldt basin (Pigière 2009). At Namur, at the Grognon site, a total of 19 herring bones have been recovered from the sieved samples from three different contexts, dating from the eighth to the first half of the ninth century (Pigière 2009). At another location on the same site, six ninth–tenth-century contexts have yielded herring remains, together with a single flatfish bone and a skeletal element of a gadid (Pigière *et al.* 2003). In northern France – at Paris, Reims and Compiègne, for example – herring bones have also been found at early medieval (eighth-century) sites (Clavel and Yvinec 2010).

If an early herring trade existed in the Meuse basin and in northern France, it most probably also did in the Scheldt basin. Herring, plaice and some gadid bones (unidentified cod family and haddock (*Melanogrammus aeglefinus*), but no cod) have also been found during the excavations of eighth–ninth-century Dorestad (The Netherlands) (Prummel 1983, table 13, table 22). Traditionally, this site is seen as a central place within the network of the so-called Frisian trade (see Jellema 1955; Lebecq 1986; 1992); this site thus provides a historical framework for the import of sea-fish to inland sites.

Even now that, in contrast with what has been put forward in earlier reviews (Ervynck *et al.* 2004; Van Neer and Ervynck 2006; 2007), the material from the *portus* at Gent can no longer be regarded as documenting the onset of the marine fish trade and inland consumption, the impression remains that it is the first site showing a real abundance of sea-fish remains in medieval Flanders. The process documented by the *portus* material thus represents a breakthrough in the trade in marine animal products. But why did this development (as far as is known today) only take place in the tenth century, and why in the *portus* of Gent while not in the neighbouring, partly contemporaneous settlement of Sint-Pieters? Most probably it relates to the fact that the *portus* is the core of later urban Gent, an independent concentration of traders and craftsmen. Such a group is unlikely to have produced all of its own food, and was thus dependent on what came in from the countryside or was imported through trade. In this period, most marine fish must have been transported in a processed form. Within this context, it is possible that sea-fish became a desirable commodity compared with freshwater fish or meat because it was a cheap and/or easily preserved product. Of course, freshwater fish can also be processed (most typically by smoking), but salting herring and drying flatfish may have resulted in more abundant and cheaper products. Additionally, the possibility exists that the inhabitants of the *portus* did not have the local fishing rights that were owned by the abbey sites, which would mean that freshwater fish could not be purchased without paying a high price. Finally, an increased demand for fish as a result of religious food rules issued by the Christian Church (that is, the prohibition on eating meat during many periods of the year) may have promoted the import of an alternative or supplement to local freshwater fish. In any case, the large numbers of sea-fish bones amongst the consumption refuse of the *portus* suggest that herring and flatfish were not a very expensive item on the market.

How, then, did the trade develop? How did a modest early-medieval import turn into a larger-scale enterprise? In sites along the coast and/or in the estuaries, production must have increased, making it vital for us to find out how this initiative started. Did independent tradesmen (the 'successors' of Frisian traders) react to growing demand from urban consumers by encouraging the development of production sites? Or was growing production and trade promoted by such authorities as the Counts of Flanders, who owned territories both at the coast and inland? Since the coastal sites belonged to domains that the counts, at that time, were strongly developing (Thoen 2004; Tys 2004; 2015), it is most likely that, in Flanders, elite authority was indeed the motor behind the development of a commercial fishery at sea (Tys and Pieters 2009). Moreover, stimulating fishing activities was not only beneficial for the general development of the coastal domains. Through tolls and taxes, it also meant making a profit from a group of consumers that were becoming more and more independent within an otherwise 'feudal' society. The possibility that the abbeys, who also owned lands near the coast, played a role in developing the trade in marine products is not supported by historical evidence; it also seems to be contradicted by the scarcity of sea-fish remains in the settlement linked with the abbey of Sint-Pieters. These observations from Flanders, if interpreted correctly, fit the general idea that the development of the fishery at sea around AD 1000 was a commercial enterprise promoted by the elite (Hoffmann 2000; Tys and Pieters 2009) and driven by the growth of towns, which represented a new group of consumers. This hypothesis, stressing social and demographic changes, has also been put forward for Britain in the tenth and eleventh centuries (Barrett *et al.* 2011), although for the northern parts (Atlantic Scotland, especially Orkney) an earlier (ninth- to tenth-century) rise in marine fish consumption is attested. This early rise is argued to have been stimulated by cultural contacts, that is, technological innovations (fishing techniques) introduced by people from further north. These shifts, which are visible in the archaeological record of individual sites and of the regions in general, led to the introduction of the concept of a 'fish event horizon' (Barrett *et al.* 2004a,b).

For the moment, Gent is the only urban site regarding which significant information about food consumption is available for its earliest phase. At Ename, near Oudenaarde, a feudal fortification has been excavated which was built towards the end of the tenth century and demolished in the eleventh century. A habitation zone was found next to the fortification, or *castrum*. This habitation can possibly be identified with a *portus* mentioned in the written sources. If this interpretation is valid, it must have had some trade functions (Callebaut 1991). Possibly, without the military intervention of the Count of Flanders, this *portus* would have developed into a town, implying that the faunal material provides a relevant comparison for the finds from Gent. Plaice, herring and even a cod bone were recovered (Van Neer and Ervynck unpublished data). But because the finds collection is small and mainly the result of hand collecting, no firm conclusions can be attached to this assemblage.

Sadly, the proposed interpretation of the onset of large-scale fishing at sea remains incomplete because of the lack

of information about production sites. So far, traces of fishermen's villages from the High Middle Ages have yet to be discovered by Flemish archaeology. A 'Dark Earth' deposit found at the former Roman *castellum* of Oudenburg (Vanhoutte 2007) cannot be used in this context because the site is not a fishermen's village and the reworked animal remains are very imprecisely dated (fourth–tenth centuries). At Leffinge, a *terp* settlement (a dwelling site upon an artificial mound erected in the coastal plain) dating from the middle of the seventh to the beginning of the eleventh century has been excavated, but the fish remains, dominated by the three-spined stickleback (*Gasterosteus aculeatus*), do not suggest that the inhabitants of this site were the pioneers of more intensive marine fishing (Ervynck *et al.* 2012).

Theoretically, it can be assumed that fishing for marine species was first carried out as a seasonal activity by coastal people who combined fishing with farming and animal herding as part of their subsistence economy. Most probably, with the growth of the inland urban markets, temporary fishing sites developed into permanent settlements, representing an evolution comparable to that documented for the coast of Devon, England (Fox 2001). On the basis of the marine fish consumed inland, it seems almost certain that fishing carried out from those coastal sites of the High Middle Ages remained restricted to coastal waters and estuaries. The dominance of flatfish clearly points to those biotopes, and the proportions of the other taxa may also reflect these fishing grounds. The numerous herring bones may come from seasonal exploitation of coastal populations. Each year, between December and March, adult herring congregate along the coast and form massive schools between Cap Griz-Nez and the Scheldt (Poll 1947). Cod, which might come from more offshore areas, are rare, whereas the other gadids – namely, whiting (*Merlangius merlangus*) and haddock – are more common. Whiting typically live in Belgian coastal waters, and juveniles are abundant in the Scheldt estuary (Poll 1947). Haddock is now extremely rare in the southern North Sea, but it used to be abundant until the nineteenth century (Poll 1947). Historical data (Egmond 1997) suggest that this species must have been quite numerous in the area during the sixteenth century.

Sea-fish consumption after *c.* AD 1000

In order to evaluate sea-fish consumption at sites younger than AD 1000, a selection has been made (Table 14.1) from a much larger database containing all identifications of fish bones from Flemish sites (the database itself is stored and maintained at the Royal Belgian Institute of Natural Sciences). The included collections were recovered through sieving and taphonomically represent ordinary consumption refuse (rather than, for example, the refuse of a specific type of fish processing). In order to mitigate against the impact

Table 14.1. Assemblages used for the diachronic comparison (Figures 14.5 and 14.6) of fish consumption in Flemish medieval and post-medieval sites.

Site	Chronology (centuries AD)*	Code(s) in Fig. 14.5 & 14.6*	N (all taxa below class)	N (identified gadids)	Reference(s) for site	Reference(s) for fish bone
Gent Sint-Pieters	9th–10th	Gent (9–10)	208	0	Van den Bremt & Vermeiren 2004	Van Neer & Ervynck unpubl.
Gent *portus* (L7-11)	10th B–11th	Gent (10B–11)	1791	3	Laleman & Stoops 1996	Van Neer & Ervynck unpubl.
Gent *portus* (L1-6)	11th–12th	Gent (11–12)	3191	107	Laleman & Stoops 1996	Van Neer & Ervynck unpubl.
Mechelen Lamot	9th–12th	Mechelen (9–12)	2462	85	Lettany 2003	Van Neer & Ervynck unpubl.
Gent Belfortstraat	13th A	Gent (13A)	220	39	Laleman *et al.* 1986	Van Neer & Ervynck 1995
Aalst Stadhuis	13th	Aalst (13)	1704	121	De Groote *et al.* 2009	De Groote *et al.* 2009
Mechelen Steen	13th B–14th A	Mechelen (13B–14A)	16265	3908	Troubleyn *et al.* 2007; 2009	Lentacker *et al.* 2007; Troubleyn *et al.* 2009
Brugge Prinsenhof	14th	Brugge (14)	1298	129	Deforce *et al.* 2007	Deforce *et al.* 2007
Mechelen Veemarkt	14th–15th	Mechelen (14–15)	4728	584	Mechelse Vereniging voor Stadsarcheologie 1995	Van Neer & Ervynck unpubl.
Aalst Stadhuis	16th A	Aalst (16A)	1236	101	De Groote *et al.* 2004	De Groote *et al.* 2004
Mechelen O.L.V.-ziekenhuis	17th	Mechelen (17)	770	168	Flemish Heritage Institute unpubl.	Van Neer & Ervynck unpubl.
Antwerpen Koolkaai	17th	Antwerpen (17)	4627	887	Van der Wee *et al.* 2000	Veeckman *et al.* 2000
Brussel Arme Klaren	17th–18th A	Brussel (17–18A)	787	165	Claes 2006	Thys *et al.* unpubl.
Gent Schepen-huisstraat	17th–18th	Gent (17–18)	2735	943	Raveschot 1991	Van Neer *et al.* unpubl.
Aalst Peperstraat	18th B	Aalst (18B)	444	172	De Groote *et al.* 2002	Van Neer & Ervynck unpubl.
Aalst Kattestraat	18th B–19th A	Aalst (18B–19A)	2499	574	De Groote & Moens 1997	Van Neer & Ervynck unpubl.
Brussels Arme Klaren	18th B–19th a	Brussel (18B–19a)	1889	398	Claes 2006	Thys *et al.* unpubl.

*A: first half of a century, B: second half of a century; a, b, c, d: first to last quarter of a century

of differing recovery methods, only sieved fractions larger than 4 mm are included here. While this will underestimate the proportion of herring, it will enhance comparability across all the assemblages studied. We focus on truly urban sites, even excluding castles and abbeys located in towns. The latter sites can rely on associated estates located in the countryside and are thus less dependent on urban markets.

For sites in the countryside, this is of course even truer, and it has been demonstrated that fish consumption differed significantly between urban and rural sites (e.g. Ervynck and Van Neer 1998). Note that the catadromous eel (*Anguilla anguilla*) is considered a 'freshwater fish' for present purposes, whereas all flatfish and anadromous species except sturgeon (*Acipenser* sp.) are classified as 'sea-fish'.

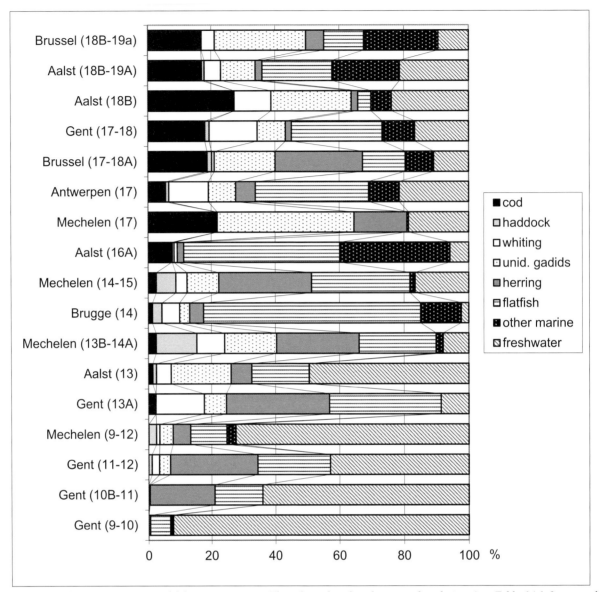

Figure 14.5. Diachronic comparison of fish consumption in Flemish medieval and post-medieval sites (see Table 14.1 for assemblage numbers and references). The three oldest assemblages, from Gent, represent material sieved to 1 mm; the other collections represent material sieved to 4 mm.

A diachronic comparison of the assemblages illustrates a clear decline in the consumption of freshwater fish in the late Middle Ages (Figure 14.5). Possibly, this was not a synchronous evolution in all of the towns documented. While a thirteenth-century context from Gent produced less than 10% freshwater fish, the proportion is much higher in contemporary Aalst (thirteenth century) and broadly contemporary Mechelen (ninth–twelfth centuries). Nevertheless, all younger contexts, from different towns, show a comparable (that is, low) proportion of freshwater fish. Without doubt, the decline must be linked to a decrease in inland fish stocks due to the pollution of waters in and near towns, due to over-fishing, and due to the negative

effects of water management works (Ervynck and Van Neer 1998; cf. Chapter 3; Chapter 15).

The diachronic comparison (Figure 14.5, Table 14.1) also illustrates the relatively low importance of gadids in sites pre-dating the seventeenth century. The high frequency of this fish family in the seventeenth-century site of Mechelen 'O.L.V.-Ziekenhuis' is striking and probably linked to high status. In contrast, the lower frequency of gadids in contemporary Antwerpen 'Koolkaai' may reflect the modest purchasing power of this household or the fact that these people, as fishermen in the estuary, only consumed leftovers from their local catch. Overall, it seems that the growing importance of gadids at the transition from the medieval to

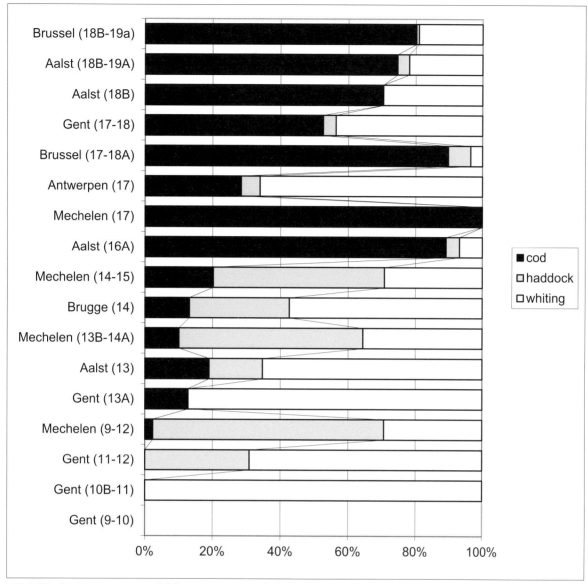

Figure 14.6. Diachronic comparison of fish consumption in Flemish medieval and post-medieval sites, identified gadid remains only (see Table 14.1 for assemblage numbers and references). The three oldest assemblages, from Gent, represent material sieved to 1 mm; the other collections represent material sieved to 4 mm.

the post-medieval period is the result of an increase in the consumption of cod. This conclusion can also be drawn from a diachronic comparison of the relative frequency of the three most important species within the gadids (Figure 14.6, Table 14.1). It again reflects the social difference between the two seventeenth-century sites (with cod clearly representing the more expensive commodity), but generally shows a rise in cod consumption through time. This change unfolded very slowly, with a near absence of cod in the sieved samples from the *portus* of Gent (a small number of specimens have been recovered by hand collecting), low frequencies in the late Middle Ages and higher numbers in the post-medieval era. Haddock, on the other hand, reached high frequencies

in the medieval sites, but disappeared almost completely in the post-medieval assemblages. Whiting was the most important species amongst the gadids in the oldest sites, decreasing significantly in the youngest contexts.

Possibly, these trends within the gadid group can be explained by a shift in fishing grounds, from catching whiting and haddock in the southern part of the North Sea to catching cod in more northern waters. This northward shift, to the rich fishing grounds off the English and Scottish coasts, is amply documented in the historical sources (Ervynck *et al.* 2004). The decline of haddock may have been caused by the disappearance of the population in the southern North Sea (due to over-fishing?). The increase

Table 14.2. Assemblages used for the evaluation of the distribution of the standard lengths (Figure 14.7) represented by cod bones from medieval Flemish inland sites.

Site	Chronology (centuries AD)*	N	Reference (site)	Reference (fish bone)
Oudenburg Dark Earth	4th–10th	1	Vanhoutte 2007	Van Neer & Ervynck unpubl.
Gent *portus*	10th B–12th	11	Laleman & Stoops 1996	Van Neer & Ervynck unpubl.
Ename *portus*	10th–11th	1	Callebaut 1991	Van Neer & Ervynck unpubl.
Mechelen Lamot	9th–12th	2	Lettany 2003	Van Neer & Ervynck unpubl.
Gent Kammerstraat	12th	5	Raveschot 1990	Van Neer & Ervynck 1995
Ename abbey	12th–13th	2	Callebaut 1991	Ervynck *et al.* 1994
Dendermonde Grote Markt	12th–13th	39	Vervoort 2007	Van Neer & Ervynck unpubl.
Aalst Stadhuis	13th	24	De Groote *et al.* 2009	De Groote *et al.* 2009
Oudenaarde Lalaing	13th	6	De Groote 1993	Van Neer & Ervynck unpubl.
Gent Belfortstraat	13th	7	Laleman *et al.* 1986	Van Neer & Ervynck 1995
Mechelen Minderbroeders-klooster	13th B–14th a	45	Troubleyn *et al.* 2006	Van Neer & Ervynck unpubl.
Brugge Prinsenhof	14th	2	Deforce *et al.* 2007	Deforce *et al.* 2007

*A: first half of a century, B: second half of a century; a, b, c, d: first to last quarter of a century

in cod consumption may also have been (partly) caused by a growing import of stockfish. Traditionally, this term refers to large, beheaded cod that were caught and dried in Arctic Norway and other parts of the North Atlantic region. Ideally, the presence of stockfish vs. fresh cod should be evaluated on the basis of the zooarchaeological material itself, but discriminating between fresh and processed cod in contexts in which both occur is not straightforward and has proven to be a more complicated exercise than is often assumed. Criteria to investigate include the presence of certain cut marks, patterns of intra-skeletal distribution (particularly the rarity of certain bones), and patterns in the reconstructed length distributions of the fish (e.g. Barrett 1997; Brinkhuizen 1994; Harland 2007; Perdikaris *et al.* 2007). However, it must be taken into account that different types of processed cod existed (Bennema and Rijnsdorp 2015; Chapter 1; Chapter 18), that specimens of different lengths were used (Brinkhuizen 1994) and that the North Atlantic monopoly was not absolute, as proven by the fact that cod was also processed in coastal settlements around the North Sea (e.g. at Raversijde: Ervynck *et al.* 2013). As a result of these three factors, beheading and/or splitting the fish was not always done in the same way, and specimens must in some cases also have remained whole.

Because of this complexity, an analysis of cut marks and of intra-skeletal distributions will not be attempted here. Differentiating between fresh and processed products on the basis of biometry is also proving problematic. An evaluation of the length distribution of the medieval material (Table 14.2) shows a uniform spread of fish sizes (Figure 14.7) comparable to the post-medieval

material (see De Groote *et al.* 2004; Van Neer and Ervynck unpublished data; Veeckman *et al.* 2000). The possibility of a diachronic trend in the length distributions of the medieval material was evaluated by arbitrarily subdividing the contexts into four phases, but this did not yield any significant results. Perhaps the only meaningful observation to be made is that the smallest specimens (30–50 cm estimated SL) all come from the oldest context (the *portus* of Gent). This may indicate that once catches of cod became part of a larger-scale marine fishery such small specimens no longer arrived at market. Most probably, these small fish were originally caught in waters near the coast.

A recently developed new approach uses measurements of stable isotope values ($\delta^{13}C$ and $\delta^{15}N$) on cod bones to reveal their possible provenance. Using a sample set of medieval cod bones from coastal settlements, representing animals of 50–100 cm total length assumed to derive from local catches, it was proven that differences in the isotopic signal occurred between different seas (e.g. Arctic Norway, eastern Baltic, Kattegat and western Baltic, northern North Sea, central and southern North Sea). By comparing the isotopic composition of cod bones from inland sites with this reference data set, it is now possible to make inferences regarding their location of catch (Barrett *et al.* 2008; 2011).

An evaluation of the Flemish isotope data from inland sites (still representing animals between 50 and 100 cm total length), allows us to put forward some conclusions. During the oldest phase of sea-fish consumption, only cod caught in the central and southern North Sea seem to have been consumed at inland sites, while northern products, perhaps from Arctic Norway, arrived during the

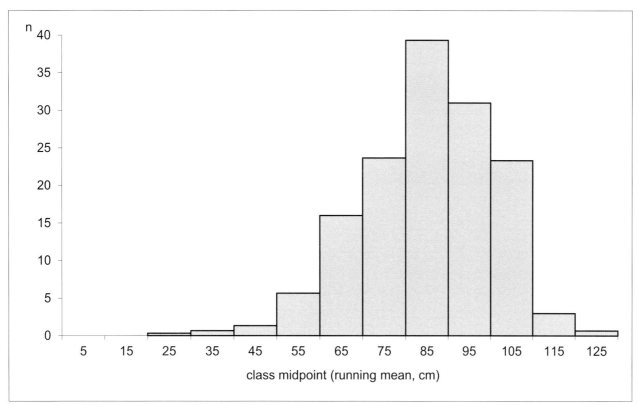

Figure 14.7. Distribution of the standard lengths (the distance from the tip of the snout to the base of the tail) represented by cod bones from medieval Flemish inland sites (hand-collected material, n = 145; see Table 14.2 for sample details and references).

late Middle Ages (Barrett *et al.* 2011; see Chapter 1). The latter, perhaps 'stockfish' of some kind, partly explain the rise in cod consumption towards the post-medieval period (Figures 14.5 and 14.6).

Conclusion

The overview presented makes it clear that the medieval fish market certainly looked different from that depicted by Beuckelaer in the sixteenth century. In fact, what was sold on the market in terms of marine products was the result of dynamic developments through time, involving the growth of an urban group of consumers, economic changes in the production sites and shifts in fishing grounds. As a result of these processes, herring and flatfish originally were the most important commodities on the fish market of the High Middle Ages, to be slightly complemented by gadids during the late Middle Ages, initially based on whiting and haddock. The real rise of gadid consumption is linked to an increase in cod consumption in post-medieval times, most probably of processed fish (stockfish and the like). In general, this evolution, as it can be reconstructed at the moment, differs markedly from what has been observed in other countries surrounding the North Sea (e.g. England and Scotland, cf. Barrett *et al.* 2004b). Instead of a 'fish event

horizon', the sites from the Scheldt basin show a gradual increase in the sea-fish trade, accelerated by urbanisation, but most probably already originating in older, possibly pre-medieval, times.

Of course, as with all archaeological interpretations, the conclusions put forward here will have to be tested against a broader data set. Certainly, assemblages from the oldest growth phases of towns other than Gent are required, as are more finds from other river basins (IJzer, Meuse). Ideally, we would like to have more data available per period, per site, and from the different social contexts within each town. The rise of sea-fish consumption in the countryside must also be taken into account in order to obtain a holistic view of this part of the food economy. Finally, the analysis needs to be redone using the finer-sieved fractions, to reach a better evaluation of the consumption of herring through time. Without doubt, the few herring depicted by Beuckelaer do not match the former economic importance of this species.

Acknowledgements

The analysis presented here is the result of a long-term project started by the Institute for the Archaeological Heritage of the Flemish Community and the Royal Museum for Central Africa, and now continued by the Flemish Heritage agency and the Royal Belgian Institute of Natural

Sciences. Many thanks go to Wim Wouters (Royal Belgian Institute of Natural Sciences) and Luc Muylaert (Flemish Heritage agency) for the treatment and study of samples from numerous Flemish sites. The contribution of Wim Van Neer to this paper presents research results of the Interuniversity Attraction Poles Programme – Belgian Science Policy.

References

Asaert, G. 1980. Scheepvaart en visserij, in D. P. Blok (ed.) *Algemene geschiedenis der Nederlanden.* Vol. 4: 128–34. Haarlem: Fibula-Van Dishoeck.

Baeteman, C. & L. Denys. 1997. Holocene shoreline and sea-level data from the Belgian coast. *Palaeoclimate Research* 21: 49–74.

Barrett, J. H. 1997. Fish trade in Norse Orkney and Caithness: a zooarchaeological approach. *Antiquity* 71: 616–38.

Barrett, J. H., A. M. Locker & C. M. Roberts. 2004a. The origins of intensive marine fishing in medieval Europe: the English evidence. *Proceedings of the Royal Society of London*, Series B 271: 2417–21.

Barrett, J. H., A. M. Locker & C. M. Roberts. 2004b. 'Dark Age Economics' revisited: the English fish bone evidence AD 600–1600. *Antiquity* 78: 618–36.

Barrett, J. H., C. Johnstone, J. Harland, W. Van Neer, A. Ervynck, D. Makowiecki, D. Heinrich, A. K. Hufthammer, I. B. Enghoff, C. Amundsen, J. S. Christiansen, A. K. G. Jones, A. Locker, S. Hamilton-Dyer, L. Jonsson, L. Lõugas, C. Roberts & M. Richards. 2008. Detecting the medieval cod trade: a new method and first results. *Journal of Archaeological Science* 35(4): 850–61.

Barrett, J. H., D. Orton, C. Johnstone, J. Harland, W. Van Neer, A. Ervynck, C. Roberts, A. Locker, C. Amundsen, I. B. Enghoff, S. Hamilton-Dyer, D. Heinrich, A. K. Hufthammer, A. K. G. Jones, L. Jonsson, D. Makowiecki, P. Pope, T. C. O'Connell, T. De Roo & M. Richards. 2011. Interpreting the expansion of sea fishing in medieval Europe using stable isotope analysis of archaeological cod bones. *Journal of Archaeological Science* 38: 1516–24.

Bennema, F. P. & A. D. Rijnsdorp. 2015. Fish abundance, fisheries, fish trade and consumption in sixteenth-century Netherlands as described by Adriaen Coenen. *Fisheries Research* 161: 384–99.

Boelmans Kranenburg, H. A. H. 1979. Visserij in de Zuidelijke Nederlanden 1650–1795, in D. P. Blok (ed.) *Algemene Geschiedenis der Nederlanden.* Vol. 8: 261–4. Haarlem: Fibula-Van Dishoeck.

Boelmans Kranenburg, H. A. H. 1980. Visserij in de Zuidelijke Nederlanden 1580–1650, in D. P. Blok (ed.) *Algemene geschiedenis der Nederlanden.* Vol. 7: 170–1. Haarlem: Fibula-Van Dishoeck.

Bogemans, F., E. Meylemans, Y. Perdaen, A. Storme & I. Verdrurmen. 2008. *Paleolandschappelijk, archeologisch en cultuurhistorisch onderzoek in het kader van het geactualiseerde Sigmaplan: Sigma-cluster Kalkense Meersen, zone Wijmeersen 2.* Brussel: Vlaams Instituut voor het Onroerend Erfgoed, at the behest of Waterwegen en Zeekanaal.

Brinkhuizen, D. C. 1994. Some notes on fish remains from the late 16th century merchant vessel Scheurrak SO1, in W. Van Neer (ed.) *Fish Exploitation in the Past*: 197–205. (Proceedings of the 7th Meeting of the ICAZ Fish Remains Working Group; Annales du Musée Royal de l'Afrique Centrale, Sciences Zoologiques 274). Tervuren: Musée Royal de l'Afrique Centrale.

Brulet, R., C. Coquelet, A. Defgnée, F. Pigière & L. Verslype. 2004. Les sites à « terres noires » à Tournai et le secteur des anciens cloîtres canoniaux: etudes archéozoologique, palynologique et contextualisation, in R. Brulet & L. Verslype (eds) *Terres noires*: 152–72. (Actes de la table ronde de Louvain-la-Neuve, 09–10 novembre 2001, Département d'archéologie et d'histoire de l'art et Centre de recherches d'archéologie nationale, Collection d'archéologie Joseph Mertens 14, Publications d'histoire de l'art et d'archéologie de l'Université catholique de Louvain). Louvain-la-neuve: Université catholique de Louvain.

Callebaut, D. 1991. Castrum, Portus und Abtei von Ename, in H. W. Böhme (ed.) *Burgen der Salierzeit. Volume 1: in den nordlichen Landschaften des Reiches*: 291–309. Mainz: Römisch-Germanischen Zentralmuseum.

Claes, B. 2006. Archeologisch onderzoek van het voormalige Arme Klarenklooster (Br.). *Archaeologia Mediaevalis* 29: 28–34.

Clavel, B. & J.-H. Yvinec. 2010. L'archéozoologie du Moyen Âge au début de la période moderne dans la moitié nord de la France, in J. Chapelot (ed.) *Trente ans d'archéologie médiévale en France: un bilan pour un avenir*: 71–87. (IXe Congrès international de la Société d'archéologie médiévale, Vincennes, 16–18 juin 2006; Publication du CRAHM). Caen: Presses Universitaires.

Decavele, J. 1989. *In Defence of a Rebellious City.* Antwerpen: Mercatorfonds.

Deforce, K., A. Ervynck, B. Hillewaert, J. Huyghe, A. Lentacker, H. van Haaster & W. Van Neer. 2007. Het archeologische onderzoek: de voorgeschiedenis van het Bourgondische hof, in B. Hillewaert & E. Van Besien (ed.) *Het Prinsenhof in Brugge*: 22–39. Brugge: Uitgeverij Van de Wiele.

De Groote, K. 1993. De middeleeuwse ambachtelijke wijk van Pamele (stad Oudenaarde, Oost-Vlaanderen): het onderzoek in het Huis de Lalaing 1. de pottenbakkersovens. *Archeologie in Vlaanderen* 3: 359–99.

De Groote, K. & J. Moens. 1997. Laat- en post-middeleeuwse bewoningssporen aan de Kattestraat te Aalst (O.-Vl.). *Archaeologia Mediaevalis* 20: 64–5.

De Groote, K., J. Moens, A. De Block & I. Zeebroek. 2002. Het afval van een laat 18de-eeuwse pataconbakker in de Peperstraat te Aalst (Oost-Vlaanderen). *Archaeologia Mediaevalis* 25: 49–50.

De Groote, K., J. Moens, D. Caluwé, B. Cooremans, K. Deforce, A. Ervynck, A. Lentacker, E. Rijmenants, W. Van Neer, W. Vernaeve & I. Zeebroek. 2004. *De Valcke, de Slotele en de Lelye*, burgerwoningen op de Grote Markt te Aalst (prov. Oost-Vlaanderen): onderzoek naar de bewoners, analyse van een vroeg-16de-eeuwse beerputvulling en de evolutie tot stadhuis. *Archeologie in Vlaanderen* 8: 281–408.

De Groote, K., J. Moens, D. Caluwé, B. Cooremans, K. Deforce, A. Ervynck, A. Lentacker & W. Van Neer. 2009. Op zoek naar de oudste middeleeuwse bewoning aan de Grote Markt te Aalst (prov. Oost-Vlaanderen): het onderzoek van afval- en beerkuilen uit de twaalfde tot de veertiende eeuw. *Relicta* 4: 135–203.

Demey, D., S. Vanhoutte, M. Pieters, J. Bastiaens, W. De Clercq, K. Deforce, L. Denys, A. Ervynck, A. Lentacker, A. Storme & W. Van Neer. 2013. Een dijk en een woonplatform uit de Romeinse periode in Stene (Oostende). *Relicta* 10: 7–70.

Dobney, K. & A. Ervynck. 2007. To fish or not to fish? Evidence for the possible avoidance of fish consumption during the Iron Age around the North Sea, in C. Haselgrove & T. Moore (eds) *The Later Iron Age in Britain and Beyond*: 403–18. Oxford: Oxbow Books.

Egmond, F. 1997. *Een bekende Scheveninger: Adriaen Coenen en zijn Visboeck van 1578.* Den Haag: Centrum voor Familiegeschiedenis van Scheveningen.

Ervynck, A. & W. Van Neer. 1998. Het archeologisch onderzoek van de voedseleconomie van laatmiddeleeuwse steden: mogelijkheden en eerste resultaten voor Leuven, in L. Bessemans & Museum Vander Kelen-Mertens (eds) *Leven te Leuven in de late Middeleeuwen*: 79–94. Leuven: Peeters.

Ervynck, A. & W. Van Neer. 1999. Dierenresten uit een waterput op de Nieuwe Beestenmarkt: een blik op de voedselvoorziening van een vroeg-middeleeuws Gent. *Stadsarcheologie. Bodem en Monument in Gent* 23(1): 5–13.

Ervynck, A., B. Cooremans & W. Van Neer. 1994. De voedselvoorziening in de Sint-Salvatorsabdij te Ename (stad Oudenaarde, prov. Oost-Vlaanderen) 3: een latrine bij de abtswoning (12de–begin 13de eeuw). *Archeologie in Vlaanderen* 4: 311–22.

Ervynck, A., A. Gautier & W. Van Neer. 1997. Import van schelpdieren en vis in een Romeinse nederzetting te Nevele. *VOBOV-info (Tijdschrift van het Verbond voor Oudheidkundig Bodemonderzoek in Oost-Vlaanderen)* 46: 24–8.

Ervynck, A., A. Lentacker & W. Van Neer. 2008. Archeozoölogisch onderzoek, in L. Meganck (ed.) *Onderzoeksbalans Onroerend Erfgoed Vlaanderen.* http://www.onderzoeksbalans.be/.

Ervynck, A., W. Van Neer & A. Lentacker. 2013. Dierenresten uit het middeleeuwse vissersdorp: een synthese, in M. Pieters *et al.* (eds) *Het archeologisch onderzoek in Raversijde (Oostende) in de periode 1992–2005*: 508–25. Brussel: Agentschap Onroerend Erfgoed.

Ervynck, A., W. Van Neer & M. Pieters. 2004. How the North was won (and lost again): historical and archaeological data on the exploitation of the North Atlantic by the Flemish fishery, in R. A. Housley & G. Coles (eds) *Atlantic Connections and Adaptations: economies, environments and subsistence in lands bordering the North Atlantic*: 230–9. (Symposia of the Association for Environmental Archaeology 21). Oxford: Oxbow Books.

Ervynck A., P. Deckers, A. Lentacker, D. Tys & W. Van Neer. 2012. 'Leffinge – Oude Werf': the first archaeozoological collection from a *terp* settlement in coastal Flanders, in D. C. M. Raemaekers, E. Esser, R. C. G. M. Lauwerier & J. T. Zeiler (ed.) *A bouquet of archaeozoological studies: essays in honour of Wietske Prummel*: 152–62. Groningen: Barkhuis & University of Groningen Library.

Fox, H. 2001. *The Evolution of the Fishing Village: landscape and society along the south Devon coast.* Oxford: Leopard's Head.

Gijzen, A. 1986. De vismarkt van Joachim Beuckelaer, in P. Verbraeken (ed.) *Joachim Beuckelaer: het markt- en keukenstuk in de Nederlanden 1550–1650*: 67–70. Gent: Gemeentekrediet.

Harland, J. 2007. Status and space in the 'fish event horizon': initial results from Quoygrew and Earl's Bu, Viking Age and medieval sites in Orkney, Scotland, in H. Hüster Plogmann (ed.) *The Role of fish in Ancient Time*: 63–8. (Proceedings of the 13th Meeting of the ICAZ Fish Remains Working Group in October, 4th–9th, Basel/Augst 2005). Rahden: Marie Leidorf.

Hoffmann, R. C. 2000. Medieval fishing, in P. Squatriti (ed.) *Working with Water in Medieval Europe: technology and resource-use*: 331–93. Leiden: Brill.

Jellema, D. 1955. Frisian trade in the Dark Ages. *Speculum* 30: 15–36.

Kavaler, E. M. 1986. Erotische elementen in de markttaferelen van Beuckelaer, Aertsen en hun tijdgenoten, in P. Verbraeken (ed.) *Joachim Beuckelaer: het markt- en keukenstuk in de Nederlanden 1550–1650*: 18–26. Gent: Gemeentekrediet.

Laleman, M. C., D. Lievois & P. Raveschot. 1986. De top van de Zandberg: archeologisch en bouwhistorisch onderzoek. *Stadsarcheologie. Bodem en Monument in Gent* 10(3): 2–61.

Laleman, M. C. & G. Stoops. 1996. Vondstmeldingen 2: Emile Braunplein. *Stadsarcheologie. Bodem en Monument in Gent* 20(3): 57–60.

Lebecq, S. 1986. Dans l'Europe du nord des VIIe/IXe siècles: commerce frison ou commerce franco-frison? *Annales. Economies, Sociétés, Civilisations* 41: 361–77.

Lebecq, S. 1992. The Frisian trade in the Dark Ages: a Frisian or a Frankish/Frisian trade? in A. Carmiggelt (ed.) *Rotterdam papers*. Vol. 7: 7–15. Rotterdam: Stad Rotterdam.

Lentacker, A., A. Ervynck & W. Van Neer. 2004. The symbolic meaning of the cock: the animal remains from the Mithraeum at Tienen (Belgium), in M. Martens & G. De Boe (eds) *Roman Mithraism: the evidence of the small finds*: 57–80. (Archeologie in Vlaanderen Monografie 5). Zellik and Tienen: Instituut voor het Archeologisch Patrimonium.

Lentacker, A., W. Van Neer, A. Ervynck & K. Desender. 2007. De dierlijke resten, in L. Troubleyn, F. Kinnaer & A. Ervynck (eds) *Het Steen en de burgers: onderzoek van de laatmiddeleeuwse gevangenis van Mechelen*: 133–54. Mechelen: Stad Mechelen.

Lentacker, A., W. Van Neer & F. Pigière. In press. L'étude archéozoologique du site du quai Marché-aux-Poissons/CV12 à Tournai, in R. Brulet & L. Verslype (eds) *L'Escaut à Tournai au fil du temps: les fouilles et surveillances archéologiques de travaux de pose de collecteurs d'eaux usées le long de l'Escaut à Tournai.* (Collection d'archéologie Joseph Mertens, Publications d'histoire de l'art et d'archéologie de l'Université catholique de Louvain). Louvain-la-Neuve: Université catholique de Louvain.

Lettany, L. (ed.) 2003. *Het ongeschreven Mechelen: archeologisch onderzoek op de Grote Markt en de Veemarkt 2001–2003.* Mechelen: Stedelijke Musea Mechelen.

Mechelse Vereniging voor Stadsarcheologie 1995. Het Carmelietenklooster op de Veemarkt te Mechelen (Ant.). *Archaeologia Mediaevalis* 18: 28.

Nicholas, D. 1992. *Medieval Flanders*. London and New York: Longman.

Perdikaris, S., G. Hambrecht, S. Brewington & T. McGovern. 2007. Across the fish event horizon: a comparative approach, in H. Hüster Plogmann (ed.) *The Role of Fish in Ancient Time*:

51–62. (Proceedings of the 13th Meeting of the ICAZ Fish Remains Working Group in October 4th–9th, Basel/Augst 2005). Rahden: Marie Leidorf.

Peters, J. 1987. De dierlijke resten uit de Romeinse nederzetting van Bredene II, in H. Thoen (ed.) *De Romeinen langs de Vlaamse kust*: 67–9. Brussel: Gemeentekrediet.

Philippart, J.-C. & M. Vranken. 1983. *Atlas des poissons de Wallonie: distribution, écologie, éthologie, pêche, conservation* (Cahiers d'éthologie appliquée à la protection et à la conservation de la vie sauvage, à la gestion et au contrôle des ressources et productions animales, Collection enquêtes et dossiers 4). Liège: Université de Liège.

Pigière, F. 2009. *Evolution de l'économie alimentaire et des pratiques d'élevage de l'Antiquité au haut Moyen Age en Gaule du nord: une étude régionale sur la zone limoneuse de la Moyenne Belgique et du sud des Pays-Bas.* (British Archaeological Report S2035). Oxford: Archaeopress.

Pigière, F., M. Udrescu, I. Boone & W. Van Neer. 2003. Etude archéozoologique, in N. Mees, R. Vanmechelen & C. Robinet (eds) Le Grognon à Namur: de l'émergence du portus au quartier des temps modernes (V^e–XVIIIe siècles). Unpublished report prepared for Région wallonne.

Plouvier, L. 1990. De 'Belgische' keuken in de vroege middeleeuwen. *De Brabantse Folklore en Geschiedenis* 267–8: 205–35.

Poll, M. 1947. *Poissons marins.* Brussel: Musée royal d'histoire naturelle de Belgique.

Prummel, W. 1983. *Early Medieval Dorestad, an Archaeozoological Study.* (Excavations at Dorestad 2, Nederlandse Oudheden 11, Kromme Rijn Projekt 2). Amersfoort: Rijksdienst voor het Oudheidkundig Bodemonderzoek.

Raveschot, P. 1990. Een middeleeuwse schoenmaker in de Kammerstraat. *Stadsarcheologie. Bodem en Monument in Gent* 14(2): 39–41.

Raveschot, P. 1991. De beerput in de Schepenhuisstraat. *Stadsarcheologie. Bodem en Monument in Gent* 15(3): 5–7.

Thoen, E. 2004. 'Social agrosystems' as an economic concept to explain regional differences: an essay taking the former county of Flanders as an example (Middle Ages–19th century), in B. J. P. van Bavel & P. Hoppenbrouwers (eds) *Landholding and Land Transfer in the North Sea Area (late Middle Ages–19th century)*: 47–66. Turnhout: Brepols.

Thoen, H. 1987. *De Romeinen langs de Vlaamse kust.* Brussel: Gemeentekrediet.

Troubleyn, L., F. Kinnaer & A. Ervynck (eds). 2007. *Het Steen en de burgers: onderzoek van de laatmiddeleeuwse gevangenis van Mechelen.* Mechelen: Stad Mechelen.

Troubleyn, L., F. Kinnaer, A. Ervynck, L. Beeckmans, D. Caluwé, B. Cooremans, F. De Buyser, K. Deforce, K. Desender, A. Lentacker, J. Moens, G. Van Bulck, M. Van Dijck, W. Van Neer & W. Wouters. 2009. Consumption patterns and living conditions inside Het Steen, the late medieval prison of Malines (Mechelen, Belgium). *Journal of the Archaeology of the Low Countries* 1(2): 5–47.

Troubleyn, L., R. Ribbens & B. Robberechts. 2006. *Het archeologisch onderzoek op de sites Minderbroedersklooster en Begijnenstraat* (Nieuwsbrief 9). Mechelen: Stad Mechelen.

Tys, D. 2004. Domeinvorming in de 'wildernis' en de ontwikkeling van vorstelijke macht: het voorbeeld van het bezit van de graven van Vlaanderen in het IJzerestuarium tussen 900 en 1200. *Jaarboek voor Middeleeuwse Geschiedenis* 7: 34–87.

Tys, D. 2015. Maritime environment and social identities in medieval coastal Flanders: the management of water and environment and its consequences for the local community and the landscape, in J. H. Barrett & S. J. Gibbon (eds) *Maritime Societies of the Viking and Medieval World*: 122–37. Leeds: Maney/Society for Medieval Archaeology.

Tys, D. & M. Pieters. 2009. Understanding a medieval fishing settlement along the southern North Sea: Walraversijde, *c.* 1200–1630, in L. Sicking & D. Abreu-Ferreira (eds) *Beyond the Catch: fisheries of the North Atlantic, the North Sea and the Baltic, 900–1850*: 91–121. Leiden and Boston: Brill.

Van den Bremt, A. & G. Vermeiren. 2004. Archeologisch vooronderzoek op het Sint-Pietersplein en aan de Tweekerkenstraat. *Handelingen der Maatschappij voor Geschiedenis en Oudheidkunde te Gent (Nieuwe Reeks)* 58: 23–58.

Van der Wee, P., M. Hendrickx & J. Veeckman. 2000. De *Groote Schalien Loove*: een laatmiddeleeuws pand op de overgang van hout- naar steenbouw, in J. Veeckman (ed.) *Berichten en Rapporten over het Antwerps Bodemonderzoek en Monumentenzorg.* Vol. 4: 27–98. Antwerpen: Stad Antwerpen.

Van Neer, W. & A. Ervynck. 1995. Gentse graten: onderzoek van archeologische visresten uit de binnenstad. *Stadsarcheologie. Bodem en Monument in Gent* 19(4): 5–11.

Van Neer, W. & A. Ervynck. 2004. Remains of traded fish in archaeological sites: indicators of status, or bulk food? in S. J. O'Day, W. Van Neer & A. Ervynck (eds) *Behaviour Behind Bones: the zooarchaeology of ritual, religion, status and identity*: 203–14. Oxford: Oxbow Books.

Van Neer, W. & A. Ervynck. 2006. The zooarchaeological reconstruction of the development of the exploitation of the sea: a *status quaestionis* for Flanders, in M. Pieters, F. Verhaeghe & G. Gevaert (eds) *Fishery, Trade and Piracy: fishermen and fishermen's settlements in and around the North Sea area in the Middle Ages and later*: 95–103. (Archeologie in Vlaanderen Monografie 6). Brussel: Flemish Heritage Institute.

Van Neer, W. & A. Ervynck. 2007. De zoöarcheologische studie van de ontwikkeling van de exploitatie van de zee: een *status quaestionis* voor Vlaanderen, in A. M. J. de Kraker & G. J. Borger (eds) *Veen-vis-zout: landschappelijke dynamiek in de zuidwestelijke delta van de Lage Landen*: 45–54. (Geoarchaeological and Bioarchaeological Studies 8). Amsterdam: Vrije Universiteit.

Van Neer, W. & A. Lentacker. 1994. New archaeozoological evidence for the consumption of locally-produced fish sauce in the northern provinces of the Roman empire. *Archaeofauna* 3: 53–62.

Van Neer, W., A. Ervynck & P. Monsieur. 2010. Fish bones and amphorae: evidence for the production and consumption of salted fish products outside the Mediterranean region. *Journal of Roman Archaeology* 23(1): 161–95.

Van Neer, W., W. Wouters, A. Ervynck & J. Maes. 2005. New evidence from a Roman context in Belgium for fish sauce locally produced in northern Gaul. *Archaeofauna* 14: 171–82.

Van Neer, W., W. Wouters, F. Vilvorder & J.-C. Demanet. 2009. Pont-à-Celles/Luttre: importation de poissons marins dans le vicus des « Bons-Villers » à Liberchies. *Chronique de l'Archéologie Wallonne* 16: 46–48.

Van Uytven, R. 1979. Visserij in de Zuidelijke Nederlanden, in D. P. Blok (ed.) *Algemene geschiedenis der Nederlanden. Volume 6*: 138–44. Haarlem: Fibula-Van Dishoeck.

van Winter, J. M. 1976. *Van soeter cokene: recepten uit de oudheid en Middeleeuwen*. Haarlem: Fibula-Van Dishoeck.

Vanderhoeven, A., G. Vynckier, A. Ervynck & B. Cooremans. 1992. Het oudheidkundig bodemonderzoek aan de Kielenstraat te Tongeren (prov. Limburg): interimverslag 1990–1993, deel 1: de vóór-Flavische bewoning. *Archeologie in Vlaanderen* 2: 89–145.

Vanhoutte, S. 2007. Het Romeinse *castellum* van Oudenburg (prov. West-Vlaanderen) herontdekt: de archeologische campagne van augustus 2001 tot april 2005 ter hoogte van de zuidwesthoek. *Relicta* 3: 199–235.

Vanhoutte, S., J. Bastiaens, W. De Clercq, K. Deforce, A. Ervynck, M. Fret, K. Haneca, A. Lentacker, H. Stieperaere, W. Van Neer, P. Cosyns, P. Degryse, W. Dhaeze, W. Dijkman, M. Lyne, P. Rogers, C. van Driel-Murray, J. van Heesch & J. P. Wild. 2009. De dubbele waterput uit het laat-Romeinse *castellum* van Oudenburg (prov. West-Vlaanderen): tafonomie, chronologie en interpretatie. *Relicta* 5: 9–141.

Veeckman, J., W. van Hoof, B. Cooremans, A. Ervynck & W. Van Neer. 2000. De inhoud van de afvalput van de *Groote Schalien Loove*: speuren naar de 17de-eeuwse bewoners, in J. Veeckman (ed.) *Berichten en rapporten over het Antwerps bodemonderzoek en monumentenzorg. Volume 4*: 115–90. Antwerpen: Stad Antwerpen.

Verbraeken, P. 1986. Catalogus, in P. Verbraeken (ed.) *Joachim Beuckelaer: het markt- en keukenstuk in de Nederlanden 1550–1650*: 113–99. Gent: Gemeentekrediet.

Vervoort, R. 2007. Archeologen kijken onder de Markt, in P. Buyse, L. Meganck, E. Vandeweghe & R. Vervoort (eds) *De Grote Markt van Dendermonde van boven tot onder bekeken* (Kleine Kultuurgidsen Provincie Oost-Vlaanderen): 26–52. Gent: Provincie Oost-Vlaanderen.

Wouters, W., L. Muylaert & W. Van Neer. 2007. The distinction of isolated bones from plaice (*Pleuronectes platessa*), flounder (*Platichthys flesus*) and dab (*Limanda limanda*): a description of the diagnostic characters. *Archaeofauna* 16: 33–95.

Fishing and Fish Trade in Medieval York: The Zooarchaeological Evidence

Jennifer F. Harland, Andrew K. G. Jones, David C. Orton and James H. Barrett

Introduction

Archaeological investigations of deep, well-stratified and well-preserved deposits in York have played a central role in the development of thinking about temporal trends in medieval fishing (e.g. Barrett *et al.* 2004; Enghoff 2000; Harland *et al.* 2008; Jones 1981; 1988a) – concurrent with a wider influence on the methods and practices of urban zooarchaeology (O'Connor 2003). Excavations at 16–22 Coppergate between 1976 and 1981 employed an unprecedented amount of sieving using fine mesh. Concurrently, the staff of York's Environmental Archaeology Unit (founded in 1975) dedicated themselves to ensuring the processing and analysis of both sieved samples and aliquots of raw sediment (Bond and O'Connor 1999; Kenward and Hall 1995; A. K. G. Jones pers. comm.; O'Connor 1989). The scale of recovery and analysis of ecofacts was revolutionary within medieval archaeology, and it set a precedent followed by subsequent excavation and post-excavation work in York – regarding such sites as Fishergate (Kemp 1996) and Blue Bridge Lane (Spall and Toop 2005). The primary zooarchaeological material for study of fishing is thus exceptional.

Conversely, publication of the fish bone from excavations in York has not previously benefited from a project funded to achieve this objective. What is known has therefore been disseminated through preliminary discussions of major trends (Harland *et al.* 2008; Jones 1988a), supplementary mention in studies of mammal bone (Bond and O'Connor 1999, 400–1; O'Connor 1988; 1989, 196; 1991, 263–7), diverse unpublished and online reports (e.g. Hall *et al.* 2002; Johnstone *et al.* 2000; Jones 1989; Large *et al.* 1999; Rowland 2005) and a survey of English fishing between AD 600 and 1600 (Barrett *et al.* 2004) – which tabulated data in archive documents, including printouts of early mainframe computer records (see below). In most of these cases it has not been possible to consider information beyond the relative abundance of different species based on the number of identified specimens (NISP). In this chapter we aim to clarify what the high-quality archaeological record from York can contribute to our understanding of changes in the role of marine and freshwater fish in medieval northern England – and by extension northwestern Europe. Our ultimate objective is to provide a more detailed understanding of the chronology and characteristics of a previously identified shift in emphasis from freshwater to marine fish during the course of the Middle Ages (Barrett *et al.* 2004; Hoffmann 1996).

Methods and materials

To achieve our objective, we have revisited the physical bone archive from two key sites, 16–22 Coppergate (hereafter just Coppergate) and Blue Bridge Lane and Fishergate House (hereafter Blue Bridge Lane), re-identifying the sieved specimens using a standardised protocol (based on Harland *et al.* 2003 and references therein) in which data regarding state of preservation, skeletal element, taxon, fish size and cut marks were recorded – after consulting records of previous identifications by one of us (AKGJ) for Coppergate and by Stephen Rowland for Blue Bridge Lane (see below). By studying changes in the representation of fish taxa, sizes and skeletal elements it is possible to infer possible trends in harvesting, transportation/trade and consumption. Moreover, by considering the state of bone preservation it is possible to qualify these interpretations in the light of possible taphonomic biases.

The bones from a third key site, Fishergate, were not available for primary study, but we consider the extant

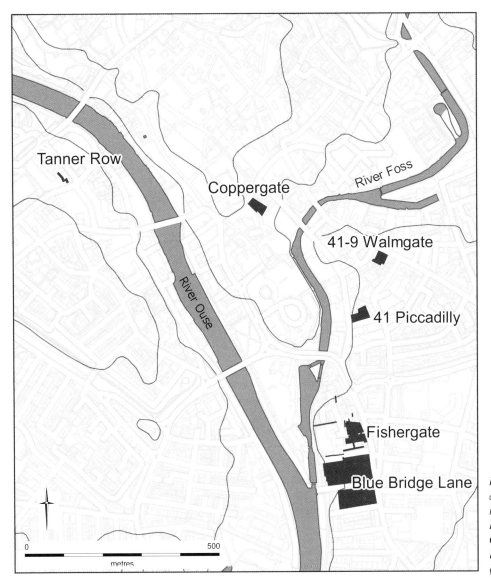

Figure 15.1. Locations of the main excavation sites in York considered in this chapter (Drawing: Jennifer Harland, using a base map © Crown Copyright and Database Right 2015. Ordnance Survey (Digimap Licence)).

species abundance data for this sieved assemblage (Jones 1989; O'Connor 1991, 263–7) along with those from 24–30 Tanner Row (O'Connor 1988, table 32), 41–9 Walmgate (Hall *et al.* 2002; Johnstone *et al.* 2000) and 41 Piccadilly (Large *et al.* 1999), which were all also recovered using sieving. Many other excavations in York have resulted in well-recovered fish bone, but the records of 33 additional assemblages we considered showed that most remain unquantified (e.g. 63–67 Micklegate and 22 Piccadilly) and that a few were not precisely phased (e.g. 1–5 Aldwark). Additional sites have also been excavated (e.g. Hungate). This material would reward future research, but the 26,663 identified sieved specimens from Coppergate and Blue Bridge Lane – dating predominantly from the seventh to sixteenth centuries – provide an important baseline data set for present purposes, particularly when combined with

the 18,246 previously identified specimens from Fishergate, 24–30 Tanner Row, 41–9 Walmgate and 41 Piccadilly (Figure 15.1).

At Coppergate, a sediment sampling regime was in place from the beginning of the excavation, and sampling was undertaken methodically, as was hand collection of bone. Bulk samples were wet sieved to 1 mm using a Siraf-type flotation tank (Jones 1982). Sample residue sorting was undertaken by a team supervised by one of us (AKGJ), typically (but not always) aided by first sieving to 2 mm (Jones 1988a). A significant portion of the hand collected and sieved bone from Coppergate was previously recorded in the 1980s, entered on punch cards and uploaded to the University of York's mainframe computer, with summary outputs forming the foundation of most previous published work. The electronic data were subsequently lost, but the original sheets have been kept, along with the data dictionaries providing definitions of numerical codes.

We re-digitised the original data sheets, creating a database including archaeological context, recovery method, quantity, taxon, element, side and limited comments. Using this database as a foundation, all sieved material from Coppergate was then re-located and examined, confirming the identifications and recording additional information: bone part (based on defined diagnostic zones), percent completeness, bone texture, fish size (based on comparison with reference specimens from fish of known total length [TL]), measurements, butchery marks and descriptive comments (see Harland *et al.* 2003). Newly sieved samples that had not previously been identified were also included. We then quantified the sieved data by NISP based on a suite of skeletal elements that are broadly comparable across taxa, including 18 cranial and appendicular elements, otoliths, all vertebrae and a few specialised osseous structures characteristic of particular groups (e.g. the first anal pterygiophore in flatfish) (Harland *et al.* 2003). During the original analysis, other fish remains, such as scales were also identified, usefully increasing the recognition of rare taxa, but also limiting the comparability of the evidence. Large numbers of perch (*Perca fluviatilis*) scales, for example, over-represented the relative abundance of this taxon. In the final NISP data presented herein, taxa identified by AKGJ only by remains not included in our systematic recording protocol are noted as present. The hand-collected assemblage of fish bone from Coppergate was substantial, but given the inherent recovery biases, it could return only limited information. Nevertheless, this material contained butchered bone, predominantly cod (*Gadus morhua*), so it was visually scanned and any cut marks were recorded. These specimens are considered qualitatively when discussing fish processing methods below, but are only quantified in Table 15.9. The overall chronology for the Coppergate fish assemblage has been constructed by matching archival context numbers and phasing information with the most recent absolute dating evidence (e.g. Hall and Hunter-Mann 2002; Hall *et al.* 2004; 2014). Details are provided in Appendix 15.1 for the benefit of future research.

All the included fish remains from Blue Bridge Lane were from sieved and sorted samples. The >2 mm and >4 mm sample fractions were re-analysed using the recording protocol explained above (Harland *et al.* 2003). One late fourteenth-century deposit composed almost entirely of herring (*Clupea harengus*) bone (contexts C1580B and C1577B in pit F125/6) has been the subject of a separate study (Keaveney 2005). It is discussed below in the context of our results, and data from a subsample appear in Table 15.8. The quantity of fish bones, predominately of herring, in the late-medieval (late fourteenth- to early sixteenth-century) contexts from Blue Bridge Lane was so large that

they were also subsampled, using a sample riffler, prior to identification. The online excavation report (Spall and Toop 2005) and archival material provided phasing and dating information.

As noted above, four additional assemblages are here considered for which reanalysis was not practicable. Fish bone from Fishergate had been identified by AKGJ in the 1980s. The samples were sieved to 2 mm. A small number of additional specimens were recovered from coarse sieving to 12 mm using experimental disaggregation with a cement mixer; these are not quantified or discussed here as they are not directly comparable with other sieve sizes or sampling strategies. An unpublished archive report (Jones 1989) and an updated published summary (O'Connor 1991, 263) include NISP data by taxon and phase, based on all possible identifications, including fish scales. The same observation applies for 24–30 Tanner Row, analysed by AKGJ after sieving to 1 mm and sorting to 2 mm (O'Connor 1988, 71–2, 103, table 32). The assemblages from 41–9 Walmgate (sieved using 1 mm mesh) and 41 Piccadilly (probably sieved using 0.5 mm and 1 mm mesh) were analysed by the former Environmental Archaeology Unit team, following on from AKGJ, using methods broadly comparable with those employed for Fishergate and 24–30 Tanner Row (Hall *et al.* 2002; Johnstone *et al.* 2000; Large *et al.* 1999).

Inter-site meta-analysis of the abundance of the most common fish taxa through time was carried out using estimated frequency density distributions (Figure 15.2). These incorporate the dating information for each relevant archaeological phase without lumping the data into coarse-grained chronological groups (e.g. the two-century blocks of Barrett *et al.* 2004). The number of relevant specimens in each site phase was divided by the length of its date range to give an estimate of frequency density across that range. These estimates were then summed across all contexts at five-year intervals to produce an overall distribution (cf. Orton *et al.* 2014). Date ranges for each site phase were converted into calendar years on a conservative basis, interpreting 'early *n*th century' as the first half of that century, 'mid-*n*th century' as the years 25–75 within that century, and 'late *n*th century' as the second half of that century.

A similar approach has been used for meta-analysis of the changing abundance of cranial bones and vertebrae through time for cod and haddock (*Melanogrammus aeglefinus*), the most abundant gadid species (Figures 15.3 and 15.4). Plotting the data in this way makes it possible to observe trends that may relate to the consumption of decapitated preserved products, such as stockfish (Orton *et al.* 2014; Chapter 16; see below). Cranial specimens are here defined as including the bones of the neurocranium, jaw apparatus, hyoid arch and gill covers. We exclude posttemporals and appendicular elements such as the supracleithra, cleithra

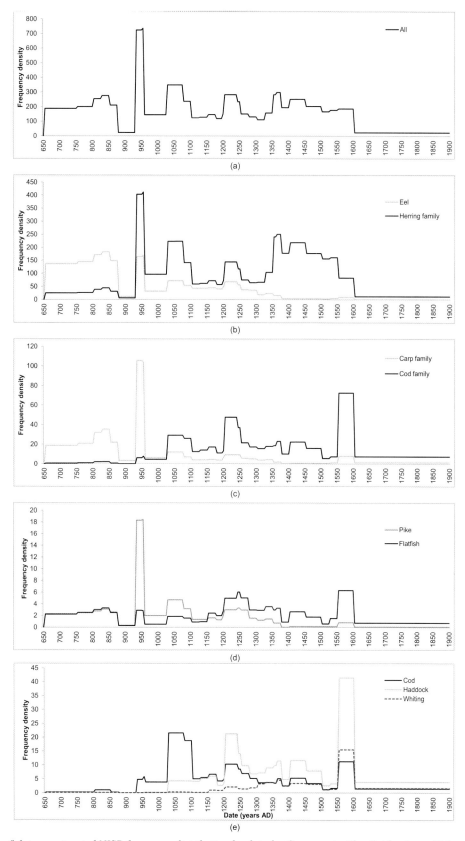

Figure 15.2. Major fish taxa estimated NISP frequency distribution by date for Coppergate, Blue Bridge Lane, Fishergate, 24–30 Tanner Row, 41–9 Walmgate and 41 Piccadilly, predominantly >2 mm sieving.

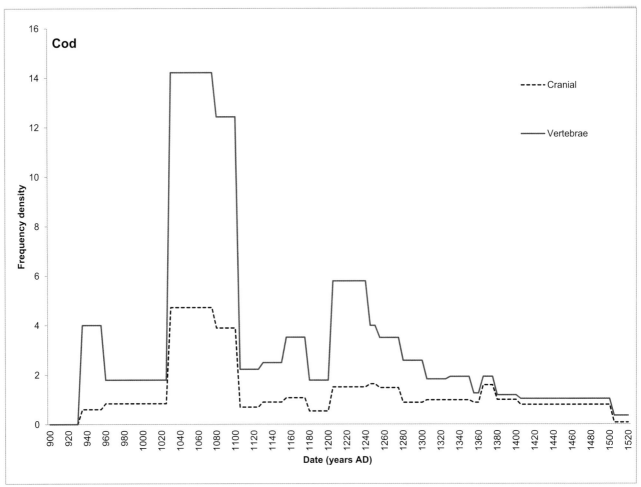

Figure 15.3. Cod estimated NISP frequency distribution of cranial elements and vertebrae by date for Coppergate and Blue Bridge Lane, predominantly >2 mm sieving.

(which support the pectoral fins, just behind the head) and scapulae because they are anatomically located where cod are often decapitated and might end up in either processing waste or a traded product.

The abundance of gadid cleithra, another potential indicator of dried fish consumption (e.g. Barrett 1997; Harland 2007; Perdikaris and McGovern 2009, 73–4), is considered separately on a site-by-site and phase-by-phase basis. The relative representation of herring anatomical elements, which can also reveal a preserved product (e.g. Enghoff 1996; Chapters 12–13), is similarly considered on a site-by-site basis.

Fish bone preservation

The fish bones from both Coppergate and Blue Bridge Lane were well preserved. Few bones showed evidence of burning (0.9% for Coppergate; 2% for Blue Bridge Lane) (Table 15.1), and the majority from both sites had surface textures described as 'good' (defined as lacking fresh appearance,

but otherwise solid, with only very localised flaky or powdery patches) (Table 15.2). Many specimens survived as substantial pieces of the original bone (Table 15.3), and in both assemblages over half of the fragments could be identified to a meaningful taxonomic group, typically species (Tables 15.4–15.5). In relative terms the Coppergate bones were slightly better preserved. This difference is most evident from the distribution of bone surface textures: 88% of the specimens from Coppergate exhibited 'good' surface textures, whereas only 64% of the Blue Bridge Lane specimens were so scored (Table 15.2). More highly fragmented specimens occur in the Coppergate assemblage (Table 15.3), but this probably reflects the inclusion of some material from the <2 mm sample fraction (see above) and the fact that even very broken bones were sufficiently well preserved to be identified. In addition to this overall difference, there is important intra-site patterning. At Coppergate, the fish bones are consistently well preserved across all phases. In no case does the frequency of burned bone exceed *c.* 2% (Table 15.1), the most common bone surface texture

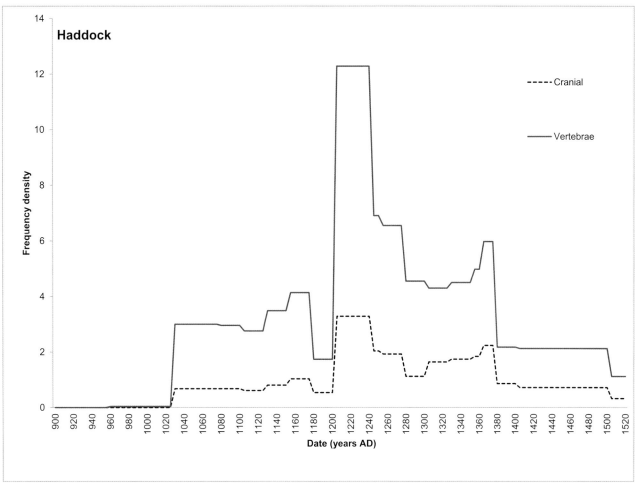

Figure 15.4. Haddock estimated NISP frequency distribution of cranial elements and vertebrae by date for Coppergate and Blue Bridge Lane, predominantly >2 mm sieving.

is 'good' in all phases (Table 15.2) and the degree of bone fragmentation is fairly consistent through time (Table 15.3). At Blue Bridge Lane, conversely, there are greater temporal differences. The fish bones from the late AD 600s to mid-800s have better bone surface textures (Table 15.2) and are much more complete (Table 15.3) than all later material from the site. Given the date of this anomalous phase, and when the Coppergate and Blue Bridge Lane chronologies meet, the implication for present purposes is that any taxonomic and/or anatomical changes observed in the ninth century need to be interpreted in light of possible preservation biases. Comparable taphonomic data do not exist for the other sites considered herein, but all the fish bone was well preserved in terms of the authors' comparative experience.

The taxa consumed

A diverse range of freshwater, migratory and marine fishes were utilised in medieval York. Lumping rather than splitting, however, it is the following taxa that account for the vast majority of specimens: herring; eel (*Anguilla anguilla*); gadids (Gadidae), mostly haddock, cod and whiting (*Merlangius merlangus*); cyprinids (Cyprinidae); flatfishes, such as flounder (*Platichthys flesus*) and plaice (*Pleuronectes platessa*); pike (*Esox lucius*); smelt (*Osmerus eperlanus*); and salmonids (Salmonidae), namely, salmon (*Salmo salar*) and trout (*Salmo trutta*). Tables 15.4–15.7 present the NISP by taxon and phase for the six sites considered here. As explained above, the estimated frequency density of the most abundant taxonomic groups across all sites is plotted through time in Figure 15.2. In this case the differences in identification methods between assemblages are minimised by excluding species, such as perch, that were sometimes identified by their scales rather than their bones. Figure 15.5, which plots the relative abundance of major taxonomic groups by feature type for Coppergate, suggests that major chronological trends transcend the vagaries of depositional context.

The first observation evident from Figure 15.2 is that there is variability in the number of fish bones identified

Table 15.1. Burnt, crushed, gnawed and acid-etched fish bone by major phase for Coppergate and Blue Bridge Lane, predominantly >2 mm sieving.

Chronology	Burnt (calcined) N	%	Burnt (charred) N	%	Crushed N	%	Other
COPPERGATE							
mid–late 800s/early 900s	5	0.6			17	2.3	
c. 930/935 c. 955/6	32	0.7	7	0.2	95	2.1	1 (gnawed)
c. 955/6	32	1.8	6	0.3	25	1.4	
c. 955/6–early/mid-1000s	25	0.7	23	0.7	67	1.9	
mid–later 1000s	17	0.5	9	0.3	77	2.4	
mid-1000s–mid-1100s	4	0.3			12	0.8	
mid-1000s–early 1200s	6	0.7	1	0.1	35	4.3	
mid-1100s–1200					2	1.8	
1200–late 1200s	4	0.2	2	0.1	39	1.9	
1275–mid-1300s	1	0.3	5	1.5			
1300–late 1300s					9	3.3	1 (gnawed)
1360–1500					1	0.1	
Total	128	0.6	53	0.3	390	1.9	4
BLUE BRIDGE LANE							
late 600s–mid-800s	8	0.1	72	1.1	204	3.1	4 (acid etched)
late 800s–mid-1000s					1	0.9	
late 1000s–late 1100s							
late 1100s–mid-1300s	6	0.2	7	0.2	9	0.3	
early–mid-1300s	1	0.1	3	0.2	36	1.9	
mid-1300s	2	0.3	9	1.4	20	3.2	
mid–late 1300s			1	6.7			
late 1300s							
late 1300s–early 1500s	249	2.9	58	0.7	17	0.2	
Total	266	1.3	150	0.7	184	0.9	4

through time, based in large part on subsampling decisions in the field and in the lab. Peaks and troughs must therefore be interpreted *vis-à-vis* the distribution of both the total fish-bone NISP through time (Figure 15.2a) and the relative frequencies, by comparing the patterns for different taxa (Figures 15.2b–15.2e). The clearest temporal trend is a shift from taxa that were definitely (e.g. cyprinids) or probably (e.g. eels) caught in fresh water to those caught in estuaries or the open sea (e.g. herring and gadids). The chronology varies by taxon, but the transition was nevertheless a tenth- to eleventh-century phenomenon (cf. Barrett *et al.* 2004; Harland *et al.* 2008). It is important to note that this progressive change is evident in the Coppergate time-series that continues beyond the ninth-century taphonomic watershed noted above (Table 15.4). Herring was the first important marine taxon. Having already been a significant minority species in the earliest phases of Blue Bridge Lane and Fishergate (both late seventh to mid-ninth century),

it rose to become the most abundant taxon – in the tenth century at Coppergate and later at other sites. Herring continued to be important in subsequent centuries at York, and were occasionally disposed of *en masse*, as at Blue Bridge Lane in the late Middle Ages. One discrete late fourteenth-century pit fill at this site, studied by Keaveney (2005), was almost entirely composed of herring bone (Figure 15.6; Table 15.8). Large-scale medieval fisheries for this species were known all along the English east coast, particularly from Scarborough to Yarmouth (Chapter 3). Starting in the fourteenth century, Baltic herring were also imported to England (Childs and Kowaleski 2000, 22; Chapter 2). The potential source (or sources) of the York herring is discussed further below.

Gadids were very rare in the earliest deposits under consideration, but they began to increase in number from the middle of the tenth century onwards. Their changing abundance through time is broken down by major taxon in

Table 15.2. Surface texture of identified fish cranial and appendicular specimens by major phase for Coppergate and Blue Bridge Lane, predominantly >2 mm sieving.

| | Surface texture | | | | | | | |
| | Excellent | | Good | | Fair | | Poor | |
Chronology	*N*	*%*	*N*	*%*	*N*	*%*	*N*	*%*
COPPERGATE								
mid–late 800s/early 900s	2	7	24	86	2	7		
c. 930/935–*c.* 955/6		0	296	94	20	6		
c. 955/6	3	3	89	87	10	10		
c. 955/6–early/ mid-1000s	2	1	125	85	20	14		
mid–later 1000s	3	2	157	89	16	9	1	1
mid-1000s– mid-1100s	6	6	84	84	9	9	1	1
mid-1000s–early 1200s			33	92	3	8		
mid-1100s–1200			11	100				
1200–late 1200s	4	3	135	91	8	5	1	1
1275– mid-1300s	3	11	20	74	4	15		
1300–late 1300s	1	6	13	81	2	13		
1360–1500	2	2	104	81	21	16	1	1
Total	34	3	1138	88	118	9	4	0
BLUE BRIDGE LANE								
late 600s– mid-800s	34	23.8	82	57.3	26	18.2	1	0.7
late 1100s– mid-1300s			45	66.2	21	30.9	2	2.9
Early– mid-1300s			41	71.9	16	28.1		
mid-1300s			19	63.3	11	36.7		
late 1300s–early 1500s			375	63.7	214	36.3		
Total	34	3.8	568	63.5	290	32.4	3	0.3

Figure 15.2e. Cod increased first, followed within a hundred years by haddock (which overtook cod itself by NISP in the thirteenth century). Haddock have more robust bones than cod, but as discussed above, no temporal differences in preservation are applicable to the relevant phases; the increase in haddock therefore cannot be attributed to preservational factors. Other gadid species were of comparatively minor quantitative importance, although the increasing amount of whiting in the late Middle Ages reflects wider English trends in the consumption of this species (Serjeantson and Woolgar 2006, 117). Based on historical sources, the English east-coast cod fishery is best known from the fourteenth century and later, when it began to compensate for competition on the herring market from Scania and The Netherlands (Childs and Kowaleski 2000; Kowaleski 2003 cf. Chapter 3). Nevertheless, it can be inferred from the present zooarchaeological evidence that the east-coast cod fishery began in the tenth century and subsequently diversified to include other cod-family species. Cod and related gadids could also have been imported as dried stockfish from Scandinavia or (after the late fourteenth century) as salted and dried fish produced by English West Country fishermen operating in a variety of in-shore and distant-water fisheries (Kowaleski 2003). These possibilities are explored below in the context of anatomical and butchery evidence.

Before (and to a lesser degree after) the rise to numerical dominance of marine taxa it is likely that many migratory fishes, such as eel, salmonids and smelt, were caught in stationary traps (e.g. Murphy 2009, 48–9; O'Sullivan 2004). Obligatory freshwater species could also have been caught in traps, and/or by use of nets and hooks (e.g. Losco-Bradley and Salisbury 1988). From approximately the twelfth century onwards, some true freshwater fish, such as pike, cyprinids and perch, may occasionally have been the produce of fish ponds established by both secular and ecclesiastical proprietors (Dyer 1988; Serjeantson and Woolgar 2006, 124–5). The King's Fishpool in York is an obvious example, but unless it was illegally poached, its produce is more likely to have provided gifts to rural magnates than food for the urban craftspeople active at sites such as Coppergate (McDonnell 1981; Rees Jones and Daniell 2002; Steane

Table 15.3. Completeness of identified fish cranial and appendicular specimens by major phase for Coppergate and Blue Bridge Lane, predominantly >2 mm sieving.

	Completeness									
Chronology	80–100%		60–80%		40–60%		20–40%		1–20%	
	N	%	N	%	N	%	N	%	N	%
COPPERGATE										
mid–late 800s/early 900s	10	33	7	23	2	7	8	27	3	10
c. 930/935–*c.* 955/6	36	11	82	26	60	19	86	27	52	16
c. 955/6	8	8	21	21	22	22	39	38	12	12
c. 955/6–early/mid-1000s	7	5	24	16	48	33	45	31	23	16
mid–later 1000s	14	8	37	21	37	21	47	27	42	24
mid-1000s–mid-1100s	13	13	10	10	23	23	31	31	23	23
mid-1000s–early 1200s	3	8	4	11	10	28	13	36	6	17
mid-1100s–1200			1	9	3	27	4	36	3	27
1200–late 1200s	18	12	35	24	32	22	33	22	30	20
1275–mid-1300s	1	4	4	15	4	15	11	42	6	23
1300–late 1300s	2	13	5	31	4	25	4	25	1	6
1360–1500	14	11	35	28	17	13	42	33	19	15
Total	135	10	274	21	275	21	383	30	228	18
BLUE BRIDGE LANE										
late 600s–mid-800s	72	50.3	35	24.5	15	10.5	21	14.7		
late 1100s–mid-1300s	4	5.9	20	29.4	17	25.0	23	33.8	4	5.9
early to mid-1300s	12	21.1	6	10.5	19	33.3	18	31.6	2	3.5
mid-1300s	7	23.3	6	20.0	7	23.3	10	33.3		
late 1300s–early 1500s	18	3.1	104	17.7	169	28.7	284	48.2	14	2.4
Total	114	12.7	172	19.2	230	25.7	359	40.1	20	2.2

1988). Regardless, freshwater fish were valued in the High Middle Ages, despite (or because of) their increasing rarity (Dyer 1988; Serjeantson and Woolgar 2006).

It has previously been noted that among the freshwater species rare identifications of grayling (*Thymallus thymallus*) and burbot (*Lota lota*) declined through time, possibly due to increasing pollution of the waters in the environs of the medieval town (O'Connor 1989, 198). Although this explanation is plausible, our ability to draw conclusions is hampered by the fact that the number of specimens of these taxa is very small at all dates. Possible human impacts on aquatic ecosystems are more clearly evidenced by the relative (and absolute) decline in the abundance of all freshwater taxa through time – and perhaps also by changes in the size distribution of species such as pike, to be discussed below.

Anatomical patterning, total length estimates, cut marks and stable isotope evidence

Having looked at broad chronological changes in the relative abundance of different taxonomic groups, we will now examine patterning within a selection of the most common species (eel, herring, cod, haddock and pike). Although cyprinid and flatfish bones were also frequent, they are difficult to identify to species routinely (e.g. Wouters *et al.* 2007) and are thus not amenable to detailed analysis. The present discussion will focus on the relative abundance of different skeletal elements, the sizes (expressed as TL) of fish represented by the bones and any cut marks evident from butchery practices. The results of stable isotope analysis of cod bones (Barrett *et al.* 2011) will also be addressed. Only the sieved assemblages from Coppergate and Blue Bridge Lane are considered in most instances, but cut marks on hand-collected bones from Coppergate are included in Table 15.9. Together, these data help indicate whether fish arrived in the town whole (perhaps fresh or lightly cured) or preprocessed (potentially implying heavily salted and/or dried products subject to long-range trade). Fish TL estimates can also reveal temporal trends that may relate to the intensity of exploitation.

Eels were the second most common taxon overall at Coppergate and Blue Bridge lane, beginning as the most abundant species in the earliest phases, dating

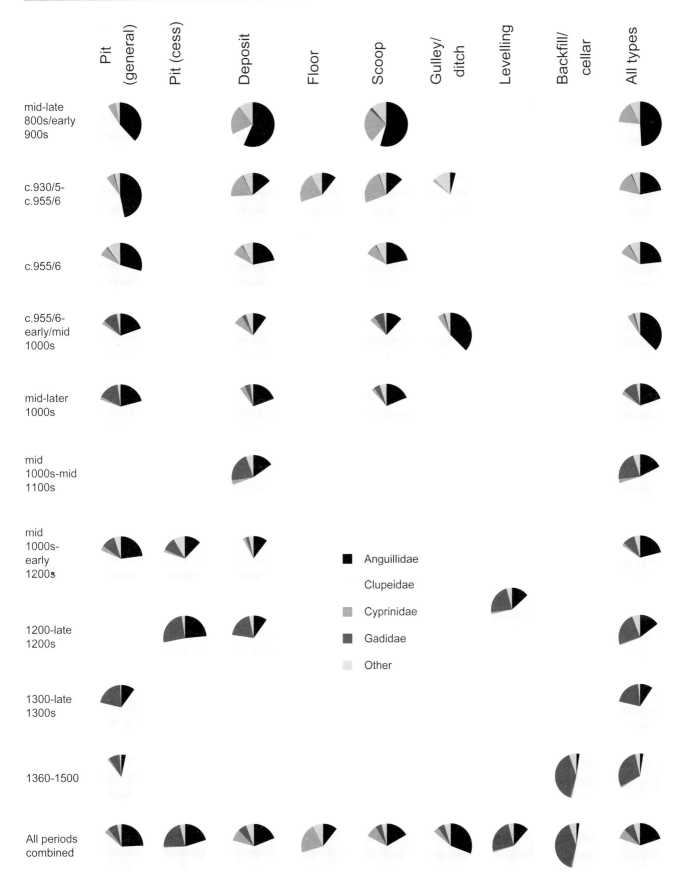

Figure 15.5. Major fish taxonomic groups NISP distribution by phase and feature type for Coppergate, predominantly >2 mm sieving.

Table 15.4. Coppergate NISP for all identified specimens, predominantly >2 mm sieving (a = noted by AKG Jones during initial analysis; p = present).

Taxon	mid- to late 800s/ early 900s	c. 930/935 to c. 955/6	c. 955/6	c. 955/6 to early/ mid-1000s	mid–later 1000s	mid–1000s to mid-1100s	mid-1000s to early 1200s	mid-1100s to 1200	1200 to late 1200s	1275 to mid-1300s	1300 to late 1300s	1360 to 1500	Other	Total
Elasmobranch				a			a		a			a	a	
Ray family				a	2	3	4		1	a				10
Thornback ray				a	1		a		3	1			a	5
Sturgeon		p	p											
Eel	265	767	327	406	479	148	132	7	189	11	13	11	176	2931
Conger eel							a		a				a	
Herring family	a	2	1	a		a	a		a	a		23	a	26
Atlantic herring	145	1982	834	1961	1550	438	393	34	683	97	93	253	366	8829
Shads	2	a	1	2	3	3	2						1	14
Allis shad							1							1
Twaite shad							1							1
Pilchard		1												1
Carp family	83	483	106	115	67	24	13	a	13	4		5	21	934
Barbel		a												
Gudgeon		4												4
Dace	a	5	3											8
Common bream				a										
Roach	a	12	3	2	2	1	a	1	a	a				21
Roach?		1												1
Rudd		2												2
Chub		2			1									3
Tench		2		1									1	4
Tench?		1												1
Pike	8	90	53	37	32	10	11		24	1			9	275
Smelt	4	66	24	27	4	5	1		6	2		4	8	151
Salmon & trout family	21	24	23	14	11	10	2		4			1	5	116
Salmon	a	a	a	a	a		a		a	a		a	a	
Trout	a	a	a	a	a				a	a				
Grayling	a	a	a						a					
Cod family	1	2	2	3	22	6	3		6	2	1	4	5	55
Cod		24	3	68	220	72	31	7	81	22	9	46	27	612
Whiting			2	1	2	3	a		6	8	13	26	9	70
Pollack										1				1
Saithe							1			3				4
Haddock				1	4	98	25	15	197	25	4	47	60	476
Burbot	2	5	4	2	9				3	1				26
Burbot?									2					2

(Continued)

Table 15.4. Coppergate NISP for all identified specimens, predominantly >2 mm sieving (a = noted by AKG Jones during initial analysis; p = present). (Continued)

Taxon	mid- to late 800s/ early 900s	c. 930/935 to c. 955/6	c. 955/6	c. 955/6 to early/ mid-1000s	mid–later 1000s	mid-1000s to mid-1100s	mid-1000s to early 1200s	mid-1100s to 1200	1200 to late 1200s	1275 to mid-1300s	1300 to late 1300s	1360 to 1500	Other	Total
Tusk										1				1
Ling						2	a	1	3	a		2	1	9
Garfish										a				
Three-spined stickleback	a	a												
Gurnard family						p								
Perch family		a	a			a	a		a			a	a	
Perch	1	18	6	8	a		2		4	1	1	1	1	43
Atlantic horse-mackerel/scad	a				2	a	1		p				2	5
Atlantic mackerel		3	1	1	2	3			1			1	2	14
Turbot family						1								1
Turbot							a							
Halibut family	7	13	5	6	11	9	5	1	24	7		9	5	102
Plaice				1		2	a		7	a	1	a	1	12
Halibut					1									1
Flounder		a		a		1			2				1	4
Total identified	539	3509	1398	2656	2425	842	625	67	1262	186	135	433	701	14778
Unidentified (QC0)	189	933	395	818	761	677	196	44	789	158	136	274	315	5685
Total	728	4442	1793	3474	3186	1519	821	111	2051	344	271	707	1016	20463

Table 15.5. Blue Bridge Lane NISP for all identified specimens, >2 mm sieving. The chronology is given in centuries AD.

Taxon	late 7th to mid-9th	late 9th to mid-11th	late 11th to late 12th	late 12th to mid-14th	early–mid-14th	mid-14th	mid-late 14th	late 14th	late 14th to early 16th	unphased	Total
Ray family	1				1	1			14		17
Thornback ray				2							2
Eel	3013	40		97	84	41	11	3	57	5	3351
Conger eel				1							1
Herring family	6										6
Atlantic herring	499	17	4	471	453	266	4	8	6067	14	7803
Allis shad/twaite shad	7				5						12
Allis shad	1										1
Carp family	74			22	5	4			10	2	117
Bleak	1										1

(Continued)

Table 15.5. Blue Bridge Lane NISP for all identified specimens, >2 mm sieving. The chronology is given in centuries AD. (Continued)

Taxon	late 7th to mid-9th	late 9th to mid-11th	late 11th to late 12th	late 12th to mid-14th	early–mid-14th	mid-14th	mid–late 14th	late 14th	late 14th to early 16th	unphased	Total
Gudgeon	1										1
Chub/ dace	1										1
Dace	1			1							2
Common bream?	1										1
Roach				1	1						2
Chub	1										1
Pike	37			12	3	2			1		55
Smelt	20	1		2		3				1	27
Salmon & trout family	6										6
Cod family	2			9	11	2		1	20		45
Cod				53	2	1			15	3	74
Cod?	1				1				2		4
Whiting	3			37	13				12		65
Haddock	1			33	62	4			62	2	164
Haddock?				3	1						4
Burbot	5			1							6
Burbot?	1										1
Ling				3					1		4
Ling?				2							2
Gurnard family	1								1		2
Perch	2			2	1						5
Flatfish order	2			1							3
Turbot family				2							2
Halibut family	14			48	3	3			7	1	76
Flounder/plaice	3										3
Plaice				13	1				1		15
Flounder	1										1
Sole family				2							2
Total identified	3706	58	4	818	646	328	15	12	6270	28	11885
Unidentified	2894	57	3	2068	1243	297	0	24	2366	26	8978
Total	6600	115	7	2886	1889	625	15	36	8636	54	20,863

between the seventh and ninth centuries. Thereafter they represented a significant minority species until declining to trivial numbers over the course of the fourteenth century. The continuing decrease in their importance through time militates against the possibility that the beginning of the shift, during the ninth century, is a taphonomic bias (see above). In all cases the vast majority of specimens were vertebrae, which is to be expected

given the large numbers of this element (110–120) per individual (Froese and Pauly 2015). All parts of the skeleton are, nevertheless, represented, indicating that the eels were used fresh or at least whole (smoked, for example), without any separation of cranial and vertebral elements. Eel TL estimates based on the sieved bones from Coppergate and Blue Bridge Lane are summarised in Figure 15.7. Numbers are small compared with the

Table 15.6. Fishergate NISP for all identified specimens, >2 mm sieving. The chronology is given in years or centuries AD.

Taxon	late 7th/early 8th to early/mid-9th	later 8th to early 9th	early 9th	late 10th to 12th	1195 to early 14th	13th to 16th	15th to 16th	c. 1538	later 16th	17th to 19th	Total
Elasmobranch				2	1	10			14	5	32
Dogfish family				2							2
Ray family				1				3	6	6	16
Thornback ray					4	3	2		19	19	47
Sturgeon					1			1			2
Eel	3140	167	275	729	305	137	66	303	58	42	5222
Conger eel						1	1	1	1	1	5
Atlantic herring	567	25	122	716	491	834	194	538	735	706	4928
Allis shad/twaite shad	50	1	1	4	7	1					64
Carp family	754	42	112	71	86	36	14	85	67	79	1346
Barbel	4			1	2						7
Gudgeon	1	1									2
Chub/dace	2				1					1	4
Roach	1					1	1	7	2	2	14
Chub	1			1	1			3			6
Pike	65	4	3	23	22	11	5	5	7	5	150
Smelt	10		6	11	10	1	1	17	2	2	60
Salmon & trout family	13	1	1	4	3	1	1	1			25
Atlantic salmon	26	1	1	4	5	1		14	1		53
Trout	14	1		2	3	1	1	1	1	1	25
Whitefish?	1										1
Grayling	9	1	2		5						17
Cod family	1	7	2	22	7	14	10	18	31	25	137
Cod	9		8	50	21	43	30	11	100	86	358
Whiting	1				15	84	32	31	145	97	405
Haddock	3			7	22	238	40	35	397	230	972
Burbot	4		1	1	2				1	1	10
Ling?						4	4	7	6	7	28
Three-spined stickleback								1	1	1	3
Gurnard family									2	2	4
Perch family	1			3	4	4	4		7	7	30
Perch	2			7	1			1			11
Atlantic horse-mackerel/ scad	1			2							3
Atlantic mackerel						4		41	5	2	52
Turbot								1			1
Halibut family	71	6	5	10	23	17	9	31	45	42	259
Plaice						6	6	1	3	3	19
Halibut						3	1		2		6
Flounder	9			1	1			2			13
Total identified	4760	257	539	1674	1043	1455	422	1159	1658	1372	14,339
Unidentified	515	39	11	321	505	851	253	400	1174	825	4894
Total	5275	296	550	1995	1548	2306	675	1559	2832	2197	19,233

Table 15.1. 24–30 Tanner Row, 41–9 Walmgate and 41 Piccadilly NISP for all identified specimens, predominantly >1 mm and >2 mm sieving.

	Tanner Row		41–9 Walmgate		41 Piccadilly	41–9 Walmgate
	1100s to 1200s	early 1200s	mid–late 1300s	early 1400s	1400s	mid–late 1500s
Ray family		2				
Thornback ray		2		1	5	3
Eel	825	159	9		9	28
Conger eel	1	1				
Herring family	6	14				
Atlantic herring	856	320	166	409	230	70
Shads	1	1				
Carp family	58	33	1			10
Roach	3					
Rudd	1					
Pike	12	1				
Smelt	18	4				
Salmon & trout family	5	1				
Salmon	2	3			1	
Trout	3					
Grayling	1					
Cod family	102	93	2	9	29	8
Cod	37	9		19	11	7
Whiting	1	1	8	3	21	3
Haddock	7	39	21	37	73	11
Burbot	1	1				
Ling					4	
Perch family	3					
Perch	2					
Sea bream family	1					
Sand eel family					10	
Flatfish order			2	8	7	14
Turbot family					1	
Halibut family	4	9				
Flounder/plaice			3			
Plaice				1	9	
Halibut						
Flounder	1					
Total identified	1951	693	212	487	410	154

overall abundance of this species because vertebrae were not attributed to size categories. After a peak in the exploitation of large individuals in the tenth to eleventh centuries at Coppergate, eels became smaller through time at both sites. This trend could have resulted from fishing pressure, an interpretation consistent with the generally decreasing numbers of eels through time. However, changing fashion in diet probably also played a role – particularly with the virtual abandonment of eel consumption by the late fourteenth century. There were no butchery marks on the eel bones from either Coppergate or Blue Bridge Lane.

Figure 15.6. A subsample (prior to sieving) of a late fourteenth-century deposit, predominately of herring bone, from Blue Bridge Lane (Photo: James Barrett).

Herring is the most abundant taxon overall in the York assemblages, overtaking eel by NISP in the tenth century. The vast majority of specimens are from fish between 15 cm and 30 cm TL regardless of chronology, with most probably within the range of 24–30 cm. These sizes are consistent with past North Sea catches (e.g. Samuel 1918, 27; Nash and Dickey-Collas 2005). The abundance of herring elements for Coppergate and Blue Bridge Lane is shown in Table 15.8, which also includes data from Keaveney's (2005) analysis of a subsample from the late fourteenth-century pit-fill at Blue Bridge Lane noted above. In the absence of ice or refrigeration, this species can be cured for short-term preservation by drying, salting and/or smoking (Cutting 1955, 53). For long-term storage, however, herring were salted and often packed in barrels with brine in the Middle Ages (Unger 1978; Chapter 2). A specific butchery practice was sometimes employed for this purpose, involving removal of part of the gut along with some bones from the hyoid, gill and appendicular regions.

This method, sometimes known as the Scania (Skåne) cure, is best known from the medieval Danish fishery, where it is archaeologically documented in thirteenth- to fourteenth-century material from Selsø-Vestby (Enghoff 1996; Chapter 13; see also Chapters 2 and 12 where it is noted earlier in other Baltic Sea contexts). It is thought to have been attempted in eastern England near the end of the fourteenth century, based on historical sources (Childs and Kowaleski 2000, 22). Later it was put to best effect by the Dutch North Sea fishery (Unger 1978; Poulsen 2008).

The remains of barrelled herring from the wreck of the Drogheda Boat in Ireland (Harland 2009) provide an illuminating comparison for the York material. The boat sank during the sixteenth century, probably when transporting a cargo to a ship anchored in the harbour. The barrels may represent a 'last' of herring, a commercial unit representing *c.* 12,000 fish (Unger 1978, 337). Anatomical and butchery patterning indicate that a consistent method of preservation was used. Bones from the appendicular

Table 15.8. Herring NISP for Coppergate and Blue Bridge Lane, predominantly >2 mm sieving. The table includes an additional sample from Blue Bridge Lane analysed by E. Keaveney. Equivalent data from the Drogheda boat assemblage of preserved herring are provided for comparison. The chronology is given in years or centuries AD.

Site	Coppergate				Blue Bridge Lane										Blue Bridge Lane (E. Keaveney)		Drogheda Boat	
Phase	1360–1500		All		Late 7th–mid-9th		Late 12th–mid-14th		Early–mid-14th		Mid-14th		Late 14th–early 16th		Late 14th		16th	
Element	NISP	%	NISP	%	NISP	%	NISP	%	NISP	%	NISP	%	NISP	%	NISP	%	NISP	%
Articular	10	4.0	76	0.9			2	0.4					78	1.3	113	1.9	137	1.1
Basioccipital	3	1.2	23	0.3			2	0.4	1	0.2	3	1.1	36	0.6	105	1.7	102	0.8
Ceratohyal	1	0.4	38	0.4	4	0.8			2	0.4	1	0.4	60	1.0	124	2.0	111	0.9
Dentary	12	4.7	102	1.2	2	0.4	1	0.2	3	0.7	4	1.5	101	1.7	101	1.7	304	2.5
Hyomandibular	4	1.6	51	0.6									43	0.7	48	0.8	132	1.1
Maxilla	19	7.5	124	1.4	2	0.4	6	1.3	5	1.1	2	0.8	101	1.7	112	1.8	444	3.6
Opercular	6	2.4	63	0.7									35	0.6	214	3.5	96	0.8
Otic Bulla	12	4.7	157	1.8	5	1.0	27	5.7	7	1.5	5	1.9	191	3.1	888	14.6	654	5.3
Parasphenoid	1	0.4	9	0.1									2	0.0	11	0.2	119	1.0
Posttemporal	3	1.2	28	0.3							1	0.4	17	0.3	20	0.3	72	0.6
Preopercular	1	0.4	16	0.2							2	0.8	13	0.2	48	0.8	54	0.4
Quadrate	3	1.2	25	0.3			2	0.4	1	0.2	2	0.8	34	0.6	52	0.9	166	1.4
Vomer									1	0.2			2	0.0	6	0.1	59	0.5
First Vertebra	8	3.2	226	2.6	16	3.2	23	4.8	20	4.4	6	2.3	282	4.6	227	3.7	386	3.1
Abdominal Vert.	99	39.1	3578	41.2	261	52.3	211	44.4	293	64.5	122	45.9	2598	42.8	1550	25.5	4294	34.9
Caudal Vertebra	65	25.7	4044	46.5	207	41.5	197	41.5	117	25.8	115	43.2	2402	39.6	2260	37.2	5042	41.0
Ultimate Vertebra	1	0.4	74	0.9	2	0.4	4	0.8	2	0.4			37	0.6	183	3.0	81	0.7
Cleithrum	3	1.2	40	0.5					1	0.2	2	0.8	23	0.4	10	0.2	4	0.0
Scapula																	1	0.0
Supracleithrum	2	0.8	15	0.2					1	0.2	1	0.4	12	0.2			35	0.3
Total	253	100	8689	100	499	100	475	100	454	100	266	100	6067	100	6062	100	12293	100

Table 15.9. Anatomical plane of butchery marks on Coppergate cod bones by phase, all recovery methods. Vertebrae groups follow Barrett (1997).

Plane	Element	c. 930/935–c. 955/6	c. 955/6	c. 955/6–early/mid-1000s	mid–later 1000s	mid-1000s–mid-1100s	mid-1000s–early 1200s	mid-1100–1200	1200–1240	1240–1275	1200–late 1200s	1275–mid-1300s	1300–late 1300s	1360–1500	1370–1720	Other	Total
Sagittal	First vertebra						1				2						3
	Abdominal vert. group 1			1			2				9			1	1	2	16
	Abdominal vert. group 2						6	2	1	2	7						18
	Abdominal vert. group 3	1			1	2	2	3	3		11	1		3		4	31
	Caudal vert. group 1							2			4				2	2	10
	Caudal vert. group 2										1						1
	Total	1		1	1	2	11	7	4	2	34	1		4	3	8	79
Transverse	First vertebra															1	1
	Abdominal vert. group 2						1										1
	Abdominal vert. group 3					2				1	2				1		6
	Caudal vert. group 1				1	1					4	1				3	10
	Caudal vert. group 2	1		1							2			1			5
	Total	1		1	1	3	1			1	8	1		1	1	4	23
Frontal	Abdominal vert. group 1						2									1	3
	Abdominal vert. group 2						3										3
	Abdominal vert. group 3					2		1								1	4
	Total					2	5	1								2	10
Combined	Abdominal vert. group 3							1	1								2
	Caudal vert. group 1					3							1			1	5
	Total					3		1	1				1			1	7

region were routinely removed, presumably along with some of the gut. Cleithra were conspicuously absent; the only four cleithrum fragments from Drogheda were all from the dorsal portion of the element, with one displaying a clear cut mark. Other elements found in abnormally low proportions at Drogheda are the scapula and supracleithra. Attempts were made to identify all urohyals (a bone not routinely recorded), but as with the scapula and the supracleithrum, they were almost absent from the assemblage. Ceratohyals, epihyals and hyomandibulars were all found in numbers comparable with other elements, however, marking a difference from herring cures like those represented at Selsø-Vestby in Denmark; in a post-medieval processing dump in Waterford, Ireland; and in a sixteenth-century Dutch shipwreck (Enghoff 1996; Lauwerier and Laarman 2008; Tourunen 2008; Chapter 13). Cleithra are key to understanding herring processing,

as this element has consistently been removed during processing across all of the above examples.

With these comparisons in mind, the large quantities of herring from Coppergate and Blue Bridge Lane do not show anatomical patterning characteristic of preparation for barrelling. Even if one focuses on the distinctive late fourteenth-century dump from Blue Bridge Lane, all parts of the herring are represented, including bones, such as the cleithrum, that would have been removed if preserved using cures like those at Selsø-Vestby or Drogheda. Moreover, the ten herring cleithra that Keaveney (2005) identified at Blue Bridge Lane only include the middle or ventral portions of the bone. In no case was the dorsal portion present, although this was the only part of the few cleithra found in the Drogheda Boat. Keaveney also identified 17 urohyal fragments in her sample. In contrast, only two were recovered from the Drogheda excavation – both in a sample

Coppergate

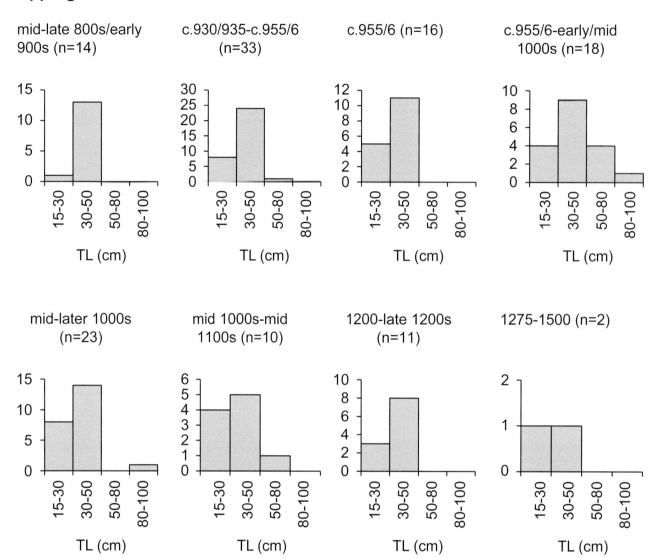

Blue Bridge Lane

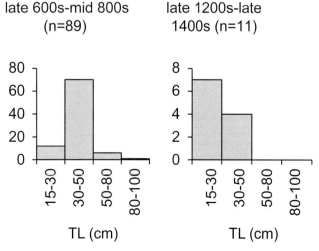

Figure 15.7. Eel total length estimates (cm) for Coppergate and Blue Bridge Lane, based on NISP of identified cranial and appendicular specimens, predominantly >2 mm sieving.

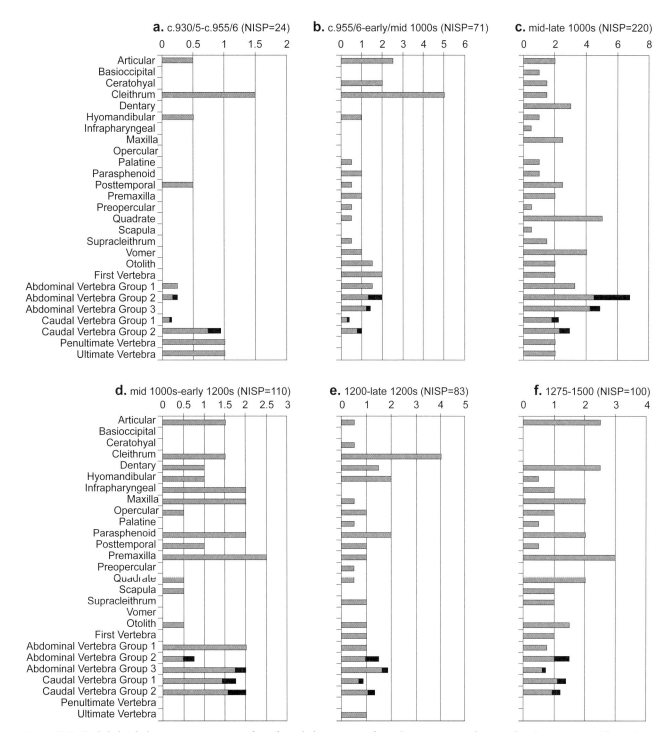

Figure 15.8. Cod skeletal element representation for selected phase groups from Coppergate, predominantly >2 mm sieving. The scale is NISP divided by number of occurrences in the body; for vertebrae, the lighter grey shows the minimum number and the darker grey the maximum, representing the fact that cod have a variable number of vertebrae per fish.

external to the main barrel contents. The urohyal was not included in the protocol used for Coppergate and most of the Blue Bridge Lane assemblage.

These results do not exclude the possibility that some of the York herring arrived salted in barrels, given that the

'Scania cure' was only attempted in England at the close of the fourteenth century. They do imply, however, that the fish are most likely to derive from relatively local catches rather than from imports of Baltic or (in the latest centuries under consideration) Dutch herring. Moreover, the material

a. c.955/6-early/mid 1000s (NISP=25)

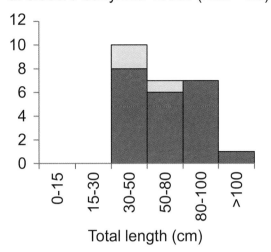

b. mid-later 1000s (NISP=66)

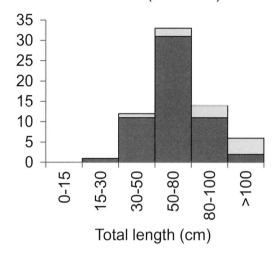

c. mid 1000s-early 1200s (NISP=40)

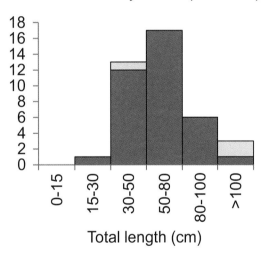

d. 1200-late 1200s (NISP=33)

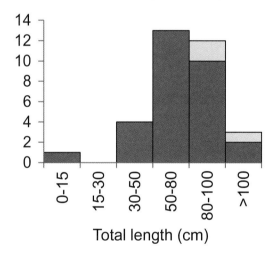

e. 1275-1500 (NISP=43)

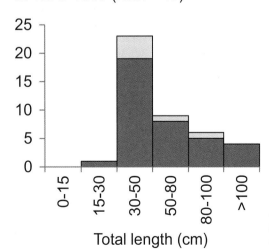

Figure 15.9. Cod total length estimates (cm) for Coppergate, based on NISP of identified cranial and appendicular specimens, predominantly >2 mm sieving. Cleithra are displayed separately in lighter grey.

Table 15.10. Cod and haddock total length estimates (cm) for Blue Bridge Lane, based on NISP of identified cranial and appendicular specimens, >2 mm sieving.

TL (cm)	late 1100s– mid- 1300s	early– mid- 1300s	mid- 1300s	late 1300s– early 1500s
Cod				
30–50	6	2		
50–80	2			2
80–100	1			
>100	3			1
Haddock				
15–30		2	1	
30–50	8	22	1	16
50–80	2			1

of late fourteenth–sixteenth-century date (dating from the time when English fishermen had adopted the 'Scania cure' for long-term storage) is perhaps more likely to represent lightly cured, unbarrelled fish, presumably from the nearby Yorkshire coast. It is not surprising that some apparently spoiled and were discarded *en masse* in a pit at Blue Bridge Lane.

A mass deposit of small clupeid (herring and sprat) bones was also found during excavations at St Mary Bishophill Junior. It has recently been described as a tenth-century herring-processing factory (Carver and Loveluck 2013, 123, citing Cramp 1967, 18–19). However, it is more likely to have been the residue of late Roman fish-sauce production, based on the stratigraphic context, the associated finds and the tiny size of the fish represented (7–11 cm TL) (Jones 1988b).

Cod family fishes are the third most abundant group overall, initially becoming numerous in the eleventh century. As noted above, the relative abundance of different skeletal elements can be used to infer whether or not preserved gadids are likely to be represented at a consumption site. In a medieval English context the most likely products were air-dried stockfish from the Scandinavian north (Nedkvitne 2014) or, particularly from the late fourteenth century onward, the salted and dried products of English fishermen working in a variety of inshore and distant waters (Candow 2009; Childs 1995; Jones 2000; Kowaleski 2003; Chapter 3). Both cures typically used decapitated fish, sometimes with appendicular elements, such as the cleithra, left in the preserved product (cf. Barrett 1997; Betts *et al.* 2014). Anterior (i.e. abdominal) vertebrae could be removed, leaving only caudal vertebrae in the finished product, or virtually all vertebrae could be left in the processed fish (cf. Chapter 1; Chapter 18).

Figure 15.8 illustrates the abundance of the identified cod skeletal elements for the major phase groups of Coppergate. In order to accurately compare between bones, the number of occurrences of each skeletal element within the body is taken into account. For example, totals of singly-occurring midline elements are unchanged, totals for paired elements are halved, and the total for each vertebral group is divided by the number of that type within the body. Where the latter represents a range from minimum to maximum, the darker bars show the difference. In this way any element that occurs in larger or smaller quantities than normal can be easily visually identified.

Patterning in these data can result from differential recovery (which is controlled for by the use of sieving in the present study), preservation (fragile bones will be less abundant, and bones that typically break into multiple robust pieces may be over-represented) and transport (the issue of present interest). Some variability will also derive from the vagaries of small sample sizes and the differing degree to which skeletal elements can be identified to species. Despite these complications, it is clear that cod cleithra are well represented in some phases at Coppergate. The cleithrum is a large bone, but for cod also a fragile one. In a sieved archaeological assemblage it should not be overrepresented unless originally deposited in larger numbers than other skeletal elements. This pattern is evident in the tenth to eleventh century and the thirteenth century. In both cases cleithra were the most abundant bone. Cod also occur in the Blue Bridge Lane assemblage, but the smaller sample size cannot support similar anatomical analysis of a single skeletal element.

Not all preserved cod products were prepared with the cleithra left in (see above). The relative abundance of cranial versus vertebral bones is thus also an important indicator. Figure 15.3 plots a time series including the relevant data from both Coppergate and Blue Bridge Lane. The method used is explained above. There are two instances in which the trends for cranial bones and vertebrae diverge – the mid-tenth century and the early thirteenth century. In both instances the abundance of vertebrae peaks and plunges without correspondingly large changes in the number of cranial specimens. It may not be a coincidence that these dates correspond with phases when cleithra represent the most abundant single skeletal element at Coppergate (Figure 15.8). Perhaps during these two periods of time some cod were brought to York in a decapitated state, increasing the number of vertebrae and cleithra *vis-à-vis* cranial bones.

Our understanding of the cod assemblage can be further enhanced by considering the estimated total lengths of the fish involved. The relevant data for cod cranial and appendicular elements from Coppergate and Blue Bridge Lane are displayed in Figure 15.9 and Table 15.10, using six ordinal TL categories. These results are based on the sieved dataset alone, thus avoiding recovery bias. Phases have been grouped into five increments to increase sample sizes.

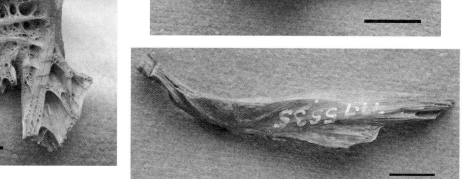

Figure 15.10. A selection of cod vertebrae and cleithra from Coppergate with butchery marks indicative of sagittal splitting, decapitation, filleting and/or removing the anterior portion of the vertebral column. Scale 1 cm (Photos: Cluny Johnstone).

Differences through time are evident, making changes in fish procurement likely. Once again the thirteenth century stands out, in this instance because the cleithra are all attributed to cod of greater than 80 cm TL. It is tempting to interpret these large specimens as indicative of preserved imports, as they are of a size typically used for some varieties of stockfish (e.g. Perdikaris and McGovern 2009, 80–1).

The hypothesis that preserved cod were imported in the thirteenth century, and perhaps also in the tenth–eleventh centuries, has been further explored by stable isotope analysis (Barrett *et al.* 2011). Of 21 vertebrae and ten cleithra studied, only four vertebrae had $\delta^{13}C$ and $\delta^{15}N$ signatures that were not consistent with an origin in local waters of the southern and central North Sea. Two vertebrae dating to AD 1040–1280 and AD 1200–1280 match isotope control data from Arctic Norway or northeastern North Atlantic sources, such as

Iceland and northern Scotland. A third example attributed to the same potential sources dates to between AD 1275 and the mid-fourteenth century. The fourth potentially non-local specimen, of the same date, was initially best matched by control data from the Kattegat and western Baltic Sea (Barrett *et al.* 2011), but it is now known that similar isotope values can be found among cod from the Northern Isles of Scotland (Hutchinson *et al.* 2015). These four specimens support the argument that some of the York cod were preserved fish from distant waters, most convincingly in the High Middle Ages (e.g. the thirteenth century). Along with the anatomical and size data, however, they also suggest that the majority were caught in relatively local waters.

Some 'locally' caught cod may also have been preserved using butchery methods analogous to those employed in distant waters. For example, five of the thirteenth-century

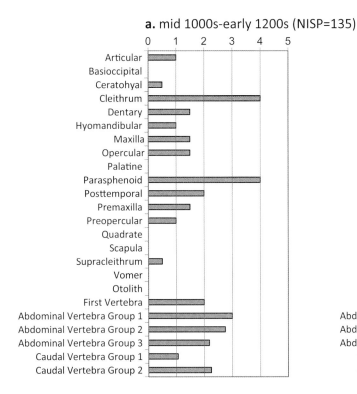

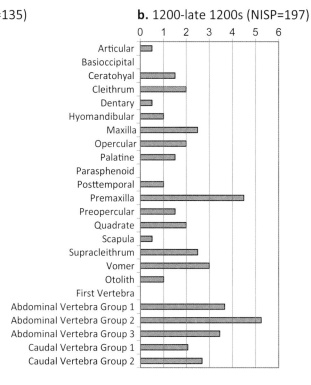

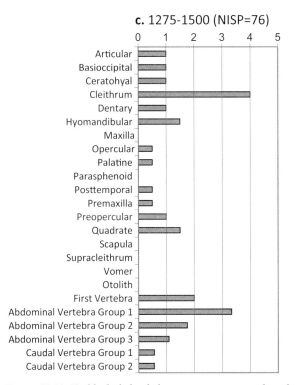

Figure 15.11. Haddock skeletal element representation for selected phase groups for Coppergate, predominantly >2 mm sieving. The scale is NISP divided by number of occurrences in the body.

a. mid 1000s–early 1200s (NISP=36)

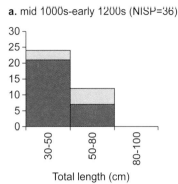

Total length (cm)

b. 1200 to late 1200s (NISP=51)

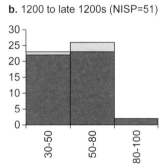

Total length (cm)

c. 1275 to 1500 (NISP=27)

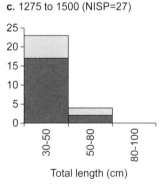

Total length (cm)

Figure 15.12. Haddock total length estimates (cm) for Coppergate, based on NISP of identified cranial and appendicular specimens, predominantly >2 mm sieving. Cleithra are displayed separately in lighter grey.

cleithra from cod between 80 cm and 100 cm TL (discussed above as possible evidence for preserved fish) yielded stable isotope values consistent with a south-central North Sea origin. Moreover, many cod bones exhibited butchery marks that could have been caused by preparation for salting and/or drying (Figure 15.10; Table 15.9; cf. Barrett 1997; Barrett *et al.* 1999; 2008). All of the Coppergate vertebrae and cleithra subjected to stable isotope analysis were chosen from among these cut specimens, but (as noted above) only four yielded results suggestive of distant-water catches.

Overall, it would seem that whole cod were routinely consumed in medieval York, but that decapitated dried and/ or salted products were also used in small numbers. Some of the latter were probably from relatively local sources in the southern or central North Sea (presumably cured by both salting and drying), whereas others were probably stockfish (dried without salt) from distant northern waters. The best evidence for these imports, from either source, relates to the thirteenth century. There are intriguing hints that a few preserved cod were also imported in the tenth to eleventh centuries, but the evidence is ambiguous. There is no case for imported preserved cod being consumed in York between the mid-eleventh century and the beginning of the thirteenth century.

The skeletal element data for haddock from Coppergate are illustrated in Figure 15.11. Haddock cleithra are naturally robust and survive better than almost all other haddock elements, in marked contrast to cod (see Harland 2006). Consequently, they would be expected in considerable quantities, as is the case. They cannot be used as a reliable guide to the presence of imported preserved fish. Overall, there is little to suggest that the haddock arrived in York in a pre-processed state. Presumably they derived from local coastal fishing. There is also no temporal trend in the size of haddock consumed, with most being 30–80 cm in TL (Figure 15.12; Table 15.10).

After cod and haddock, pike are the next most abundant taxon identified to the species level. It is notable that large pike of 50–80 cm TL are represented until the mid-eleventh century, but almost completely absent thereafter (Figure 15.13). It seems highly plausible that this decrease in maximum TL reflects fishing pressure, perhaps combined (in some instances) with a shift in source from natural watercourses to fish ponds. There is no clear butchery evidence, from cut marks or the representation of different skeletal elements, to suggest a cured product (cf. Hoffmann 2009; Jonsson 1986).

Conclusions

Detailed analysis of the fish-bone evidence from York confirms a rapid shift in emphasis from freshwater and migratory species (such as eel, cyprinids and pike) to marine taxa (specifically herring and gadids) between the mid-tenth and mid-eleventh centuries – followed by ongoing increases in the relative importance of sea fish through the Middle Ages. The initial shift was concurrent with a brief increase in, and subsequent near-total disappearance of, the bones of very large eel and pike (greater than 50 cm TL). Among the gadids, cod became abundant first, with haddock increasing in the eleventh century and whiting in the late twelfth to thirteenth centuries. Overall, the evidence suggests a major increase in demand for fish of diverse kinds from the turn of the

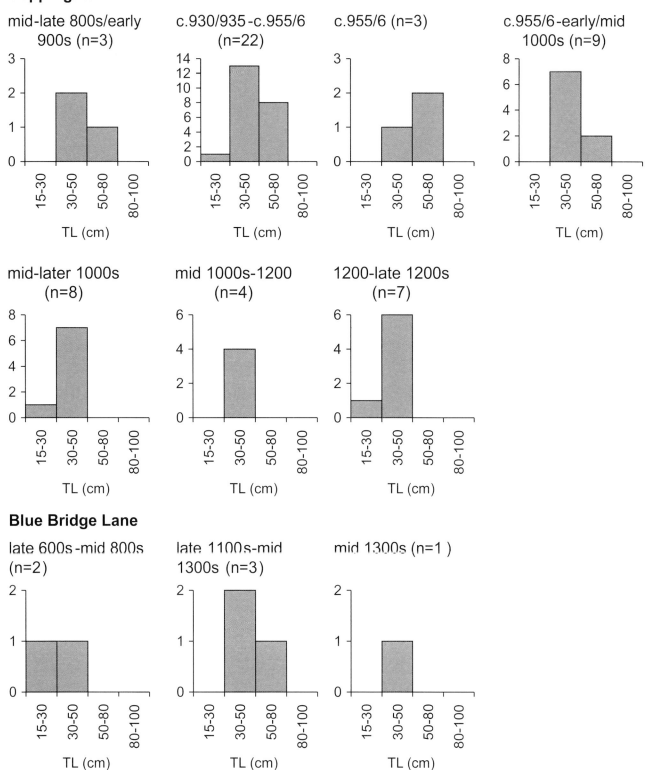

Figure 15.13. Pike total length estimates (cm) for Coppergate and Blue Bridge Lane, based on NISP of identified cranial and appendicular specimens, predominantly >2 mm sieving.

first/second millennia AD – sufficient to initiate new sea fisheries and to put pressure on local freshwater species, leading to decreases in their maximum size at catch (cf. Barrett *et al.* 2004; 2011; Harland *et al.* 2008; Hoffmann 1996). A subsequent further decline in the representation of freshwater and migratory taxa during the fourteenth century may reflect continued fishing pressure, increasing monopolisation by the elite, or both.

The evidence of anatomical representation, fish size estimates, cut marks and (in the case of cod) stable isotope analysis all converges on the interpretation that most of the herring, cod and haddock consumed in York were probably fished in relatively local waters of the North Sea. None of the herring assemblages show convincing evidence of the gutting practice that characterised barrelled Baltic imports of the Middle Ages, and was adopted in England towards the end of the fourteenth century. Some may have arrived as barrelled English herring, before the 'Scania cure' was adopted locally, but it is equally likely that most had been preserved in other ways.

Turning to the main gadids, the haddock bones show no convincing inconsistencies in skeletal element representation or fish size estimates that might imply the presence of a product processed for storage and/or long-range trade (cf. Harland 2007, 64–5). Most cod probably also arrived in a fairly fresh state. In the case of this species, however, some preserved (dried or dried and salted) fish were probably consumed – from both relatively local fisheries and distant waters of the North Atlantic region. The evidence includes patterning in the abundance of cleithra and vertebrae *vis-à-vis* other skeletal elements, inconsistencies in the sizes of fish represented between phases and skeletal elements, the presence of distinctive butchery marks and stable isotope analysis. The preserved cod inferred on these bases are not represented in all phases. They only appear in the High Middle Ages, especially the thirteenth century, and much more equivocally, also the tenth to eleventh centuries. The former of these patterns matches what is now known about the chronology of London's consumption of preserved cod (Barrett *et al.* 2011; Orton *et al.* 2014; Chapter 16). The latter possibility opens interesting questions for future research.

Acknowledgements

Access to the Coppergate assemblage and associated data was kindly organised by Christine McDonnell of the York Archaeological Trust (YAT). Ailsa Mainman was generous with her time in answering questions about excavations conducted and/or published by YAT. Cecily Spall and Nicky Toop of Field Archaeology Specialists kindly provided access to the fish bone from Blue Bridge Lane. Rachel Parks assisted with bone identifications. Image credits are provided in the figure captions. The research was funded by the Leverhulme Trust.

References

Barrett, J. H. 1997. Fish trade in Norse Orkney and Caithness: a zooarchaeological approach. *Antiquity* 71: 616–38.

Barrett, J. H., R. A. Nicholson & R. Cerón-Carrasco. 1999. Archaeo-ichthyological evidence for long-term socioeconomic trends in northern Scotland: 3500 BC to AD 1500. *Journal of Archaeological Science* 26: 353–88.

Barrett, J. H., A. M. Locker & C. M. Roberts. 2004. 'Dark Age Economics' revisited: the English fish bone evidence AD 600–1600. *Antiquity* 78: 618–36.

Barrett, J. H., C. Johnstone, J. Harland, W. Van Neer, A. Ervynck, D. Makowiecki, D. Heinrich, A. K. Hufthammer, I. B. Enghoff, C. Amundsen, J. S. Christiansen, A. K. G. Jones, A. Locker, S. Hamilton-Dyer, L. Jonsson, L. Lõugas, C. Roberts & M. Richards. 2008. Detecting the medieval cod trade: a new method and first results. *Journal of Archaeological Science* 35(4): 850–61.

Barrett, J. H., D. Orton, C. Johnstone, J. Harland, W. Van Neer, A. Ervynck, C. Roberts, A. Locker, C. Amundsen, I. B. Enghoff, S. Hamilton-Dyer, D. Heinrich, A. K. Hufthammer, A. K. G. Jones, L. Jonsson, D. Makowiecki, P. Pope, T. C. O'Connell, T. de Roo & M. Richards. 2011. Interpreting the expansion of sea fishing in medieval Europe using stable isotope analysis of archaeological cod bones. *Journal of Archaeological Science* 38: 1516–24.

Betts, M. W., S. Noël, E. Tourigny, M. Burns, P. E. Pope & S. L. Cumbaa. 2014. Zooarchaeology of the historic cod fishery in Newfoundland and Labrador, Canada. *Journal of the North Atlantic* 24: 1–21.

Bond, J. M. & T. P. O'Connor. 1999. *Bones from medieval deposits at 16–22 Coppergate and other sites in York*. (Archaeology of York Fascicule 15/5). York: Council for British Archaeology.

Candow, J. E. 2009. Migrants and residents: the interplay between European and domestic fisheries in northeast North America, 1502–1854, in D. J. Starkey, J. T. Thór & I. Heidbrink (eds) *A History of the North Atlantic Fisheries. Volume 1: from early times to the mid-nineteenth century*: 416–52. Bremen: H. M. Hauschild.

Carver, M. & Loveluck, C. 2013. Early medieval, AD 400 to 1000, in J. Ransley & F. Sturt (eds) *People and the Sea: a maritime archaeological research agenda for England*: 113–137. (Council for British Archaeology Research Report 171). York: Council for British Archaeology.

Childs, W. R. 1995. England's Icelandic trade in the fifteenth century: the role of the port of Hull, in P. Holm, O. U. Janzen, O. Uwe & J. Thor (ed.) *Northern Seas Yearbook, 1995* (Fiskeri- og søfartsmuseets studieserie 5): 11–31. Esbjerg: Fiskeri- og Søfartsmuseet.

Childs, W. & M. Kowaleski. 2000. Fishing and fisheries in the Middle Ages, in D. J. Starkey, C. Reid & N. Ashcroft (eds) *England's Sea Fisheries: the commercial sea fisheries of England and Wales since 1300*: 19–28. London: Chatham.

Cutting, C. L. 1955. *Fish Saving: a history of fish processing from ancient to modern times*. London: Leonard Hill.

Dyer, C. C. 1988. The consumption of fresh-water fish in medieval England, in M. Aston (ed.) *Medieval Fish, Fisheries and Fishponds in England*: 27–38. (British Archaeological Report 182). Oxford: British Archaeological Reports.

Enghoff, I. B. 1996. A medieval herring industry in Denmark and the importance of herring in eastern Denmark. *Archaeofauna* 5: 43–7.

Enghoff, I. B. 2000. Fishing in the southern North Sea region from the 1st to the 16th century AD: evidence from fish bones. *Archaeofauna* 9: 59–132.

Froese, R. & D. Pauly (eds) 2015. *FishBase*. Available at: www.fishbase.org.

Hall, A., H. Kenward, D. Jaques, J. Carrott & S. Rowland. 2002. Technical report: biological remains from excavations at 41–9 Walmgate, York (site code: 1999.941). Report prepared by Palaeoecology Research Services, York.

Hall, R. A. & K. Hunter-Mann. 2002. *Medieval Urbanism in Coppergate: refining a townscape.* (Archaeology of York Fascicule 10/6). York: Council for British Archaeology.

Hall, R. A., D. W. Rollason, M. Blackburn, D. N. Parsons, G. Fellows-Jensen, A. R. Hall, H. K. Kenward, T. P. O'Connor, D. Tweddle, A. J. Mainman & N. S. H. Rogers (eds) 2004. *Aspects of Anglo-Scandinavian York.* (Archaeology of York). York: Council for British Archaeology.

Hall, R. A., D. T. Evans, K. Hunter-Mann & A. J. Mainman. 2014. *Anglo-Scandinavian occupation at 16–22 Coppergate: defining a townscape.* (Archaeology of York Fascicule 8/5). York: Council for British Archaeology.

Harland, J. 2006. Zooarchaeology in the Viking Age to medieval Northern Isles, Scotland: an investigation of spatial and temporal patterning. Unpublished PhD dissertation, University of York.

Harland, J. 2007. Status and space in the 'fish event horizon': initial results from Quoygrew and Earl's Bu, Viking Age and medieval sites in Orkney, Scotland, in H. H. Plogmann (ed.) *The Role of Fish in Ancient Time*: 63–8. Rahden: Marie Leidorf.

Harland, J. 2009. Technical report: fish remains from the Drogheda boat, Ireland. Report prepared by Centre for Human Palaeoecology, Department of Archaeology, University of York.

Harland, J., C. Johnstone & A. Jones. 2008. A case study from the medieval origins of commercial sea fishing project: zooarchaeological results from York (UK), in P. Béarez, S. Gronard & B. Clavel (eds) *Archéologie du poisson: 30 ans d'archéo-ichtyologie au CNRS*: 15–26. Antibes: Éditions APDCA.

Harland, J. F., J. H. Barrett, J. Carrott, K. Dobney & D. Jaques. 2003. The York system: an integrated zooarchaeological database for research and teaching. *Internet Archaeology* 13. Available at: http://intarch.ac.uk/journal/issue13/harland_index.html.

Hoffmann, R. C. 1996. Economic development and aquatic ecosystems in medieval Europe. *American Historical Review* 101: 631–69.

Hoffmann, R. C. 2009. *Strekfusz*: a fish dish links Jagiellonian Kraców to distant waters, in P. Górecki & N. Van Deusen (eds) *Central and Eastern Europe in the Middle Ages: a cultural history*: 116–24. London: I. B. Tauris.

Hutchinson, W. F., M. Culling, D. C. Orton, B. Hänfling, L. Lawson Handley, S. Hamilton-Dyer, T. C. O'Connell, M. P. Richards & J. H. Barrett. 2015. The globalization of naval provisioning: ancient DNA and stable isotope analyses of stored cod from the wreck of the *Mary Rose*, AD 1545. *Royal Society Open Science* 2: 150199.

Johnstone, C., J. Carrott, A. Hall, H. Kenward & D. Worthy. 2000. Assessment of biological remains from 41–40 Walmgate York (site code 1999.941). Report 2000/04 prepared by Environmental Archaeology Unit, York.

Jones, A. K. G. 1981. Reconstruction of fishing techniques from assemblages of fish bones, in I. B. Enghoff, J. Richter & K. Rosenlund (eds) *Fish Osteoarchaeology Meeting: Copenhagen 28th–29th August 1981*: 4–5. Copenhagen: Danish Zoological Museum.

Jones, A. K. G. 1982. Bulk-sieving and the recovery of fish remains from urban archaeological sites, in A. R. Hall & H. Kenward (eds) *Environmental Archaeology in the Urban Context*: 79–85 (Council for British Archaeology Research Report 43). London: Council for British Archaeology.

Jones, A. K. G. 1988a. Provisional remarks on fish remains from archaeological deposits at York, in P. Murphy & C. French (eds) *The Exploitation of Wetlands*: 113–27. (British Archaeological Report 186). Oxford: British Archaeological Reports.

Jones, A. K. G. 1988b. Fish bones from excavations in the cemetery of St Mary Bishophill Junior, in T. P. O'connor (ed.) *Bones from the General Accident site, Tanner Row*: 126–30. (Archaeology of York Fascicule 15/2). London: Council for British Archaeology.

Jones, A. K. G. 1989. Fish remains and excrements from 46–54 Fishergate, York. Report prepared by Environmental Archaeology Unit, York.

Jones, E. T. 2000. England's Icelandic fishery in the early modern period, in D. J. Starkey, C. Reid & N. Ashcroft (eds) *England's Sea Fisheries: the commercial sea fisheries of England and Wales since 1300*: 105–10. London: Chatham.

Jonsson, L. 1986. Finska gäddor och Bergenfisk – ett försök att belysa Uppsalas fiskimport under medeltid och yngre Vasatid, in N. Cnattingius & T. Neréus (eds) *Uppsala stads historia. Volume 7: från Östra Aros till Uppsala: en samling uppsatser kring det medeltida Uppsala*: 122–39. Uppsala: Almqvist and Wiksell.

Keaveney, E. 2005. Fish Trade in York: bones from Blue Bridge Lane and Fishergate House. Unpublished MSc dissertation, University of York.

Kemp, R. L. 1996. *Anglian Settlement at 46–54 Fishergate.* (Archaeology of York Fascicule 5/1). York: Council for British Archaeology.

Kenward, H. K. & A. R. Hall 1995. *Biological Evidence from Anglo-Scandinavian Deposits at 16–22 Coppergate.* (Archaeology of York Fascicule 14/7). York: Council for British Archaeology.

Kowaleski, M. 2003. The commercialization of the sea fisheries in medieval England and Wales. *International Journal of Maritime History* 15(2): 177–231.

Large, F., A. Hall, C. Johnstone, D. Worthy & J. Carrott. 1999. Assessment of biological remains from 41 Piccadilly, York (sitecode: 1998.15). Report 99/45 prepared by Environmental Archaeology Unit, York.

Lauwerier, R. C. G. M. & F. J. Laarman. 2008. Relics of 16th-century gutted herring from a Dutch vessel. *Environmental Archaeology* 13: 135–42.

Losco-Bradley, P. M. & C. R. Salisbury. 1988. A Saxon and a Norman fish weir at Colwick, Nottinghamshire, in M. Aston (ed.) *Medieval Fish, Fisheries and Fishponds in England*:

329–51. (British Archaeological Report 182). Oxford: British Archaeological Reports.

McDonnell, J. 1981. *Exploitation of Inland Fisheries in Early Medieval Yorkshire 1066–1300*. York: University of York Borthwick Institute of Historical Research.

Murphy, P. 2009. *The English Coast: a history and prospect*. London: Continuum.

Nash, R. D. M. & M. Dickey-Collas. 2005. The influence of life history dynamics and environment on the determination of year class strength in North Sea herring (*Clupea harengus* L.). *Fisheries Oceanography* 14: 279–91.

Nedkvitne, A. 2014. *The German Hansa and Bergen 1100–1600*. Köln: Böhlau.

O'Connor, T. P. 1988. *Bones from the General Accident Site, Tanner Row*. (Archaeology of York Fascicule 15/2). London: Council for British Archaeology.

O'Connor, T. P. 1989. *Bones from Anglo-Scandinavian levels at 16–22 Coppergate*. (Archaeology of York Fascicule 15/3). London: Council for British Archaeology.

O'Connor, T. P. 1991. *Bones from 46–54 Fishergate*. (Archaeology of York Fascicule 15/4). London: Council for British Archaeology.

O'Connor, T. P. 2003. *The Analysis of Urban Animal Bone Assemblages: a handbook for archaeologists*. York: York Archaeological Trust/Council for British Archaeology.

Orton, D. C., J. Morris, A. Locker & J. H. Barrett. 2014. Fish for the city: meta-analysis of archaeological cod remains and the growth of London's northern trade. *Antiquity* 88: 516–30.

O'Sullivan, A. 2004. Place, memory and identity among estuarine fishing communities: interpreting the archaeology of early medieval fish weirs. *World Archaeology* 35: 449–68.

Perdikaris, S. & T. H. McGovern. 2009. Viking Age economics and the origins of commercial cod fisheries in the North Atlantic, in L. Sicking & D. Abreu-Ferreira (eds) *Beyond the Catch: fisheries of the North Atlantic, the North Sea and the Baltic, 900–1850*: 61–90. Leiden: Brill.

Poulsen, B. 2008. *Dutch Herring: an environmental history, c. 1600–1860*. Amsterdam: Amsterdam University Press.

Rees Jones, S. & C. Daniell. 2002. The King's Pool, in R. A. Hall & K. Hunter-Mann (eds) *Medieval Urbanism in Coppergate: refining a townscape*: 696–98. (Archaeology of York Fascicule 10/6). York: Council for British Archaeology.

Rowland, S. 2005. Artefacts & environmental evidence: the animal & fish bone, in C. A. Spall & N. J. Toop (eds) Blue Bridge Lane & Fishergate House, York: report on excavations: July 2000 to July 2002. Report prepared by Archaeological Planning Consultancy Ltd. Previously available at: http://www.archaeologicalplanningconsultancy.co.uk/mono/001/rep_bone_animal.html

Samuel, A. M. 1918. *The Herring: its effect on the history of Britain*. London: John Murray.

Serjeantson, D. & C. M. Woolgar. 2006. Fish consumption in medieval England, in C. M. Woolgar, D. Serjeantson & T. Waldron (eds) *Food in Medieval England: diet and nutrition*: 102–30. Oxford: Oxford University Press.

Spall, C. A. & N. J. Toop. (eds) 2005. Blue Bridge Lane & Fishergate House, York: report on excavations: July 2000 to July 2002. Report prepared by Archaeological Planning Consultancy Ltd.

Available at: http://www.archaeologicalplanningconsultancy.co.uk/mono/001/index.html

Steane, J. M. 1988. The royal fishponds of medieval England, in M. Aston (ed.) *Medieval Fish, Fisheries and Fishponds in England*: 39–68. (British Archaeological Report 182). Oxford: British Archaeological Reports.

Tourunen, A. 2008. Gutted and salted: a fish bone assemblage from John Street, Waterford, Ireland. *Archaeofauna* 17: 139–45.

Unger, R. W. 1978. The Netherlands herring fishery in the late Middle Ages: the false legend of Willem Beukels of Biervliet. *Viator* 9: 335–56.

Wouters, W., L. Muylaert & W. Van Neer. 2007. The distinction of isolated bones from plaice (*Pleuronectes platessa*), flounder (*Platichthys flesus*) and dab (*Limanda limanda*): a description of the diagnostic characters. *Archaeofauna* 16: 33–95.

Appendix 1. Phasing and dating of the Coppergate fish-bone assemblage

Below we provide a simplified concordance of the published numerical period codes for Coppergate and their associated dates as used in this study, followed by a detailed concordance of period codes, their calendrical dates (at varying levels of lumping) and the individual contexts they include. Calendrical dates (rather than period codes) have been used throughout the chapter. The general periodization is based on previous Coppergate publications (Hall and Hunter-Mann 2002, table 21; Hall 2014, table 25), with new dates for Anglo-Scandinavian phases 4B, 5A and 5B provided by Hall *et al.* (2014). Comprehensive database files were also kindly provided by York Archaeological Trust, listing context numbers, tenements, period codes and phases. Using these, we were able to correlate each individual context with a specific phase of activity within each tenement. Thus it was possible to subdivide the originally lengthy Period 6. Moreover, Hall and Hunter-Mann (2002, fig. 397) detail the various 'phases of activity' within each of the tenements (A, B, C and D) through time, making it possible to approximate the date range of each context.

Period 1, of Roman date, was represented by only 35 fish bones and has not been included in our study (the species found were predominantly freshwater, including 20 eel and eight cyprinid bones). No fish were found from Period 2 deposits. Period 4A was very poorly represented, with only three fish bones; it was therefore grouped with Period 3 in the mid–late ninth–early tenth-century phase. Similarly, Periods 5Cf and 5Cr were grouped into the mid–later eleventh-century phase because Period 5Cr contained only 168 fish bones, compared with more than 3000 for Period 5Cf. Lastly, zooarchaeological data from a few of the original periods (e.g. 4/5) have not been included in this chapter because the date range was broad and there were only small quantities of fish bone.

Concordance of original period attributions and calendrical date ranges used in this study

Original period	Date range
1	Not included (Roman)
2	Not included (no fish recovered)
3	Mid- to late 800s–early 900s
4A	Mid- to late 800s–early 900s
4B	*c.*930/935–*c.*955/6
5A	*c.*955/56
5B	*c.*955/6–early to mid-1000s
5Cf	Mid- to later 1000s
5Cr	Mid- to later 1000s
6 (A6c/B6a, B6a, B6a/C6a, B6a/C6c, B6u, B6u/C6v, C6c)	Mid-1000s–mid-1100s
6 (B6a/C6d, C6c/D6a, C6d/D6a, D6a)	Mid-1000s–early 1200s
6 (A6c, A6d, B6b, B6b/C6d, C6x, C6d)	Mid-1100s–1200
6 (B6c, B6w)	1200–1240
6 (D6c, D6d)	1240–1275
6 (A6e, B6c/C6e, B6d, B6d/C6e, B6e/C6e, C6e/D6d, C6e)	1200–late 1200s
6 (A6f, B6e, C6f, D6e)	1275–mid-1300s
6 (A6g, B6f)	1300–late 1300s
6 (A6h, C6g)	1360–1500
6 (A6i, A6j, B6g, B6y, C6h, C6i, D6g, D6h, D6j, D6m)	1370–1720

Concordance of phase codes, individual context numbers, detailed date ranges and date categories used in this study

Broad period or tenement-period-phasing code	Detailed date range	Date range used in this study	Contexts
1	late 1st–late 4th or later	late 1st–late 4th or later	30904, 31741, 31929, 31973, 33121, 33131, 33145, 34150
2	5th–mid-9th	5th–mid-9th	30982, 31762
3	mid–late 800s/early 900s	mid–late 800s/early 900s	19739, 20523, 20894, 25990, 26016, 26019, 26630, 26631, 26715, 26718, 26721, 26732, 26946, 27194, 27813, 27868, 28033, 28384, 28408, 30549, 30629, 30642, 30652, 30693, 30704, 30710, 30735, 30747, 30748, 30750, 30751, 30801, 30808, 30826, 30835, 30841, 30842, 30843, 30864, 30882, 30902, 30922, 30925, 30929, 30936, 30954, 30969, 30973, 30974, 31595, 31637, 31644, 32265, 32980, 33072, 33094, 34789, 34814, 36154
4A	late 800s/early 900s–*c.* 930/5	mid–late 800s/early 900s	26263, 26610, 26635, 27943, 30039, 30063, 30352
4B	*c.* 930/935–*c.* 955/6	*c.* 930/935–*c.* 955/6	8574, 8627, 8731, 8856, 15897, 15897/15894, 15902, 18529, 19626, 19738, 20462, 20476, 20747, 20779, 20982, 21430, 21451, 22259, 22415, 22421, 22452, 22490, 22505, 22523, 22524, 22547, 22560, 22586, 22595, 22636, 22659, 22670, 22713, 22714, 22745, 22760, 22767, 22797, 22808, 22843, 22845, 22868, 22911, 23080, 23243, 23437, 23456, 23543, 23753, 24004, 24064, 24399, 24560, 24599, 25065, 25084, 25104, 25110, 25257, 25340, 25371, 25380, 25391, 25466, 25750, 25934, 26012, 26157, 26249, 26250, 26636, 26992, 27017, 27018, 27203, 27368, 27503, 27504, 28087, 28904, 28967, 29222, 29528, 29844, 29885, 29926, 31424, 32026, 32185, 32189, 32243, 32571, 34185, 34289, 34290, 34292, 34377, 34412,

(Continued)

Broad period or tenement-period-phasing code	Detailed date range	Date range used in this study	Contexts
			34413, 34424, 34483, 35086, 35323, 35519, 35560, 35673, 35674, 35675, 35677, 35678, 35679
4/5	late 800s–later 1000s		3520, 3661
5A	*c.*955/6	*c.*955/6	2446, 2677, 8056, 8169, 8170, 8253, 8258, 8282, 8448, 8454, 8455, 8491, 8523, 8572, 8799, 8800, 8801, 8844, 8845, 8847, 8874, 8996, 14466, 14787, 14790, 14874, 14882, 14883, 14928, 14963, 18286, 18344, 18594, 20062, 20105, 20143, 20181, 20185, 20186, 20719, 20746, 20767, 20808, 20877, 20970, 20990, 22044, 22090, 22104, 22122, 22154, 22154A, 22154B, 22209, 22256, 22267, 22309, 22339, 22376, 22382, 26953, 26958, 27388, 34295
5B	*c.* 955/6–early/mid-1000s	*c.* 955/6–early/mid-1000s	1076, 1091, 1473, 1541, 1611, 2081, 2104, 2224, 2321, 2367/82, 2403, 2409, 2410, 2470, 2476, 2534, 2566, 2571, 2578, 2682, 2686, 2723, 2730, 2732, 2733, 2768, 2778, 2789, 2794, 2812, 2875, 2974, 5247, 5321, 5416, 5419, 5424, 5435, 5438, 5439, 5588, 5641, 5650, 5651, 5663, 5673, 5714, 5772, 5775, 5804, 5848, 5870, 5885, 5887, 5889, 5893, 5980, 6284, 6287, 6347, 6425, 6429, 6433, 6434, 6444, 6472, 6531, 6532, 6535, 6578, 6787, 6798, 6866, 6879, 6927, 7232, 7251, 7252, 7257, 7258/7204/7258, 7361, 7362, 7369, 7377, 7379, 7390, 7405, 7406, 7467, 7483, 7488, 7553, 7589, 7595, 7672, 7974, 8033, 8044, 8051, 8106, 8107, 8109, 8122, 8195, 8225, 8444, 8526, 8528, 8730, 9630, 9726, 13093, 13716, 13875, 14005, 14184, 14187, 14254, 14296, 14297, 14350, 14418, 14433, 14434, 14529, 14548, 14581, 14595, 14630, 14659, 14666, 14767, 14890, 14941, 14973, 15131, 15177, 15189, 15193, 15195, 15314, 15332, 15361, 15382, 15387, 15401, 15471, 15483, 15493, 15561, 15608, 15622, 15628, 15645, 15663, 15745, 15999, 16881, 16882, 18139, 18712, 19245, 19285, 19307, 19313, 19320, 19353, 19448, 19620, 19622, 19623, 19625, 20052, 20164, 20170, 20221, 20231, 20234, 20269, 20273, 20286, 20301, 20303, 20310, 20314, 20318, 20322, 20342, 20441, 20670, 21140, 21197, 21204, 21244, 21252, 21257, 21265, 21270, 21322, 21381, 21404, 21463, 21478, 21510, 21674, 21680, 21746, 21747, 21766, 21886, 22103, 22107, 24870, 26015
C6v	1040–1100	mid–later 1000s	1489, 1585, 7320
5CF	mid–later 1000s	mid–later 1000s	3463, 3464, 3467, 3507, 3529, 3543, 6379, 6570, 6573, 6774, 6781
5CR	mid–later 1000s	mid–later 1000s	6795, 6903, 6926, 7862, 7863, 7868, 7954, 7966, 9798, 14069, 14133, 14182, 14183, 14186, 15311, 16720, 16763, 16865, 16877, 16887, 16889, 17742, 19090, 19120, 19138, 19141, 19165, 19166, 19192, 19197, 19212, 19247, 19255, 19267, 19268, 19269, 19270, 19271, 19272, 19283, 19310, 20162, 20166, 20178, 20179, 20205, 21008, 21075, 21088, 21141, 21433
A6c/B6a	1040–1160	mid-1000s–mid-1100s	13665
B6a/C6a	1040–1160	mid-1000s–mid-1100s	17890
B6a/C6c	1040–1160	mid-1000s–mid-1100s	13698, 17333, 17695, 17697, 18331
B6u/C6v	1040–1160	mid-1000s–mid-1100s	2096, 2159
B6a	1040–1160	mid-1000s–mid-1100s	13243, 13245, 13523, 13554, 13568, 13571, 13658, 13785, 15124, 15285, 16044, 16549, 16557, 16596, 16653, 16735, 16803, 17008, 17253, 17332, 17397, 17457, 18256, 18329, 18366
B6u	1040–1160	mid-1000s–mid-1100s	2290, 2303, 2306, 13147, 13287
C6c	1100–1140	mid-1000s–mid-1100s	13673, 17103, 17322, 17452, 17463, 17551, 17574, 17599, 17627, 17699, 17740, 17760, 17786, 17787, 5000, 5398, 5826, 5975, 6257

(Continued)

Broad period or tenement-period-phasing code	Detailed date range	Date range used in this study	Contexts
B6a/C6d	1040–1200	mid-1000s–early 1200s	16949, 16951, 17421, 17586, 18338
C6c/D6a	1040–1220	mid-1000s–early 1200s	16734, 5231, 5232, 5395, 5415, 5719, 5906, 5981, 6258, 6291, 17418, 17482, 17532, 17814, 5724, 5838,
C6d/D6a	1040–1220	mid-1000s–early 1200s	16361, 16533, 16639, 16640
D6a	1040–1220	mid-1000s–early 1200s	5230, 5331, 5333, 5485, 5541, 5545, 5607, 5655, 5666, 5755, 5776, 5777, 5811, 5977, 5978, 5993, 5996, 6016/6026, 6111, 6245, 6246, 6281, 6285, 6297, 6339, 6357, 9302, 9305, 9362, 9394, 9396, 9397, 9481, 9570, 9572, 9574, 9793, 9794, 9812, 9815, 12548, 12561, 12759, 16114, 16129, 16130, 16170, 16404, 16409, 16410, 16456, 16464, 16465, 16517, 16522, 16523, 16525, 16530, 16535, 16544, 16551, 16590, 16603, 16608, 16612, 16643, 16688, 16724, 16870, 16876, 16886, 16970, 17040, 17059, 17084, 17087, 17090, 17129, 17526
A6n/B6a	1040–1280		15136
A6n/B6c	1040–1280		13445, 13532, 13539
C6e/D6a	1040–1280		5238, 5345, 5348, 5372, 5484, 5494, 5499
A6z/B6u	1040–1300		13759
C6x	1130–1200	mid-1100s–1200	1331, 1350, 1382, 1440, 1441, 1442, 1445, 1455, 1475, 7854, 7973
A6c	1140–1160	mid-1100s–1200	3403, 3419, 3443, 3446, 3453, 3493, 3496, 3558
B6b/C6d	1140–1200	mid-1100s–1200	16442, 16512
C6d	1140–1200	mid-1100s–1200	5814, 5976, 16006, 16112, 16123, 16153, 16311, 16331, 16401, 16462, 16463, 16487, 16490, 16597, 16642, 16699, 16731, 16758, 16785, 16790, 16797, 16853, 16890, 16891, 16892, 16893, 16923, 16972, 17045, 17070, 17137, 17162, 17184, 17235, 17241, 17272, 17275, 17328, 17360, 17375, 17396, 17478, 17513, 17515, 17528, 17535, 17537, 17539, 17540, 17593, 17616, 17626, 18668
A6d	1160–1200	mid-1100s–1200	3435, 13663
B6b	1160–1200	mid-1100s–1200	16604, 16605
B6c	1200–1240	1200–1240	11416, 11458, 11514, 11713, 11818, 13042, 13058, 13191, 13315, 13336, 13363, 13364, 13389, 13410, 13434, 13465, 13501, 13525, 13527, 13528, 13541, 13579, 13591, 13638, 13686, 13740/13746, 13802, 13900, 13902, 13949, 13964A, 15008, 15040, 15183, 15185, 16174, 16362, 16374, 16443, 18047, 18073, 18074, 18076, 18079, 18119, 18120, 18132, 18133, 18134, 18153, 18171, 18172, 18189, 18193, 18194
B6w	1200–1240	1200–1240	2056, 2130, 2259, 2260, 2296, 2587, 2589, 2605, 2606, 2607, 2738, 2742, 2830, 13212
D6c	1240–1250	1240–1275	5261, 5262, 5263, 5264, 9248
D6d	1250–1275	1240–1275	5240, 5241, 5245, 5289, 9249
A6e	1200–1280	1200–late 1200s	3362, 3364, 3377, 3397, 3401, 3404, 3407, 3409, 3414, 3440, 3442, 3494
B6c/C6e	1200–1280	1200–late 1200s	13800, 16145
B6d/C6e	1200–1280	1200–late 1200s	5308
C6e/D6d	1200–1280	1200–late 1200s	5104
C6e	1200–1280	1200–late 1200s	4175, 4184, 4184a, 4213, 4540, 4583, 4585, 4586, 4597, 4600, 4605, 4620, 4636, 4648, 4693, 4695, 4745, 4756, 4757, 4788, 4850, 4915, 5021, 5032, 5037, 5051, 5052, 5064, 5140, 5177, 5186, 5215, 5239, 5246, 5370, 5373, 5394, 5401, 5403, 5406, 5412, 5432, 5441, 5442, 5464, 5510, 5535, 5536, 5539, 5573, 5585, 5586, 5605,

(Continued)

Broad period or tenement-period-phasing code	Detailed date range	Date range used in this study	Contexts
			5611, 5645, 5668, 5668A, 5671, 5693, 5715, 5717, 5786, 5803, 5869, 6249, 9224, 9229, 9276, 11017, 11020, 11021, 12389, 12412, 12762, 16024, 16072, 16100, 16359, 16381, 16636
B6e/C6e	1200–1300	1200–late 1200s	13492
C6e/D6e	1200–mid-1300s		1502, 10735
B6d	1240–1280		13454, 13546
D6e	1275–mid-1300s	1275–mid-1300s	1114, 1115, 1118, 1119, 1506, 1549, 1550, 1552, 1604, 1605, 1609, 1622, 1624, 1702, 1703, 1706, 1944, 9091, 9213, 9215, 9219, 9252, 9269, 9641, 12147, 12274
B6e	1280–1300	1275–mid-1300s	13005, 13521
A6f	1280–1350	1275–mid-1300s	3361, 3382, 3386, 3439
C6f	1280–1360	1275–mid-1300s	4097, 4550, 4704, 4804, 4885, 5110, 5181, 10739
B6f	1300–1370	1300–late 1300s	11927, 11948, 11953, 13299, 13514
A6g	1350–1360	1300–late 1300s	3233, 3235, 3258, 3344, 3349
A6h	1360–1420	1360–1500	3371, 3473
C6g	1360–1500	1360–1500	4239, 4290, 4426, 4440, 4442, 4548, 4589, 4604, 4630, 4658, 4689, 4724, 4752, 4758, 4763, 4790, 4797, 4808, 4829, 4830, 4867, 4874, 4879, 5163, 9277, 10464, 10511, 10923
B6g	1370–1720	1370–1720	10604, 10617, 10711, 10721, 10758, 11513, 11616, 11656, 11687, 11712, 11797, 13577, 15298
A6i	1420–1450	1370–1720	3237, 3256
A6j	1450–1650	1370–1720	3047, 3051, 3054, 3061, 3155, 3224, 3225, 3345
D6g	1500–1540	1370–1720	9030
C6h	1500–1670	1370–1720	1034, 1214, 1263, 1267, 4007, 4009, 4010, 4061, 4089, 4129, 4153, 4209, 4463
D6h	1540–1580	1370–1720	1133, 1773
D6j	1600–1640	1370–1720	1797
C6i	1670–1720	1370–1720	1011, 1030, 1396, 4002, 4012, 4054, 4060, 4083, 4085, 4141, 4289b, 4385
B6y	1680–1700	1370–1720	11998, 13185
D6m	early/mid-1600s–1720	1370–1720	1742
C6j	1720–1780		4005, 4015, 4042, 4049, 4053
A6z	1040–1300 and 1440–1480		2185, 2191, 2192, 2193, 2194, 2195, 2231, 2237, 2238, 2239, 2243, 2266, 2268, 2336, 2338, 2621, 2635, 2636, 2841, 2844, 8304, 8305, 13119
C6e–g	1200–1500		10962
C6z	1230–1320 and 1500–1720		1096, 1101, 1226, 1229, 1283, 1297, 1298, 1338, 1346, 1347, 4043, 4059, 4125, 4164, 4181, 7791
B6f/C6g	1300–1500		13037
D6y	late 1100s–1320		1010, 1085, 1359, 1404, 1523, 1567, 1570, 7782, 9800, 9801
6	later 1000s–1500s		12393, 12741, 17412
?6?	later 1000s–1500s		12741, 17412
D6f	mid-1300s–early/mid-1600s		1111, 1186, 1697, 1701, 1842
B/CMODERN	modern		1017, 1019, 1070, 1265, 1295, 2006, 2092, 2134, 4000, 4162, 4193, 5184, 7788

16

Fish for London

David C. Orton, Alison Locker, James Morris and James H. Barrett

Introduction and background

Previous synthesis of zooarchaeological remains from English medieval sites has shown that marine fish consumption increased, from a very low baseline, around AD 1000, a phenomenon dubbed the fish event horizon (FEH) by Barrett *et al.* (2004a). This change appears initially to have been linked primarily to urban sites, with widespread marine fish consumption at inland rural settlements argued to be a slightly later development (Barrett *et al.* 2004b; but see also Chapter 17 regarding near-coastal elite settlements). That early towns and cities led the way in the expansion of marine resource use is perhaps unsurprising, given that urban settlements almost by definition involve a concentration of food consumers rather than producers and hence require a significant hinterland to meet demand. Turning to marine resources is one way to expand this resource base.

Stable isotope provenancing has shown that this FEH initially involved relatively locally caught fish, with imports from northern waters only becoming common during the thirteenth to fourteenth centuries (Barrett *et al.* 2011). To the extent that the shift towards marine resources was driven by demand from urban populations, one might also expect expanding cities to have been at the forefront of the eventual development of long-distance trade in fish. Apart from the increasing strain that must have been placed on local hinterlands (both terrestrial and marine) by growing cities such as London – whose population is estimated to have climbed from around 20,000 in the twelfth century to around 80,000 in the thirteenth century (Campbell *et al.* 1993, 24) – trade functions were in any case central to medieval urbanism (Astill 2009; Biddle 1976). Indeed, in this context it is worth noting that significant consumption of herring (*Clupea harengus*) seems to have pre-dated the FEH, specifically at early medieval proto-urban trading centres (Barrett *et al.* 2004b; see also Chapters 14–15).

London is thus a good case study through which to explore the onset and development of long-distance trade in fish to England, using cod (*Gadus morhua*) as our example. Apart from being one of the largest medieval cities in the kingdom, its historic core has also been subject to intensive, high-quality excavation over the past 40 years. Moreover, stable isotope provenancing results suggest that London came to rely on imported cod relatively early, within its southern North Sea context (Barrett *et al.* 2011), although sample size – and hence representativeness – is ultimately restricted by the practicalities of destructive laboratory analysis.

This chapter presents the results of a meta-analysis of cod remains from 95 sites across London, aimed at detecting changes over time in the contribution of imports, complementing and placing in context the isotopic results. It builds on an earlier publication (Orton *et al.* 2014), adding assessments of changes in fish preservation techniques and of the spatial distribution of the bone finds. Late and post-medieval trends are also given more attention in the present chapter than in the previous study.

Data and methods

The dataset used here is that of Orton *et al.* (2014), consisting of a total of 2827 reasonably well-dated cod remains from 95 excavation sites, along with associated context information and dating. Specimens with date ranges greater than 300 years were not used and are not included in this number. The vast majority of sites are in London's historic core, while a few from what is today known as Greater London can be considered part of the (post-)medieval city's wider economic catchment. Data were gathered from a range of published and unpublished sources, but principally the database of MOLA – a commercial excavation company that grew out of the Museum of London's field archaeology unit – and co-author

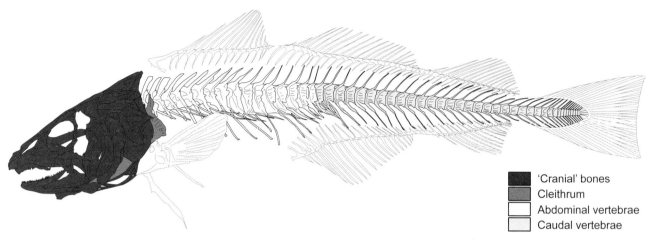

Figure 16.1. Skeleton of cod, showing anatomical categories used in this study. (Drawing: David Orton using a base image by Michel Coutureau and Benoît Clavel, ©ArchéoZoo.)

Alison Locker's personal archive. Further details can be found in the online supplementary information to Orton *et al.* (2014).

Our analysis relies on the assumption that, prior to the use of ice and/or refrigeration in modern times, cod were typically decapitated before drying and/or salting for long-distance transport (e.g. Barrett 1997; Candow 2009; Perdikaris and McGovern 2008). Thus cranial elements will usually represent relatively locally caught fish. Postcranial elements, by contrast, could derive either from local catches or from imports. There are exceptions to this rule, but they represent atypical examples (e.g. Bennema and Rijnsdorp 2015; Jonsson 1986). 'Cranial' is here defined to include the neurocranium, jaw apparatus, hyoid arch and gill covers, while postcranial refers to the cleithra (paired bones that support the pectoral fins, just behind the head) and the vertebrae. The vertebrae are subdivided into abdominal and caudal vertebrae where specified in the original data (Figure 16.1). Other postcranial bones, including supracleithra, postcleithra and scapulae, are excluded from analysis for the present purpose (leaving 2827 specimens for study, from an original total of 3034). The few known exceptions to the practice of decapitating cod prior to drying and/or salting may reduce the visibility of imports (that is, fish from some sources) but are unlikely to alter the overall picture.

Changes in the relative frequency of postcranial versus cranial bones can thus be used as a proxy for shifts in the contribution of imported cod. Moreover, because the inclusion of abdominal vertebrae in preserved cod varies according to technique and tradition, changes in relative frequencies *within* the postcranial category may reveal shifts in the types of products imported, and hence hint at shifts in the relative importance of different sources. For example, abdominal vertebrae and cleithra were left in *rundfisk*, one of the most common varieties of unsalted Norwegian stockfish (Chapter 18). Conversely, anterior vertebrae (but not cleithra) were typically removed during the production of *råskjær* (another variety of

stockfish, made in Norway, Iceland and the Northern Isles of Scotland), and in the salted and dried products prepared, for example, in the early Newfoundland fishery (Candow 2009; Harland and Barrett 2012; Perdikaris and McGovern 2008; Chapter 18).

Relative frequencies of each bone category over time are compared using estimated frequency distributions, which are constructed using a simple, two-step procedure:

1. Divide the number of relevant specimens in each context by the length of that context's date range to create an estimated frequency density across that range. Note that this assumes a uniform probability distribution for the true date of deposition, within the limits provided by the context dating.
2. Sum the frequency density from all contexts at five-year intervals. This is effectively equivalent to calculating the aoristic sum with five-year bins (see Crema 2012).

In order to identify the sites and areas of London in which early imports seem to have been consumed and deposited, we also plot relative frequencies of cranial specimens, vertebrae, and cleithra by geographical location on a century-by-century basis, using the mid-points of each context's date range. Results are only plotted where 10 or more specimens from a given site fall within the relevant century.

Results

Figure 16.2 shows the estimated frequency distributions for all cod specimens over the course of the city's history. Changes through time should be treated with caution due to the possibility (indeed probability) of date-correlated biases in research intensity. Nonetheless, the near-total absence of cod specimens between the abandonment of the Roman city and *c.* AD 1000 is striking. The paucity of early medieval cod specimens is unlikely to represent research bias. Fish

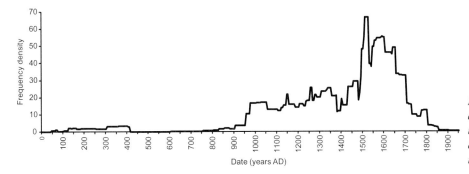

Figure 16.2. Estimated NISP frequency distribution for 3034 dated cod bones recovered from London (all anatomical elements). See text for explanation of methodology.

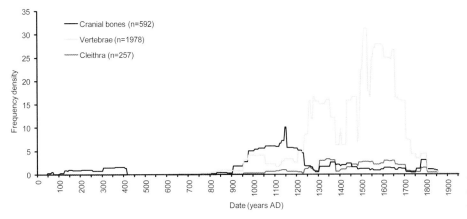

Figure 16.3. Estimated NISP frequency distributions for London cod by anatomical group.

remains from Saxon Lundenwic have certainly been studied, but they are mostly of species caught in fresh water, such as cyprinids (Cyprinidae), pike (*Esox lucius*) and eel (*Anguilla anguilla*) (e.g. Locker 1988).

Moving to the end of the sequence in the late seventeenth to nineteenth centuries, a marked decline in the frequency of cod specimens is less likely to represent a genuine fall-off in consumption. Rather, it may reflect a combination of changing depositional practices – reducing the chances of fish bones being deposited in well-dated urban contexts – and relatively limited archaeological interest. In ongoing work, these hypotheses are being tested by studying trends in the relationship between archaeological chronology and the number of environmental samples processed in London.

The estimated frequency distribution is broken down into the three main specimen types (cranial bones, cleithra and vertebrae) in Figure 16.3. Changes in the relative frequencies of different skeletal elements should be more robust than trends in overall frequencies, although possible sources of bias are considered below. Numerous small-scale fluctuations in relative frequency should probably be considered noise, emanating in large part from the vagaries of date brackets. Nevertheless, several clear trends are evident:

1. Specimens from Roman London are overwhelmingly cranial. This pattern is likely to represent an identification bias and is discussed below.

2. The re-appearance of cod specimens in the medieval city initially involves both cranial and postcranial bones in significant numbers, lending support to isotopic evidence for locally caught fish during this period.

3. This pattern changes in the early thirteenth century, when there is a sudden increase in vertebrae and a concurrent sharp decline in cranial specimens, suggesting the onset of a significant import trade in processed fish. The frequency of cleithra also starts to climb at this point, albeit more gradually. At this date the most likely product is stockfish from the North Atlantic, particularly Norway (see Chapters 4–5). Cranial specimens subsequently remain rare for the remainder of the sequence.

4. There is a marked dip in both vertebrae and cleithra in the decades around AD 1400, with no parallel for cranial elements, suggesting a temporary decline in imports. Given the timing, this might – tenuously – be linked to changes in consumption and trade resulting from the Black Death. Prices of Norwegian stockfish substantially increased and, based on English customs records, less stockfish was imported than in the early fourteenth century (Nedkvitne 2014; Chapter 5).

5. A recovery in the mid-fifteenth century and a dramatic increase in vertebrae at around AD 1500 suggest further rises in the contribution of imports. These shifts coincide with the historically documented growth of English fisheries and fish trade in Iceland (see Chapters 7–8), and ultimately in Newfoundland (Candow 2009). The extent

to which the early Newfoundland trade supplied English markets is a matter of debate. It appears to have been small; English vessels probably played a minor role in the Newfoundland fishery until the late sixteenth century (Starkey and Haines 2001, 8) and even then they primarily supplied continental Europe (Gray and Starkey 2000, 97). On the other hand, the Iceland fisheries did provide for English consumption (Childs 1995; Jones 2000).

Since the dataset includes both hand-collected specimens and bones recovered by the sieving of sediment samples, and since certain anatomical elements are larger and more robust than others, it is necessary to check for possible

recovery biases. Figure 16.4 compares results between hand-collected specimens and those recovered by sieving. Whilst there are clear differences between Figures 16.4a and 16.4b, they are broadly consistent in terms of the temporal trends noted above. The most notable difference between the two is that the significant increase in vertebrae seen during the thirteenth century for the hand-collected bones – and for the combined dataset – is not apparent until around AD 1500 in the sieved dataset. Moreover, the hand-collected vertebrae show a marked trough at AD 1500 rather than a further increase to match their sieved counterparts.

The reasons for these inconsistencies become clear, however, when vertebrae are broken down into subgroups

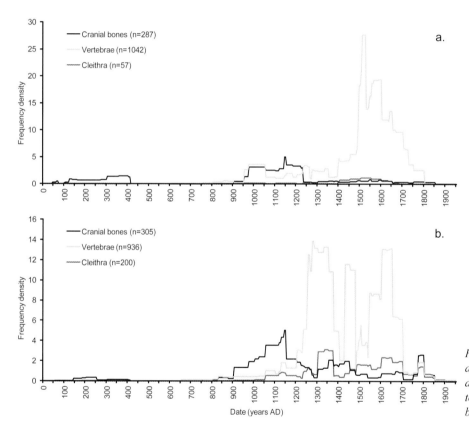

Figure 16.4. Estimated NISP frequency distributions for London cod by anatomical group, separated according to recovery method. a: sieved samples, b: hand-collected specimens.

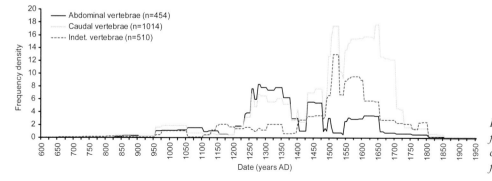

Figure 16.5. Estimated NISP frequency distributions for different classes of cod vertebrae recovered from London.

(Figure 16.5). The initial thirteenth-century increase involves both abdominal and caudal vertebrae in roughly equal numbers, and frequencies of the two groups track each other very closely over the following two-and-a-half centuries, through the *c.* AD 1400 trough and subsequent recovery. However, they part company at the end of the fifteenth century; the *c.* AD 1500 surge in vertebrae is caused entirely by caudal specimens, while abdominal vertebrae actually decline at this point. In fact, an examination of the 'indeterminate' vertebrae data suggests that this trend may have started somewhat earlier, in the early fifteenth century. The reasoning is as follows: cod have between 51 and 55 vertebrae, and those classified as 'indeterminate' in terms of position in the vertebral column (at least in the case of the Locker data) are most likely to be approximately midway, around vertebrae 22–30, and hence technically caudal (Fjelldal *et al.* 2013). It is therefore likely that the 'indeterminate' group includes specimens from this part of the body, plus perhaps damaged vertebrae from farther into the caudal group which could not be definitively identified as such. They might thus be expected to be well represented even where preservation techniques involving the removal of anterior vertebrae were practised (see discussion below), a point that is reinforced by the broad similarity in the observed frequency curves for caudal and 'indeterminate' specimens.

Returning to Figure 16.4, it is this *c.* AD 1500 increase in the frequency of caudal vertebrae that is picked up by the sieved dataset but not the hand-collected one, which makes sense given that caudal vertebrae are smaller than their abdominal counterparts and hence less likely to be recovered without sieving. This can be confirmed by re-plotting the abdominal–caudal comparison with the datasets separated (Figure 16.6). The increase in caudal vertebrae shows up much more clearly in the sieved dataset, although it is also visible in the hand-collected dataset. It is harder to explain the fact that the initial thirteenth-century increase in vertebrae of both types is not apparent in the sieved dataset, but it may be that larger specimens spotted in the field were often collected prior to separation of sediment samples for sieving – an understandable digression from technically correct sampling protocol. It should be borne in mind throughout this comparison that any changes in the typical *size* of fish caught and/or imported will also affect the relationship between hand-collected and sieved datasets, over and above shifts in anatomical representation. It would thus be valuable for future work to consider the size of the bones present, based on osteometry and/or comparison with reference specimens from fish of known total length.

While a systematic date-correlated bias in frequency of sieving is unlikely, it should be noted that the majority of specimens – 501 out of 727 – contributing to the sixteenth-century peak in caudal vertebrae actually derive from only three of the 21 assemblages represented in this period. This peak should thus be treated with a degree of caution since we cannot be sure how representative these three assemblages are of the wider picture. On the other hand, since there is

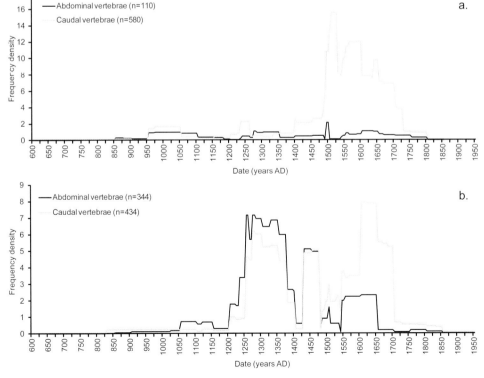

Figure 16.6. Estimated NISP frequency distributions for different classes of cod vertebrae recovered from London, separated according to recovery method. a: sieved samples, b: hand-collected specimens.

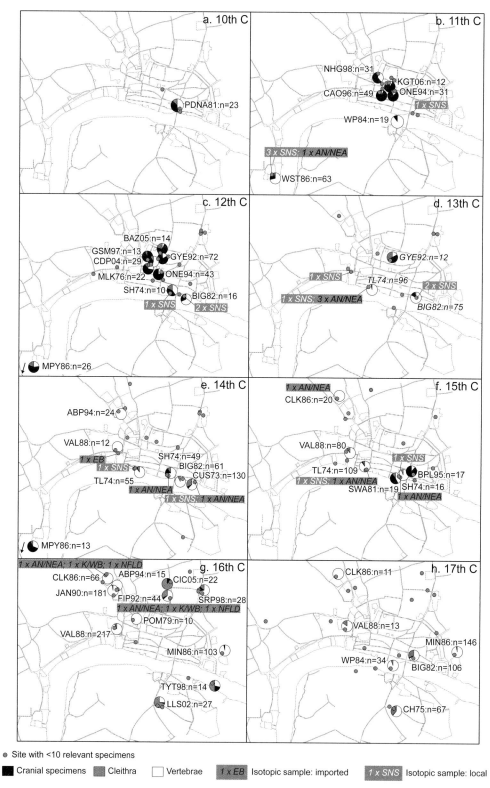

a. 10th C

PDNA81:n=23

b. 11th C

NHG98:n=31
CAO96:n=49
KGT06:n=12
ONE94:n=31
1 x SNS
WP84:n=19

3 x SNS; 1 x AN/NEA

WST86:n=63

c. 12th C

BAZ05:n=14
GSM97:n=13
CDP04:n=29
GYE92:n=72
MLK76:n=22
ONE94:n=43
SH74:n=10
1 x SNS
BIG82:n=16
2 x SNS

↓ MPY86:n=26

d. 13th C

GYE92:n=12

1 x SNS
TL74:n=96
2 x SNS
1 x SNS; 3 x AN/NEA
BIG82:n=75

e. 14th C

ABP94:n=24
VAL88:n=12
1 x EB
SH74:n=49
BIG82:n=61
TL74:n=55
1 x SNS
CUS73:n=130
1 x AN/NEA
1 x SNS; 1 x AN/NEA

↓ MPY86:n=13

f. 15th C

1 x AN/NEA
CLK86:n=20
VAL88:n=80
1 x SNS
TL74:n=109
BPL95:n=17
1 x SNS; 1 x AN/NEA
SWA81:n=19
SH74:n=16
1 x AN/NEA

g. 16th C

1 x AN/NEA; 1 x K/WB; 1 x NFLD
CLK86:n=66
ABP94:n=15
CIC05:n=22
JAN90:n=181
FIP92:n=44
SRP98:n=28
1 x AN/NEA; 1 x K/WB; 1 x NFLD
POM79:n=10
VAL88:n=217
MIN86:n=103
TYT98:n=14
LLS02:n=27

h. 17th C

CLK86:n=11
VAL88:n=13
MIN86:n=146
WP84:n=34
BIG82:n=106
CH75:n=67

● Site with <10 relevant specimens

■ Cranial specimens   ▨ Cleithra   □ Vertebrae   1 x EB Isotopic sample: imported   1 x SNS Isotopic sample: local

Figure 16.7. Maps showing relative frequencies of cranial specimens, vertebrae and cleithra at individual London sites, by century. Specimens are allocated to centuries based on the mid-points of their dating ranges, so the precise groupings should be treated with caution. Where available, tentative isotopic provenancing results (from Barrett et al. *2011) are also marked. SNS = southern North Sea, AN/NEA = Arctic Norway/northeast Atlantic Ocean, EB = eastern Baltic Sea, K/WB = Kattegat/western Baltic Sea, NFLD = Newfoundland (Drawing: David Orton and James Morris using a base map courtesy of MOLA (Museum of London Archaeology)).*

no shortage of fish bone data from this period, the relative paucity of cod abdominal vertebrae and especially cranial specimens is likely to be meaningful. Taking the vertebra pattern at face value, it would suggest a change in the type of cod products imported, and hence either a shift in source regions, a diachronic change in the preservation techniques used in those regions, or both.

Identification biases must also be assessed, since analysts vary regarding which elements they routinely identify to the taxonomic level of species. Comparison between specimens recorded (a) by Alison Locker and (b) by in-house specialists at MOLA indicate that the latter were much more conservative about identifying cod vertebrae to species. However, there does not appear to have been any date-correlated bias by analyst except in the Roman period versus all other periods. The bulk of Roman assemblages were studied by MOLA specialists, which explains the predominance of cranial bones in this period (see Orton *et al.* 2014, figs 6 and 7).

London is not, of course, a homogeneous settlement, and it is worth assessing from where within the city and its environs the various types of cod bones derive. Figure 16.7 shows the sites from which cod data were taken within each century, based on the mid-point of date ranges for relevant contexts. It should be noted that chronological resolution for specific contexts is often greater than one century, and that therefore the chronological groupings in the maps should be considered indicative rather than firm. The extent of overlap in date ranges for specimens is illustrated by Figure 16.8. Where the sample from a given site and century is ten specimens or more, relative frequencies of cranial bones, vertebrae, and cleithra are plotted on Figure 16.7. Sites have also been annotated with the results of stable isotope research, where available, with the single most likely source listed for each sample (Barrett *et al.* 2011).

The situation in the tenth century is hard to assess due to a profusion of very small samples. The single reasonably large sample, from Pudding Lane (PDN81), has a balance of elements.

Most of the eleventh-century data come from sites in the City itself and are dominated by cranial bones. The exceptions are Westminster Abbey (WST86, included as eleventh century here but actually dated AD 1050–1150) and a group of pits on the site that would later become Winchester Palace (WP84), both of which are, interestingly, already dominated by vertebrae. Four vertebrae from the former site were included in the isotope study, and one of these had a probable Arctic Norway/northeastern Atlantic signature. If some of the remains at WST86 indeed represent early imports, the site's monastic status is likely to be significant. A single vertebra sampled from New Fresh Wharf (St. Magnus, SM75) had a local isotopic signature.

Contexts in the City core dated to the twelfth century show more variation, but are still dominated by cranial

bones overall. Postcranial bones make up more than 50% only at Milk Street (MLK76) and at two riverfront sites with relatively small samples: Billingsgate (BIG82; located, appropriately enough, on the site that would become Old Billingsgate fish market) and Seal House (SH74). Three isotope samples from these latter sites all gave local stable isotope signatures. Meanwhile, cranial bones also make up a little over half of the cod assemblage from Merton Priory (MPY86), *c.* 13 km south of the City.

Most of the thirteenth-century data come from three sites in the City, and particularly from Trig Lane (TL74) and Billingsgate on the waterfront. These sites have greater than 50% postcranial bones, but Trig Lane in particular is heavily dominated by vertebrae. Four of these were sampled for isotopes, of which three are probable northern imports. By contrast, two specimens from BIG82 and one from Ludgate Hill Car Parks (PWB88) are probably local.

A larger number of reasonable-sized samples are available for the fourteenth century, but these are biased towards (a) the waterfront and (b) the Fleet Valley and Clerkenwell, outside the City's western walls. This may be a real pattern rather than research bias, since fourteenth-century contexts have certainly also been excavated in the City's core, and smaller numbers of cod bones have been reported from them. All of the waterfront assemblages are dominated by postcranial bones, but those beyond the walls are exclusively composed of vertebrae. It may be significant that Albion Place (ABP94), which is dominated by vertebrae, represents part of the outer precinct of the Priory of St. John of Jerusalem, part of the monastic landscape to the west of the City. It can be contrasted with the cranial-dominated assemblage from Merton Priory (MPY86) in London's southern hinterland, suggesting the consumption of fresh fish at the latter site. Stable isotope results from the fourteenth century all come from sites on or near the waterfront. From a total of five specimens, two are probable northern imports, while a third, from PWB88, has a surprising eastern Baltic Sea signature. This is not inconceivable given the evidence for the production of dried cod around Gdańsk starting in the fourteenth century (cf. Nedkvitne 2014, 130; Orton *et al.* 2011; forthcoming), and the contemporary growth of trade between England and the towns of the Baltic littoral (e.g. Lloyd 1991, 91–2).

The picture in the fifteenth century is broadly similar to that in the fourteenth. The sample from St. Mary's Nunnery, Clerkenwell (CLK86), consists entirely of vertebrae – one of which is thought to be a probable import based on isotope analysis – while Fleet Valley (VAL88) and the waterfront sites of Trig Lane (TL74) and Seal House (SH74) each have more than 80% vertebrae. Three out of four isotope samples from the latter two sites have northern signatures. Swan Lane (SWA81), by contrast, which is also on the waterfront, yielded a cod bone assemblage almost evenly split between cranial and postcranial bones, with a single

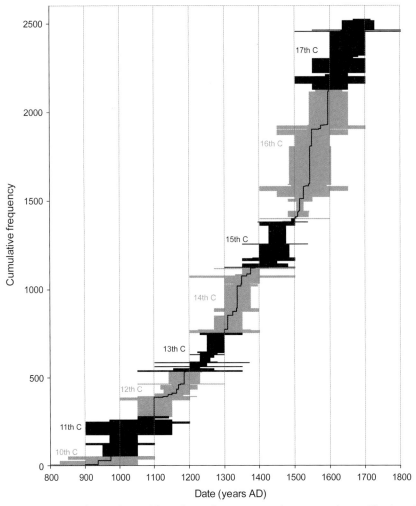

Figure 16.8. Cumulative frequency curve for London cod from the tenth to seventeenth centuries (central line), with associated date ranges for specimens (shaded area). Century groupings as used in Figure 16.7 are shaded in alternating black and grey to show the potential extent of overlap between them.

isotope sample having a southern North Sea signature. The cranial-dominated assemblage from Botolph Place (BPL95) is associated with what was probably a merchant's house.

The distribution of cod remains changes considerably in the sixteenth century. There are still good samples from the Fleet Valley and Clerkenwell, but none from within the City walls. Instead, data are available from sites around the fringes of London, including Islington (FIP92, CIC05) and Spitalfields (SRP98) in the north, Southwark in the south (TYT98, LLS02), and the site of the Royal Mint (MIN86) in the east – the latter a Royal Navy victualling yard at this time. Cranial bones are rare throughout, but cleithra are frequent at several sites north and south of the City. This may suggest changes in how traded cod were processed, and potentially in their source, so it is interesting that the various sites in Clerkenwell (and the west in general) retain the vertebrae-dominated profile seen in previous centuries. On the other hand, the limited available provenancing data do

not support any interpretation in terms of differing sources: Clerkenwell (CLK86) and Finsbury Pavement (FIP92) have rather different anatomical signatures but very similar isotopic results, both having single specimens attributed to each of the following three regions: Arctic Norway/the northeastern Atlantic, the Baltic, and Newfoundland (the latter being particularly tentative, however). The variation may thus have to do with either the very general geographical assignments that are possible through isotope analysis or differing fragmentation, recovery, and identification of the large but easily broken cleithra. We should also note that the samples with large percentages of cleithra are generally smaller than those with an overwhelming dominance of vertebrae, and that the map thus over-represents their frequency in London as a whole.

Finally, the seventeenth-century map shows a broadly similar pattern to that from the sixteenth century, but with fewer reasonably sized samples, perhaps due to a drop off in the intensity of environmental sampling on sites dating from the post-medieval period and/or changing refuse disposal practices in the past. No isotope analysis was conducted on seventeenth-century or later material from London.

Discussion and conclusions

While changes over time in the overall frequency of cod bones must be treated with considerable caution due to possible research biases, the patterns of relative anatomical representation can provide more reliable information on shifts in how London was provisioned. The drop in cranial bones and increase in postcranial bones in the thirteenth century is interpreted as evidence of a change in supply away from fresh (whole) locally caught cod from the North Sea to (decapitated) preserved cod from more distant waters, perhaps Norwegian stockfish (Orton *et al*. 2014). This inference is supported by the available stable isotope data (Barrett *et al*. 2011). The subsequent fluctuations in the abundance of postcranial bones in the fourteenth and fifteenth centuries may tentatively be related to the decline and then recovery in stockfish production and trade following the Black Death (see Chapter 5). Until the late fifteenth century, anterior (abdominal) vertebrae and caudal vertebrae are both well represented. This pattern suggests that most imported fish had been headed, but retained their vertebral columns intact. Conversely, from the end of the fifteenth century there was a sharp rise in the representation of caudal vertebrae, whereas abdominal specimens dropped off, remaining low through the sixteenth and seventeenth centuries.

Whereas the initial thirteenth-century change in the relative abundance of cranial and postcranial bones suggests an increasing supply of preserved cod to London, the subsequent sixteenth-century shift in the representation of caudal versus abdominal vertebrae implies the introduction of a different processing technique. The change need not have been absolute, but the body portion representation data nevertheless imply a quantitative trend. If the main source of imports had remained Norwegian stockfish, as was probably being consumed in the thirteenth century based on historical evidence (e.g. Nedkvitne 2014), the shift would be best interpreted as a preference for *råskjær* over *rundfisk* (see above and Chapter 18). However, over the course of the fifteenth and early sixteenth centuries preserved cod from a variety of new sources entered the English market. These included stockfish (perhaps *råskjær*?) brought directly from Iceland by English merchants, as well as cod that was both salted and dried – by English fishermen operating around southwestern England, Ireland, Iceland and (from *c*. AD 1502) Newfoundland. Most of these salt cod were dried by being laid flat on coastal rocks or wooden platforms, methods which entailed removing the anterior vertebrae (e.g. Candow 2009; Kowaleski 2000, 439; 2003). It may be relevant that cleithra were also frequent at several sites of sixteenth-century date.

This paired skeletal element was also left in the new salted and dried cures (e.g. Betts *et al*. 2014). In sum, the anatomical shift around AD 1500 is consistent with a switch in London's supply of preserved cod, from stockfish (perhaps a mixture of *råskjær* and *rundfisk*, but with *rundfisk* dominant) to salted and dried products of the expanding long-distance sea fisheries of late-medieval England.

Turning to the spatial distribution of the data, the evidence for dried or salted and dried cod consumption occurs across the City of London and its hinterland. Ecclesiastical institutions were among the consumers, but not exclusively so. Vertebrae and/or cleithra were also over-represented at secular sites, and some monastic institutions produced assemblages with good representation of both cranial and postcranial bones (e.g. Merton Priory). When subdivided by site and century, the sample sizes are small, but such as they are, the data suggest a diverse market for imported cod across the medieval and post-medieval city and its hinterland. In terms of chronological trends, the changes in anatomical representation, and thus probably cod imports, of the thirteenth/fourteenth and sixteenth centuries are also evident at the level of individual sites.

Acknowledgements

We thank the Museum of London Archaeology staff, especially Alan Pipe, Natasha Powers, David Bowsher and Stephanie Ostrich. Philip Armitage and Hannah Russ generously provided additional unpublished data. The research was funded by the Leverhulme Trust and The Fishmongers' Company.

References

Astill, G. 2009. Medieval towns and urbanization, in R. Gilchrist & A. Reynolds (eds) *Reflections: 50 years of Medieval Archaeology, 1957–2007*: 255–70. Leeds: Maney.

Barrett, J. H. 1997. Fish trade in Norse Orkney and Caithness: a zooarchaeological approach. *Antiquity* 71: 616–38.

Barrett, J. H., A. M. Locker & C. M. Roberts. 2004a. The origin of intensive marine fishing in medieval Europe: the English evidence. *Proceedings of the Royal Society B* 271(1556): 2417–21.

Barrett, J. H., A. M. Locker & C. M. Roberts. 2004b. 'Dark Age Economics' revisited: the English fish bone evidence AD 600–1600. *Antiquity* 78: 618–36.

Barrett, J. H., D. Orton, C. Johnstone, J. Harland, W. Van Neer, A. Ervynck, C. Roberts, A. Locker, C. Amundsen, I. B. Enghoff, S. Hamilton-Dyer, D. Heinrich, A. K. Hufthammer, A. K. G. Jones, L. Jonsson, D. Makowiecki, P. Pope, T. C. O'Connell, T. de Roo & M. Richards. 2011. Interpreting the expansion of sea fishing in medieval Europe using stable isotope analysis of archaeological cod bones. *Journal of Archaeological Science* 38: 1516–24.

Bennema, F. P. & A. D. Rijnsdorp. 2015. Fish abundance, fisheries, fish trade and consumption in sixteenth-century Netherlands as described by Adriaen Coenen. *Fisheries Research* 161: 384–99.

Betts, M. W., S. Noël, E. Tourigny, M. Burns, P. E. Pope & S. L. Cumbaa. 2014. Zooarchaeology of the historic cod fishery in Newfoundland and Labrador, Canada. *Journal of the North Atlantic* 24: 1–21.

Biddle, M. 1976. Towns, in D. M. Wilson (ed.) *The Archaeology of Anglo-Saxon England*: 99–150. London: Methuen.

Campbell, B. M. S., J. A. Galloway, D. Keene & M. Murphy. 1993. *A Medieval Capital and its Grain Supply: agrarian production and distribution in the London region c. 1300*. London: Centre for Metropolitan Research.

Candow, J. E. 2009. The organisation and conduct of European and domestic fisheries in northeast North America, 1502–1854, in D. J. Starkey, J. T. Thór & I. Heidbrink (eds) *A History of the North Atlantic Fisheries. Volume 1: from early times to the mid-nineteenth century*: 387–415. Bremen: H. M. Hauschild.

Childs, W. R. 1995. England's Icelandic trade in the fifteenth century: the role of the port of Hull, in P. Holm, O. U. Janzen, O. Uwe & J. Thor (eds) *Northern seas yearbook, 1995*: 11–31. (Fiskeri- og søfartsmuseets studieserie 5). Esbjerg: Fiskeri- og Søfartsmuseet.

Crema, E. R. 2012. Modelling uncertainty in archaeological analysis. *Journal of Archaeological Method and Theory* 19: 440–61.

Fjelldal, P. G., G. K. Totland, T. Hansen, H. Kryvi, X. Wang, J. L. Sondergaard & S. Grotmol. 2013. Regional changes in vertebra ontogeny reflect the life history of Atlantic cod (*Gadus morhua* L.). *Journal of Anatomy* 222: 615–24.

Gray, T. & D. J. Starkey. 2000. The distant-water fisheries of south west England in the early modern period, in D. J. Starkey, C. Reid & N. Ashcroft (eds) *England's Sea Fisheries: the commercial sea fisheries of England and Wales since 1300*: 96–104. London: Chatham.

Harland, J. F. & J. H. Barrett. 2012. The maritime economy: fish bone, in J. H. Barrett (ed.) *Being an Islander: production and identity at Quoygrew, Orkney, AD 900–1600*: 115–38. (McDonald Institute Monograph). Cambridge: McDonald Institute for Archaeological Research.

Jones, E. T. 2000. England's Icelandic fishery in the early modern period, in D. J. Starkey, C. Reid & N. Ashcroft (eds) *England's Sea Fisheries: the commercial sea fisheries of England and Wales since 1300*: 105–10. London: Chatham.

Jonsson, L. 1986. Finska gäddor och Bergenfisk: ett försök att belysa Uppsalas fiskimport under medeltid och yngre Vasatid, in N. Cnattingius & T. Neréus (eds *Uppsala stads historia. Volume 7, Från Östra Aros till Uppsala: en samling uppsatser kring det medeltida Uppsala*: 122–39. Uppsala: Almqvist and Wiksell.

Kowaleski, M. 2000. The expansion of the southwestern fisheries in late medieval England. *Economic History Review* 53: 429–54.

Kowaleski, M. 2003. The commercialization of the sea fisheries in medieval England and Wales. *International Journal of Maritime History* 15(2): 177–231.

Lloyd, T. H. 1991. *England and the German Hanse 1157–1611*. Cambridge: Cambridge University Press.

Locker, A. 1988. The fish bones in two Middle Saxon occupation sites: excavations at Jubilee Hall and 21– 22 Maiden Lane. *Transactions of the London & Middlesex Archaeological Society* 39: 149–50.

Nedkvitne, A. 2014. *The German Hansa and Bergen 1100–1600*. Köln: Böhlau.

Orton, D. C., J. Morris, A. Locker & J. H. Barrett. 2014. Fish for the city: Meta-analysis of archaeological cod remains and the growth of London's northern trade. *Antiquity* 88: 516–30.

Orton, D., E. Rannamäe, L. Lõugas, D. Makowiecki, S. Hamilton-Dyer, A. G. Pluskowski, T. O'Connell & J. H. Barrett. forthcoming. The Teutonic Order's role in the development of a medieval eastern Baltic cod fishery, in A. G. Pluskowski (ed.) *The Ecology of Crusading, Colonisation and Religious Conversion in the Medieval Eastern Baltic: Terra Sacra II*. Leiden: Brepols.

Perdikaris, S. & T. H. McGovern. 2008. Codfish and kings, seals and subsistence: Norse marine resource use in the North Atlantic, in T. C. Rick & J. M. Erlandson (eds) *Human Impacts on Ancient Marine Ecosystems: a global perspective*: 187–214. Berkeley: University of California Press.

Starkey, D. J. & M. Haines. 2001. The Newfoundland fisheries c. 1500–1900, in P. Holm, T. Smith & D. J. Starkey (eds) *The Exploited Seas: new directions for marine environmental history*: 1–11. St John's: International Maritime Economic History Association/Census of Marine Life.

The Social Complexities of Early Marine Fish Consumption: New Evidence from Southeast England

Rebecca Reynolds

Introduction

The trading of what were to become high-bulk commodities – cod (*Gadus morhua*) and herring (*Clupea harengus*) – in England is believed to have begun around AD 1000 (Barrett *et al.* 2004), based on comparison of archaeological fish-bone assemblages. Most examples from the middle (AD 600–800) and late (AD 800–1000 for present purposes) Anglo-Saxon periods have, with a few exceptions, come from *wics* (urban or proto-urban trading centres) and other trading places, and, as a consequence, the rise of fishing and fish consumption is thought to be closely linked to the development of medieval economies and the growth of towns.

Whilst long-distance trade of preserved fish became the norm in the later medieval period, recent stable isotope studies of archaeological fish bones have revealed that the earliest cod consumed in medieval England were fished from local waters (Barrett *et al.* 2011), probably satisfying local demands and representing a combination of direct and indirect subsistence, the latter being controlled by elites (Hoffmann 1996; Tsurushima 2007). For the centuries before AD 1000, however, due to the paucity of evidence from relevant rural settlements, there has been little zooarchaeological opportunity to explore the role that the elite may have played in the catching and consumption of marine fish. Until recent years one of the few examples has been Flixborough, near the River Trent in North Lincolnshire, which yielded very few sea-fish (Dobney *et al.* 2007). The Anglo-Saxon situation in southern England has thus been uncertain. However, recent excavations at two high-status sites near the south coast – the middle to late Anglo-Saxon *thegnly* settlement of Bishopstone, East Sussex, and the Anglo-Saxon monastery at Lyminge, Kent – have revealed substantial amounts of sea-fish bone. These sites provide a rare opportunity to explore the social complexities that may also have played a role in establishing a taste for marine fish in Anglo-Saxon England (Figure 17.1).

Bishopstone

The fish remains from Bishopstone, dating from the late eighth–late tenth/early eleventh century AD, are mostly of marine taxa (Table 17.1). A few cyprinids (Cyprinidae) are the only freshwater fish. Eel (*Anguilla anguilla*) remains and one possible salmonid (Salmonidae) vertebra are the only evidence for migratory fish. The marine fish include herring, cod, whiting (*Merlangius merlangus*), flatfish (Pleuronectidae), mackerel (*Scomber scombrus*), horse mackerel (*Trachurus trachurus*) and small numbers of

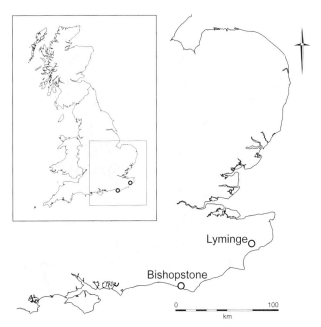

Fig. 17.1. The main fish-bone assemblages discussed in the text (Drawing: Vicki Herring).

Table 17.1. Number of identified specimens of fish taxa from Bishopstone recovered by wet sieving (>2 mm and >4 mm mesh), dry sieving (>4 mm mesh) and hand collecting.

Taxon	Wet sieved	Dry sieved/ hand collected	Total
Ray family	176	10	186
Eel	594		594
Conger eel	19	17	36
Herring family	1		1
Herring	749		749
Herring?	2		2
Cyprinid	8		8
Salmonid		1	1
Gadid	53	26	79
Cod	80	344	424
Cod?	1	1	2
Whiting	359	55	414
Bib	4	1	5
Haddock		3	3
Haddock?	1		1
Ling		1	1
Ling?	1		1
Garfish	3		3
Garfish?	3		3
Sand smelt?		1	1
Gurnard family	10	2	12
Gurnard family?	3		3
Seabass	83	26	109
Seabass?	2	1	3
Perch	2		2
Horse mackerel	39	34	73
Horse mackerel?	2		2
Sea bream family	4		4
Mullet family?	2		2
Atlantic mackerel	165	48	213
Tuna?		1	1
Flatfish order	4	1	5
Turbot/brill		2	2
Plaice/flounder	325	120	445
Plaice	1		1
Flounder		1	1
Unidentified fish	3890	525	4415
Total	6586	1221	7807

specimens of other species. The bones came from a variety of features, including cesspits and rubbish dumps. The overall depositional trend observed at Bishopstone was that the bones of bigger fish were generally found with mammal remains, while those of smaller fish were found on their own. This is not surprising, as the bones of smaller fish may be ingested, whereas those of bigger fish are more likely to be discarded as part of kitchen preparation or table waste. Many of the fish remains demonstrate evidence of (presumably human) digestion, but stable isotope results from the human burials indicate that fish did not make a significant contribution to the diet (Thomas 2010; see Chapter 20). At Bishopstone, the nearby English Channel thus seems to have provided an occasional addition to a varied diet.

Poole's (2010a) study of the wider zooarchaeological assemblage has indicated a considerable frequency of pig and wild animals, including wild fowl, in the Bishopstone assemblage, traits indicative of a high-status site. This interpretation is supported by the presence of several bones from large marine mammals. Contemporary textual and other archaeological sources suggest that marine mammals were highly prized by the elite, both as a foodstuff and as a source of raw material (Gardiner 1997). Regardless of the size of the fish, deposits which held high numbers of fish bones were generally sparse in bird remains despite being abundant in mammal remains (Reynolds 2008). There were thus different attitudes to the treatment of these foodstuffs.

Lyminge

Excavation at Lyminge, Kent, has revealed a sequence of occupation spanning the early Anglo-Saxon period through to the establishment and use of a middle Anglo-Saxon double monastery (Thomas and Knox 2012; Thomas 2013). The excavations have revealed a large faunal assemblage (Knapp forthcoming) accompanied by many artefacts. Only a fraction of the fish assemblage has been analysed, but it has proven rich in sea-fish remains (Table 17.2; Reynolds 2015). Much of the material studied to date has been hand collected and dry sieved and therefore favours the recovery of bones of larger species, such as cod. Nevertheless, some environmental samples from the phases believed to date from the later part of the early Anglo-Saxon period have revealed the presence of herring and a member of the family Pleuronectidae, either plaice (*Pleuronectes platessa*) or flounder (*Platichthys flesus*) (Table 17.3). A basic assessment of the faunal material has revealed a similar depositional pattern to that at Bishopstone. The bones of fish and domestic fowl were rarely found together, suggesting different treatment and consumption (Reynolds 2009).

Table 17.2. Number of identified specimens of fish taxa from Lyminge recovered by wet sieving (>2 mm and >4 mm mesh), dry sieving (>4 mm mesh) and hand collecting.

Taxon	Wet sieved	Dry sieved/ hand collected	Total
Ray family	81	3	84
Eel	1512	3	1515
Conger eel		4	4
Herring family	6		6
Herring	244	9	153
Herring?	1		1
Cyprinid	10		10
Salmonid	10		10
Cod family	24	85	109
Cod family?	2		2
Cod	63	562	625
Whiting	114	55	169
Haddock	2	11	13
Haddock?		1	1
Gurnard family	15	12	27
Seabass?	2		2
Perch family		72	72
Horse mackerel	319	74	393
Sea bream family		1	1
Sea bream family?		9	9
Blackspot/red sea bream	34	34	68
Atlantic mackerel	110	43	153
Tuna?		2	2
Flatfish order		11	11
Turbot/brill		2	2
Plaice/flounder	143	116	259
Plaice/flounder?	1	1	2
Unidentified fish	2189	2232	4421
Total	4882	3342	8124

Provisioning and anatomical representation

The skeletal element representation at Bishopstone and Lyminge indicates that whole fish were brought to the sites, although both settlements were near rather than immediately adjacent to the coast. It is thus significant that textual evidence may link the minster of Lyminge with a coastal settlement at Sandtun, West Hythe, Kent, which has been identified as a seasonal trading and landing place that may also have acted as a fishing site (Gardiner *et al.* 2001). A recent re-examination of the charters of Lyminge suggests that the relevant grant of land may in fact have been for Lympne, not Lyminge (Kelly 2006), so establishing a definite

Table 17.3. Number of identified specimens of fish taxa from the early Anglo-Saxon phases of Lyminge, recovered from wet-sieved environmental samples (>2 mm and >4 mm mesh).

Taxon	Wet sieved
Eel	8
Herring	118
Cyprinid	2
Horse mackerel	1
Plaice/flounder	27
Unknown fish	66
Total	222

link between the two sites is problematic. Nevertheless, one can imagine that both Lyminge and Bishopstone may have been provisioned with sea-fish from coastal outstations *like* Sandtun. The presence of both vertebrae and cranial elements at Bishopstone and Lyminge makes it unlikely that fish were decapitated for drying at landing places (cf. Chapter 1; Chapter 18), but coastal sites may still have been providing whole fish and/or access to the sea. Toll privileges issued to minsters in Kent during the eighth century are prolific, and it is likely that ecclesiastical centres held a vested interest in trade (Kelly 1992). However, what the relevant commodities were, and whether they included fish, is harder to establish.

The methods and equipment used for fishing can seldom be inferred from available evidence. An important exception is provided by the remains of weirs, such as those on the Blackwater estuary in Essex and in the River Thames at London. Their impressive size suggests that a great deal of time and effort went into their construction and upkeep, possibly implying that they were controlled by ecclesiastical or secular elites (Strachan 1998). Initial construction by small communities is also conceivable (O'Sullivan 2003), with control of the weirs only later being acquired by overlords. Textual sources from the late Anglo-Saxon period indicate that elites increasingly sought to control their environment and food sources, and many weirs and fisheries are listed in charters as being very valuable (Hooke 2007; Tsurushima 2007). The weirs on the Blackwater estuary, constructed in the middle Anglo-Saxon period, were being used and repaired for several centuries thereafter (Strachan 1998). Some of those on the River Thames date back to the early Anglo-Saxon period (Cohen 2011; Cowie and Blackmore 2008). No weirs have yet been found near Lyminge or Bishopstone, but later textual sources, such as Domesday Book, note Kent in particular as having important herring and eel fisheries. Thus the possibility that the minster of Lyminge and/or the settlement of Bishopstone may have controlled one cannot be entirely discounted. Tsurushima (2007) argues that the fishing of herring around the eleventh century was highly controlled by manors and overlords (including the king). It was thus probably centrally managed. Fishing in earlier periods may have been similarly organised.

The social context of fish consumption

The presence of herring at both proto-urban settlements and some elite sites before AD 1000 may help our understanding of how fish were perceived and how different social groups within Anglo-Saxon England provisioned themselves. Received wisdom regarding fish consumption in the middle to late Anglo-Saxon period is that eels were dominant, with cyprinids also being represented in high frequencies, whereas marine fish were seldom abundant (e.g. Barrett *et al.* 2004; Serjeantson and Woolgar 2006). Rare exceptions to this pattern include the settlement at Sandtun (see above), where the assemblage was dominated by cod, whiting and flatfish, and hardly any eel were identified (Hamilton-Dyer 2001), and the site of Fishtoft, Lincolnshire, where a fishery specialising in flatfish coincided with salt making (Locker 2012). A tiny sixth–early eighth-century AD assemblage of 44 identified fish bones from the near-coastal settlement of Bloodmoor Hill, Suffolk, also included sea-fish (Parks and Barrett 2009). At Flixborough in North Lincolnshire, the middle to late Anglo-Saxon assemblage was dominated by freshwater and migratory species. Only 14 remains of cod and 11 of herring were found (Dobney *et al.* 2007, 40).

At Lyminge, the middle Anglo-Saxon phases of occupation seem to be richest in sea-fish remains, though some small numbers have also been found in the early Anglo-Saxon phases. At Bishopstone, the middle and late Anglo-Saxon fish bone could not be divided, but it remains notable that marine fish were abundant. Taken together, the evidence discussed in this chapter may suggest that, until the wider adoption of sea-fish consumption in tenth- and eleventh-century England, marine fishing was associated with *wics* (see above), settlements near the shore (Fishtoft, Bloodmoor Hill and Sandtun) and some elite sites (Lyminge and Bishopstone, both near the southern coast). Thus some coastal regions were already developing a taste for fish, perhaps supported by elite demand, in the middle Anglo-Saxon period (Reynolds 2015).

We now know that marine fish can be found on both urban (that is, *wic*) and elite sites during the middle Anglo-Saxon period. It is therefore worthwhile to compare the dynamics of consumption. Bishopstone and Lyminge have richer and more diverse assemblages of sea-fish than do *wics*, where herring predominate (Barrett *et al.* 2004; see Chapter 15). This wide selection may be a correlate of high status or affluence (e.g. Ervynck *et al.* 2003). Some of the cod whose bones were excavated at both Lyminge and Bishopstone were also large (sometimes more than 1 m in total length). This in itself may imply labour-intensive offshore fishing and associated elite investment.

Marine fish remains have been found on only a very small number of early to middle Anglo-Saxon settlements – particularly Lyminge and Bloodmoor Hill (see above), both of which show evidence of social stratification (Scull 2009; 2010; Thomas 2013). This early foray into marine-fish consumption then blossomed in the later part of the middle Anglo-Saxon period at Lyminge and probably also Bishopstone. Although thus-far a feature of southern England, the comparatively large amounts of marine fish found at these two high-status sites may suggest that elites had a role in initiating a greater taste for marine fish, rather than (or in addition to) people living in the urban settlements emphasised by Barrett *et al.* (2004; 2011). Indeed, it has been argued that middle Anglo-Saxon *wics* were dependent on elites for their food supply (Hamerow 2007; Holmes 2014, 102; O'Connor 2001). Furthermore, as the Anglo-Saxon period progressed, urban centres will also have been home to wealthy secular and ecclesiastical residents.

The presence of fish on monastic sites, such as Lyminge, has often been explained by the Rule of St Benedict, which forbids the consumption of the flesh of quadrupeds except by the sick or elderly. Monastic settlements of later periods in Flanders do show higher amounts of fish and avian remains (Ervynck 1997), and in England such settlements are also characterised by a preference for flatfish (Dobney *et al.* 2007, 231–3). The Rule was written in the sixth century, but assessing its influence on fish consumption in Anglo-Saxon England is problematic (Frantzen 2014, 232–45). Most scholars agree that the period after the conversion of England (following the arrival of St Augustine in AD 597) was characterised by what is known as a *regula mixta*, where practices were guided by the abbot or by the secular rulers who had established the religious house (Blair 2005; Foot 2006; Mayr-Harting 1976). However, after the religious reforms of the later tenth century (Gem 1997, 12), fish remains do increase in numbers on such monastic sites as Westminster Abbey (Locker 1997).

The nature of early monastic settlements is not well understood, especially *vis-à-vis* secular high-status settlements. There is a strong possibility that their archaeological signatures are very similar (Loveluck 2011; 2013). Fish remains were found throughout the phases at Flixborough, with no discernible increase in their abundance during the identified monastic phase of the ninth century (Dobney *et al.* 2007). Moreover, remains of wild mammals and birds (which might be associated with aristocratic tastes and pastimes) have been found on ecclesiastical sites (Poole 2010b). Thus a distinct religious signature has not yet been recognised. While textual sources attest to the presence of a minster at Lyminge, it is also possible that it had been a local secular administrative centre with a royal *vill* (Chadwick Hawkes 1982). Given that sea-fish remains have been found at a few middle Anglo-Saxon secular sites (see above), an explanation entirely founded on the observance of a religious rule is unlikely.

Conclusion

Identifying the exact role elites played in marine-fish consumption in the period leading up to AD 1000 is far from simple. The scarcity of fish remains for the early Anglo-Saxon period is notable, but its significance is not yet fully understood. The finds from Bishopstone and Lyminge have shown that elites of the middle and late Anglo-Saxon periods were consuming marine fish that were being transported whole to the settlements. These two sites are characterised by significant quantities of cod and other gadids, and it would thus seem that cod may originally have had high-status associations. These elites were also consuming a variety of other marine species. The fact that marine fish have been found on a few high-status sites may support the theory that elites had some role in the early growth of English sea-fish exploitation. Concurrently, the increase in marine fish on urban settlements was probably due to a variety of factors, including the provisioning actions of elites, emulation of what started off as a prestigious diet, and of course economic and demographic development (Reynolds 2015). The situation witnessed after AD 1000 points to how important marine fish became, as both an economic commodity and a foodstuff. Thus it is important to study the origins of sea-fish consumption not just from an economic point of view, but also from a social one (Reynolds 2015).

Acknowledgements

I would like to thank James Barrett for inviting me to contribute and for his help throughout my research. James Rackham and Naomi Sykes have given suggestions and patiently read previous drafts, for which I am very grateful. This paper derives from PhD research conducted at the University of Nottingham.

References

Barrett, J. H., A. M. Locker & C. M. Roberts. 2004. 'Dark Age Economics' revisited: the English fish bone evidence AD 600–1600. *Antiquity* 78: 618–36.

Barrett, J. H., D. Orton, C. Johnstone, J. Harland, W. Van Neer, A. Ervynck, C. Roberts, A. Locker, C. Amundsen, I. B. Enghoff, S. Hamilton-Dyer, D. Heinrich, A. K. Hufthammer, A. K. G. Jones, L. Jonsson, D. Makowiecki, P. Pope, T. C. O'Connell, T. de Roo & M. Richards. 2011. Interpreting the expansion of sea fishing in medieval Europe using stable isotope analysis of archaeological cod bones. *Journal of Archaeological Science* 38: 1516–24.

Blair, J. 2005. *The Church in Anglo-Saxon Society*. Oxford: Oxford University Press.

Chadwick Hawkes, S. 1982. *Anglo-Saxon Kent, c. 425–725*, in P. Leach (ed.) *Archaeology in Kent to AD 1500*: 425–75. London: Council for British Archaeology.

Cohen, N. 2011. Early Anglo-Saxon fish traps along the River Thames, in S. Brookes, S. Harrington & A. Reynolds (eds) *Studies in Early Anglo-Saxon Art and Archaeology: papers in*
honour of Martin G. Welch (British Archaeological Reports British Series 527): 131–8. Oxford: Archaeopress.

Cowie, R. & L. Blackmore. 2008. *Early and Middle Saxon Rural Settlement in the London Region*. London: Museum of London Archaeological Services.

Dobney, K., D. Jaques, J. H. Barrett & C. Johnstone. 2007. *Farmers, Monks and Aristocrats: the environmental archaeology of an Anglo-Saxon estate centre at Flixborough, north Lincolnshire, UK*. Oxford: Oxbow Books.

Ervynck, A. 1997. Following the rule? Fish and meat consumption in monastic communities in Flanders (Belgium), in G. De Boe & F. Verhaeghe (eds) *Environment and Subsistence in Medieval Europe* (Papers of the Medieval Brugge 1997 Conference): 67–81. Zellik: Instituut voor het Archeologisch Patrimonium Rapporten.

Ervynck, A., W. Van Neer, H. Hüster-Plogmann & J. Schibler. 2003. Beyond affluence: the zooarchaeology of luxury. *World Archaeology* 34: 428–41.

Foot, S. 2006. *Monastic Life in Anglo-Saxon England, c. 600–900*. Cambridge: Cambridge University Press.

Frantzen, A. J. 2014. *Food, Eating and Identity in Early Medieval England*. Woodbridge: Boydell Press.

Gardiner, M. 1997. The exploitation of sea-mammals in medieval England: bones and their social context. *Archaeological Journal* 154: 173–95.

Gardiner, M., R. Cross, N. Macpherson-Grant & I. Riddler. 2001. Continental trade and nonurban ports in mid-Anglo-Saxon England: excavations at Sandtun, West Hythe, Kent. *Archaeological Journal* 158: 255–61.

Gem, R. 1997. Introduction, in R. Gem (ed.) *St. Augustine's Abbey Canterbury*: 9–14. London: B. T. Batsford/English Heritage.

Hamerow, H. 2007. Agrarian production and the emporia of mid Saxon England, ca. AD 650–850, in J. Henning (ed.) *Post-Roman Towns, Trade and Settlement in Europe and Byzantium. Volume 1: the heirs of the Roman West*. Berlin: Walter de Gruyter.

Hamilton-Dyer, S. 2001. Bird and fish remains, in M. Gardiner, R. Cross, N. Macpherson-Grant & I. Riddler, Continental trade and non-urban ports in middle Anglo-Saxon England: excavations at Sandtun, West Hythe, Kent. *Archaeological Journal* 158: 255–61.

Hoffmann, R. C. 1996. Economic development and aquatic ecosystems in medieval Europe. *American Historical Review* 101: 631–69.

Holmes, M. 2014. *Animals in Saxon and Scandinavian England: backbones of economy and society*. Leiden: Sidestone Press.

Hooke, D. 2007. Uses of waterways in Anglo-Saxon England, in J. Blair (ed.) *Waterways and Canal-building in Medieval England*: 37–54. Oxford: Oxford University Press.

Kelly, S. 1992. Trading privileges from eighth-century England. *Early Medieval Europe* 1: 3–28.

Kelly, S. 2006. Lyminge Minster and its early charters, in S. Keynes & A. P. Smyth (eds) *Anglo-Saxons: studies presented to Cyril Hart*: 98–113. Dublin: Four Courts Press.

Knapp, Z. *forthcoming*. The zooarchaeology of the Anglo-Saxon Christian conversion: Lyminge, a case study. Unpublished PhD dissertation, University of Reading.

Locker, A. 1997. The fish bones, in P. Mills *Excavations at the Dorter Undercroft, Westminster Abbey* (Transactions of the London and Middlesex Archaeological Society 46): 111–3.

Locker, A. 2012. The fish assemblage, in P. Cope-Faulkner *Clampgate Road, Fishtoft: archaeology of a middle Saxon island settlement in the Lincolnshire fens*: 98–102. Sleaford: Lincolnshire Archaeology and Heritage Report Series.

Loveluck, C. 2011. Problems of the definition and conceptualisation of early medieval elites, AD 450–900: the dynamics of the archaeological evidence, in F. Bougard, H.-W. Goetz & R. Le Jan (eds) *Théorie et pratique des elites du Haut Moyen Age. conception, perception et réalisation sociale*: 22–67. Turnhout: Brepols.

Loveluck, C. 2013. *Northwest Europe in the Early Middle Ages, c. AD 600–1150*. Cambridge: Cambridge University Press.

Mayr-Harting, H. M. R. E. 1976. *The Venerable Bede, the Rule of St. Benedict and Social Class* (Jarrow Lecture 1976). Newcastle-upon-Tyne: J. and P. Bealls.

O'Connor, T. 2001. On the interpretation of animal bone assemblages from *Wics*, in D. Hill & R. Cowie (eds) *Wics: the early medieval trading centres of northern Europe*: 54–60. Sheffield: Sheffield Academic Press.

O'Sullivan, A. 2003. Place, memory and identity among estuarine fishing communities: interpreting the archaeology of early medieval weirs. *World Archaeology* 35: 449–68.

Parks, R. L. & J. H. Barrett. 2009. Fish bone, in S. Lucy, J. Tipper & A. Dickens (eds) *The Anglo-Saxon Settlement and Cemetery at Bloodmore Hill, Carlton Colville, Suffolk*: 304–306. Cambridge: Cambridge Archaeological Unit.

Poole, K. 2010a. Mammal and bird remains, in G. Thomas *The Later Anglo-Saxon Settlement at Bishopstone: a downland manor in the making*: 146–57. York: Council for British Archaeology.

Poole, K. 2010b. The nature of Anglo-Saxon society. Unpublished PhD dissertation, University of Nottingham.

Reynolds, R. V. 2008. The dynamics of late Anglo-Saxon fish consumption: the evidence from the site of Bishopstone, East Sussex. Unpublished BA dissertation, University of Nottingham.

Reynolds, R. V. 2009. Surf 'n' turf: the components and influences of a middle Anglo-Saxon diet from Lyminge, Kent. Unpublished MSc dissertation, University of Nottingham.

Reynolds, R. V. 2015. Food for the soul: the social dynamics of marine fish consumption along the southern North Sea coast from AD 700 to AD 1200. Unpublished PhD dissertation, University of Nottingham.

Scull, C. 2009. The human burials, in S. Lucy, J. Tipper and A. Dickens (eds) *The Anglo-Saxon Settlement and Cemetery at Bloodmoor Hill, Carlton Colville, Suffolk*: 385–424. Cambridge: Cambridge Archaeology Unit.

Scull, C. 2010. Social transactions, gift exchange, and power in the archaeology of the fifth to seventh centuries, in H. Hamerow, H. Hinton and S. Crawford (eds) *The Oxford Handbook of Anglo-Saxon Archaeology*: 848–864. Oxford: Oxford University Press.

Serjeantson, D. & C. M. Woolgar. 2006. Fish consumption in medieval England, in C. M. Woolgar, D. Serjeantson & T. Waldron (eds) *Food in Medieval England: diet and nutrition*: 102–30. Oxford: Oxford University Press.

Strachan, D. 1998. Inter-tidal stationary fishing structures in Essex: some C14 dates. *Essex Archaeology and History* 29: 274–82.

Thomas, G. 2008. *The Later Anglo-Saxon Settlement at Bishopstone: a downland manor in the making* (Council for British Archaeology Research Report 163). York: Council for British Archaeology.

Thomas, G. 2013. Life before the Minster: the social dynamics of monastic foundation at Anglo-Saxon Lyminge, Kent. *Antiquaries Journal* 93: 109–45.

Thomas, G. & A. Knox. 2012. A window on Christianisation: transformation at Anglo-Saxon Lyminge, Kent, England. *Antiquity Bulletin* 86: 334.

Tsurushima, H. 2007. The eleventh century in England through fish-eyes: salmon, herring, oysters, and 1066, in C. P. Lewis (ed.) *Anglo-Norman Studies* (XXIX Proceedings of the Battle Conference): 193–213. Woodbridge: Boydell Press.

Fish Trade in Norway AD 800–1400: Zooarchaeological Evidence

Anne Karin Hufthammer

Introduction

This paper explores the medieval exploitation and trade of marine resources, in particular fish, in Viking Age and medieval Norway. It focuses on the differences among regions, between rural and urban settlements and between production and domestic consumption of stockfish (defined as air-dried cod: *Gadus morhua*). Its method is zooarchaeology and its main focus is cod (being the most important source of stockfish, and also one of the most important fish for local consumption, during the centuries under consideration). The paper first deals with the role of fish in the diet based on the relative abundance of bones from domestic animals, birds, marine mammals and fish. It then considers how zooarchaeological evidence can be used to recognise the spread of stockfish consumption (and therefore at least regional trade) to the Norwegian towns of Trondheim, Bergen, Stavanger, Tønsberg and Oslo. Given that stockfish were exported from medieval Norway, the results have significant implications for understanding the growth of commodity trade in medieval Europe more generally.

The rise of Norwegian trading centres and towns in the eleventh and twelfth centuries must have brought about a household economy in these densely populated places that differed significantly from that of rural areas. It is likely that rural households were largely dependent on local resources, whereas urban populations could choose from a range of goods, including those imported from inland and abroad. The growth of long-range trade in fish and marine mammal products is well known from twelfth-century and later historical sources, and it may be inferred from more anecdotal eleventh-century textual evidence (Nedkvitne 2014; Nielssen 2009; see Chapters 4 and 5). However, there is only one known household menu regarding medieval Norway: in Trondheim in AD 1532, different kinds of meat and fish, as well as cheese and bread, were expected by

the archbishop (Nordeide 2003, 298). For the rest of the population, and at other times and places, zooarchaeological evidence is required.

Fishing was an important means of subsistence in Norway in the past and continues to be so in the present. The zoologist Jens Rathke, who travelled along the Norwegian coast in AD 1795–1802, claimed that fishing was one of the most important means of subsistence in the country (Rathke 1907). An earlier source, a letter from the Pope to the Archbishop in Nidaros (Trondheim) in AD 1276, claimed that people consumed mainly milk products and fish (Grøn 1926, 27). In a country with one of the longest coastlines in Europe and some of the most productive fish-spawning areas (Nielssen 2009), it is a reasonable starting assumption that fishing also played an important role in Viking Age and medieval times and that marine fish constituted a significant component of the diet.

Remains of marine fish (mostly of cod family species [Gadidae]) are very abundant from Mesolithic and most Neolithic coastal sites along the entire Norwegian coast (e.g. Degerbøl 1951; Helskog 1983; Hufthammer 1992; 1998; Olsen 1967; 1976). But there are exceptions; at the Neolithic site Auve on the south coast of Norway the number of fish bones is very low, illustrating that fish played a minor role in the economy compared to marine and terrestrial mammals (Hufthammer 1997). Bronze Age and Iron Age sites are less well known, but these also produce marine-fish bone (e.g. Brinkmann and Shetelig 1920). Marine fish bones have even been found in the mountains far away from the coast, at the Bronze Age site Skrivarhelleren, 713 m above sea level (Prescott 1993). In the Lofoten area, the main district for stockfish production, a stable economy based on animal husbandry and fishing was maintained from at least the Migration Period to the Middle Ages (Johansen 1982; Nielssen 2009). It was here that winter spawning grounds for the migratory northeastern Arctic (Barents Sea) cod co-occurred with ideal climatic conditions for drying fish without the need of salt. Freshwater fish (particularly Atlantic salmon [*Salmo salar*] and trout [*Salmo trutta*]) also occur in Norwegian zooarchaeological assemblages, but in small numbers (e.g. Barrett *et al.* 2007; 2015). However, a comparison between faunal remains from the Archbishop's Palace and urban sites indicate that salmonid fishes were more common amongst the upper classes than amongst the ordinary people in medieval Trondheim (Hufthammer 2000).

There have been a number of scholarly investigations regarding the roles of fishing and the stockfish trade in the Norwegian economy (e.g. Nedkvitne 1977; 1988; 2014; Nielssen 2009; Solhaug 1976). However, except for some small publications based on zooarchaeology (Hufthammer 2000; 2003), little has been written about the significance of fish in the diet. One of the goals of the present paper is, therefore, to explore this significance, even if it is only possible in a comparative way (see below).

Stockfish production was an important means of subsistence in the north, and it became the basis of growth and wealth in the south, particularly in Bergen, the main transhipment port for export. Stockfish itself, mainly of cod, is a distinctive Norwegian product that was (based on historical evidence) exported in bulk to other Scandinavian towns, continental Europe and Britain in the Middle Ages. As noted above, the earliest historical evidence for this trade relates to the eleventh and (particularly) twelfth centuries. Earlier export, on a smaller scale, is also possible based on zooarchaeological and stable isotope analyses of ninth- to eleventh-century bones from Haithabu/Hedeby (Barrett *et al.* 2008; Heinrich 2006; Lepiksaar and Heinrich 1977).

The production sites for stockfish were (and are) mainly in the three northernmost counties: Finnmark, Troms and Nordland. In the Middle Ages the dried cod was transported southwards to transhipment ports (usually Bergen, but also Trondheim), from which it was exported abroad. There are no Norwegian records of the amount of stockfish exported in the Middle Ages. However, based on toll accounts from eastern English ports relating to the early fourteenth century, it has been estimated that dried cod constituted *c.* 80% of the value of the Norwegian imports (Helle 1982, 306; Nedkvitne 1976, 250; Nielssen 2009, 104). In some of these fourteenth-century documents, 12 different words were used to characterise the stockfish (Nedkvitne 1977, 39). This level of specialisation and implied quality control suggests an advanced trade. Between AD 1650 and AD 1654, 11.3 million dried cod were exported from Bergen and another 1.2 million from Trondheim (Tande and Tande 1986). Quantitative estimates for earlier centuries are smaller, but the amounts are unlikely to have been trivial from the perspective of contemporary producers, merchants and consumers (see Chapter 5).

Despite the large numbers of terms used for dried cod, the stockfish known to have been exported from medieval Norway are categorised into two broader categories, as either *rundfisk* or *råskjær*. Nedkvitne (1977, 113) estimated that in AD 1577 the export from Bergen was 698 tons of *rundfisk* and 871–1058 tons of *råskjær*. The characteristics of each, and thus their expected zooarchaeological signatures, are discussed further below.

Stockfish was also exported from the north for consumption in the Norwegian towns themselves. There are very few medieval historical sources dealing with this internal trade. Nevertheless, it is reasonable to suggest that the growth of internal trade to transhipment ports such as Trondheim and particularly Bergen was probably concurrent with the growth of export trade. Therefore, the second aim of this paper is to analyse the spread of stockfish consumption in the domestic market.

Materials and methods

Using zooarchaeological evidence, we can evaluate the relative abundance (and hence potential significance to human diet) of the remains of fish, birds, terrestrial mammals and marine mammals recovered from archaeological sites. Here we compare the number of identified specimens (NISP) from each taxonomic group in nine bone assemblages. Four are from rural settlements, and five are from urban sites (Figure 18.1). From the towns, where the absolute quantity of material is very large, selected bone samples from well-dated contexts have been chosen for consideration. The sites range in date from the Viking Age to the late Middle Ages (Figure 18.1, Table 18.1). The northernmost is from the island of Helgøy in Troms (70°06'46"N), and the southernmost is from the town of Stavanger in Rogaland (58°58'12"N) (Figure 18.1). All the assemblages, except for the majority of the Kaupang material, have been analysed at the osteological laboratory of the University Museum of Bergen, limiting the error from inter-analyst variability. All of the material, with the exception of that from Kaupang, is very well preserved. The sample sizes (by NISP) range from 580 to 16,793 (see Table 18.1).

The rural sites are Helgøy, Storvågan and Blomsøy in northern Norway. The urban settlements are Oslo (Mindets tomt), Tønsberg (Baglergaten 3), Stavanger (Skagen 3), Bergen (Dreggen) and Trondhein (Televerkstomten) (Table 18.1, Figure 18.1). Kaupang differs from the other sites in being of earlier date. For present purposes it is grouped with the rural settlements, although it is accepted here that it also had urban functions (Skre 2008). Storvågan, a medieval collection point for northern stockfish, also had both rural and urban characteristics (Bertelsen 1985).

Almost all of the material was recovered by sieving, in general with 5 mm mesh. The main exceptions are Skagen

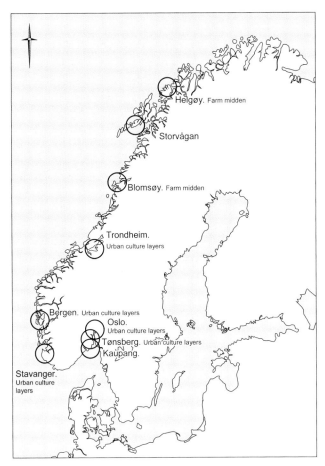

Figure 18.1. Zooarchaeological assemblages discussed in the text (Drawing: Anne Karin Hufthammer and Vicki Herring).

in Stavanger, Dreggen in Bergen and Helgøy. All three sites were excavated in the 1970s, when sieving had not yet become routine and material was hand-collected. Thus the bones of fish, and small skeletal elements in general, may be under-represented at these sites.

Given that a cod bone and a cow bone obviously represent very different amounts of potential food (not to mention the importance of secondary products from livestock), it is meaningless to try to measure the precise significance of fish, birds and mammals to human diet based on the archaeological assemblages. However, comparisons between the different sites can form the basis for observing general trends. The relative abundance (as numbers of identified specimens) of each grouping (fish, bird, terrestrial mammal and marine mammal) has been visually summarised using Principal Component Analysis (PCA) (Leps and Šmilauer 2004) (Figure 18.2). To help interpret the resulting patterns, the relative abundances of domestic versus wild terrestrial mammals and of different fish species within the same assemblages are also explored (Figures 18.3 and 18.4).

Our ability to identify the production and trade of stockfish using zooarchaeological methods depends on an understanding of how stockfish was made. Norwegian stockfish was (and still is) traditionally prepared either as *rundfisk* or as *råskjær*. With *rundfisk* ('round fish' in English), the head and intestines are removed, and two fish are tied together by a string around their tails before being hung on a drying rack. To tie two fish together in this way is called to *sperre* the fish. This is presently the most common preparation method for stockfish.

Table 18.1. Overview of the nine zooarchaeological assemblages used in this study. In most cases only bones from selected archaeological layers have been included.

Site	County	Type	Date (centuries AD)	NISP	University Museum of Bergen No.	Reference(s)
Helgøy	Troms	Farm midden	14th–15th	8320	JS 568	Holm-Olsen 1977
Storvågan	Nordland	Settlement midden	9th–15th, mainly 13th	3086	JS 747	Bertelsen 1977; pers. comm.
Blomsøy	Nordland	Farm midden	7th–13th	580	JS 818	Berglund 1995
Trondheim, Televerkstomten	Sør-Trøndelag	Town sediments	11th–13th	10,288	JS 632	Marthinussen 1992
Bergen, Dreggen	Hordaland	Town sediments	11th–13th	6212	JS 630	Undheim 1985
Oslo, Mindets tomt	Oslo	Town sediments	11th–13th	16,793	JS 537	Lie 1988
Tønsberg, Baglergate 3	Vestfold	Town sediments	11th–13th	4479	JS 644	Brendalsmo 1983
Kaupang	Vestfold	Village sediments	9th–10th	2081	JS 1144, JS 1490	Pilø 2007
Stavanger, Skagen 3	Rogaland	Town sediments	12th–13th	5060	JS 519	Lillehammer 1970

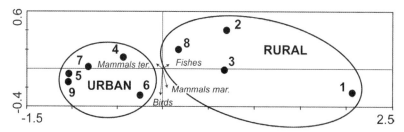

Figure 18.2. Principal component analysis (PCA) illustrating the relative abundance of terrestrial mammal, marine mammal, fish and bird bone in nine Norwegian zooarchaeological assemblages of Viking Age to medieval date. The original quantification was by number of identified specimens (NISP). Rural sites, Storvågan and Kaupang: – 1: Helgøy, 2: Storvågan, 3: Blomsøy, 8: Kaupang. Urban sites – 4: Televerkstomten in Trondheim, 5: Dreggen in Bergen, 6: Mindets tomt in Oslo, 7: Baglergate 3 in Tønsberg, 9: Skagen 3 in Stavanger. Additional details are provided in Table 18.1.

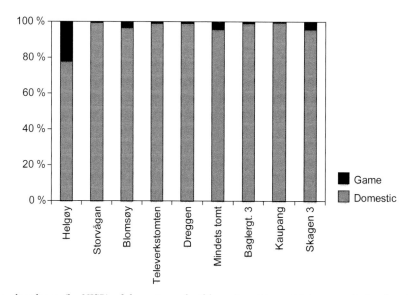

Figure 18.3. Relative abundance (by NISP) of domestic and wild species of terrestrial mammals for the nine assemblages summarised in Table 18.1.

With *råskjær*, in addition to removing the head and intestines, the fish is split to the anal fin, and the anterior part of the vertebral column is removed, as far posterior as three vertebrae behind the anus. This method is traditionally used for large fish or in warm weather, when there is a danger of infestation with fly larvae. By removing the anterior (precaudal) part of the vertebral column, the visceral artery is removed. Thus the probability of spoilage and/or infestation is reduced. Single *råskjær* fish are also hung on a drying rack, one split half on each side of the stock.

In the University of Bergen laboratory, approximately 94 bones of a cod skeleton are routinely identified: 33 from the head, eight from the appendicular ('shoulder') region and 53 from the 'tail' (all vertebrae) (Table 18.2). Thus a finished *rundfisk* should have no cranial bones, eight bones from the 'shoulder' region and approximately

53 vertebrae. Conversely, a *råskjær* fish should contain no cranial bones, eight bones from the appendicular region and approximately 30 vertebrae (all caudal) (Table 18.2). Based on these observations, the representation of different skeletal elements in an archaeological assemblage (based on NISP) can be used to interpret the relative abundance of dried and fresh cod. In Figure 18.5, we employ PCA to summarise variability in the frequency of cranial bones, appendicular bones and vertebrae between a selection of archaeological cod-bone assemblages. The sites discussed above are included, but for this purpose it is also possible to consider additional sieved and dated medieval material from two sites in Trondheim, one in Stavanger, two in Tønsberg and two in Oslo (Table 18.2). One post-medieval site from Oslo (Revierstredet) has also been included to provide a comparison with earlier material from this town.

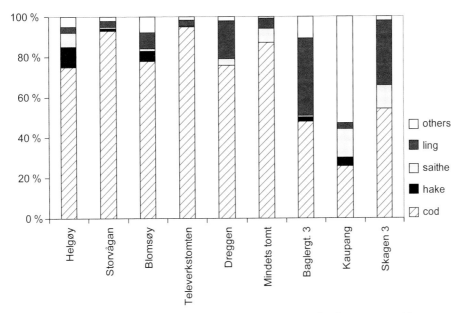

Figure 18.4. Relative abundance (by NISP) of the most common fish species for the nine assemblages summarised in Table 18.1.

Results and discussion

Overall, fish and marine mammal bones are most abundant in the rural assemblages and bird bones are very infrequent in both the rural and the urban collections (Figure 18.2). The only two sites at which birds seem to have been of some importance are Helgøy in northern Norway and Mindets tomt in Oslo. Furthermore, except for Mindets tomt (with its relatively high frequency of birds), the assemblages from the medieval towns cluster together in the PCA results. The relative abundance of the different vertebrate groups is very similar for all the towns, with terrestrial mammals most numerous.

Even for the rural sites and Kaupang, the main protein source was typically terrestrial mammals – in particular cattle and sheep or goat, with game animals playing only a minor role (Figure 18.3). The exception is Helgøy, where more than 90% of the identified bone specimens were of fish. Furthermore, at this site marine mammals also played an important role in the diet, constituting more than 20% of the mammal bones.

Despite the statistical probability that more rare species will be identified in larger zooarchaeological assemblages, there is little correlation between the number of fish bones and the number of fish species identified (Figure 18.6). Nevertheless, except for Baglergaten 3 in Tønsberg and Skagen 3 in Stavanger, the frequency of fish species seems in general to be higher at the rural sites (Helgøy, Storvågen, Blomsøy and Kaupang) than in the towns Trondheim, Bergen and Oslo (Televerkstomten, Dreggen, Mindets tomt I, Mindets tomt III). Thus it appears that more fish species were utilised in the countryside than in urban areas and

that some fish species were only consumed locally, rather than being exported to the towns. For example, among the 3282 fish bones in the Mindets tomt III assemblage, only four fish species have been identified.

In six out of the nine assemblages employed for taxonomic comparison, more than 75% of the identified fish bones were of cod (Figure 18.4). In all except one – namely, Kaupang, with 48% herring (*Clupea harengus*) – cod was by far the dominant fish species. Ling (*Molva molva*) is typically the second most abundant species, having been a significant part of the menu, particularly in Bergen, Stavanger and Tønsberg.

Herring is the most abundant fish species at Kaupang, with cod second in importance. In Norway, the main historical herring fishery focused on spawning populations that moved inshore in late winter and spring. The appearance of spring herring in Norwegian coastal waters is known to be cyclical (e.g. Alheit and Hagen 1997; Devold 1963; see Chapter 2). Their inconsistent archaeological presence may thus relate partly to their natural availability. However, herring bones are also very small, and their recovery is particularly dependent on sampling methods. Although most of the assemblages in this investigation have been sieved, none have been recovered as meticulously as in the recent excavations at Kaupang. Therefore any comparison of the significance of small species, such as herring, between Kaupang and the other sites will probably be biased. The presence of herring at Kaupang merely demonstrates that herring was indeed important there. The absence or low frequency of herring in the other sites could, conversely, be due to several factors: sampling methods, the abundance of the species, or differing dietary preferences.

Table 18.2. The relative abundance (by % NISP) of cranial, appendicular and vertebral bones in complete and preserved cod, and in 17 zooarchaeological assemblages. The chronology varies from Table 18.1 in cases where material from different subsamples has been analysed.

	Head (%)	Shoulder (%)	Vertebrae (%)	No. cod bones	University Museum of Bergen No.	Reference(s)
Complete fish	35	8.5	54.5	94		
Dried cod, Rundfisk	0	13	87	61		
Dried cod, Råskjær	0	21	79	38		
Trondheim, Erkebispegården, AD 1250–1532	3	32	65	144	JS 845	Hufthammer 1999
Trondheim, Folkebibliotekstomten, AD 1000–1125	3	10	87	1137	JS 765	Lie 1989
Trondheim, Televerkstomten, AD 900–1350	0.5	1.5	98	1671	JS 632	Marthinussen 1992
Bergen, Dreggen, AD 1170–1330	16	40	44	184	JS 639	Undheim 1985
Stavanger, Skagen 3, AD 1100–1272	52	12	36	119	JS 519	On file at University Museum of Bergen, Natural History (UMBNH)
Stavanger, Stavanger torg, medieval	24	5	71	150	JS 1398	On file at UMBNH
Tønsberg, Baglergate 3, c. AD 1200–1350	6	14	80	153	JS 644	On file at UMBNH
Tønsberg, Storgate 35, AD 1100–1200	2	2	96	116	JS 563	On file at UMBNH
Tønsberg, Storgate 24/26, AD 1100–1500	31	4	64	245	JS 637	On file at UMBNH
Oslo, Mindets tomt, AD 1075–1225	6	1	93	218	JS 537	Lie 1988
Oslo, Kansler gate, c. AD 1250	16	5	79	192	JS 768	Lie 1991
Oslo, Oslogate 4, AD 1000–1100	19	5	76	224	JS 798	Lie 1991
Oslo, Revierstredet, AD 1624–1730	11	27	62	71	JS 600	Lie 1981
Helgøy, 14th & 15th centuries	50	11	39	375	JS 568	On file at UMBNH
Storvågan, mainly 13th century	42	11	47	202	JS 747	On file at UMBNH
Blomsøy, 13th century	4	6	90	193	JS 818	Hufthammer 1993
Kaupang, Vestfold. 9th–10th centuries	25	2	73	156	JS 1144 JS 1490	Barrett *et al.* 2007; Hufthammer & Bratbak 2000

Cod was probably by far the preferred fish on the menu in medieval Norway (Figure 18.4), but this does not indicate whether it was fresh or dried, locally caught or traded. Assuming, however, that *rundfisk, råskjær* and fresh cod can be identified by the frequency of different skeletal elements (cranial, appendicular and vertebral), it is possible to illustrate the distinctions between assemblages comprised of different fish products using PCA (Figure 18.5). As discussed above, a larger number of sites can be included in this analysis than have been considered thus far (Table 18.2).

Three assemblages from medieval towns have a typical *rundfisk* 'signature': two from Trondheim (Erkebispegården and Folkebibliotekstomten) and one from Tønsberg (Baglergaten 3). Moreover, the assemblages from Bergen Dreggen and Oslo Revierstredet also resemble a stockfish signature, but with more head bones and (in particular at Dreggen) a high frequency of appendicular elements. Based on the zooarchaeological evidence alone, these samples look like what would be expected of assemblages combining *råskjær* and locally caught fish – perhaps both imported dried and local fresh cod, with stockfish being most abundant.

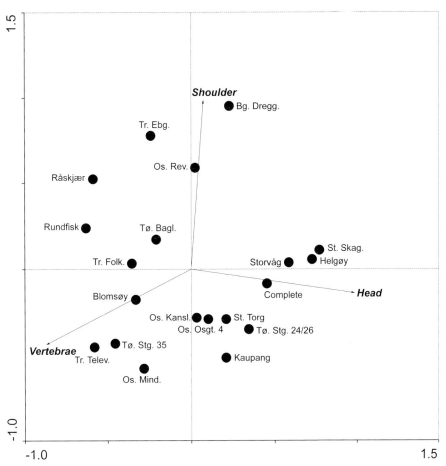

Figure 18.5. Principal component analysis (PCA) based on the frequency (by NISP) of cod cranial (head), appendicular ('shoulder') and vertebral bones present in a complete modern cod, a råskjær *and a* rundfisk, *and at selected archaeological sites: the rural sites Helgøy and Blomsøy; Kaupang and Storvågan; and urban sites in Oslo (Mindets tomt, Kanslergate 10, Oslogate 4, Revierstredet), Tønsberg (Baglergate 3, Storgaten 35, Storgaten 24/26), Stavanger (Skagen 3, Stavanger torg), Bergen (Dreggen), and Trondheim (Televerkstomten, Folkebibliotekstomten, Erkebispegården). Additional details are provided in Table 18.2.*

Cranial bones are left behind when both *rundfisk* and *råskjær* are made. Therefore the fish bone assemblages from Helgøy and Storvågan, with their high relative frequencies of cranial elements, are characteristic of production sites. The percentage of head bones is higher than for a complete cod, as would be expected if the material comprised both fresh fish used for local consumption and heads left over from stockfish production.

Surprisingly, one town assemblage, Skagen 3 from Stavanger, shows bone-element distributions similarly characteristic of a combined production and consumption site. However, in this case the explanation probably relates to sample bias. The assemblage includes only 118 cod bones and was not sieved. Therefore large appendicular and skull bones will be over-represented. When recovery bias is taken into account, the Skagen 3 assemblage is better interpreted as evidence for the consumption of only locally caught cod.

The rural site of Blomsøy (part of the Tjøtta/ Alstahaug complex) (Berglund 1995) and the urban sites of Televerkstomten in Trondheim, Storgaten 35 in Tønsberg and Mindets tomt in Oslo all give mixed signals. Vertebrae are over-represented at these sites, as would be expected if stockfish (in particular *rundfisk*) were present, but skull bones (potentially from locally caught cod) are also abundant. Lastly, the assemblages from Kaupang, from Kanslergata and Oslogate 4 in Oslo, from Storgaten 24/26 in Tønsberg and from Stavanger torg in Stavanger show little or no evidence for the presence of stockfish (Figure 18.5).

Turning to the issue of chronology, Folkebibliotekstomten in Trondheim (dating from AD 1000–1125) is the oldest of the urban assemblages to exhibit a typical stockfish signature, in this case *rundfisk*. Next in date is Dreggen in Bergen (AD 1170–1330), with its combination of *råskjær* and locally caught fish, followed by probable *rundfisk* from Baglergaten

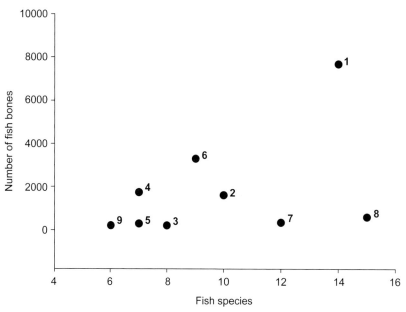

Figure 18.6. The relationship between sample size (NISP) and the number of identified species for the nine assemblages summarised in Table 18.1.

3 in Tønsberg (AD 1200–1350). Moreover, mixed stockfish and fresh-fish signatures were observed in assemblages dating to the twelfth and thirteenth centuries at Tønsberg (Storgaten 35) and Oslo (Mindets tomt). Viking Age (mainly ninth century) Kaupang and the eleventh-century assemblage from Oslogate 4 lacked zooarchaeological evidence of stockfish. The same is true of Kanslergata in Oslo (*c.* AD 1250) and the more broadly dated assemblages Storgaten 24/26 in Tønsberg and Stavanger torg in Stavanger. The seventeenth- to eighteenth-century assemblage from Oslo Revierstredet provides clear zooarchaeological evidence for preserved cod. Appendicular elements are over-represented, suggesting the presence of *råskjær*. At this date, however, it could also be *klippfisk*, a salted and dried product butchered in a similar way that began to be produced in the 1690s (Nielssen 2009, 90).

Conclusions

Marine fish bones are abundant in Viking Age and medieval sites from Norway (with the caveat that bone is not usually well preserved at inland sites, where this pattern could differ). The rural sites considered have more fish bone *vis-à-vis* mammal and bird remains than do the medieval towns, with Kaupang being intermediate between the two (Figure 18.2). Cod is typically the most important species, often followed by ling and saithe (*Pollachius virens*). Only at Kaupang, where herring bones are comparatively well represented, is a species other than cod the most abundant fish. Marine mammals are present at all sites except

Kaupang, but typically in very small numbers. They are most common in Helgøy, the northernmost site.

The anatomical distributions within the fish bone assemblages suggest that stockfish (in the form of *rundfisk* and/or *råskjær*) was produced at the northern Norwegian sites. Although the chronology is not tight, this pattern is certainly evident by the thirteenth century and can probably be extrapolated back to the Viking Age or earlier (Table 18.1). In Trondheim, the only town in central Norway, it seems that stockfish (presumably transported from the north) were important already in the eleventh to twelfth centuries and continued to be so. They constituted a significant component of all three assemblages from this town included in the present investigation. The mixed signal from between AD 1100 and AD 1200 in Tønsberg (Storgaten 35) illustrates that stockfish were also an early element in the diet there. The eleventh–fourteenth-century material from Dreggen in Bergen is not so tightly dated, but provides a clear anatomical picture of what is probably *råskjær*. Overall, no zooarchaeological evidence of stockfish clearly pre-dates the twelfth century in any of the towns of southern Norway. Thus it would seem that stockfish became a widespread element in the diet of Norwegian town dwellers in the twelfth–thirteenth centuries. There are, however, exceptions. No stockfish are represented in the two medieval settlements in Stavanger, for example, where demand was presumably met with locally caught fish. The post-medieval (seventeenth–eighteenth-century) assemblage from Oslo Revierstredet included either stockfish or *klippfisk*.

Acknowledgements

This work was initiated by James Barrett which is gratefully acknowledged. I also thank Olaug Flatnes Bratbak and Tore Fredriksen for technical assistance, Kari Loe Hjelle for running the Principal Component Analysis and an anonymous reviewer for comments on the manuscript and language.

References

Alheit, J. & E. Hagen. 1997. Long-term climate forcing of European herring and sardine populations. *Fisheries Oceanography* 6(2): 130–9.

Barrett, J. H., A. K. Hufthammer & O. Bratbak. 2015. Animals and animal products at the Late Iron Age settlement of Bjørkum, Lærdal: the zooarchaeological evidence. Report on file, University Museum of Bergen.

Barrett, J. H., A. Hall, C. Johnstone, H. Kenward, T. O'Connor & S. Ashby. 2007. Interpreting the plant and animal remains from Viking-age Kaupang, in D. Skre (ed.) *Kaupang in Skiringssal*: 283–319. (Norske Oldfunn 22, Kaupang Excavation Project Publication Series). Aarhus: Aarhus University Press.

Barrett, J. H., C. Johnstone, J. Harland, W. Van Neer, A. Ervynck, D. Makowiecki, D. Heinrich, A. K. Hufthammer, I. B. Enghoff, C. Amundsen, J. S. Christiansen, A. K. G. Jones, A. Locker, S. Hamilton-Dyer, L. Jonsson, L. Lõugas, C. Roberts & M. Richards. 2008. Detecting the medieval cod trade: a new method and first results. *Journal of Archaeological Science* 35(4): 850–61.

Berglund, B. 1995. Tjøtta-riket: en arkeologisk undersøkelse av maktforhold og sentrumsdannelse på Helgelandskysten fra Kr.f. til 1700 e.Kr. Unpublished PhD dissertation, University of Trondheim.

Bertelsen, R. 1977. Lofotfisket og Vågan i vikingtid og mellomalder, in S. Høgmo (ed.) *Lofoten i går, idag, i morgen. Volume 2*: 41–51. Tromsø: Universitetet i Tromsø.

Bertelsen, R. 1985. The medieval Vågan: an Arctic urban experiment. *Archaeology and Environment* 4: 49–56.

Brendalsmo, J. 1983. *Rapport over de arkeologiske utgravningene i Baglergaten 3, Tønsberg, 1979–80*. Riksantikvaren: Utgravningskontoret for Tønsberg.

Brinkmann, A. & H. Shetelig. 1920. *Ruskenesset en stenalders jaktplass* (Norske Oldfund 3). Kristiania: Norske Arkeologiske Selskap.

Degerbøl, M. 1951. Det osteologiske materialet, in H. E. Lund (ed.) *Fangst-boplassen i Vistehulen på Viste i Randaberg, Nord-Jæren*: 52–93. Stavanger: Stavanger Museum.

Devold, F. 1963. The life history of the Atlanto-Scandian herring, in *Rapports et procès-verbaux des réunions*. Vol. 154–5: 98–108. Copenhagen: Conseil permanent international pour l'exploration de la mer.

Grøn, F. 1926. *Om kostholdet i Norge indtil aar 1500*. (1984 facsimile). Oslo: Kildeforlaget.

Heinrich, D. 2006. Die Fischreste aus dem Hafen von Haithabu – handaufgelesene funde, in U. Schmölcke, K. Schietzel, D. Heinrich, H. Hüster-Plogmann & K. J. Hüser, *Untersuchungen an Skelettresten von Tieren aus dem Hafen von Haithabu*: 157–93 (Berichte über die Ausgrabungen in Haithabu 35). Neumünster: Wachholtz Verlag.

Helle, K. 1982. *Bergen bys historie. Volume 1: kongesete og kjøpstad: fra opphavet til 1536*. Bergen, Oslo, Tromsø: Universitetsforlaget.

Helskog, E. T. 1983. *The Iversfjord Locality: a study of behavioural patterning during the late Stone Age of Finnmark, north Norway*. (Tromsø Museums Skrifter vol. 19). Tromsø: Universitetet i Tromsø/Tromsø Museum.

Holm-Olsen, I. M. 1977. Foreløpig innberetning om utgravning av gårdshaug på Helgøy, gnr. 33, br. nr 3, Karlsøy kommune, Troms, 1975 og 1976. Report on file, University of Tromsø.

Hufthammer, A. K. 1992. De osteologiske undersøkelsene fra Kotedalen, in K. L. Hjelle, A. K. Hufthammer, P. E. Kaland, A. B. Olsen & E. C. Soltvedt (eds) *Kotedalen – en boplass gjennom 5000 år. Bind 2: Naturvitenskaplige undersøkelser*: 9–64. Bergen: Historisk Museum, University of Bergen.

Hufthammer, A. K. 1993. Beinmateriale fra gårdshaugene på Helgeland. Report on file, University Museum of Bergen.

Hufthammer, A. K. 1997. The vertebrate faunal remains from Auve – a palaeoecological investigation, in E. Østmo, B. Hulthén, S. Isaksson, A. K. Hufthammer, R. Sørensen, S. Bakkevik and M. S. Thomsen (eds) *Auve bind II. Tekniske og naturvitenskapelige undersøkelser*: 43–59. (Norske Oldfunn 17). Oslo: Institutt for arkeologi, kunsthistorie og numismatikk, Universitetets Oldsaksamling.

Hufthammer, A. K. 1998. The use of vertebrate fauna remains in the interpretation of subsistence strategy and settlement patterns, with emphasis on fish and bird bones: a case study from Kotedalen, western Norway. *Archaeologica Baltica* 3: 109–19.

Hufthammer, A. K. 1999. *Utgravningene i Erkebispegården i Trondheim: kosthold og erverv i Erkebispegården: en osteologisk analyse*. (Norsk Institutt for Kulturminneforskning Temahefte 17). Oslo: Norsk Institutt for Kulturminneforskning.

Hufthammer, A. K. 2000. Kosthold hos overklassen og hos vanlige husholdninger I middelalderen. En sammenligning mellom animalosteologisk materiale fra Trondheim og Oslo, in A. Dybdahl (ed.) *Osteologisk materiale som historisk kilde*: 163–87. Trondheim: Tapir Akademisk Forlag.

Hufthammer, A. K. 2003. Med kjøtt og fisk på menyen, in O. Skevik (ed.), *Middelaldergården i Trøndelag. Foredrag fra to seminarer*: 182–96. Stiklestad Nasjonale Kultursenter AS.

Hufthammer, A. K. & O. F. Bratbak. 2000. Bones from the year 2000 excavation at the Kaupang, Tjølling site. Report on file, University Museum of Bergen.

Johansen, O. S. 1982. Viking Age farms: estimating the number and population size. *Norwegian Archaeological Review* 15(1–2): 45–69.

Lepiksaar, J. & D. Heinrich. 1977. *Untersuchungen an Fischresten aus der frühmittelalterlichen Siedlung, Haithabu*. (Berichte über die Ausgrabungen in Haithabu 10). Neumünster: Wachholtz Verlag.

Leps, J. & P. Šmilauer. 2004. *Multivariate Analysis of Ecological Data using CANOCO*. Cambridge: Cambridge University Press.

Lie, R. W. 1981. Dyrebein, in E. Schia (ed.) *Fra Christianias bygrunn: arkeologiske utgravninger i Revierstredet 5–7*: 257–71. Øvre Ervik: Alvheim & Eide.

Lie, R. W. 1988. Animal bones, in E. Schia (ed.) *De arkeologiske utgravninger i Gamlebyen, Oslo*. Vol. 5: 153–96. Øvre Ervik: Universitetsforlaget.

Lie, R. W. 1989. *Dyr i byen: en osteologisk analyse.* (Meddelelser fra Prosjektet Fortiden i Trondheim Bygrunn, Folkebibliotekstomten 18). Trondheim: Riksantikvaren.

Lie, R. W. 1991. Dyrebein fra Oslogt. 4 og Kanslergt. 10, in E. Schia & T. Wiberg (eds) *De arkeologiske utgravninger i Gamlebyen, Oslo. Volume 10: grøftegravninger:* 75–84. Øvre Ervik: Alvheim & Eide.

Lillehammer, A. 1970. Melding om den arkeologiske utgravninga på tomta Skagen 3 i Stavanger sommaren 1968. Report on file, Archaeological Museum, Stavanger.

Marthinussen, K. L. 1992. Et osteologisk materiale fra Televekstomten: undersøkelse av et utgravd beinmateriale fra middelalderens Trondheim (Nordre gt 1). Unpublished Cand. Scient. thesis, University of Bergen.

Nedkvitne, A. 1976. Omfanget av tørrfiskeksporten fra Bergen på 1300-tallet. *Historisk Tidskrift* 33: 340–55.

Nedkvitne, A. 1977. *Handelssjøfarten mellom Norge og England i høymiddelalderen.* (Norwegian Yearbook of Maritime History 1976). Bergen: Sjøfartsmuseum.

Nedkvitne, A. 1988. *Mens bønderne seilte og jægterne for: nordnorsk og vestnorsk kystøkonomi 1500–1730.* Oslo: Universitetsforlaget.

Nedkvitne, A. 2014. *The German Hansa and Bergen 1100–1600.* (Quellen und Darstellungen zur Hansischen Geschichte, Neue Folge 70). Vienna and Cologne: Böhlau.

Nielssen, A. R. 2009. Norwegian fisheries *c.* 1100–1850, in D. J. Starkey, J. T. Thór & I. Heidbrink (eds) *A History of the North Atlantic Fisheries. Volume 1: from early times to the midnineteenth century.* Bremen: H. M. Hauschild.

Nordeide, S. W. 2003. *Erkebispegården i Trondheim: beste tomta i by'n.* Trondheim: Norsk Institutt for Kulturminneforskning.

Olsen, K. 1967. *Osteologiske materiale: innledning – fisk – fugl.* (Varanger-funnene 4). Bergen, Oslo, Tromsø: Universitetsforlaget.

Olsen, K. 1976. Skipshelleren – osteologisk materiale. Report on file, Zoologisk Museum, University of Bergen.

Pilø, L. 2007. The settlement: extent and dating, in D. Skre (ed.) *Kaupang in Skiringssal: 161–78. (Norske Oldfunn 22, Kaupang Excavation Project Publication Series).* Aarhus: Aarhus University Press.

Prescott, C. 1993. From the Stone Age to Iron Age: a study from Sogn, Western Norway. Unpublished PhD dissertation, Historical Museum, University of Bergen.

Rathke, J. 1907. *Afhandling om de norske fiskerier og beretninger om reiser i årene 1795–1802 for at studere fiskeriforhold m.v.* Bergen: Selskabet for de Norske Fiskeriers Fremme.

Skre, D. 2008. Post-substantivist towns and trade: AD 600–1000, in D. Skre (ed.) *Means of Exchange: dealing with silver in the Viking Age:* 327–41. Aarhus: Aarhus University Press.

Solhaug, T. 1976. *De norske fiskeriers historie: 1815–1880.* Bergen: Universitetsforlaget.

Tande, T. & T. Tande. 1986. *Norsk tørrfisknærings historie.* Askim: Norges Tørrfiskeksportørers Landsforening.

Undheim, P. 1985. Osteologisk materiale fra Dreggen: en økologisk studie fra middelalderens Bergen. Unpublished Cand. Scient. thesis, University of Bergen.

Exploring the Contrasts: Fish-Bone Assemblages from Medieval Ireland

Sheila Hamilton-Dyer

Introduction

Fish-bone assemblages from medieval Ireland (defined as fifth to sixteenth centuries for present purposes) vary greatly in size, from small collections from coastal ringforts (enclosed rural settlements) to groups of many thousands of specimens from major urban centres. Few assemblages are from sites away from the coast (Table 19.1; see Figure 9.1 in Chapter 9 for site locations). The taxa represented are also varied. This chapter provides a brief survey of the most significant discoveries. The distribution of the main taxonomic groups (by number of identified bone specimens) at the sites to be discussed is given in Table 19.1, and a breakdown of the different gadid (cod family) species is provided in Table 19.2. For the sake of convenience, the term gadid here includes the closely related hake (*Merluccius merluccius*).

For the early centuries (fifth–twelfth) of present concern, the only inland assemblage currently available is from the monastery of Clonmacnoise on the River Shannon in Co. Offaly. This important site is situated in the very centre of Ireland and has produced the only fish-bone assemblage examined that does not have any obligate marine species. Salmon (*Salmo salar*), together with eel (*Anguilla anguilla*) and shad (*Alosa* sp.), were the only taxa identified (Hamilton-Dyer 2011). All of these species are migratory and could have been caught from the river.

Other early medieval assemblages are from sites excavated on the western, that is, Atlantic, coast of Ireland. These include ringforts and monasteries. The fish-bone assemblages are often small, and those from older excavations are not reported in detail. All have a wide variety of species represented, but sea bream (*Sparidae*), wrasse (Labridae), Atlantic horse mackerel/scad (*Trachurus trachurus*) and other local inshore fish are typically most abundant. Gadids, apart from some pollack (*Pollachius pollachius*) and whiting (*Merlangius merlangus*), tend to be absent or rare. Omey and Doonloughan are typical of the small coastal sites where sieving has been undertaken. The ninth-century AD deposits at Doonloughan contained mainly wrasse, sea bream and gurnard (Triglidae) (Hamilton-Dyer 2012), while Omey (*c.* AD 1000) produced mainly Atlantic horse mackerel/scad and wrasse (Hamilton-Dyer 1994a). In recent years excavation of the island monastery of Illaunloughan on the southwest coast has produced a much larger faunal assemblage of mainly eighth–ninth-century date (Hamilton-Dyer 2005). The fish remains recovered are mainly of sea bream and wrasse, but the 18 species present also include examples such as hake, pollack, Atlantic horse mackerel/scad, seabass, conger (*Conger conger*) and cod (*Gadus morhua*) (Table 19.3). Complete fish are indicated by the anatomical distribution, where numbers are sufficient to allow a judgment in this regard. Most of the 'gadid' remains are of pollack and hake; cod is represented by only a few bones. The monastery was not self-sufficient, but the fact that the food refuse includes remains of birds from the local seabird colonies could suggest that the fish supply was local too. Some of the cod bones are of fish over 1 m in length, but as these are head bones, they are unlikely to be of imported, that is, processed, cod. The eremitic island settlement on Skellig Michael produced a small assemblage also mainly of pollack, whiting and sea bream (Hamilton-Dyer 2011b).

At the slightly later (eleventh–twelfth-century) monastic site of Staad Abbey on the northwest coast, wrasse are the most frequent taxa in the small sample available (Hamilton-Dyer 1994b). The seven species present do include four 'gadids': cod, ling (*Molva molva*), whiting and hake. While the two bones of ling are from large fish, the single cod vertebra is from a small individual and, again, all the fish are likely to have been sourced locally. Pollack is the main species in a twelfth–thirteenth-century deposit from St Thomas' Church, Rathlin Island, off the northeast coast of Ireland (Hamilton-Dyer 2008). This species favours rocky coasts and is also likely to indicate exploitation of a purely local resource.

Table 19.1. Number of identified specimens by taxon for a selection of fish-bone assemblages from Ireland (see text for references and Figure 9.1 in Chapter 9 for site locations).

Site	Herring	Salmonids	Cod & large Gadidae	Pollack & whiting	Gadidae indet.	Gurnard, scad, sea bream, wrasse	Other species	Total identified	Indeterminate
Clonmacnoise		145					423	568	230
Doonloughan		8		17		315	36	376	2381
Omey			3	8		341		352	334
Illaunloughan		2	139	99	183	1987	118	2528	2142
Skellig Michael			1	19	4	56		80	148
Staad Abbey			9	10		76	5	100	231
Rathlin Island	18			33	236	43	3	333	257
Galway City medieval	154	6	1008	548	357	870	512	3455	3180
Galway City post-medieval	16	4	311	45	57	42	30	505	383
Dublin Arran Quay			692		397	1	19	1109	
Dublin other sites			150		7	4	24	185	34
Trim Castle	8	3	342	31	114	17	82	597	563
Maynooth Castle		1	116			33	68	218	133
Knowth late Christian		18	3					21	34
Knowth Anglo-Norman		4	12					16	8
Cork medieval sites	104	2	504	59		37	165	871	?
Cork Christ Church			766	45		3	33	847	52
Wexford Viking	8		116	22	84	11	9	250	?
Wexford medieval	168		1284	85	348	107	56	2048	?
Dungarvan			499			10	1	510	221
Total	476	193	5955	1021	1787	3953	1584	14,969	10,331

The majority of urban sites investigated are ports. Mention should be made at this point of the ninth/tenth–twelfth-century Hiberno-Scandinavian settlements in Ireland. Major trading settlements were established at Dublin, Wexford, Waterford and Cork on the eastern/southeastern seaboard and at Limerick in the west. To date, few excavations of Scandinavian-period occupations have resulted in fish-bone analyses. Exceptions include a small sieved assemblage from Bride Street, Wexford, which yielded mainly cod and pollack, and early medieval contexts of excavations at Dublin Castle, which also yielded sea-fish bones (McCarthy 1998, 61; pers. comm.). A glimpse of the fish species consumed in Hiberno-Scandinavian Dublin during the tenth and eleventh centuries AD is also provided by bones incidentally identified during the analysis of botanical samples from Fishamble Street (Geraghty 1996). These included herring, saithe and ling.

On the western seaboard lies the urban port of Galway City, at the mouth of the River Corrib. Excavations at Courthouse Lane recovered a large assemblage of medieval date (c. AD 1100–c. 1350). Cod is much more in evidence than in the earlier, non-urban, assemblages on the west coast, but bones of whiting, gurnards and scad are also common finds from this site (Hamilton-Dyer 2004a). It is interesting to note that the post-medieval assemblage contained a much higher proportion of cod and hake (Table 19.4). At a similar latitude, but on the east coast, Dublin has also yielded assemblages of medieval and post-medieval date. Most are small assemblages that were mainly recovered by hand collection, which likely has introduced a bias against small and rare taxa. Thus it may not be surprising that they are all dominated by bones of gadids, mainly cod, ling and haddock (Hamilton-Dyer 1993; 1996; 1997). These remains represent large fish, the cod and ling often well over 1 m in length, unlike most of those from the early medieval sites discussed above, and may thus imply the emergence of fishing in deeper water (see Chapter 9).

Table 19.2. Number of identified specimens of cod family taxa and hake for a selection of fish-bone assemblages from Ireland (see text for references).

Site	Hake	Cod	Haddock	Ling	Pollack	Saithe	Whiting	Indet. Gadidae	Total Gadidae	Total identified fish	% of identified fish
Clonmacnoise									0	115	0
Doonloughan				17					17	376	4.5
Omey		2		1			8		11	352	3.1
Illaunloughan	120	17	1	88	1		11	183	301	2528	11.9
Skellig Michael		1		12			7	4	24	80	30.0
Staad Abbey	6	1		2			10		13	100	13.0
Rathlin Island				33				236	269	332	81.0
Galway City medieval	50	639	30	289	208		340	357	1863	3455	53.9
Galway City post-medieval	88	166	15	42	33		12	57	325	505	64.4
Dublin Arran Quay	1	148		543				397	1088	1109	98.1
Dublin Back Lane	5	14	2	17					33	58	56.9
Dublin Bridge St		25	10	8					43	44	97.7
Dublin Cornmarket		3	3	6					12	14	85.7
Dublin Nicholas St	3	11	3	40				7	61	69	88.4
Trim Castle	32	190	26	94			31	114	455	597	76.2
Maynooth Castle	96	9	11						20	218	9.2
Knowth late Christian		2	1						3	21	14.3
Knowth Anglo-Norman		11	1						12	16	75.0
Cork medieval sites	244	134	24	102	2		57		319	871	36.6
Cork Christ Church	544	162	2	58	45				267	847	31.5
Wexford Viking	9	101	3	3	19		3	84	213	250	85.2
Wexford medieval	240	918	6	117	48	3	37	348	1477	2048	72.1
Dungarvan	485	14							14	510	2.7
Total	1923	2568	138	1322	505	4	516	1787	6840	10,774	

At Arran Quay in Dublin, a late-medieval (fifteenth–early sixteenth-century) deposit was excavated from adjacent to the stone quay wall (Finbar McCormick pers. comm.). It consists largely of ling and cod heads (Hamilton-Dyer 2004b). Abdominal vertebrae (those nearer the head) are present but underrepresented. Caudal (tail) vertebrae and cleithra (a bone that supports the pectoral fin, behind the head) are almost entirely absent (Table 19.5). These remains may have been from fish prepared for either local use or export. The town of Chester in northwest England was one destination for fish from Dublin and nearby Drogheda (Vanes 1979, 14).

At a distance of some 40 km inland from Dublin lies Trim, a major high-status Anglo-Norman castle. Fish-bone assemblages dating to the late thirteenth–early fourteenth and mid-fourteenth–mid-fifteenth centuries AD show several interesting characteristics (Table 19.6). They are dominated by cod, and several other gadid species are represented among the 14 taxa identified (Hamilton-Dyer 2011c). In this aspect, they are similar to the Dublin assemblages. There is, however, a difference in the gadid representation between the two phases of the site. Cod and ling are most frequent in both phases, with ling approximately half as abundant as cod. However, there are fewer bones of other 'gadids' – including haddock, whiting and hake – in the later phase. The cod, ling and haddock bones are mainly of large fish. Butchery marks include cuts across cleithra and precaudal (abdominal) vertebrae consistent with beheading. This

Table 19.3. Distribution of taxa and anatomical elements (by number of identified specimens) for the eighth–ninth-century fish-bone assemblage from Illaunloughan.

Element	Conger	Hake	Cod	Pollack	Other Gadidae	Indet. Gadidae	Seabass	Atlantic horse mackerel/ scad	Sea bream	Wrasse	Total
Basioccipital			3	8					15	2	28
Supraoccipital									19	1	20
Posttemporal									5	10	15
Parasphenoid	1		1	6					42	4	54
Vomer	2		1	2					3	9	17
Pharyngeal										64	64
Premaxilla	3	10	2	4			2		30	63	114
Maxilla		2		5	1				19	31	58
Dentary	4	4		8	1		4	1	20	47	89
Articular	1	4		23					37	33	98
Ceratohyal	5	1							16	1	23
Urohyal	1									5	6
Quadrate			2	2					12	10	26
Hyomandibular				2					24	11	37
Preopercular									3	7	10
Opercular	2								14	4	20
Subopercular										1	1
Other cranial	1								12	32	45
Scute								6			6
Cleithrum	2			2					3	2	9
Other pectoral				1							1
Abdominal vertebra	13	72	2	13		84	7	4	251	131	577
Caudal vertebra		27	6	12		58	27	70	708	189	1097
Vertebral fragment						60					60
Total	35	120	17	88	2	202	40	81	1233	657	2475

does not, however, indicate the presence or manufacture of preserved fish at this site, as both head bones and vertebrae are frequent. Instead, large, complete fish were being delivered to Trim's kitchens. All the fish were probably fresh, as the castle is close enough to Drogheda for rapid transport up the River Boyne. The fish remains also included three bones of pike (*Esox lucius*), a significant find. Ireland has an extremely limited suite of native freshwater fish, and pike is one of a number of species present today that were almost certainly introduced. These finds are, to date, the earliest evidence for this species in Ireland and underline the high status of this castle.

Maynooth is another high-status Anglo-Norman castle in the area, in this case 24 km to the southwest of Dublin. A thirteenth–fourteenth-century sample from a well at Maynooth was dominated by bones of hake rather than cod (Hamilton-Dyer 2000). Cod bones were few, and they were outnumbered by those of gurnards and flatfish (Pleuronectidae). The mammal remains, although indicating high-status occupation, also differed from those at Trim.

Also in this area is Knowth, sister site to the famous prehistoric site of Newgrange. The medieval samples available from Knowth are rather small, but they offer a significant contrast between the two phases (Hamilton-Dyer 2007). The 'later Early Christian' phase (tenth century–AD 1169) is dominated by salmon and contains just a few bones of cod and haddock, while the situation is reversed for the later Anglo-Norman (post-AD 1169) phase. Apart from the earlier assemblage from Clonmacnoise, this is the only site with significant amounts of salmon. This is of interest because medieval documents might lead one to assume that salmon were of paramount importance. Legal

Table 19.4. Number of identified specimens by taxon for medieval and post-medieval fish-bone assemblages from Galway.

Taxon	Medieval	Post-medieval	Total
Sharks & rays	406	20	426
Eel	31	1	32
Conger	12	1	13
Herring	154	16	170
Shad		1	1
Salmon	6	4	10
Gadidae indeterminate	357	57	414
Cod	639	166	805
Whiting	340	12	352
Pollack	208	33	241
Haddock	30	15	45
Ling	289	42	331
Hake	50	88	138
Gurnard	683	35	718
Atlantic horse mackerel/scad	131	2	133
Sea bream		1	1
Mullet	7		7
Wrasse	56	4	60
Mackerel	12		12
Flatfish	44	7	51
Total identified	3455	505	3960
Indeterminate	3180	383	3563
Total	6635	888	7523

Table 19.5. Distribution of anatomical elements (by number of identified specimens) for a fifteenth- to early sixteenth-century fish-bone assemblage from Arran Quay in Dublin.

Element	Hake	Cod	Ling	Gadidae	Total
Basioccipital		1	8		9
Supraoccipital			5		5
Posttemporal		3	6		9
Parasphenoid		3	17		20
Frontal		1	23		24
Vomer		2	5		7
Premaxilla		9	17		26
Maxilla		13	28		41
Dentary		22	50		72
Articular		11	54		65
Ceratohyal		9	45		54
Other hyal		6	19	2	27
All branchial				286	286
Quadrate		1	10		11
Hyomandibular		8	16		24
Preopercular		9	21		30
Opercular		6	13		19
Other opercular		8	11		19
Other cranial		8	47	100	155
Otolith		1			1
Cleithrum		1			1
Supracleithrum		4	13		17
Atlas			8		8
Abdominal vertebra		25	131		156
Caudal vertebra	1	2			3
Total	1	151	549	388	1089

tracts often discuss fishing rights and control of fish weirs, mainly concerning salmon (Kelly 1998). Marine fish are very rarely mentioned in these secular documents, nor in the ecclesiastical ones (Murray *et al.* 2004), yet the evidence shows that they, and not fish caught in fresh water, appear to have been the major resource.

In the south, Cork has urban assemblages of eleventh–early fourteenth-century date that are again dominated by 'gadids'. Hake is the most frequent species in this case, followed by cod and then ling (McCarthy 2003). At Christ Church, a mid-thirteenth-century assemblage represents the waste from a single household. In this hand-collected sample hake account for 60% of the identified total and cod for much of the remainder. The use of whole fish is indicated by the anatomical distribution (McCarthy 1997). At Wexford, cod is the most frequent species, with hake second (McCarthy pers. comm.). Hake was clearly available off the southern and western coasts of Ireland, and a favoured catch.

A great deal of the medieval hake trade was carried out by fishermen from Devon and Cornwall, who took salt with them to process the fish before returning home (Kowaleski 2000; 2003). Shore camps are indicated by documentary evidence, in Dingle Bay, for example. The fishermen were not only supplying their home ports, but also markets further afield. Bristol, in England, also had significant trade with all the ports of Ireland's southern coast (e.g. Wexford, Waterford, Cork, Kinsale, Youghal and Dungarvan), and fish was the major commodity (Childs and O'Neill 1993; Vanes 1979). From the end of the fifteenth century Continental fishermen (from Iberia and Brittany, for example) also became heavily involved in fishing Irish waters (Chapter 9). The amount of hake and other fish being taken, and the potential loss of income for the English, can be gauged by

Table 19.6. Number of identified specimens by taxon for medieval fish-bone assemblages from Trim Castle.

Taxon	late 13th to early 14th	mid-14th to mid-15th	Total
Shark & ray		1	1
Conger	8	10	18
Herring	1	7	8
Pike	2	1	3
Salmon		3	3
Gadidae	42	72	114
Cod	82	107	189
Whiting	23	8	31
Haddock	18	8	26
Ling	38	56	94
Hake	24	8	32
Gurnard	3	4	7
Sea bream	5	1	6
Wrasse	4		4
Flatfish	20	40	60
Indeterminate	230	333	563
Total	500	659	1159

the imposition of various regulations, including one in AD 1516 requiring foreign fishermen to land at least a third of their catch in Ireland (Kowaleski 2003). The continuing importance of hake is indicated by the post-medieval assemblage from Dungarvan, Co. Waterford, in which this species makes up 95% of the identified fish remains in the sieved samples (McCarthy 1995). The availability of hake probably fluctuated, dropping during the eighteenth century in Waterford Bay, for example (Smith 1746). It remains a (sometimes overexploited) target of commercial fishing today (cf. FAO 1997; 2011).

Where the bones of cod and other large gadids are found in any quantity it is illuminating to examine the pattern of anatomical distribution. With the exception of Arran Quay, Dublin, where heads predominate, it is mostly whole fish that are represented. Vertebrae (especially the smaller caudal ones) are under-represented, but this appears to be at least partly due to recovery bias, as several of these assemblages were mostly hand collected, with little or no sieving. Where butchery marks indicate beheading of fish by cutting through the cleithrum, both the cranial and the caudal portions are present. This implies that many of these fish were prepared and used in the local area rather than representing the production or consumption of traded fish (cf. Barrett 1997). A few remains of prepared traded fish may be included in some assemblages, but these are not obvious from the collections examined so far.

Finally, mention must be made of herring. This small but economically very important species is found at some Irish sites. It has not, however, been discussed in detail because at many sites its abundance and element distribution are unlikely to reflect what was in the ground immediately prior to excavation. Many of the assemblages were hand collected, and the small bones of herring would mostly have been missed. However, even where sieving was undertaken, herring has only been found with any regularity in Viking Age Dublin (based on Fishamble Street) and at major High-medieval and post-medieval settlements such as Galway City, Cork, Waterford, Wexford and Trim Castle. Even at these sites, herring has not been found in large quantities – with the exception of a water-sieved sample (1 mm mesh) from a seventeenth–eighteenth-century dump relating to herring curing at John Street in Waterford (Tourunen 2008). Herring was fished for off the southern Irish coast as well as in the Irish Sea and further north, but thus far there is little fish-bone evidence for its widespread consumption in Ireland itself. This pattern is probably a recovery bias in part, but much herring fishing was carried out by non-Irish boats, often from the same home ports as were used for hake fishing. Boats from Scotland are recorded as trading herring in Dublin and Drogheda in AD 1306 (O'Neill 1987), and much herring was imported into Chester, Bristol, Minehead and Padstow (Kowaleski 2003). Many fishermen also came from France and Spain (Chapter 9). Presumably much of this fish only passed through the Irish ports. The exceptional find of a sixteenth-century wreck in the River Boyne at Drogheda contained barrels of herring, probably being ported to a larger ship (Harland 2009; Chapter 15). Local herring processing in more recent centuries is demonstrated by the John Street, Waterford, assemblage noted above (Tourunen 2008).

Conclusions

In summary, the zooarchaeological evidence suggests that fishing for cod and other large gadids in medieval Ireland was initially of relatively low intensity, but that this intensity increased through time. The paucity of analysed Viking Age assemblages is a serious lacuna. It would be most interesting to know whether these were consistently dominated by gadids. The small collection from Wexford is dominated by cod and pollack, and sediment samples from Fishamble Street in Dublin contain both gadids and herring. Nevertheless, Irish fish exploitation during the Viking Age cannot be extrapolated from the limited available evidence. There is more fish-bone evidence from urban contexts post-dating the Norman invasion. Late- and post-medieval sites do seem to have more gadids, but even in these cases the assemblages are limited in size and geographical spread. At present, the broad range and late dates of many urban deposits examined limit the possibilities of detecting subtle changes in fish exploitation and trade in the medieval period. One notable spatial pattern is the high proportion of small, local species in medieval Galway and the limited evidence for cod and other large gadids from this town. Customs duties for AD 1276–1333 indicate that the majority of Ireland's trade went through New Ross and Waterford, with a further 42% going through Cork, Drogheda and Dublin. The remaining ports, including Galway, accounted for only 7.5% of trade (Graham 1987). Thus it is possible that limited options for export influenced the strategies of fishermen supplying Galway.

The position regarding herring is also unclear. On the current zooarchaeological evidence, its exploitation appears to have been less important for local consumption than for export. However, this interpretation may have to be revised once more sieved assemblages are analysed. There are also many areas of Ireland that have not yet had any substantial fish-bone assemblages analysed or published, or that have gaps by site type and period. There is great potential for further research.

Acknowledgements

The majority of the results surveyed in this paper derive from research commissioned in the context of professional contract archaeology. I thank the project funders, and I am particularly grateful to Finbar McCormick and Emily Murray of Queen's University Belfast, and to Heather King, for facilitating this work and providing invaluable information regarding the archaeological contexts of the finds. Thanks are also due to Margaret McCarthy for her reports on material from Cork and Wexford.

References

Barrett, J. H. 1997. Fish trade in Norse Orkney and Caithness: a zooarchaeological approach *Antiquity* 71: 616–38.

Childs, W. & T. O'Neill. 1993. Overseas Trade, in A. Cosgrove (ed.) *A New History of Ireland. Volume 2: medieval Ireland 1169–1534*: 492–524. Oxford: Clarendon Press.

FAO Food and Agriculture Organization of the United Nations. 1997. *Review of the State of World Fishery Resources: marine fisheries*. (FAO Fisheries Circular No. 920 FIRM/C920). Rome: Food and Agriculture Organization of the United Nations.

FAO Food and Agriculture Organization of the United Nations. 2011. *Review of the State of World Marine Fishery Resources*. (FAO Fisheries and Aquaculture Technical Paper 569). Rome: Food and Agriculture Organization of the United Nations.

Geraghty, S. 1996. *Viking Dublin: botanical evidence from Fishamble Street*. Dublin: Royal Irish Academy.

Graham, B. J. 1987. Urban genesis in early medieval Ireland. *Journal of Historical Geography* 13(1): 3–16.

Hamilton-Dyer, S. 1993. Bird and fish bones from Back Lane, Dublin. Report prepared for F. McCormick, Queen's University Belfast.

Hamilton-Dyer, S. 1994a. Fish and marine invertebrates from Omey, Galway. Report prepared for F. McCormick, Queen's University Belfast.

Hamilton-Dyer, S. 1994b. Bird, fish and marine invertebrates from Staad Abbey, Co Sligo. Report prepared for F. McCormick, Queen's University Belfast.

Hamilton-Dyer, S. 1996. Bird and fish bones from Cornmarket and Bridge Street, Dublin. Report prepared for F. McCormick, Queen's University Belfast.

Hamilton-Dyer, S. 1997a. Birds, fish and marine invertebrates from site G, in C. Walsh (ed.) *Archaeological Excavations at Patrick, Nicholas & Winetavern Streets, Dublin*: 220–1. Dingle: Brandon Books/Dublin Corporation.

Hamilton-Dyer, S. 2000. Fish and bird bones from Maynooth Castle, Co. Kildare. Report prepared for F. McCormick, Queen's University Belfast.

Hamilton-Dyer, S. 2004a. Bird and fish bone, in E. Fitzpatrick, M. O'Brian & P. Walsh *Archaeological investigations in Galway City, 1987–1998*: 609–26. Bray: Wordwell.

Hamilton-Dyer, S. 2004b. Fish bones, in A. Hayden (ed.) *Excavation of the Medieval River Frontage at Arran Quay* (Medieval Dublin 5): 235–8. Dublin: Four Courts Press.

Hamilton-Dyer, S. 2005. Fish bone, in J. White Marshall & C. Walsh (edn) *Illaunloughan Island: an early medieval monastery in Co. Kerry*. Bray: Wordwell.

Hamilton-Dyer, S. 2007. Bird and fish remains, in F. McCormick & E. Murray (eds) *Knowth and the Zooarchaeology of Early Christian Ireland*: 71–3. (Excavations at Knowth 3, Royal Irish Academy Monographs in Archaeology). Dublin: Royal Irish Academy.

Hamilton-Dyer, S. 2008. Fish and bird bones from St Thomas' Church, Rathlin Island, Co. Antrim. Report prepared for E. Murray, Queen's University Belfast.

Hamilton-Dyer, S. 2011a. Bird and fish bones from Clonmacnoise. Report prepared for H. King & J. Soderberg.

Hamilton-Dyer, S. 2011b. Bird and fish bones, in E. Bourke, A. R. Hayden & A. Lynch (eds) Skellig Michael, Co. Kerry: the monastery and South Peak archaeological stratigraphic report: excavations 1986–2010 (E338; 90E34; 93E195): 437–49. Report on file, Department of Arts, Heritage and the Gaeltacht, Ireland.

Hamilton-Dyer, S. 2011c. Fish and bird bones, in A. Hayden *Trim Castle, Co. Meath: Excavations 1995 1998*: 411–18. (Archaeological Monograph Series 6). Dublin: Stationery Office.

Hamilton-Dyer, S. 2012. Appendix C: The fish bones, in E. V. Murray & F. McCormick, *Doonloughan: a seasonal settlement site on the Connemara coast. Proceedings of the Royal Irish Academy* 112C: 1–52.

Harland, J. 2009. Technical report: fish remains from the Drogheda boat, Ireland. Report prepared by Centre for Human Palaeoecology, Department of Archaeology, University of York.

Kelly, F. 1998. *Early Irish Farming*. (Early Irish Law Series 4). Dublin: Dublin Institute for Advanced Studies.

Kowaleski, M. 2000. The expansion of the south-western fisheries in late medieval England. *Economic History Review* 53: 429–54.

Kowaleski, M. 2003. The commercialization of the sea fisheries in medieval England and Wales. *International Journal of Maritime History* 15(2): 177–231.

McCarthy, M. 1995. Dungarvan – a faunal report. *Tipperary Historical Journal* 28: 202–6.

McCarthy, M. 1997. Faunal remains: Christ Church, in R. M. Cleary, M. F. Hurley & E. Shee Twohig (eds) *Skiddy's*

Castle and Christ Church, Cork: excavations 1974–77 by D. C. Twohig: 349–59. Cork: Cork Corporation.

McCarthy, M. 1998. Archaeozoological studies and early medieval Munster, in M. A. Monk & J. Sheehan (eds) *Early Medieval Munster*: 59–64. Cork: Cork University Press.

McCarthy, M. 2003. The faunal remains, in R. M. Cleary & M. F. Hurley (eds) *Cork City Excavations 1984– 2000*: 375–89. Cork: Cork City Council.

Murray, E., F. McCormick & G. Plunkett. 2004. The food economies of Atlantic island monasteries: the documentary and archaeo-environmental evidence. *Environmental Archaeology* 9: 179–88.

O'Neill, T. 1987. *Merchants and Mariners in Medieval Ireland.* Sallins: Irish Academic Press.

Smith, C. 1746. *The Ancient and Present State of the County and City of Waterford: being a natural, civil, ecclesiastical, historical and topographical description thereof.* Waterford: Mercier Press.

Tourunen, A. 2008. Gutted and salted: a fish bone assemblage from John Street, Waterford, Ireland. *Archaeofauna* 17: 139–45.

Vanes, J. 1979. *Documents Illustrating the Overseas Trade of Bristol in the Sixteenth Century.* (Bristol Records Society 31). Bristol: Bristol Records Society.

Marine Fish Consumption in Medieval Britain: The Isotope Perspective from Human Skeletal Remains

Gundula Müldner

Introduction

When considering the archaeological evidence for the early history of marine fishing, we need to accord a central role to isotope data. While zooarchaeological studies of fish-bone assemblages reflect the availability of marine resources in greater detail – to the level of individual species – carbon and nitrogen stable isotope analysis of human skeletal remains is the only direct method for reconstructing the actual level of human consumption of marine foods. Stable isotope analysis thus enables us to explore if, when and where marine fish became a dietary staple and by whom such fish were preferentially consumed. Therefore a comprehensive approach to the question of the early fishing industries should always combine zooarchaeological and isotopic methods (see Barrett et al. 2001).

Although some of the earliest applications of stable isotope analysis in European archaeology included skeletons of medieval date (Bocherens et al. 1991; Johansen et al. 1986; Tauber 1981), it was only in the late 1990s that Mays (1997) published a study with the explicit aim of exploring variation in marine food use in the medieval and post-medieval periods. The size of Mays' sample, 67 individuals from five sites in northeastern England, was very large for the time and allowed for differences between populations to be observed, both geographically, between coastal and inland locations, and socially, between different social groups buried in the same cemetery. Since Mays' study, the amount of stable isotope data available from medieval and post-medieval sites both in Britain and continental Europe has increased greatly. This dataset enables us to draw at least preliminary conclusions about the impact of the expanding fisheries on everyday subsistence. This paper aims to review the currently available carbon and nitrogen stable isotope evidence for marine fish consumption in medieval England and southern Scotland with a special emphasis on chronology – at what point in time did marine fish begin to make a regular contribution to human diet? – and on characterising the groups in society that were preferentially involved in its consumption. As will become apparent in this chapter, the available evidence is still patchy and leaves much to be desired in terms of our ability to address specific questions. Nevertheless, it is derived from a sufficient variety of site types and chronological phases to allow us to begin to outline some wider trends and formulate agendas for future research. Isotope data offer a different perspective on the history of the early sea fisheries and their effects on society than do the more traditional, zooarchaeological methods.

Carbon and nitrogen stable isotope data as a reflection of marine food consumption

There is now abundant evidence from field studies and controlled-feeding experiments to demonstrate that the carbon and nitrogen stable isotope composition of bone collagen, the main organic constituent of bone, reliably reflects that of the main protein sources in an individual's diet (see Ambrose 1993; Kelly 2000; Schwarcz and Schoeninger 1991).

Isotopes are atoms of the same element with slightly different atomic masses. Carbon has two stable isotopes, C-12 (^{12}C), with an atomic mass of 12, and the heavier C-13 (^{13}C). In contrast to the widely known radiocarbon (C-14), these two carbon isotopes do not decay radioactively over time; in other words, they are stable. The relative abundance of the stable isotopes of carbon in nature, or the ratio of

C-13 to C-12 ($^{13}C/^{12}C$, usually expressed as $\delta^{13}C$ or delta-13-C), varies significantly between different environments, especially between plants of different photosynthetic pathways (so-called C_3 and C_4 plants) as well as between C_3 plant-dominated terrestrial and marine ecosystems. These differences are preserved in all foods produced in these environments, and they can also be traced in the body tissues of their human consumers. In temperate northwestern Europe, where C_4 plants were largely absent until modern times, carbon stable isotope analysis is therefore the method of choice for distinguishing between the terrestrial and marine components in past human diet (Chisholm *et al.* 1982; Tauber 1981).

Nitrogen also has two stable isotopes, N-14 and N-15. The abundance of the 'heavy' N-15 in relation to N-14 (i.e. the $^{15}N/^{14}N$ ratio, or $\delta^{15}N$) in consumer tissues increases by 3–5‰ with each step up the food chain – a mechanism called the trophic level effect. $\delta^{15}N$ values are therefore useful to assess the trophic position of an organism and hence the importance of plant versus animal protein in human diet (DeNiro and Epstein 1981; Hedges and Reynard 2007). Nitrogen isotope ratios of marine animals are usually several per mil higher than those of terrestrial organisms, mainly because of the longer food-chains in aquatic ecosystems. The bone collagen of consumers of marine fish should consequently be enriched, not only in 'heavy' ^{13}C but also in ^{15}N. Both the $\delta^{13}C$ and $\delta^{15}N$ values of consumers of marine fish are therefore normally higher than those of organisms consuming an exclusively terrestrial diet (Richards and Hedges 1999; Schoeninger and DeNiro 1984).

It has to be understood that stable isotope data give only very general information about diet. Bone collagen, the preferred material for archaeological investigations, reflects a long-term average of foods eaten over at least several years to several decades of an individual's life

(Hedges *et al* 2007; Sealy *et al.* 1995). The isotopic signal obtained from collagen is heavily biased towards the protein component of the diet, meaning that foods with relatively little protein, such as plants, fats or carbohydrate-rich foods will be underrepresented or even invisible in comparison with protein-rich items, such as lean meat or fish (see Ambrose 1993; Jim *et al.* 2006). For this reason, and especially where the varied diets eaten by humans are concerned, stable isotope values are often best interpreted in relative terms, i.e. whether an individual consumed relatively more or less marine protein than other individuals, rather than truly quantitatively, even if stable isotope mixing models that seek to fully quantify consumption are becoming continuously more sophisticated (see Phillips *et al.* 2014).

Stable isotope data reflect only very broadly defined groups of foods, especially terrestrial and marine and plant or animal protein. They do not, for example, allow us to distinguish among different products of the same animal, such as meat or dairy; nor do they easily allow us to detect the consumption of different species of marine fish (O'Connell and Hedges 1999; Richards and Hedges 1999). The method is also not particularly sensitive to small variations in dietary intake. Even though the large isotopic differences between C_3 terrestrial and marine foods make these foods particularly suitable for this type of analysis, humans still have to derive a substantial ($\geq$~20%) portion of their dietary protein from marine resources (a figure which may increase further in certain low-protein diets) before these can be traced with certainty (see Hedges 2004; Schwarcz 2000). Even then, it is usually essential to establish local 'baseline' values typical for the different foods available to a population, by processing bone from fish and terrestrial animals from a site alongside the human samples, before small contributions of marine protein can be inferred with confidence (Figure 20.1; see Müldner and

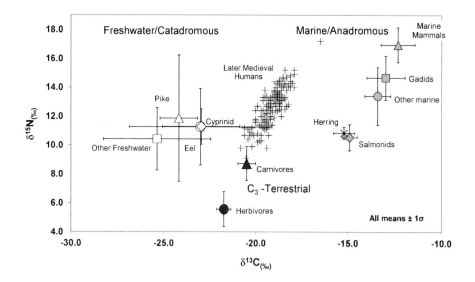

Figure 20.1. Carbon and nitrogen stable isotope ratios for fauna from northern England, illustrating the differences in isotope values between terrestrial, marine and freshwater animals as well as by trophic level. Also shown are human stable isotope data from the later medieval (thirteenth–early sixteenth-century AD) Gilbertine priory at Fishergate in York, which reflect a range of diets with varying proportions of marine protein (Data: Müldner and Grimes 2007, 201–2; Müldner and Richards 2005; 2007a,b).

Richards 2007b). Small-scale or irregular consumption of marine foods by humans may well go undetected.

Finally, some consideration must be given to our ability to distinguish between freshwater and marine fish by means of stable isotope analyses. The nitrogen isotope ratios of freshwater animals are similar to those of marine organisms of comparable trophic level, at least where there is no heavy load of sewage or other pollutants (France 1995). $\delta^{15}N$ values therefore have little potential for distinguishing between freshwater and marine foods in the diet. Carbon stable isotope values of freshwater fauna, on the other hand, are notoriously variable and can show significant overlap with the ranges of both terrestrial and marine habitats (Dufour *et al.* 1999; Katzenberg and Weber 1999). Nevertheless, especially in rivers and shallow lakes where there is a large input of terrestrial plant material to the nutrient pool, animal $\delta^{13}C$ values often resemble or are more negative than those of terrestrial fauna (Fry 1991).

Numerous isotope measurements on fish-bone collagen from medieval contexts in northern England allow us to characterise the ranges of freshwater and marine species relatively well (Figure 20.1; Müldner and Grimes 2007, 201–2; Müldner and Richards 2005; 2007b). They demonstrate that the two are completely separated by their carbon stable isotope signatures. Contributions of freshwater and marine protein to human diet should therefore not normally be confused; however, if both are being eaten in some quantity, a 'mixing effect' might obscure the isotopic signals of either (Müldner and Richards 2005). Also, recognising small contributions of freshwater foods to an otherwise terrestrial diet is even more difficult than for marine protein, as they may not be distinguishable from a number of other (terrestrial) foods that have high $\delta^{15}N$ values (Müldner and Richards 2007b; Privat *et al.* 2002). Importantly, the isotope values of migratory fish reflect those of the ecosystems where they spend most of their life-cycle (and do most of their feeding) (Kline *et al.* 1998). While life-cycles are species-specific and may also vary geographically (see Fuller *et al.* 2012), anadromous species (mainly salmonids [Salmonidae]) from British medieval contexts have been shown to have $\delta^{13}C$ most similar to marine fish, while catadromous fish (especially eel [*Anguilla anguilla*]) plot with the freshwater taxa (see Figure 20.1). When human consumption of 'marine' or 'freshwater' foods are referred to in this paper, it should therefore be understood that these categories may include migratory species.

An isotope perspective on the 'fish event horizon'

One of the key questions regarding England's early maritime economy is when the sea fisheries became suitably commercialised to supply fish in large numbers to the English markets. Historical evidence pre-dating the later medieval period is relatively scarce, although the Domesday Book

indicates that productive herring fisheries were established in southeastern England, and especially East Anglia, by the later eleventh century (Campbell 2002; Kowaleski 2003; Chapter 3). Barrett and colleagues' (2004) survey of the English fish-bone evidence suggested that a landmark change occurred somewhat earlier, in the decades around AD 1000, when marine species, initially herring (*Clupea harengus*) and later increasingly cod (*Gadus morhua*) and similar species, began to dominate the assemblages. So how is this 'fish event horizon' (Barrett *et al.* 2004, 621) reflected in the human stable isotope data?

The stable isotope evidence certainly suggest a dietary change on a significant scale. Available isotope data from Anglo-Saxon England and early medieval southern Scotland suggest that marine protein contributed very little, if anything, to the diet (Hull and O'Connell 2011; Lightfoot *et al.* 2009; Lucy *et al.* 2009; Mays and Beavan 2012; Modzelewski 2008; Privat *et al.* 2002), while isotope values for most later medieval humans are visibly different from those of earlier time periods and indicate that marine protein was a regular part of subsistence (Lakin 2010; Lamb *et al.* 2012; Müldner and Richards 2005; 2007a,b; Müldner *et al.* 2009). Although the contribution to the diet made by marine protein was arguably still only minor in absolute terms – the medieval isotope values are entirely different from those of coastal hunter-gathers, for example (Coltrain *et al.* 2004; Richards and Hedges 1999) – consumption of sea-fish in medieval Britain must have been much more substantial than in preceding periods in order to effect such clear isotopic differences between early and later medieval populations.

Differences between late/post-medieval and earlier populations, which probably reflect a shift towards increased marine consumption, can also be observed in isotope data from other European countries. Apart from northern Scotland, which is relatively well researched (Barrett and Richards 2004; Barrett *et al.* 2000; 2001; Richards *et al.* 2006; Curtis-Summer *et al.* 2014), changes through time have also been suggested based on palaeodietary data from Belgium (Polet and Katzenberg 2003), northeastern Germany (Peitel 2006; Schäuble 2006) and Italy (Salamon *et al.* 2007). Nevertheless, in other isotopic data sets no diachronic trend appears to register, even where it might be expected based on other archaeological sources (Yoder 2010; Ervynck *et al.* 2013). More detailed investigations are therefore needed before any general European trends can be postulated.

Pinning down when the transition in England occurred in absolute chronological terms is more difficult, for two reasons. First, as noted above, only a limited number of populations have been analysed. Second, dating later Anglo-Saxon and medieval cemeteries to a meaningful precision is notoriously difficult, as accompanying dateable artefacts are scarce. The time-frames that can be assigned to different phases of a cemetery are therefore often quite wide, even where additional stratigraphic information is available (e.g. Hadley 2001; Kemp and Graves 1996). Although direct dating of human remains by radiocarbon analysis is carried out more and more frequently (e.g. Bayliss *et al.* 2013; Pollard *et al.* 2012), interpreting the results for the specific question of early fish consumption is complicated by the 'marine reservoir effect', which makes individuals with marine foods in their diet appear chronologically older than those with a terrestrial diet (Stuiver *et al.* 1998; see Barrett *et al.* 2000). While a number of different methods for marine reservoir effect correction are available, these unavoidably introduce additional error, especially since the exact contribution of marine protein cannot necessarily easily be quantified (Ambrose *et al.* 1997; Arneborg *et al.* 1999; Barrett and Richards 2004). The dating programme on human skeletal remains from the medieval Timberhill cemetery in Norwich (Bayliss *et al.* 2004) illustrates the potentially significant consequences of the marine reservoir effect for dating burials in and around the 'fish event horizon'. Although Bayesian statistics were employed to improve the precision of the dates, application of different algorithms for calculating marine contributions to the diet resulted in archaeologically very significant variation in the absolute dates assigned to the burials. Possible dates ranged from a short pre-Conquest sequence (without reservoir correction) to the later twelfth and thirteenth centuries (with correction) (Bayliss *et al.* 2004). Even though the chronological resolution of the evidence discussed below could therefore no doubt be significantly improved by a number of strategically placed radiocarbon dates, it would probably take a much larger programme to settle the questions of chronology.

Starting with the early Anglo-Saxon period, isotope data are now available from a significant number of sites in southern England and East Anglia (Hull and O'Connell 2011; Lightfoot *et al.* 2009; Lucy *et al.* 2009; Mays and Beavan 2012; Privat *et al.* 2002). These data indicate that marine foods played a role in supplementing human diet in coastal settlements, where they could easily be obtained (Mays and Beavan 2012); however, at the overwhelming majority of fifth–seventh-century AD sites, marine protein made no measurable contribution to subsistence. Evidence from the remainder of the Anglo-Saxon period is much scarcer, but several publications either report no measurable marine-fish consumption (Buckberry *et al.* 2014; Müldner and Richards 2007b) or use a possible marine component in the diet as further evidence to identify Scandinavian incomers on British soil (Chenery *et al.* 2014; Pollard *et al.* 2012). Still, Hull and O'Connell (2011) note a significant shift in carbon and nitrogen isotope ratios between their large sample of early Anglo-Saxons and individuals at three later Anglo-Saxon sites in Norfolk which very likely reflect marine consumption. Two of the sites, Caister-on-Sea and Burgh Castle, are in the immediate vicinity of

Great Yarmouth. They would date the beginning of the East-Anglian fisheries to the eighth to tenth centuries AD, and possibly even as far back as the seventh century, although it is unclear whether relatively early radiocarbon dates obtained on the human remains from Burgh Castle are in need of reservoir correction (Hull and O'Connell 2011; see Johnson 1983, 111–12; Rodwell 1993, 252).

This early evidence from Norfolk is of great interest. However, there seems to be significant regional variation in the introduction of marine fish to the diet, just as the historical evidence for the development of the fisheries suggests (Kowaleski 2003; Chapter 3). Results of the palaeodietary analysis of eighth–late ninth-century burials from Bishopstone, near Brighton in East Sussex, suggest that any contribution of marine foods to the diet of these individuals was below the threshold where it can be confidently identified in the stable isotope signal (Thomas 2010). This threshold is commonly assumed to be 20% of the dietary protein, but it may be higher in certain low-protein diets (see above). The results from Bishopstone are surprising, given the large proportion of marine species in Bishopstone's fish-bone assemblage (Reynolds 2011; Chapter 17), which makes the site one of the few Anglo-Saxon estates in England known to have actively exploited marine resources (Thomas 2013, see Chapter 17 and references therein). A considerable number of the recovered fish bones showed signs of digestion, and stable isotope values for the bones of a cat, radiocarbon dated to the same horizon as the human burials, suggest that marine fish waste was lying around openly in the settlement for animals to scavenge (Reynolds 2010; Thomas 2010).

For the northeast, stable isotope evidence from the later seventh–eighth-century cemetery at Belle Vue House/Lamel Hill in York equally shows no indication of marine consumption on a measurable scale (Müldner and Richards 2007b), and neither do a small number of human values from ninth-century Coppergate (York). Palaeodietary data of seventh–eleventh-century date from rural Masham, near Ripon in North Yorkshire, and from the coastal cemetery of Lundin Links, in Fife in southeast Scotland, which has been dated to the fifth–seventh centuries, also allow for only small marine components, below the detection limits of the method (Buckberry *et al.* 2014; Modzelewski 2008). York is the only site in England for which we have a long-term diachronic sequence of isotope data, from the Roman to the post-medieval period, although, regrettably, the important ninth–tenth centuries are so far represented by only three samples from Coppergate (Buckberry *et al.* 2014; Müldner and Richards 2007b). Isotope evidence from the parish cemetery of St Andrew, Fishergate, suggests that a shift towards the consumption of marine fish in measurable quantities was only just in progress in Period 4b, dated to the mid-/late eleventh or early twelfth centuries (Kemp and Graves 1996). A significant minority of the people who were eventually buried in this cemetery (all young males) seem to have adopted marine foods, while the majority continued to eat the terrestrial diet that had been characteristic of the Anglian period (Figure 20.2; Müldner 2009; Müldner and Richards 2007a). By the thirteenth century (Period 6a/b), represented by the large sample from the Gilbertine priory at Fishergate, the situation had completely reversed. The great majority of the population now consumed marine protein on a regular basis, with only few individuals registering little or no marine protein in their bones (Müldner and Richards 2007b).

Finally, an example from southern Scotland allows us to glimpse the chronology on the west coast of Britain. A small group of radiocarbon-dated burials from Whithorn Cathedral Priory on the Machars Peninsula (Dumfries and Galloway) suggests a transition to measurable quantities of marine protein in the diet as late as the thirteenth century (Montgomery et al. 2009). However, the isotope data also demonstrate large differences in diet according to social status, which is likely to have biased the results (see Müldner et al. 2009, and below). Unfortunately, few identifiable fish-bone fragments were retrieved during the excavations at Whithorn, and the faunal assemblage therefore has extremely limited value for deducing temporal trends (Hamilton-Dyer 1997, 602–3). The observation that most marine species are not present before Period V/1–3 (from c. AD 1260) must therefore be treated with caution, although it may support the relatively late date for dietary change suggested by the isotope data.

In summary, the English and southern-Scottish isotope evidence, although still very incomplete, suggests significant regional variation in the adoption of marine fish into the diet. The currently available data span from mid-/late Anglo-Saxon beginnings in East Anglia, to a probably post-Conquest transition in York and an even later dietary shift on the southwest coast of Scotland.

When trying to reconcile these results with the fish-bone record, it is apparent that there are several instances when the isotope evidence for marine consumption post-dates the arrival of marine species in the fish-bone assemblages by some time. At Anglo-Saxon Bishopstone, marine fish has been recorded in significant numbers throughout the occupation sequence (Thomas 2010; see Chapter 17), and in York, the relative abundance of herring increased from the mid-tenth century (Harland et al. 2008; O'Connor 1991, 263–7; see Chapter 15).

These data are of course not irreconcilable. A certain lag of time between zooarchaeological and isotopic evidence is actually to be expected, since fish-bone assemblages register more subtle trends and will pick up intensification of marine exploitation earlier than will stable isotope data. The latter will register major change and will therefore only reflect when sea-fish became a significant part of human diet. With this in mind, a transition in York from the eleventh century onwards, as can be argued from the Fishergate isotope data (assuming an early mid-eleventh-century date for Period 4b and taking into account that the individuals with marine diets were all between c. 20 and 30 years old when they died; see Müldner 2009), actually fits well with the fish-bone evidence, as it is only then that marine species began to truly dominate the assemblages (Barrett et al. 2004; see Chapter 15).

An additional factor for seeming disagreement between isotope and fish-bone data may be that the most abundant species in the early 'fish event horizon' is herring (Barrett et al. 2004). Herring occupy a relatively low trophic level in the marine food chain, and their carbon and nitrogen isotope values are therefore generally lower (and less distinct from

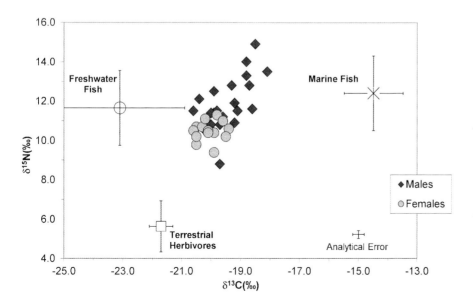

Figure 20.2. Stable isotope data from Period 4b (mid-eleventh–twelfth century) of the parish cemetery of St Andrew Fishergate in York, showing a population in transition between the 'traditional' terrestrial diet that appears to have been typical of the Anglo-Saxon period and the mixed marine-terrestrial diet that would become characteristic for Fishergate in the later medieval phase. For discussion of gender differences see below and Müldner (2009) (Data: Müldner and Richards 2007b).

terrestrial foods) than are those of top predators, such as cod (Figure 20.1). Consumption of herring in smaller quantities will therefore not register as strongly in consumer collagen; in other words, it would take relatively more herring protein in the diet to effect a clearly recognisable 'marine' signature in humans than it would cod protein. This issue is yet to be fully explored, especially since the isotope values of different fish species can vary significantly according to their geographical origin (see Barrett *et al.* 2008). Nevertheless, cod and related taxa appearing in greater numbers from the eleventh century onwards (Barrett *et al.* 2004; Serjeantson and Woolgar 2006) may have made any marine component become more visible in the human isotope data.

These technical issues aside, we also have to consider social variation as a reason for any seeming divergence of fish-bone evidence and isotope data. An important difference between the two methods is that animal-bone assemblages are unspecific regarding the identity of the human consumers behind them. Their associations can rarely even be narrowed down further than to a single household. In contrast, stable isotope values are specific to individuals, and allow levels of consumption to be assessed at least in relative terms (see Gumerman 1997). They therefore enable us to investigate in much more detail the different groups in society that partook in fish consumption, and how this consumption may have changed over time. The social dynamics of marine food use in medieval England will be explored in the next section.

Social variation in fish consumption in medieval England

Fish, its substantial nutritional value aside, had special significance in medieval diet as a fasting food. It could be eaten during Lent and on the numerous other days in the Church calendar on which the faithful were to abstain from meat as a sign of penitence and religious worship. References to the numerous associations of Christ and the Apostles with fish and fishing illustrate that fish consumption was indeed more than just a convenient 'loophole' in the Rule of St Benedict (which, strictly speaking, forbade only the 'flesh of quadrupeds'); it was imbued with spiritual value in itself (Serjeantson and Woolgar 2006; Woolgar 2000). Possibly fuelled by this attitude, fish in large quantities or species that were particular costly to produce or obtain became signs of social distinction. They were exchanged as gifts and adorned the tables of the rich as items of conspicuous consumption (Dyer 1988; Woolgar 2001). The value attributed to different fish species changed over time according to how readily available they were. While in the later Anglo-Saxon period any type of marine fish, including herring, may have carried special status (Fleming 2000, 5–6), it was larger or exotic specimens and costly pond-reared freshwater fish which were most highly regarded and priced in later medieval England (Dyer 1988; Woolgar 2000;

see also Gardiner 1997). Because of the special qualities attributed to fish – qualities that go beyond its nutritional value – investigating variation in fish consumption in medieval society is a particularly fruitful approach towards understanding the success of the early fishing industries.

The timing of when fish became common in monastic and lay fasts is a subject of special significance for the early history of medieval sea fishing. The key question is whether the obligation to abstain from meat created great demand for fish early on in the medieval period and was therefore a significant driving force behind the rapid expansion of the marine fisheries (see Barrett *et al.* 2004). While numerous references to fasting regulations are found in Anglo-Saxon documents, especially of the tenth century AD, it is disputed whether fish was widely accepted as appropriate for consumption on fast days prior to the Norman Conquest (Barrett *et al.* 2004, 629–30; Frantzen 2014, 232–45; Serjeantson and Woolgar 2006, 104; Woolgar 2000: 36; see Hagen 1992, 393–408).

It has been suggested that the marine signal in the majority of later-medieval isotope data can be seen as a reflection of how the fasting requirements had transformed diet on a large scale (Müldner and Richards 2005). If this interpretation is accepted, then the absence of an obvious marine (or freshwater) component in the pre-Conquest isotope data could indicate that fish was not yet a regular substitute for meat on fast days. Unfortunately, however, although work on later Anglo-Saxon populations to date has not found much evidence for marine fish consumption (with the notable exception of the sites from Norfolk discussed above), stable isotope data from the crucial ninth to early eleventh centuries are currently too sparse to infer any bigger picture.

The possibility of profound social variation in diet, at a time when marine fish were not yet widely available, is a key issue in this respect. The main consumers of fish in later Anglo-Saxon England may simply elude us, since isotope evidence from monastic communities, early towns and high-status contexts of such an early medieval date is still sparse or missing entirely. Interestingly, both Burgh Castle and Caister-on-Sea, two of the Norfolk sites with early, pre-Conquest evidence of marine foods consumption, have been discussed as possible monastic settlements – although in both cases the main argument is the suggested identification with Cnobheresburgh, a religious house mentioned by Bede in the eighth century (Darling and Gurney 1993; Hull and O'Connell 2011; Johnson 1983). Similarly, it is currently unclear whether the cemetery of Belle Vue House/Lamel Hill in York, where no evidence of marine food consumption was found, was serving the population of the Anglian town, an important ecclesiastical and economic centre at the time, or rather that of an agricultural settlement nearby (Tweddle *et al.* 1999, 176). Moreover, the three individuals with a terrestrial diet sampled from ninth-century Coppergate, York's proto-urban Anglo-Scandinavian settlement, are arguably unusual, having been deposited in pits rather than afforded a proper burial (Buckberry *et al.* 2014). At Bishopstone in East Sussex, only the periphery of the Anglo-Saxon churchyard has been excavated. Burials of the highest-ranking members of the community, which were often situated inside or in close proximity to the church, may therefore have been missed. If marine fish was at the time indeed mainly served at the tables of the elite, we would therefore not necessarily see its consumption reflected in the Bishopstone isotope data (Reynolds 2010). The chronologically later case-study from Whithorn Cathedral Priory in Scotland illustrates this point well. Here, isotope data for a number of very high-ranking individuals, including several bishops of Galloway, were contrasted with data for a group of lower-status burials. The results demonstrate significant differences in diet between the two groups. In fact, any contributions from marine foods to the diet of the lower-status individuals were too small to be detectable (Müldner *et al.* 2009).

While the Whithorn data provide excellent evidence that the quantities of fish consumed indeed increased with socio-economic status, this is not necessarily a universal trend. At the later medieval Gilbertine Priory at Fishergate, there were no systematic differences between individuals buried in designated high- and low-status areas or those reserved for the monastic community. However, a group of men and women buried in the church crossing, one of the most prestigious locations for lay burial the house offered, stood out by having consumed significantly *less* marine protein than individuals from other areas (Figure 20.3; Müldner and Richards 2007a). These results correspond to observations of

unusually low counts of fish-bone at a number of medieval high-status sites (Serjeantson and Woolgar 2006, 128–9; Sykes 2007, 60–1). They suggest that some aristocratic families sought to distinguish themselves by avoiding what was by now a popular and widely available food, rather than by the conspicuous consumption of it. A closer look at the documentary evidence indicates that preference for fish and its acceptance as suitable fare on fast days was indeed often a matter of personal choice, at least among those who could afford it (Woolgar 1995; 2006).

In several instances, stable isotope data have been able to demonstrate gender differences in the consumption of marine food. These are discussed in detail elsewhere (Müldner 2009) and shall only be briefly touched upon here. The differences between males and females among the early consumers of sea-fish at pre-monastic St Andrew, Fishergate, in York are particularly striking (Figure 20.2). Here, only males ate marine protein in measurable quantities. Given the associations of the Fishergate area with the fishing trade, it seems that the first regular consumers of marine fish, at least in a relatively low-status parish like St Andrew, were those directly involved in its production (Kemp and Graves 1996, 95–6; Müldner 2009).

Finally, let us now briefly consider the isotope evidence for variation between town and country. The large isotopic differences between the peasant population of Wharram Percy in the Yorkshire Wolds and other medieval assemblages have been discussed before (Müldner and Richards 2005; 2006). The Wharram Percy humans exhibit slightly elevated carbon isotope ratios, which could be interpreted as evidence for marine food consumption (Mays 1997; 2007, 94–5). However, the lack of correlation between $\delta^{13}C$ and $\delta^{15}N$ values, a correlation that is expected if a group of individuals is consuming marine protein in

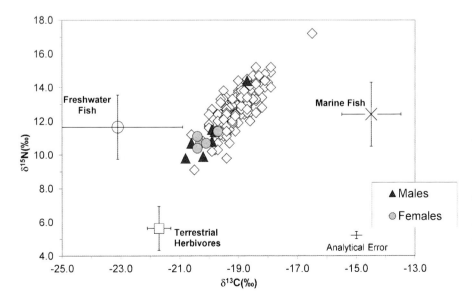

Figure 20.3. Stable isotope data for humans buried at the later medieval (thirteenth– early sixteenth-century AD) Gilbertine priory at Fishergate in York. A group of males and females buried in prominent position in the crossing of the priory church, probably in the early thirteenth century, stands out by having consumed significantly less marine protein than usual for the time (Data: Müldner and Richards 2007a).

varying quantities, and also the uncharacteristically low δ^{15}N values, suggest that diet at Wharram Percy was very different than at the other medieval sites, where a marine contribution is more evident. Nevertheless, being in the countryside and having limited economic means alone did not mean that there could not be a regular supply of fish for the table. This is evidenced by the isotope data from the small rural hospital of St Giles, near Richmond in North Yorkshire (Müldner and Richards 2005). As a hospital, St Giles was effectively a monastic institution. Again, social context is likely the key to understanding patterns of fish consumption at different sites. In this context and more generally, the fact that most available isotope data from the later medieval period are from monastic sites, and may therefore be skewed towards higher fish consumption, is a problem that future research will have to address (Müldner 2009; Quintelier *et al.* 2014).

Conclusions

The current survey of the stable isotope evidence for fish consumption in medieval England has shown that the expansion of the marine-fishing industry was accompanied by a significant change in the diet of large segments of the population. The isotope evidence is still sketchy and, at the moment, only affords spotlights on individual cemeteries from across England and southern Scotland. Still, many of the available data resonate with evidence from other sources. They suggest substantial geographical variation in the introduction of marine fish as a dietary staple. So far, only the results from East Anglia give convincing isotopic evidence for recognisable marine-food consumption before the turn of the first millennium AD (Hull and O'Connell 2011). They correspond to historical sources for the early development of the eastern fisheries, especially in the Great Yarmouth area (Kowaleski 2003; Chapter 3). Elsewhere, major changes do not seem to occur until the mid-eleventh century or later. In some areas this may have been as late as the thirteenth century, as is suggested by the small data set from Whithorn in southwest Scotland (Müldner *et al.* 2009). Nevertheless, stable isotope data from the ninth to eleventh centuries, which would be critical for exploring when and for whom sea-fish first became available in significant quantities, are still woefully scarce. These results should therefore be seen as preliminary. Social variation in fish consumption has been shown to be clearly reflected in the stable isotope data, especially those from the later medieval period. This observation reminds us that the socio-economic context of each cemetery has to be considered before stable isotope values can be interpreted in terms of continuity and change of medieval diet in general. Context is all-important, and in this respect stable isotope data are no different from any other archaeological evidence.

Future research should therefore seek not only to increase the number of samples available from the later Anglo-Saxon period, but also to deliberately target different settlement types, such as high and low status, urban and monastic, in order to establish which groups in society first made use of sea-fish as a new resource. Exploring geographic variation in the prominence of marine foods in the diet of medieval humans seems another fruitful avenue for future work. Stable isotope data for the bishops of Whithorn in the thirteenth and fourteenth centuries indicate that they ate significantly more fish than their local contemporaries, a special diet befitting their high office (Müldner *et al.* 2009). However, if their isotope values are compared with those of humans from later medieval York, the bishops' consumption of marine protein appears no more than average, hinting that substantial regional diversity existed irrespective of social status (Montgomery *et al.* 2009).

Overall, stable isotope studies on diet in the medieval period are still in their early stages. In time, they will hopefully be considered to be as important as more traditional archaeological approaches to the early history of marine fishing in medieval Europe.

References

Ambrose, S. H. 1993. Isotopic analysis of paleodiets: methodological and interpretive considerations, in M. K. Sandford (ed.) *Investigations of Ancient Human Tissue: chemical analyses in anthropology*: 59–130. Langthorne: Gordon & Breach.

Ambrose, S. H., B. M. Butler, D. B. Hanson, R. L. Hunter-Anderson & H. W. Krueger. 1997. Stable isotopic analysis of human diet in the Marianas Archipelago, western Pacific. *American Journal of Physical Anthropology* 104: 343–61.

Arneborg, J., J. Heinemeier, N. Lynnerup, H. L. Nielsen, N. Rud & A. E. Sveinbjörnsdóttir. 1999. Change of diet of the Greenland Vikings determined from stable carbon isotope analysis and ^{14}C dating of their bones. *Radiocarbon* 41(2): 157–68.

Barrett, J. H. & M. P. Richards. 2004. Identity, gender, religion and economy: new isotope and radiocarbon evidence for marine resource intensification in early historic Orkney, Scotland, UK. *European Journal of Archaeology* 7(3): 249–71.

Barrett, J. H., R. P. Beukens & D. R. Brothwell. 2000. Radiocarbon dating and marine reservoir correction of Viking Age Christian burials from Orkney. *Antiquity* 74: 537–43.

Barrett, J. H., R. P. Beukens & R. A. Nicholson. 2001. Diet and ethnicity during the Viking colonization of northern Scotland: evidence from fish bones and stable carbon isotopes. *Antiquity* 75: 145–54.

Barrett, J. H., A. M. Locker & C. M. Roberts. 2004. 'Dark Age economics' revisited: the English fish bone evidence AD 600–1600. *Antiquity* 78: 618–36.

Barrett, J. H., C. Johnstone, J. Harland, W. Van Neer, A. Ervynck, D. Makowiecki, D. Heinrich, A. K. Hufthammer, I. B. Enghoff, C. Amundsen, J. S. Christiansen, A. K. G. Jones, A. Locker, S. Hamilton-Dyer, L. Jonsson, L. Lõugas, C. Roberts & M. Richards. 2008. Detecting the medieval cod trade: a new method and first results. *Journal of Archaeological Science* 35(4): 850–61.

Bayliss, A., E. Shepherd Popescu, N. Beavan-Athfield, C. Bronk Ramsey, G. T. Cook & A. Locker. 2004. The potential significance of dietary offsets for the interpretation of radiocarbon dates: an archaeologically significant example from medieval Norwich. *Journal of Archaeological Science* 31(5): 563–75.

Bayliss, A., J. Hines, K. Hoilund Nielsen, G. McCormac & C. Scull. 2013. *Anglo-Saxon Graves and Grave Goods of the 6th and 7th centuries AD: a chronological framework.* London: Society for Medieval Archaeology.

Bocherens, H., M. Fizet, A. Mariotti, C. Olive, G. Bellon & D. Billiou. 1991. Application de la biogéochimie isotopique (^{13}C, ^{15}N) a la détermination du régime alimentaire des populations humaines et animales durant les périodes antique et médiévale. *Archives des Sciences (Geneva)* 44(3): 329–40.

Buckberry, J., J. Montgomery, J. Towers, G. Müldner, M. Holst, J. Evans, A. Gledhill, N. Neale & J. Lee-Thorp. 2014. Finding Vikings in the Danelaw. *Oxford Journal of Archaeology* 33: 413–34.

Campbell, J. 2002. Domesday herrings, in C. Harper-Bill, C. Rawcliffe & R. G. Wilson (eds) *East Anglia's History: studies in honour of Norman Scarfe*: 5–17. Woodbridge: Boydell Press.

Chenery, C. A., J. A. Evans, D. Score, A. Boyle & S. R. Chenery 2014. A boat load of Vikings? *Journal of the North Atlantic, Special Volume* 7: 43–53.

Chisholm, B. S., D. E. Nelson & H. P. Schwarcz. 1982. Stable-carbon isotope ratios as a measure of marine versus terrestrial protein in ancient diets. *Science* 216: 1131–2.

Coltrain, J. B., M. G. Hayes & D. H. O'Rourke. 2004. Sealing, whaling and caribou: the skeletal isotope chemistry of Eastern Arctic foragers. *Journal of Archaeological Science* 31(1): 39–57.

Curtis-Summer, S., J. Montgomery & M. Carver. 2014. Stable isotope evidence for dietary contrast between Pictish and medieval populations at Portmahomack, Scotland. *Medieval Archaeology* 58: 21–43.

Darling, M. J. & D. Gurney. 1993. *Caister-on-Sea, excavations by Charles Green 1951–1955.* (East Anglian Archaeology 60). Dereham: Norfolk Archaeological Unit/Norfolk Museums Service.

DeNiro, M. J. & S. Epstein. 1981. Influence of diet on the distribution of nitrogen isotopes in animals. *Geochimica et Cosmochimica Acta* 45(3): 341–51.

Dufour, E., H. Bocherens & A. Mariotti. 1999. Palaeodietary implications of isotopic variability in Eurasian lacustrine fish. *Journal of Archaeological Science* 26(6): 617–27.

Dyer, C. C. 1988. The consumption of freshwater fish in medieval England, in M. Aston (ed.) *Medieval Fish, Fisheries and Fishponds in England:* 27–38 (British Archaeological Report 182). Oxford: British Archaeological Reports.

Ervynck, A., M. Boudin, T. Van den Brande & M. Van Strydonck. 2013. Dating human remains from the historical period in Belgium: diet changes and the impact of marine and freshwater reservoir effects. *Radiocarbon* 56(2): 779–88.

Fleming, R. 2000. The new wealth, the new rich and the new political style in late Anglo-Saxon England. *Anglo-Norman Studies* 23: 1–22.

France, R. 1995. Stable nitrogen isotopes in fish: literature synthesis on the influence of ecotonal coupling. *Estuarine, Coastal and Shelf Science* 41(6): 737–42.

Frantzen, A. J. 2014. *Food, Eating and Identity in Early Medieval England.* Woodbridge: Boydell Press.

Fry, B. 1991. Stable isotope diagrams of freshwater food webs. *Ecology* 72(6): 2293–7.

Fuller, B. T., G. Müldner, W. Van Neer, A. Ervynck & M. P. Richards. 2012. Carbon and nitrogen stable isotope ratio analysis of freshwater, brackish and marine fish from Belgian archaeological sites (1st and 2nd millennium AD). *Journal of Analytical Atomic Spectrometry* 27: 807–20.

Gardiner, M. 1997. The exploitation of sea-mammals in medieval England: bones and their social context. *Archaeological Journal* 154: 173–95.

Gumerman, G. 1997. Food and complex societies. *Journal of Archaeological Method & Theory* 4(2): 105–39.

Hadley, D. M. 2001. *Death in Medieval England.* Stroud: Tempus.

Hagen, A. 1992. *A Handbook of Anglo-Saxon food: processing and consumption.* Frithgard: Anglo-Saxon Books.

Hamilton-Dyer, S. 1997. The bird and fish bones, in P. Hill (ed.) *Whithorn and St Ninian: the excavation of a monastic town 1984–91:* 601–05. Stroud: Whithorn Trust/Sutton.

Harland, J., C. Johnstone & A. Jones. 2008. A case study from the medieval origins of commercial sea fishing project: zooarchaeological results from York (UK), in P. Béarez, S. Grouard & B. Clavel (eds) *Archéologie du poisson: 30 ans d'archéo-ichtyologie au CNRS:* 15–26. Antibes: Éditions APDCA.

Hedges, R. E. M. 2004. Isotopes and red herrings: comments on Milner *et al.* and Lidén *et al. Antiquity* 78: 34–7.

Hedges, R. E. M. & L. M. Reynard. 2007. Nitrogen isotopes and the trophic level of humans in archaeology. *Journal of Archaeological Science* 34: 1240–51.

Hedges, R. E. M., J. G. Clement, C. D. L. Thomas & T. C. O'Connell. 2007. Collagen turnover in the adult femoral mid-shaft: modeled from anthropogenic radiocarbon tracer measurements. *American Journal of Physical Anthropology* 133(2): 808–16.

Hull, B. D. & T. C. O'Connell. 2011. Diet: recent evidence from analytical chemical techniques, in H. Hamerow, D. A. Hinton & S. Crawford (ed.) *The Oxford Handbook of Anglo-Saxon Archaeology:* 667–87. Oxford: Oxford University Press.

Jim, S., V. Jones, S. H. Ambrose & R. P. Evershed. 2006. Quantifying dietary macronutrient sources of carbon for bone collagen synthesis using natural abundance stable carbon isotope analysis. *British Journal of Nutrition* 95(6): 1055–62.

Johansen, O. V., S. Gulliksen & R. Nydal. 1986. δ^{13}C and diet: analysis of Norwegian human skeletons. *Radiocarbon* 28: 754–61.

Johnson, S. 1983. *Burgh Castle: excavations by Charles Green, 1958–61* (East Anglian Archaeology 20). Dereham: Norfolk Archaeological Unit/Norfolk Museums Service.

Katzenberg, M. A. & A. W. Weber. 1999. Stable isotope ecology and palaeodiet in the Lake Baikal region of Siberia. *Journal of Archaeological Science* 26(6): 651–9.

Kelly, J. F. 2000. Stable isotopes of carbon and nitrogen in the study of avian and mammalian trophic ecology. *Canadian Journal of Zoology* 78(1): 1–27.

Kemp, R. L. & C. P. Graves. 1996. *The Church and Gilbertine Priory of St. Andrew, Fishergate.* (Archaeology of York 7/1). York: Council for British Archaeology.

Kline, T. C., W. J. Wilson & J. J. Goering. 1998. Natural isotope indicators of fish migration at Prudhoe Bay, Alaska. *Canadian Journal of Fisheries and Aquatic Sciences* 55(6): 1494–502.

Kowaleski, M. 2003. The commercialization of the sea fisheries in medieval England and Wales. *International Journal of Maritime History* 15(2): 177–231.

Lakin, K. E. 2010. Diet in Medieval London: Stable Isotope Analysis of Human and Faunal Remains. Unpublished PhD dissertation, University of Reading.

Lamb, A. L., M. Melikian, R. Ives & J. Evans. 2012. Multi-isotope analysis of the population of the lost medieval village of Auldhame, East Lothian, Scotland. *Journal of Analytical Atomic Spectrometry* 27(5): 765–77.

Lightfoot, E., T. C. O'Connell, R. E. Stevens, J. Hamilton, G. Hey & R. E. M. Hedges. 2009. An investigation into diet at the site of Yarnton, Oxfordshire, using stable carbon and nitrogen isotopes. *Oxford Journal of Archaeology* 28(3): 301–22.

Lucy, S., R. Newman, N. Dodwell, C. Hills, M. Dekker, T. O'Connell, I. Riddler & P. Walton Rogers. 2009. The burial of a princess? The later seventh-century cemetery at Westfield Farm, Ely. *Antiquaries Journal* 89: 81–141.

Mays, S. A. 1997. Carbon stable isotope ratios in mediaeval and later human skeletons from northern England. *Journal of Archaeological Science* 24(6): 561–7.

Mays, S. A. 2007. The human remains, in S. Mays, C. Harding & C. Heighway (eds) *Wharram: a study of settlement on the Yorkshire Wolds. Volume 11: the churchyard*: 77–192. (York University Archaeological Publications 13). York: University of York.

Mays, S. & N. Beavan. 2012. An investigation of diet in early Anglo-Saxon England using carbon and nitrogen stable isotope analysis of human bone collagen. *Journal of Archaeological Science* 39(4): 867–74.

Modzelewski, K. 2008. Palaeodietary analysis of an early medieval coastal community in southern Scotland: the cemetery of Lundin Links, Fife. Unpublished MA dissertation, University of Reading.

Montgomery, J., G. Müldner, G. Cook, A. Gledhill & R. Ellam. 2009. Isotope analysis of bone collagen and tooth enamel, in C. Lowe (ed.) *'Clothing for the soul divine': burials at the tomb of St Ninian: excavations at Whithorn Priory 1957–67*: 63–80. Edinburgh: Historic Scotland: Edinburgh.

Müldner, G. 2009. Investigating medieval diet and society by stable isotope analysis of human bone, in R. Gilchrist & A. Reynolds (eds) *Fifty Years of Medieval Archaeology*: 327–46. Leeds: Maney.

Müldner, G. & V. Grimes. 2007. Stable isotope analysis of cetaceans and fish from Flixborough, in K. Dobney, D. Jaques, J. Barrett & C. Johnstone (eds) *Farmers, Monks and Aristocrats: the environmental archaeology of Anglo-Saxon Flixborough*: 201–6 (Excavations at Flixborough 3). Oxford: Oxbow Books.

Müldner, G. & M. P. Richards. 2005. Fast or feast: reconstructing diet in later medieval England by stable isotope analysis. *Journal of Archaeological Science* 32(1): 39–48.

Müldner, G. & M. P. Richards. 2006. Diet in medieval England: the evidence from stable isotopes, in C. M. Woolgar, D. Serjeantson & T. Waldron (eds) *Food in Medieval England: diet and nutrition*: 228–38. Oxford: Oxford University Press.

Müldner, G. & M. P. Richards. 2007a. Diet and diversity at later medieval Fishergate: the isotopic evidence. *American Journal of Physical Anthropology* 134(2): 162–74.

Müldner, G. & M. P. Richards. 2007b. Stable isotope evidence for 1500 years of human diet at the City of York, U. K. *American Journal of Physical Anthropology* 133(1): 682–97.

Müldner, G., J. Montgomery, G. Cook, R. Ellam, A. Gledhill & C. Lowe. 2009. Isotopes and individuals: diet and mobility among the medieval bishops of Whithorn. *Antiquity* 83: 1119–33.

O'Connell, T. C. & R. E. M. Hedges. 1999. Investigations into the effect of diet on modern human hair isotopic values. *American Journal of Physical Anthropology* 108(4): 409–25.

O'Connor, T. P. 1991. *Bones from 46–54 Fishergate.* (Archaeology of York 15/4). York: Council for British Archaeology.

Peitel, D. 2006. Rekonstruktion der Ernährung und weiterer Subsistenzgrundlagen dreier frühneuzeitlicher Bevölkerungen anhand der Analyse stabiler Isotope und Spurenelemente. Unpublished PhD dissertation, Freie Universität Berlin.

Phillips, D. L., R. Inger, S. Bearhop, A. L. Jackson, J. W. Moore, A. C. Parnell, B. X. Semmens & E. J. Ward. 2014. Best practices for use of stable isotope mixing models in food-web studies. *Canadian Journal of Zoology* 92: 823–35.

Polet, C. & M. A. Katzenberg. 2003. Reconstruction of the diet in a mediaeval monastic community from the coast of Belgium. *Journal of Archaeological Science* 30(5): 525–33.

Pollard, A. M., P. Ditchfield, E. Piva, S. Wallis, C. Falys & S. Ford. 2012. 'Sprouting like cockle amongst the wheat': The St Brice's Day Massacre and the isotopic analysis of human bones from St John's College, Oxford. *Oxford Journal of Archaeology* 31(1): 83–102.

Privat, K. L., T. C. O'Connell & M. P. Richards. 2002. Stable isotope analysis of human and faunal remains from the Anglo-Saxon cemetery at Berinsfield, Oxfordshire: dietary and social implications. *Journal of Archaeological Science* 29(7): 779–90.

Quintelier, K., A. Ervynck, G. Müldner, W. Van Neer, M. P. Richards & B. T. Fuller. 2014. Isotopic examination of links between diet, social differentiation, and DISH at the post-medieval Carmelite Friary of Aalst, Belgium. *American Journal of Physical Anthropology* 153: 203–13.

Reynolds, R. 2010. Fish remains, in G. Thomas (ed.) *The Later Anglo-Saxon Settlement at Bishopstone: a downland manor in the making*: 157–64. (Council for British Archaeology Research Report 163). York: Council for British Archaeology.

Reynolds, R. 2011. Anglo-Saxon fish remains from Lyminge, Kent. Report on file, Reading University. http://www.reading.ac.uk/web/FILES/archaeology/LymingeReynolds2011.pdf.

Richards, M. P. & R. E. M. Hedges. 1999. Stable isotope evidence for similarities in the types of marine foods used by Late Mesolithic humans at sites along the Atlantic coast of Europe. *Journal of Archaeological Science* 26(6): 717–22.

Richards, M. P., B. T. Fuller & T. I. Molleson. 2006. Stable isotope palaeodietary study of humans and fauna from the multi-period (Iron Age, Viking and Late Medieval) site of Newark Bay, Orkney. *Journal of Archaeological Science* 33(1): 122–31.

Rodwell, K. 1993. The cemetery, in M. J. Darling & D. Gurney (eds) *Caister-on-Sea: excavations by Charles Green 1951–1955*:

252–55. (East Anglian Archaeology 60). Dereham: Norfolk Archaeological Unit/Norfolk Museums Service.

Salamon, M., A. Coppa, M. McCormick, M. Rubini, R. Vargiu & N. Tuross. 2007. The consilience of historical and isotopic approaches in reconstructing the medieval Mediterranean diet. *Journal of Archaeological Science* 35(6): 1667–72.

Schäuble, A. 2006. Ernährungsrekonstruktion dreier mittelalterlicher Bevölkerungen anhand der Analyse stabiler Isotope und Spurenelemente. Unpublished PhD dissertation, Freie Universität Berlin.

Schoeninger, M. J. & M. J. DeNiro. 1984. Nitrogen and carbon isotopic composition of bone collagen from marine and terrestrial animals. *Geochimica et Cosmochimica Acta* 48: 625–39.

Schwarcz, H. P. 2000. Some biochemical aspects of carbon isotopic palaeodiet studies, in S. H. Ambrose and M. A. Katzenberg (ed.) *Biogeochemical Approaches to Palaeodietary Analysis*: 189–209. New York: Kluwer Academic/Plenum.

Schwarcz, H. P. & M. J. Schoeninger. 1991. Stable isotope analyses in human nutritional ecology. *Yearbook of Physical Anthropology* 34(S13): 283–321.

Sealy, J., R. Armstrong & C. Schrire. 1995. Beyond lifetime averages: tracing life histories through isotopic analysis of different calcified tissues from archaeological human skeletons. *Antiquity* 69: 290–300.

Serjeantson, D. & C. M. Woolgar. 2006. Fish consumption in medieval England, in C. M. Woolgar, D. Serjeantson & T. Waldron (eds) *Food in Medieval England: diet and nutrition*: 102–30. Oxford: Oxford University Press.

Stuiver, M., P. J. Reimer & T. F. Braziunas. 1998. High-precision radiocarbon age calibration for terrestrial and marine samples. *Radiocarbon* 40(3): 112–51.

Sykes, N. 2007. *The Norman Conquest: a zooarchaeological perspective.* (British Archaeological Report S1656). Oxford: Archaeopress.

Tauber, H. 1981. [13]C evidence for dietary habits of prehistoric man in Denmark. *Nature* 292: 332–3.

Thomas, G. 2010. *The Later Anglo-Saxon Settlement at Bishopstone: a downland manor in the making* (Council for British Archaeology Research Report 163). York: Council for British Archaeology.

Thomas, G. 2013. Life before the Minster: the social dynamics of monastic foundation at Anglo-Saxon Lyminge, Kent. *Antiquaries Journal* 93: 109–45.

Tweddle, D., J. Moulden & E. Logan. 1999. *Anglian York: a survey of the evidence* (Archaeology of York 7/2). York: Council for British Archaeology.

Woolgar, C. M. 1995. Diet and consumption in gentry and noble households: a case study from around the Wash, in R. E. Archer & S. Walker (eds) *Rulers and Ruled in Late Medieval England: essays presented to Gerald Harriss*: 17–31. London: Hambledon Press.

Woolgar, C. M. 2000. 'Take this penance now, and afterwards the fare will improve': seafood and late medieval diet, in D. J. Starkey, C. Reid & N. Ashcroft (eds) *England's Sea Fisheries: the commercial sea fisheries of England and Wales since 1300*: 36–44. London: Chatham.

Woolgar, C. M. 2001. Fast and feast: conspicuous consumption and the diet of the nobility in the fifteenth century, in M. Hicks (ed.) *Revolution and Consumption in Late Medieval England*: 7–25. Woodbridge: Boydell Press.

Woolgar, C. M. 2006. Group diets in late medieval England, in C. M. Woolgar, D. Serjeantson & T. Waldron (eds) *Food in Medieval England: diet and nutrition*: 191–200. Oxford: Oxford University Press.

Yoder, C. 2010. Diet in medieval Denmark: a regional and temporal comparison. *Journal of Archaeological Science* 37: 2224–36.

Medieval Sea Fishing, AD 500–1550: Chronology, Causes and Consequences

James H. Barrett

Introduction

Major changes in the importance of sea fishing and marine fish consumption during the Middle Ages were both punctuated and incremental. The trade in sea-fish was neither trivial nor static for European communities around the northern North Atlantic Ocean and the Baltic, North and Irish seas. Its volume could exceed that of the ever-important wine trade (Chapter 2). Its foci of production and transhipment shifted through time and space, with concomitant social, economic and political ramifications. Major regional differences were influenced by natural species distributions and a complex mix of socio-economic factors. Yet many temporal developments also followed similar trajectories in different locations. Others had very widespread influence, integrating much or all of northern and western Europe into the well-known exchange networks of the High Middle Ages – be it the trade of East Anglian herring (*Clupea harengus*) for Gascon wine or of Arctic Norwegian stockfish for English cloth, German beer and Baltic rye.

As is noted in Chapter 1, investigating the ebb and flow of different fisheries over the *longue durée* and large geographical distances comes with methodological complexities. Systematic and quantitative historical records only emerge in the fourteenth century – near the end of the study's chronological range – and not all regions are equally well documented in all periods. Archaeological data from fishing gear (e.g. tidal traps), human skeletal remains (e.g. stable isotope values) and, most important, fish bones suffer less chronological bias. Nevertheless, the interpretation of this source material is complicated by a wide variety of other confounding factors, such as how well bones were preserved and how methodically they were recovered during excavation. The degree to which seemingly esoteric finds such as fish bones have been the focus of study and publication is also highly variable. In sum, it is likely that our archaeological evidence is based on only a tiny and biased sample of what was initially consigned to the earth and that many aspects of fishing (including most activities conducted at sea) will not have entered the ground in the first place.

Accepting these obvious limitations, this chapter aims to compare and contrast what we do know, based on the wealth of information from the contributions to this volume, in order to infer a plausible outline of the archaeology and history of medieval fishing in the north of Europe from AD 500 to 1550. In so doing, it may provide a starting point for a more comprehensive future understanding of the complex relationships between human societies and the sea.

Fifth to mid-seventh centuries AD

During the earliest years of the Middle Ages, considered prehistory in many parts of northern Europe, there is no reliable documentary evidence regarding sea fishing. The availability of relevant archaeological evidence is highly variable, but this evidence is nevertheless sufficient to show differing regional traditions. The recovery of sea-fish remains is rare among the former Roman provinces of the North Sea coast, in terms of both the number of known sites and the number of marine fish bones recovered. For example, Carlton Colville, a rural settlement of the sixth–early eighth centuries AD in Suffolk, England, yielded only 44 identified fish bones, of which nine were from obligate marine species (Parks and Barrett 2009). An additional 15 were of flatfish, probably plaice (*Pleuronectes platessa*) or flounder (*Platichthys flesus*). These flatfish may represent coastal fishing, but, as noted in Chapter 1, flounder can be caught in both fresh and salt water. A small, early Anglo-Saxon (sixth to early seventh century) assemblage from Lyminge in Kent is also notable, having 118 herring bones among its 156 identified specimens (Chapter 17; Gabor Thomas pers. comm.). The neighbouring continental evidence tells a similar story. In Belgium, France and the Netherlands, marine fish bones are rarely found on settlements of this early date. When they do occur, it is either in coastal locations (as in a seventh-century context from The Hague: Magendans and Waasdorp 1989) or in assemblages where there is a strong possibility that they represent residual material of Roman date (as in sixth- to seventh-century contexts from Tournai: Chapter 14).

The limited importance of sea fishing during the fifth–mid-seventh centuries AD is also a feature of, on the one hand, coastal Scotland and Ireland and, on the other, the eastern Baltic lands of Poland and the Baltic states. In coastal Scotland there was some use of sea-fish, shellfish and cetaceans at this early date (e.g. Barrett *et al.* 1999; 2001; Cerón-Carrasco 2005; Hardy 2015; Harland 2006; Mulville 2002; Nicholson 2010; Sharples *et al.* 2015, 251–2), but their dearth in comparison with the very large quantities of herring (in the west) and cod family (Gadidae) fish (in the west, north and east) known from the Viking Age (c. 790–1050) and later is striking (e.g. Harland and Barrett 2012; Sharples *et al.* 2015). Assessment of human diet based on stable isotope analysis of human burials from Orkney in northern Scotland tells a similar story – the Pictish population of pre–Viking Age Scotland consumed marine protein on only a modest scale (Barrett and Richards 2004; Chapter 20). In Ireland, no marine fish bone assemblages of specifically fifth- to mid-seventh-century date are known (Chapter 19), but estuarine fish traps of this age suggest that at least migratory species were exploited (e.g. O'Sullivan 2001, 138; Chapter 9). Coastal shell middens of this early date are also known, although they are better represented in the following centuries (O'Sullivan *et al.* 2014, 116–19).

In the eastern Baltic, there is even less evidence for marine fish consumption in the middle centuries of the first millennium AD. Although sea fisheries had been important earlier in prehistory, especially during the Neolithic, fish such as cod (*Gadus morhua*) and herring do not reappear in the archaeological record of Poland until the eighth to ninth centuries AD (Chapter 12) and of Estonia until after AD 1000 (Chapter 11).

The reasons for this widespread avoidance of sea fishing must differ given that it occurs in locations as disparate as (for example) Estonia, Poland, Belgium, England, Scotland and Ireland. In the eastern Baltic, a reduction in salinity associated with postglacial changes in relative sea level may have contributed to the end of the Neolithic fishing boom (Chapter 11). In some regions bordering the North Sea it has been suggested that there was a culturally driven avoidance of fish amongst indigenous Iron Age peoples (Dobney and

Ervynck 2007; Chapter 14), a tradition that may have been reinvented in the wake of collapsing Roman influence in the west. Dio's *Roman History* of the early third century AD actually claims that the Caledonians (residing in what is now Scotland) did not eat fish (Cary 1969, 263; but cf. Russ *et al.* 2012). Important factors in post-Roman Europe must have included limited urban and elite demand, the low cargo capacities of contemporary ships and a relative emphasis on the exchange of high-value rather than staple goods (Chapter 4; cf. Barrett *et al.* 2004a; Loveluck 2013; Wickham 2005). In Ireland, one might speculate that the key importance of pastoralism – particularly cattle husbandry – diverted labour from the sea (see McCormick 2008; 2014). Ultimately, however, all such interpretations are limited by the fact that they are arguments from absence, built on the paucity of available fish bone evidence.

Scandinavia provides a distinct contrast to the widespread pattern of limited sea fishing in the fifth–mid-seventh centuries AD. Here there is evidence that marine fishing, for cod family species (especially cod itself), herring, or both, was practiced during the Migration Period (AD 400–570), and in some cases even as early as the Bronze Age. Cod and related species were the most important taxa in Norway, whereas both gadids and herring (together, or at separate sites) played significant roles in the Danish and Swedish islands of the western Baltic (Chapter 13; Chapter 18). A particularly striking example is Sorte Muld on Bornholm, where more than 13,000 herring bones dating to the sixth–seventh centuries AD were recovered from a central-place settlement (Chapter 13). Looking west and north, seasonal fishing settlements dating to the Migration Period have been excavated at coastal locations in Hordaland in western Norway (Johannessen 1998, 43) and on the island of Flakstad in the Lofoten archipelago, a later focus of commercial stockfish production (Wickler and Narmo 2014, 84–5). Diverse environments and cultural contexts are represented by this early Scandinavian sea fishing. Nevertheless, they all shared an insular or coastal setting, access to abundant stocks of marine fish and – perhaps most important – a maritime-oriented culture, as known from material culture and rock art in addition to fish bone evidence (e.g. Crumlin-Pedersen 2010; Lødøen and Mandt 2010).

The 'long eighth century' AD

There is evidence that patterns of sea fishing began to change in what may usefully be called the 'long eighth century', from the middle of the 600s to the early years of the 800s. This period shows emerging – albeit still small-scale – use of sea-fish in a number of regions where it was previously very difficult to detect. In England, the evidence comes from the earliest post-Roman towns (sometimes referred to collectively as *wics*, a term of convenience that will be adopted here) at what are now York, Ipswich, London and

Southampton (Barrett *et al.* 2004a); a salt-production site at Fishtoft in Lincolnshire (Locker 2012); a beach market at Sandton on the English Channel coast (Hamilton-Dyer 2001); and – in the case of southern England – high-status central places (Chapter 17). It is also relevant that radiocarbon dates on intertidal fish traps, especially along the Essex, Suffolk and Norfolk coasts of eastern England, cluster between AD 600 and 900 (Murphy 2009, 48). Moreover, assessment of human diet using stable isotope analysis suggests small-scale consumption of marine protein as early as the seventh–tenth centuries AD on the eastern coast of Norfolk, a region famous for its abundant herring in later centuries (Hull and O'Connell 2011; Chapter 20).

A variety of marine taxa were caught in England during this developmental phase. In most cases, however, bones of flatfish and herring are the most abundant finds, with smaller numbers of other species, such as cod and/or whiting (*Merlangius merlangus*). The significance of the flatfish is a little ambiguous. Many were presumably caught in coastal and estuarine waters to be used as food. As is noted above, however, flounder are also found in fresh water. Moreover, small flounder are often eaten by eels (Maitland and Campbell 1992, 248), a migratory species that is abundantly represented in English sites of the late seventh–early ninth centuries AD (and later). Thus some of the small flatfish bones that comprise the majority of potentially marine fish remains at eighth–ninth-century Fishtoft could conceivably have arrived onsite in the stomachs of eels – given that the latter are the most abundant taxon in the assemblage (Locker 2012). A similar explanation may apply to the smallest flatfish recovered from broadly contemporary phases of the elite settlement of Flixborough near the Humber estuary, where very few other bones of marine fish were found despite extensive sieving with fine mesh (Dobney *et al.* 2007, 40–1).

Regardless of what proportion of flatfish remains represent fish caught in the sea rather than in fresh water, and the food of people rather than eels, it is clear that the herring and (in lesser quantities) cod bones found in England's earliest trading sites represent the emergence of a small-scale fishery supplying town-dwellers. It is not yet possible to tell whether this development represents commercialisation, or instead indicates provisioning by elite patrons using renders of sea-fish from coastal estates. This is an important question at the core of debate regarding the nature of England's early medieval trading settlements (cf. Crabtree 2010; Hammerow 2007; Holmes 2014, 15; O'Connor 2001). Regardless, it is probable that the sea-fish found at elite centres such as Bishopstone and Lyminge, near the Channel coast of England, derived from estate holdings – possibly even Sandtun in the case of Lyminge, although there is uncertainty regarding the location indicated by the relevant charter (Chapter 17). By the mid-ninth century, coastal fisheries belonging to ecclesiastical estates elsewhere

in England are also known from the textual record (e.g. Fox 2001, 47).

The neighbouring continental evidence tells a similar story. In Belgium, France and the Netherlands, bones of sea-fish (especially herring) begin to occur in very small numbers from the eighth century onwards, at urban or elite sites upriver from their marine source. Examples include herring bones of eighth-century date from Paris, Reims and Compiègne in northern France (Clavel and Yvinec 2010), of eighth–ninth-century date from Namur in Belgium and of eighth–ninth-century date at Dorestad in the Netherlands (Chapter 14). Small numbers of plaice and haddock (*Melanogrammus aeglefinus*) bones were also found at the latter site (Prummel 1983, 90, 94). A very small number of cod bones dating to the eighth century have also been recovered from an excavation along the Oude Rijn (Old Rhine) river near Utrecht (Beerenhout 2009). All of these species were probably caught at coastal settlements that combined diverse economic activities; specialist fishing villages are not know at this date (Chapter 14).

In the case of both the continent and England, the earliest, small-scale growth of medieval sea fishing may have been driven by the demand of urban populations and elite central places. Conversely, the situation in Ireland and Scotland during the 'long eighth century' may show more continuity than change. Neither region had adopted urbanism, coinage or other aspects of incipient commercialisation at this date. In Ireland, the main focus of economic intensification in the 'long eighth century' may have been innovations in arable agriculture – as evidenced by the distribution of radiocarbon and dendrochronological dates on water mills, for example (Kerr *et al.* 2009; McCormick 2014). Nevertheless, substantial coastal fish traps continued to be built, as at Chapel Island in Strangford Lough (Chapter 9). No fish bone assemblages are dated exclusively to the 700s, but eighth- to ninth-century material from the monastery of Illaunloughan, on an island off the southwest coast, included a diverse range of 18 taxa. The most abundant were sea bream (Sparidae), wrasse (Labridae) and hake (*Merluccius merluccius*), with lesser numbers of gadids, such as pollack (*Pollachius pollachius*) and cod. All parts of the skeleton were represented. This fishery was not specialised and probably entailed local activity.

In Scotland, where preservation conditions have seldom favoured tree-ring dating, it is difficult to differentiate between developments of the eighth and ninth centuries using archaeological methods. Nevertheless, by the mid-AD 800s there was clearly Scandinavian influence in northern and western Scotland, not least on diet and subsistence practices, including a marked increase in the importance of sea fishing (see below). There remains a transitional archaeological period, however, for which we cannot yet be certain whether this Scandinavian influence was preceded by a small-scale increase in indigenous use of marine resources (e.g. Barrett *et al.* 2001, 148; Nicholson 2007).

The eastern Baltic region has resonances with contemporary circumstances in Ireland and Scotland. Coastal Estonians continued to eschew towns and marine fish at this date (cf. Chapter 11; Mägi 2015). In contrast, areas such as the Polish and German Baltic coasts participated in both the emerging urbanism of the 'long eighth century' and the first beginnings of medieval sea fishing. The earliest herring bones from medieval Poland, dating to the first half of the eighth century, are from the near-coastal trading site at Truso, part of the network to which the North Sea *wics* also belonged (Makowiecki 2012; Chapter 12). Cod bones, conversely, are not known from Poland until later (see below). Near the western limits of Baltic Slavonic settlement, at the eighth- to early ninth-century trading site of Groß Strömkendorf on the Wismar Bight in Germany, herring is the most common fish taxon, and both cod and whiting are also well represented (Schmölcke 2004, 30; see Müller-Wille and Tummuscheit [2004, 37] regarding the chronology of the site). The abundance of cod is atypical for a Slavonic settlement of this date, raising the possibility of Scandinavian influence in addition to the emergence of urban demand (Müller-Wille and Tummuscheit 2004, 36–7; Schmölcke 2004, 119).

In Scandinavia, conversely, the 'long eighth century' was characterised by a continuity of fishing patterns established by the Migration Period. Nevertheless, there may also have been some intensification of activity associated with elite and proto-urban demand. Continuity of herring fishing in the Baltic region can be assumed based, for example, on the eighth–ninth-century phase of Selsø Vestby on Zealand and perhaps also on the fifth–eighth-century phase of Eketorp on Oland (Enghoff 1999, 57; Chapter 13). Two ling (*Molva molva*) bones from the early phase of Eketorp (geographically beyond the typical range of this species) – and the presence of a few large cod vertebrae without associated cranial bones – may imply some trade of preserved gadids at a very early date (Enghoff 1999, 56–8, 63–4). Alternatively, these specimens could be intrusive from the overlying eleventh- to fourteenth-century phase, from which almost 60,000 fish bones were recovered. Regardless, we know gadid fishing was practiced during the 'long eighth century' in southern Scandinavia, with (albeit imprecisely dated) cod of a range of sizes being represented in finds from central places, such as Tissø and Gammel Lejre (Chapter 13). There were also cod and haddock remains in the eighth-century phase of the trading settlement of Ribe on the North Sea coast of Jutland. They were of small individuals – 35–65 cm total length (TL) in the case of cod – and included all parts of the skeleton, implying local fishing. Unlike the occupants from many other early *wic* sites – from Truso in the east to Hamwic (Southampton) in the west – those of Ribe seem not to have consumed herring. Instead, the gadids were found together with larger numbers of flatfish bones.

Norwegian fishing – mostly for cod and related species – showed some evidence of intensification within an overall pattern of continuity. In Hordaland there was greater use of fishing stations from the later part of the Merovingian Period (AD 570–800) onwards (Johannessen 1998, 44). In Lofoten, seasonal fishing continued and may have expanded. It is evidenced by excavations at Flakstad and Borgvær (Wickler and Narmo 2014). It is unlikely to be a coincidence that both Lofoten sites were within the political sphere of the Migration Period to Viking Age central place at Borg in Lofoten (Wickler and Narmo 2014), where the largest longhouse known in Scandinavia (*c.* 83 m) was built around the seventh century AD (Näsman and Roesdahl 2003). Thus it has been argued that the early (pre-commercial) expansion of Norwegian fishing was driven by the demands of chiefly patrons (Chapters 4–6).

AD 850 to 1050: the 'fish event horizon'

In the years around the turn of the first and second millennia AD there was an increase in the importance of sea-fish across most of the geography of present interest. The initially tongue-in-cheek label the 'fish event horizon' (Barrett *et al.* 2004a) has acquired some currency as a convenient shorthand for this complex phenomenon (e.g. Frantzen 2014, 233; Perdikaris and McGovern 2008, 203). The character, scale and precise chronology of change varied with location. Nevertheless, there were clear inter-regional resonances. More bones of marine fish (in both absolute numbers and in relative terms *vis-à-vis* freshwater and migratory taxa) are found at more sites and/or at greater distances from the sea.

The English evidence derives from a diversity of sites (e.g. Barrett *et al.* 2004a and references therein), but the time-series from York is one of the most informative. It is large, well dated, well preserved, well recovered and (with minor exceptions) identified using comparable laboratory methods (Chapter 15). Here we see a major increase in the abundance of herring during the middle of the tenth century, when it replaced eel as the most frequent taxon by number of identified specimens (NISP). At the same time, gadid fish (in this case mainly cod itself) first began to appear in more than trace numbers. Gadids became frequent by the middle of the eleventh century, outnumbering freshwater and migratory species other than eels by NISP. The story from London is similar. Far more herring remains are known from around the turn of the millennium than in previous centuries. Gadids such as cod and whiting appear in meaningful numbers for the first time and flatfish also occur in large numbers (Locker 1997; Orton *et al.* 2014; Chapter 16). In the south of England, marine fish (mostly herring) replaced freshwater and migratory species (mostly eel) as the most abundant fish in Hamwic/Southampton by AD 1030 (cf. Colley 1984; Coy 1977; Hamilton-Dyer 1997). These examples are sufficient to illustrate a general trend.

It is also relevant that the marine fish represented in zooarchaeological assemblages from earlier English sites, particularly southern English sites, such as Sandton, Bishopstone and Lyminge, rarely translate into evidence for the consumption of measureable amounts of marine protein based on stable isotope analysis of human bones (Chapter 20). With the exception of probable Scandinavian migrants (e.g. Pollard *et al.* 2012), only burials from Caister-on-Sea and Burgh Castle near the Norfolk coast show even limited evidence of a marine diet before the eleventh century (Hull and O'Connell 2011). At York, the first probable locals from medieval England to have clearly adopted a marine diet were a minority of young men who died in the mid–late eleventh or early twelfth centuries and were buried in the parish cemetery of St Andrew, Fishergate. It seems reasonable to infer, as Müldner (2009; Chapter 20) has, that these men had been directly involved in the growing sea fishery.

The organisation of the growing English fisheries of *c.* AD 950–1050 can be inferred by back-projecting from Domesday Book of AD 1086 and later sources. In Domesday we observe the role of herring as renders from both rural estates and coastal ports (Chapter 3), such as Dunwich, an important Suffolk town that now lies under the North Sea (Sear *et al.* 2011). Campbell (2002) has estimated that the Domesday renders alone may imply an annual catch of at least 3,298,000 herring. During the centuries under consideration, these fish were probably salted and dried or salted and smoked, rather than barrelled in brine (Chapter 1; Chapter 3). Coastal, river and inland trade must have brought the fish to centres such as York and London, where their bones have been excavated.

It remains an open question to what degree English herring were serving distant markets during the 'fish event horizon'. Given the centrality of English herring exports to the later (thirteenth–fourteenth centuries) wine trade with Gascony (Littler 1970, 203, 224–5; see below), it may be reasonable to speculate that many preserved English catches were sent up the Rhine and the Seine, which were the earlier conduits of continental wine to England (see Rose 2011, 17, 61). If we accept the possibility of some continuity into later centuries, it is also relevant that fish were the first item in Henry of Huntingdon's list of English exports to Germany *c.* 1130 (Oksanen 2013, 179). One might speculate that in the tenth and eleventh centuries, England's abundant herring were traded for both Rhenish wine and the influx of German silver thought to have influenced the coinage of late Anglo-Saxon England (cf. Sawyer 2013, 28, 98).

Unlike herring, gadid fish are invisible in Domesday and other historical sources of a similarly early date (Chapter 3). Yet bones of cod do occur in tenth- and eleventh-century assemblages from England (e.g. Barrett *et al.* 2004a; Chapters 15–16). At present, stable isotope analysis of fish bones suggests that very few dried cod from the Scandinavian North Atlantic region reached England in the ninth–eleventh centuries. Instead there was an increase in local cod fishing, probably driven by the demand of expanding urban populations (Barrett *et al.* 2011).

The apparent discrepancy between historical and archaeological evidence regarding cod fishing around AD 1000 can probably be understood in light of the much later accounts of the Dunwich fishery preserved in the Bailiffs' *Minute Book* of 1404–30 (Bailey 1992). Here we see that the main seasonal fishery was for herring and the related sprat (*Sprattus sprattus*), but that this focus was supplemented by opportunistic fishing for cod (and other taxa) (Bailey 1992, 16). For many English fishermen – of the tenth–eleventh centuries and later in the Middle Ages – gadid fish and other marine taxa probably represented important secondary catches in a calendar (and taxation system) focused mainly on herring.

From the perspective of environmental history, a final key aspect of the English 'fish event horizon' is whether a decline in freshwater resources was one cause of the move to the sea (Barrett *et al.* 2004b; Hoffmann 1996; 2002). Opinion is divided (cf. Chapter 3; Chapter 15). It is clear that freshwater and migratory fish did continue to be consumed in England after AD 1000. Nevertheless, the large and well-recovered zooarchaeological dataset from York allows us to observe important trends that may be obscured elsewhere (Chapter 15). The absolute abundance of freshwater fish remains declined through time, and the average TL of one of the most abundant obligate freshwater species, pike (*Esox lucius*), also decreased. Large individuals (50–80 cm TL) were well represented until the middle of the eleventh century but almost absent after that. These patterns suggest the possibility of overfishing. In addition, it has been suggested that the disappearance of grayling (*Thymallus thymallus*) and burbot (*Lota lota*) from the York fauna was due to increasing water pollution (O'Connor 1989, 198). What is not yet known is the degree to which we can generalise from the picture of a single town.

Crossing the English Channel, we see that the chronology of increased sea-fish consumption is similar in Belgium (Chapter 14). At present, the key assemblage is from the trading centre, or *portus*, of Gent (Ghent), where sea-fish appear in layers of mid-tenth-century date. The most abundant marine taxa are herring and flatfish. Some of the latter may represent flounder caught in freshwater, but plaice (which does not migrate upriver) were identified among the remains of this group despite the difficulty of attributing flatfish bones to species (Wouters *et al.* 2007; Chapter 14). Gadids are not represented among these earliest sea-fish remains from Ghent, only appearing (and then in very small numbers) in the eleventh–twelfth centuries. In this instance, the relevant species was haddock. Another important assemblage comes from a short-lived settlement at Ename, dating to the late tenth and eleventh centuries. It is a small and unrepresentative collection, having been recovered by hand rather than sieving with mesh, but it did include herring, plaice and a single cod bone. The settlement, which developed near a fortification, is thought to have had trade functions, reinforcing the link between early sea fishing and emergent urbanisation. Van Neer and Ervynck (Chapter 14) rightly caution that earlier evidence of marine fishing may yet be found in Belgium, pointing out the occasional eighth-century continental finds mentioned above. Yet on the present evidence the assemblage from Ghent provides a synchronous parallel for those from English towns such as York or, more appropriately (given the relative abundance of flatfish bones), London.

Synthesis of the medieval fish bone record from the Netherlands is still in progress (Inge van der Jagt pers. comm.). Based on existing evidence, it appears there was some limited use of sea-fish, including gadids and flatfish, at terp sites (raised settlements) of the Wadden Zee (Wadden Sea) area throughout the first millennium AD (Prummel and Heinrich 2005). There were also small numbers of herring, flatfish and gadid bones inland, at Dorestad and near Utrecht in the 'long eighth century' (Beerenhout 2009; Prummel 1983, 94, 108; see above). Nevertheless, marine fish are rare finds in the Netherlands prior to the turn of the tenth and eleventh centuries AD, based on the records of the on-line Dutch zooarchaeological database (BoneInfo nd; Lauwerier and de Vries 2004). The regular appearance of species such as cod in the faunal record begins with examples from such settlements as Vlaardingen *c.* AD 1000–1050 (Buitenhuis *et al.* 2006, 67) and Deventer between the ninth and twelfth centuries (IJzereef and Laarman 1986, 453). If sea fishing had been important in earlier centuries, one would expect to see the evidence at a site like Domburg, a coastal stronghold and market centre with a peak in activity during the ninth and tenth centuries. Yet this settlement has produced very few fish bones despite excellent preservation conditions; there were seven from the most recent excavation, of which only one (a cod) was unambiguously marine. There were no herring bones in either the hand collected material or the environmental samples (Buitenhuis 2011).

In parts of Scotland, there is unambiguous evidence for a major increase in the importance of sea fishing during the ninth and tenth centuries. Areas of the country influenced by the Scandinavian diaspora of the Viking Age saw holistic changes in material culture and subsistence at this time (Barrett 2003). Among the new cultural practices were catching and consuming more and larger cod family fish, particularly cod, saithe (*Pollachius virens*) and ling (Barrett *et al.* 2001). This new development is most evident in the Northern and Western Isles, at sites such as Old Scatness (Nicholson 2010), Pool (Nicholson 2007), Quoygrew (Harland and Barrett 2012), Bornish (Sharples *et al.* 2015) and Bostadh (Cerón-Carrasco 2011). By the eleventh–twelfth centuries, there was yet further intensification of

marine fishing – evident from zooarchaeology (Barrett 1997; 2012; Barrett *et al.* 2004a) and the assessment of human diet using stable isotope analysis (Barrett and Richards 2004; Chapter 20). By this time it is likely that unsalted dried gadids (stockfish) were being produced for the payment of local renders and possibly also for export trade (Barrett 1997; 2012; Harland 2007). The emergence of large-scale sea-fish consumption in the extensive areas of Scotland without overt Scandinavian influence is discussed below in the context of developments in the High Middle Ages.

Within the Scandinavian-influenced zone of northern and western Scotland, there is a contrast between the north, where fishing for cod family species was the main focus, and the west, where herring were of greater importance (cf. Barrett *et al.* 1999; Cerón-Carrasco 2011; Serjeantson 2013; Sharples *et al.* 2015). This difference cannot be explained based on the distribution of the relevant species. It may instead relate to differing communities of practice in the Northern and Western Isles. There has been some discussion regarding whether western herring fishing developed in part for trade – to the Hiberno-Scandinavian town of Dublin, for example. Although not inconceivable, it remains to be established where, at this date, Hebridean islanders could have procured the large supplies of salt usually necessary to preserve this species for long-range trade (Sharples *et al.* 2015, 253).

There is not yet abundant fish bone evidence for Ireland from the ninth to eleventh centuries. Like their antecedents (see above), coastal settlements of the west continued to use a diverse range of marine species – perhaps indicating expedient rather than specialised fishing. Moreover, taxa such as wrasse suggest a focus on inshore activity at settlements like Doonloughan (Chapter 19). However, large gadids may have been favoured in newly established Hiberno-Scandinavian towns, such as Dublin and Wexford (McCarthy 1998, 61; Chapter 9; Chapter 19). Herring may also have been important in these urban settlements, based on finds noted in botanical samples of tenth–early eleventh-century date from Fishamble Street in Dublin (Geraghty 1996, 55). Much research regarding fish bone from Ireland's Hiberno-Scandinavian towns remains for the future.

Turning to the eastern Baltic region, we see that in Poland the late ninth to eleventh centuries witnessed the growth of sea fishing for herring and the first evidence for inland trade of this fish (Chapter 12). Herring bones became abundant in coastal sites during the tenth century, at the trading site of Wolin and the strongholds of Kołobrzeg-Budzistowo and Sopot. In the tenth–eleventh centuries, they appeared far inland for the first time, at Grzybowo, a stronghold of the Piast dynasty, 250 km from the coast. By the eleventh century, they occurred at a number of inland sites – Wrocław, Poznań, Kruszwica and Kałdus 2 – all of which served as elite strongholds at this date.

It is clear from this inland transport alone that the early catches of herring from the Baltic waters of Poland were being cured. The method may be indicated by the eighth-century deposit of herring bones from the trading site of Truso (introduced above) and an assemblage from Kołobrzeg-Budzistowo dating to between the ninth and thirteenth centuries. In both cases, the cleithra are missing (Chapter 12). This is one of the characteristics of the gutted and pickled cure made famous by later Scanian and Dutch fisheries (cf. Chapter 2; Chapter 13; Chapter 15).

In Chapter 12 it is argued that this growth of herring fishing and trade was probably related to Polish state formation under the Piast dynasty in the tenth century, combined with the concurrent adoption of Christianity (with its fasting practices that encouraged fish consumption). It is also likely to be relevant that the Piasts conquered both the trading settlement of Wolin (with its herring fishery) and the stronghold of Kołobrzeg (with its salt springs). A new socio-economic network combined with increased elite demand and a new Christian ideology to create conditions favourable for an increase in sea fishing around the turn of the first and second millennia AD.

The chronology of growing herring fishing and trade seen in Poland has also been observed in the previously Slavonic areas of the Baltic coast of Germany – at ninth–tenth-century Menzlin, for example – and perhaps also on the island of Rügen, which appeared in the textual record as a major centre of herring fishing and trade by the twelfth century (Benecke 1982, 286; 1987; Enghoff 1999; Chapter 2). Conversely, at the eastern limits of the Baltic region, marine fish do not seem to have had either a social or an economic role until later in the Middle Ages. Unlike Poland and Germany, Estonia is distinctive in that even at the Viking Age harbour site of Tornimäe, on Saaremaa Island, there is no evidence of sea fishing among the many bones of a well-preserved and sieved assemblage (Chapter 11).

Unlike most areas discussed, Scandinavia showed much continuity in sea fishing during the Viking Age. Nevertheless, there are some hints of increases in the intensity of fishing and the distances over which fish were transported. Some herring were transported to relatively local inland centres, such as the town of Viborg Søndersø in Jutland (Chapter 13). Herring were particularly abundant at Haithabu (in modern Germany, but part of the kingdom of Denmark when occupied in the Viking Age), where they could have been easily trapped in the adjacent Schlei fjord (Lepiksaar and Heinrich 1977, 36; Radtke 1977). This site is also notable for having yielded bones from saithe, ling and halibut (*Hippoglossus hippoglossus*), all species more likely to have been caught in the Skagerrak, North Sea or North Atlantic than in the Kattegat or Baltic Sea (Heinrich 2006, 186–9; Lepiksaar and Heinrich 1977, 90–4, 106–7; Schmölcke and Heinrich 2006, 220–9). These finds co-occur with small numbers of cod bones that may also be non-local catches based on carbon and

nitrogen isotope analysis (Barrett *et al.* 2008; cf. Becker and Grupe 2012). It is tentatively proposed that the cod arrived as stockfish imported from Norway, an explanation that could equally apply to the saithe and the ling. Moreover, cured halibut is known to have been another (albeit rarer) Norwegian export in later centuries (Nedkvitne 2014, 364). The remains of these four species have been interpreted as provisions of mariners trading more precious cargoes (Heinrich 2006, 188–9; Lepiksaar and Heinrich 1977, 118). Nevertheless, it is possible that they provide the earliest evidence for a pattern of transport that was later commercialised.

Very small numbers of herring and cod bones at the trading town of Birka in Sweden, where remains of freshwater fish dominated the assemblage, could conceivable also represent trade goods (Chapter 11). Elsewhere in Sweden, Viking Age fishing is evidenced at a variety of coastal sites, with herring, gadids (especially cod) and flatfish all well represented. These remains are consistent with local catches (Enghoff 1999), although ongoing stable isotope studies may ultimately revise this interpretation (Sabine Sten pers. comm.).

Few Viking Age fish bone assemblages are known from the Skagerrak. A small collection (mainly from waterlogged pits) of ninth–tenth-century date from the trading site of Kaupang confirms the widespread role of herring, while also including smaller numbers of gadids, such as cod. As at Birka, we do not know whether these remains indicate local fishing or exchange (Barrett *et al.* 2007). Fish bones from Oslogate 4, dating to the early (eleventh-century) years of Oslo's existence as a town, represent whole cod and thus probably local fishing rather than trade (Chapter 18). Moving to the North Sea coast, the trading site of Ribe has not yet produced an analysed fish bone assemblage dating between AD 850 and 1050. Its early ninth-century phase yielded no herring, and at this date the settlement instead relied mainly on small haddock and cod, probably caught locally.

Viking Age fishing in western and northern Norway showed much continuity from the previous Merovingian Period. Ephemeral, probably seasonal, fishing structures remained the norm in the Lofoten islands (at Borgvær, for example), where the harvest of the sea must have continued to be a contributor to the wealth of chieftains at such centres as Borg. The main target species was cod, with small numbers of saithe and other taxa also being caught (Wickler and Narmo 2014, 81). Fishing booths like those in Lofoten also continued to see use in Hordaland, in western Norway (Johannessen 1998, 44). Nevertheless, there is evidence for an increase in fish consumption, at least in northern Norway, from the Merovingian Period to the Viking Age, based on stable isotope analysis of human skeletal remains (Naumann *et al.* 2014).

The consumption of cod and other marine taxa reached deep into Norway's inner fjords and perhaps even its mountainous interior. At eighth–tenth-century Bjørkum, *c.* 12 km from the innermost reach of the Søgnfjord system, small numbers of herring, cod, saithe, pollack, and tusk (*Brosme brosme*) (be they from the nearby fjord or the distant open sea) were found together with more numerous salmonid bones representing catches available from the immediately adjacent Lærdal river (Barrett *et al.* 2015). Much farther inland, a few bones of herring, cod, saithe, haddock and other sea-fish were also recovered from Vesle Hjerkinn, a royal farm and mountain lodge for travellers over the Dovre mountains in central Norway (Lie and Fredriksen 2007, 164). The Vesle Hjerkinn chronology is ambiguous because there is no differentiation between Viking Age and later (eleventh–fourteenth-century) bones from the site. It is an important example, however, because both cranial and postcranial cod bones were recovered at a location to which only preserved sea-fish could realistically have been transported. Thus some Norwegian stockfish were made without decapitating the fish first – or some fish heads were also dried for long-range transport. The existence of occasional travelling fish heads obviously limits the resolution of the main methods used to recognise traded gadids – anatomical patterning and stable isotope analysis (Chapter 1). Some imports may thus go unnoticed based on archaeological evidence. However, it is relevant to note that this observation does not undermine cases in which processed and/or isotopically non-local fish *are* recognised.

As is noted above, the Scandinavian practice of harvesting cod and other sea-fish spread to west-Norse settlements in Scotland and Ireland during the Viking Age. It was also an important component of the socio-economic package transported to Iceland during its permanent settlement in the late ninth and tenth centuries. Here, as in Norway, cod and other gadid fish were air dried without salt for later consumption. Some were transported to interior sites, such as Sveigakot in Mývatn, as part of a complex Viking Age provisioning network that long preceded Iceland's entry into the international stockfish trade in the thirteenth century (Perdikaris and McGovern 2009; Chapter 7). In this instance, zooarchaeological evidence shows that these transported preserved fish were invariably decapitated (Perdikaris and McGovern 2009, 85).

The High Middle Ages: AD 1050 to 1350

The centuries between AD 1050 and 1350 were characterised by the increasing importance of sea fishing in areas where it was already practiced, the spread of this activity to those places in the north of Europe where it had not previously been adopted, and the emergence of a pan-European fish-trade network. These changes entailed both incremental and punctuated developments. They also involved both shifting foci of production and consumption and shifting networks of exchange. The end of the period was characterised by the well-documented and widespread trade of salted herring from the western Baltic and dried stockfish from Norway,

largely monopolised by member towns of the German Hansa under the leadership of Lübeck. However, this apogee of the medieval fish trade emerged from diverse origins, some driven by rural suppliers as much as by urban merchants and consumers.

Beginning in England, we know from written sources that the herring fishery of the east and south coasts was increasing in scale during the eleventh century (Chapter 3). In addition to the seasonally abundant shoals of fish, it was the surprisingly large-scale salt production (based on boiling sea water), especially in East Anglia, that made this possible (Campbell 2002). The catch was cured by various combinations of salting, drying and smoking, not yet involving barrels (Chapter 1; Chapter 3).

This herring fishery continued to be conducted from both rural estates and ports. The latter shifted in relative importance through time. Most famously, the central role of Dunwich was overtaken by the comparatively recent settlement of Great Yarmouth in the course of the twelfth and thirteenth centuries. The Yarmouth fair was seasonally attended by many continental fishermen who landed their fish for curing. Forty German vessels are mentioned in a twelfth-century hagiography (Campbell 2002, 6; Chapter 3), and sources from the early fourteenth century refer to between *c.* 500 and *c.*1000 fishing boats (Saul 1982, 78). In 1295, Edward I issued an order protecting fishermen from Holland, Zeeland and Friesland at Yarmouth (Unger 1978, 341). King's Lynn and Boston, important new coastal centres of the late eleventh to early twelfth centuries, were more ports than fishing harbours, albeit very significant importers of Scandinavian stockfish (Carus-Wilson 1962–3; Nedkvitne 2014, 31, 55; Chapter 5). To the north, Grimsby developed as a centre of herring and cod fishing and fish trade in the twelfth and thirteenth centuries, only to be challenged by the growth of Hull during the early thirteenth century and the short-lived efflorescence of Ravenserodd (situated only a few kilometres away, across the Humber estuary, near what is now Spurn Head) between the 1230s and the mid-fourteenth century (Beresford 1988, 513–14; Rigby 1993, 7–12, 31–3). Yet farther north, fishing was conducted from (for example) Bridlington, Whitby and Scarborough (Chapter 3). To the south, there were large-scale herring fisheries in Kent and Sussex (Tsurushima 2006; Chapter 3).

In all instances there was a hierarchy of major and minor fishing centres jostling for relative advantage. Thus, for example, Dunwich contended with local rivals Blythburgh and Walberswick (Bailey 1992, 11–12) in addition to Great Yarmouth. Most of the fishermen themselves probably resided in rural estates, putting to sea as a seasonal activity, but (smuggling aside) fish were often landed in ports. It can be assumed that this was partly to facilitate salting, and partly related to tithe and market regulations. Contemporary sources record how renders, such as tithe, should be paid and which market regulations (intended to prevent practices such as forestalling) must be adhered to (Rigby 1993, 11; Chapter 3)

Both zooarchaeological and textual evidence informs us that the harvests of these fisheries were consumed across England, in towns, monasteries and rural settlements (e.g. Barrett *et al.* 2004a; Colson 2014; Serjeantson and Woolgar 2006). As is argued above, herring may also have been exchanged for continental wine in the eleventh and twelfth centuries. This was certainly the case by the time England's wine trade shifted from Rouen and La Rochelle to Gascony, early in the thirteenth century (Littler 1970, 203, 224–5; Rose 2011, 63).

The English herring fisheries of the High Middle Ages were associated with important secondary fisheries for cod and related species. Some of these fish may have reached market while still fresh or lightly salted, but many must have been preserved for longer storage. Cod bones from late twelfth–mid-fourteenth-century Cartergate in Grimsby provide evidence for local manufacture of a *råskjær*-like product (with head and anterior vertebrae removed on-site) (Russ 2011; cf. Chapter 1; Chapter 18). Given the southerly latitude of Grimsby, it is likely that these fish were both dried and salted, creating a cure like those better documented from the activities of later medieval and post-medieval fishermen of the English West Country (cf. Kowaleski 2000, 439). It is unlikely to be a coincidence that the consumption of both whole and pre-processed (decapitated) cod can be observed in the thirteenth-century fish bone record from York, a journey upriver from Grimsby (Chapter 15).

Some specialised fishing ports (such as Grimsby and Ravenserodd) also became key centres for the import of North Atlantic stockfish. In other instances, such as King's Lynn and Boston, ports arose that focused on the transhipment of diverse English and imported goods, including preserved fish. Yet other centres, such as London and Bristol, were foci of urban population that imported large quantities of fish for local consumption (Colson 2014; Kowaleski 2003, 217). These examples represent a continuum from major metropolis (London) to short-lived fishing port (Ravenserodd), along which other settlements are harder to categorise. For example, the Cinque Ports of southern England held major rights in the East Anglian herring fishery – by charter from the thirteenth century and in practice already from the twelfth or even the eleventh century (Chapter 3).

Turning to fish *imports*, can we tell when, and to what degree, imported preserved fish (especially salted herring and dried gadids) from distant waters entered the medieval English market? Related to this, can we tell when English fishermen first sought out distant fisheries? The growth of English long-distance fisheries for cod and related species was predominantly a development of the late 1300s and subsequent centuries (discussed below). Similarly, although

English imports of barrelled herring from the Baltic Scanian fishery are first documented in 1308, large-scale import of this iconic product only became common later, in the second half of the fourteenth century (Nedkvitne 2014, 517–18). English fishermen also made efforts to engage directly in the Scanian fishery, although their success was ultimately limited by the protectionist policies of the Hansa (Jahnke 2009, 177). The absence of fish bone evidence for the distinctive Scanian cure (in which specific parts of the skeleton are removed during processing) from consumer settlements such as York is consistent with these observations based on the written record (e.g. Chapter 15).

In contrast, there is plentiful evidence for the import of stockfish to England from Norway and perhaps elsewhere in the North Atlantic region before AD 1350. The first anecdotal historical references are from the late eleventh and early twelfth centuries (Chapters 4–5). However, zooarchaeological evidence from London suggests that the major shift from consumption of locally caught (whole) to traded (decapitated) cod may not have occurred until the early thirteenth century (Orton *et al.* 2014; Chapter 16). Moreover, stable isotope analysis of a sample of English cod bones shows that most specimens predating the thirteenth century have δ^{13}C and δ^{15}N values consistent with relatively local catches, in either the southern North Sea/English Channel or the Irish Sea (Barrett *et al.* 2011; see Hutchinson *et al.* 2015 for comparative data regarding the isotope values to be expected in the Irish Sea). It is not until the thirteenth to fourteenth centuries that specimens from London better match North Atlantic or Arctic Norwegian sources. A few thirteenth- to fourteenth-century cod bones from York also have stable isotope signatures characteristic of stockfish-producing regions, but this town seems to have continued to rely on relatively local fish products to a greater degree than London (Barrett *et al.* 2011; Chapter 15).

It is difficult to pinpoint the precise source of the earliest stockfish imported to England. By AD 1294, before systematic English custom records began, the Norwegian crown had successfully implemented Bergen as a staple transhipment port through which all northern exports should flow (DN vol. 5, 23). Based on more detailed fourteenth-century sources, it is clear that the town's economic monopoly was intended to include both northern Norway and the Norwegian 'taxlands' of the North Atlantic (e.g. Iceland and the Northern Isles of Scotland) (Keyser and Munch 1849, 181–2). Thus the origin of stockfish can seldom be traced farther than its transhipment in Bergen using historical sources. Moreover, the precision of stable isotope evidence is not sufficient to separate Iceland, northern Scotland and northern Norway with confidence (Barrett *et al.* 2011; Hutchinson *et al.* 2015).

What we do know is that Arctic Norway was a major producer of stockfish from at least the turn of the eleventh and twelfth centuries (Chapters 4–5), that the medieval town of Borgund in Sunnmøre's stockfish-producing area was thriving by the twelfth century (Chapter 6), and that Iceland first began to export stockfish at some point in the thirteenth century (Edvardsson 2010; Thór 2009, 329; Chapter 7). In addition, it is likely that the Scandinavianised region of northern Scotland, ruled by the earls of Orkney, was producing stockfish for trade by the eleventh–twelfth centuries (Barrett 1997; 2012; Barrett *et al.* 1999; Harland 2007; Harland and Barrett 2012). In this last context, it may be relevant that Orcadians are noted as visiting Grimsby in the early twelfth century (Guðmundsson 1965, 130) and that one of the parishes dominated by London's medieval stockfishmongers was dedicated to Magnus the Martyr, patron saint of Orkney (Colson 2014). As the stockfish trade became increasingly commercialised in the thirteenth and fourteenth centuries, it is likely that some small-scale producers – including those in Orkney – were marginalised *vis-à-vis* Arctic Norway (Barrett 2012). A similar process of 'globalisation', entailing increased regional specialisation of labour, may also explain the decline between AD 1100 and 1200 of the cod fishery in Hordaland, western Norway (Johannessen 1998, 43; Wickler and Narmo 2014, 85; see above).

Shifting focus to the Low Countries, we see that the years from AD 1050–1350 were characterised by growth in the importance of sea fishing and sea-fish trade. Based on the zooarchaeological evidence from urban sites in Belgium, the relative importance of marine species *vis-à-vis* freshwater fish steadily increased between the eleventh and fourteenth centuries; bones of the former switched from being a small minority to the majority of identified fish remains (Chapter 14). It is argued that this evidence probably implies a reduction in the availability of inland fish stocks due to human impacts, such as habitat changes and overfishing, perhaps combined with limited urban access to increasingly controlled freshwater fisheries (Ervynck and Van Neer 1998; Chapter 14; cf. Hoffmann 1996).

The pre-existing emphasis on herring and flatfish in Belgian sites (see above) continued in the eleventh–fourteenth centuries. Gadid fish slowly increased in relative importance, reaching over 40% of the identified fish bones in late thirteenth–early fourteenth-century Mechelen. In the eleventh to twelfth centuries, gadid fish from Flemish sites were mostly whiting and haddock. Cod first became common (although not dominant) in the thirteenth and fourteenth centuries (Chapter 14). The earliest stable isotope evidence for possible imports of preserved cod from beyond the North Sea in Belgium is from thirteenth–fourteenth-century Mechelen, but at this date only one of five studied specimens yielded a potentially non-local signature (Barrett *et al.* 2011).

The expansion of the local Flemish fishery in the twelfth and thirteenth centuries is also well documented in historical sources, having emerged out of earlier initiatives by the counts of Flanders to extract marketable goods from their holdings within newly embanked coastal landscapes (Tys 2015; see also Ervynck *et al.* 2004). Estates paid renders in herring, the

coastal plain became heavily populated, and by the middle of the twelfth century, fish markets were established in coastal comital towns such as Nieuwpoort and Grevelingen. These ports served as continental parallels to Dunwich and Great Yarmouth in the sense that they, too, were intermediaries between fishermen from dispersed rural estates and inland centres of population and consumption (Tys 2015).

That these fisheries entailed shore-based processing of local catches is evidenced by the linkage of herring and salt production (by burning peat) in historical sources (Tys 2015). Similar fisheries along the coast of the Netherlands are also known from twelfth-century sources (Unger 1978, 340). The barrelling of salted herring at sea was a subsequent development, of the late fourteenth and fifteenth centuries (see below). Nevertheless, fishermen from the Low Countries were also active in the East Anglian and Scanian herring fisheries (Campbell 2002, 6). As is discussed above, catches off the English coast were landed and cured near the fishing grounds before transhipment to both local and distant markets. Regarding Scania, Unger (1978, 339) suggests that fishermen from Kampen and Harderwijk (in what is now the Netherlands) were already participating by the end of the twelfth century. Merchants from around the North Sea were certainly active at the Scanian fairs in the thirteenth century (Jahnke 2009, 172–3). As a result, some Øresund herring were shipped west. One might speculate whether twelfth- and thirteenth-century herring bones from Bremen, for example, derived from this source (Küchelmann 2015, 268).

The importance of sea fishing and fish trade also increased in Scotland during the High Middle Ages. The intensification of fishing for cod and related species in the Scandinavianised highland and island regions of northern Scotland in the eleventh–twelfth centuries has been introduced above. At fishing settlements such as Quoygrew in Orkney much of the archaeological sediment is made up of gadid bones and marine molluscs, the latter probably having served as bait for the fishery (Barrett 2012). There is reasonable circumstantial evidence that stockfish was being produced (probably without salt) at sites like Quoygrew and Robert's Haven, for export to destinations farther south, both directly and via Bergen in Norway, by the twelfth century at the latest (Barrett 1997; 2012). This initially booming fishery may have been in decline by the 1200s, although it revived episodically in later centuries beyond the chronological range of this chapter.

Fish bone evidence suggests that a cured gadid product also began to be manufactured in coastal settlements of lowland Scotland in the course of the High Middle Ages. A thirteenth-century assemblage of cranial bones and abdominal vertebrae from large cod and ling recovered at Leith suggests that a *råskjær*-like product was manufactured on-site, be it for consumption in nearby Edinburgh or export (Stronach 2002; cf. discussion of Cartergate, Grimsby, above). Similar anatomical patterning is evident in a cod and ling assemblage of thirteenth–fourteenth-century date from the burgh of Eyemouth (Dixon 1986, 83). Large cod and ling are also well represented in thirteenth–fourteenth-century deposits in Aberdeen (Harland 2010). Here, however, the relatively even representation of different skeletal elements is consistent with local consumption. A few groups of bones from specific archaeological contexts may represent the remains of preserved cod that had been decapitated elsewhere. Otherwise bones from all parts of the skeleton show that whole fish were brought to the town (Harland 2010, 32) – a pattern similarly noted in twelfth–fourteenth-century deposits from Perth (Hodgson *et al.* 2011, 53–7).

Zooarchaeology also demonstrates that herring were an important catch in some areas of Scotland in the High Middle Ages. They continued to play a major role in the Western Isles, as in previous centuries (e.g. Ingrem 2005, 192–3), but were also caught in the Firth of Forth (Chapter 3) and consumed in the newly established burghs of the east (e.g. Hodgson *et al.* 2011; Stronach 2002). Salmon (*Salmo salar*), another of Scotland's major traditional resources (albeit a migratory rather than truly marine species), is surprisingly poorly represented in the zooarchaeological record. Scottish customs were first applied to fish exports in the late fourteenth and fifteenth centuries (see below), making inferences regarding the scale of earlier trade impossible. Nevertheless, external sources record cargoes destined for both England and the continent in the thirteenth century (e.g. Littler 1970, 202; Rorke 2001, 183).

Our understanding of the chronology of the expanding sea fisheries in Ireland is limited by the present paucity of published fish bone evidence from its Hiberno-Scandinavian towns and the late medieval date of the first customs accounts to record Irish fish exports (Chapter 9; Chapter 19). Nevertheless, Trim castle reveals a thirteenth–fourteenth-century fishery for cod, and to a lesser degree other marine species, that were transported inland. The anatomical evidence suggests that these fish were whole, and thus perhaps fresh or only lightly cured (Chapter 19). A few cod and haddock bones were also traded inland as early as the tenth–twelfth centuries, with bones having been recovered from deposits of this date at Knowth (Chapter 19). In the south, as at eleventh–fourteenth-century Cork, hake was the most frequent species, followed by cod and ling. This pattern prefigures an export fishery for these taxa known from late and post-medieval sources. The relevance of herring fishing is particularly hard to gauge, due to recovery bias, but one can perhaps interpolate its importance based on the earlier (tenth–early eleventh-century) consumption waste from Fishamble Street in Dublin discussed above and from later medieval archaeological and historical evidence for its export from Ireland (see below). Target species such as hake, ling and large cod imply open-water fishing, but coastal fish traps also continued to be built – with eleventh–thirteenth/fourteenth-century examples known from both the west (County Clare) and east (Strangford Lough) (Chapter 9).

In Scandinavia, as was already implied by the discussion above of inter-regional trade, the years between AD 1050 and 1350 witnessed the large-scale expansion of commercial fisheries, particularly those for cod (dried as stockfish) in the north and for herring (pickled in brine) in the western Baltic Sea. There is less evidence for changes in local consumption of fish, although the relative importance of different taxa did shift in specific contexts.

The date of the emergence of the stockfish trade from Arctic and western Norway has occasionally been a matter of debate. As is discussed above, some stockfish may have been exchanged within Scandinavia (or at least used as travellers provisions) during the Viking Age – as implied by finds from Haithabu (in modern-day Germany), for example. On the present evidence, however, initial expansion of the trade to include English and continental markets can best be attributed to the decades around AD 1100 (Chapters 4–5; Chapter 18). The Lofoten Islands proved to be the most consistently important centre, but production also occurred elsewhere: farther north in northern Troms and Finnmark (Chapter 4), around Borgund in western Norway (Chapter 6) and, as already discussed, probably also in the Scandinavian earldom of Orkney (which controlled Shetland and the northern mainland of Scotland in addition to Orkney itself). At some point in the thirteenth century, Icelandic stockfish were drawn into the trade as well (Chapter 7).

One consequence of the stockfish trade was the growth of Bergen as a major urban centre in the late eleventh and (particularly) twelfth centuries (Hansen 2005; 2015). Another was the development of permanent fishing villages along the coast of Arctic Norway (beyond the limits of cereal cultivation) by the mid-thirteenth century (Chapter 4). Yet another was the growth of trading ports in eastern England – such as King's Lynn, Boston, Grimsby and Ravenserodd – that relied in part on the exchange of English cloth and grain for Norwegian stockfish (Carus-Wilson 1962–3; Nedkvitne 2014; Rigby 1993). Many of these fish were sent on to urban consumers in centres of population such as London (Colson 2014; Orton *et al.* 2014; Chapter 16).

What began as a modest trade, of importance to northern producers but perhaps less so to southern consumers (Barrett 2012, 276–82), rapidly expanded in the middle decades of the thirteenth century. The evidence derives from diverse sources. In England, the earliest recorded formal trade agreement with Norway dates to AD 1223 (Helle 1968). In the fish bone record, London shows a rapid shift from consumption of whole cod to decapitated cod at much the same time (Chapter 16). Stable isotope analysis also suggests that imported cod first became common in England – particularly in London, but to a lesser degree also in other centres – in the thirteenth–fourteenth centuries (Barrett *et al.* 2011; Chapters 15–16). Historical sources indicate that direct Anglo–Norwegian trade was initially frequent, gradually being superseded by middlemen of the

German Hansa during the first half of the fourteenth century (Nedkvitne 2014).

Another consequence of the stockfish trade (and of the herring trade to be considered below) was the growth of the German Hansa itself. As in England, this development emerged from pre-existing trade. Towns of the lower Rhine region may have been one important early market. We know, for example, that Norwegians were trading at Utrecht by 1122 (Nedkvitne 2014, 33). Another locus of initial consumption was probably the western Baltic. Finds of decapitated large cod from late eleventh–twelfth-century Schleswig (Lübeck's economic predecessor) may derive from imported stockfish, continuing a trend first proposed *vis-à-vis* neighbouring Haithabu (Heinrich 1987). This evidence has been central to the inference that emerging urbanism (rather than Christian fasting practices, for example) played a major role in the growth of fish trade and fish consumption in medieval Germany (Lampen 2000, 210–11). Following these possible antecedents, it was in AD 1278 that German merchants, led by Lübeck, first secured distinctive trading privileges in Norway (Nedkvitne 2014, 51). The influence of Lübeck and other Wendish towns increased thereafter, with occasional setbacks, and culminated in the establishment of a powerful Hanseatic Kontor at Bergen in the 1360s. Hanseatic ships carried some stockfish into the Baltic, but many were transported to England as part of a triangle trade that saw Baltic rye and German beer taken to Norway and English cloth transported to the east (Nedkvitne 2014; Unger 2002).

In southern Scandinavia and neighbouring areas of northern Germany, herring fishing grew into an export industry of major international importance in the years between AD 1050 and 1350. The most well-known fishery, in the Øresund, and adjacent western-Baltic waters around Scania, came to dominate the European herring trade by the late thirteenth century and continued to do so until the sixteenth century (Chapter 2). Its focus was the paired settlements of Skanør and Falsterbo. These were both fishing settlements and international markets, under the protection of the Danish kings and (from the thirteenth century) with officially recognised communities of German and Dutch merchants.

Before the dominance of Scania there was already a commercial herring fishery in the Western Baltic Sea. It was conducted from Rügen in Germany during the twelfth century (Jahnke 2009, 168–70). The first of a series of episodically-productive herring fisheries from the Norwegian (now Swedish) coast of Bohuslen, (Bohuslän) at the eastern limit of the Skagerrak, was also recorded in the 1100s (Alheit and Hagen 1997; Chapter 2). Moreover, herring may have been cured in Poland and northeastern Germany for small-scale inland trade already in the tenth and eleventh centuries, and this species comprised a high proportion of the fish bone recorded from the Viking-Age trading sites of Haithabu (see above), Wolin and Truso

(Chapter 12). Thus the large-scale Baltic herring industries of the High Middle Ages emerged from pre-existing practices.

Although also rooted in earlier traditions, the Scanian herring fishery was of unprecedented scale and significance. Its branded, high-quality, barrels served Baltic and North Sea consumers and also reached distant markets of the continental interior. Holm (Chapter 2) estimates that, at *c.* 35,000 tonnes, the Scanian herring trade exceeded the volume of the important wine trade between Bordeaux and England by the early fifteenth century. Like that of North Atlantic stockfish, the trade of this commodity came increasingly under the control of the German Hansa. Lübeck was particularly influential, being well placed to control access to abundant supplies of salt mined at nearby Lüneburg and to widely distribute the finished product (Chapter 2).

Moving to the eastern Baltic, as already noted, we can see that herring had begun to be cured and traded in Poland by the tenth–eleventh centuries at the latest. The number of sieved medieval fish bone assemblages is not sufficient to observe how the importance of this species may have ebbed or flowed over the following 300 years. The changing use of cod can be observed, however, as this species grows larger and thus has bones that are more easily recovered by hand collecting. With very minor exceptions (one bone from Sopot and three from Kołobrzeg–Budzistowo), remains of cod only begin to occur in Poland (and then in small numbers) in assemblages of the thirteenth–fourteenth centuries. The find contexts are the towns of Gdańsk, Kołobrzeg and Elbląg and the monastery of Stargard Szczeciński. The earliest chronology of these finds may be the second half of the thirteenth century, based on the most closely dated specimens (Chapter 12). The majority have stable isotope values inconsistent with an origin in the eastern Baltic Sea region. They are instead likely to represent preserved imports – stockfish or analogous products – from beyond the Kattegat (Orton *et al.* 2011; see also Orton *et al.* forthcoming). It is also striking that some of these bones are from individuals of greater than 100 cm estimated TL, larger than is typical for fish taken from the Baltic cod population (Chapter 12). Interestingly, their overall length distribution is bimodal, with peaks around 50–60 cm and 70–100 cm, perhaps implying stockfish were categorised by size, as is known from historical sources (e.g. Wubs-Mrozewicz 2009, 191).

The chronology of medieval cod finds in Estonia supports the Polish evidence (Chapter 11). In Tallinn, the best-dated examples occurred in contexts of the thirteenth and/or early fourteenth centuries. These finds were from excavations at 4 Rahukohtu Street (which revealed a residence of Danish conquerors within Toompea Castle) and at 10 Viru Street and 10 Sauna Street (which both represent more typical occupation in the lower town). The above-mentioned finds were all of vertebrae and cleithra, bones typically left in preserved cod, such as stockfish. The same applies to cod found during excavations of broadly dated (thirteenth–sixteenth-century) material at Vanemuise Street in the inland town of Tartu. In both Tallinn and Tartu, some of the bones were from fish of between 95 cm and 125 cm estimated TL, larger than is typical for Baltic cod. Moreover, like the early cod from medieval Poland, stable isotope analysis of a sample of these finds shows that they are unlikely to have been caught in eastern Baltic waters. Sources in Arctic Norway, elsewhere in the North Atlantic and/or in the North Sea are more likely (Orton *et al.* 2011; forthcoming).

Two sites of thirteenth-and/or fourteenth-century date, namely, 1 Vabaduse Square in Tallinn and Karksi castle in southern Estonia, include trace numbers of cranial bones (no more than two per site) in assemblages that are otherwise also of postcranial specimens (Rannamäe and Lõugas forthcoming; Lembi Lõugas pers. comm.). These may be the first Estonian indications of medieval cod fishing in the eastern Baltic. The emergence of a local fishery for this species is clearly evidenced by later (fourteenth–sixteenth-century) finds of small cod, including cranial bones, from the coastal Estonian town of Pärnu (Chapter 11) and from Tartu Street in Tallinn (Russow *et al.* 2013; Lembi Lõugas pers. comm.). Cod from the inland town of Valmiera, upriver from the Gulf of Riga in Latvia, probably belong to this later development. They are broadly dated to the Middle Ages (Chapter 11).

Unlike in Poland, the earliest finds of herring in Estonia are contemporary with, not earlier than, the thirteenth–early fourteenth-century introduction of cod. Herring bones are represented in several assemblages from Tallinn, although they were not found in the Vanemuise Street excavations at Tartu (Chapter 11). The number of recovered specimens from medieval Estonian sites is still small, but because many of the relevant collections were hand collected, the importance of this species is probably underestimated on present evidence.

Although many of the earliest medieval finds of cod in the eastern Baltic region may be imports, a local fishery for this species did develop in the course of the fourteenth century, with its own distinctive traditions of butchery and curing (Chapter 12). One term, *strekfusz*, was probably applied to both dried pike and dried cod. Its first mention, in 1328, relates to a purchase by the Teutonic Knights at Goldingen near Elbląg (Hoffmann 2009, 119). However, most evidence for cod fishing and processing in the eastern Baltic post-dates AD 1350. One catalyst may have been the preceding trade of imported stockfish, which developed concurrent with Scandinavian and German colonization (Chapters 11–12).

The late Middle Ages: AD 1350 to *c.* 1550

Although their volume is difficult to quantify, Europe's established sea fisheries reached a peak in the decades around AD 1300. Thus they tracked the apogee of High medieval demographic and economic expansion, prior to the economic, climatic and epidemiological challenges of the mid-fourteenth century (e.g. Campbell 2010). Some fisheries declined, temporarily or permanently, after the mid-1300s. Others prospered in a new period of expansion that culminated in the major (Dutch) North Sea herring and (Iberian, French and English) cod fisheries off Newfoundland of the sixteenth and seventeenth centuries.

The English 'particular' customs accounts provide unique, semi-quantifiable glimpses of the Norwegian stockfish trade during the changes of the 1300s. Imports may have reached 2000 metric tonnes per annum in the early fourteenth century, but probably fell to less than half of this after the famine of 1315–22 and the plague of 1346–53 significantly reduced Europe's population (Chapter 5). Nevertheless, prices rose. Problems of supply presumably exceeded the reduction in demand, particularly in a post-plague world where the surviving consumers had more disposable income. Fishermen became comparatively wealthy, and migrants of diverse origin flocked to new fishing villages along Norway's Arctic coast (Chapter 4). Direct imports of stockfish from Bergen to England (albeit mostly via Hanseatic ships by the mid-fourteenth century) continued until the late fifteenth century (Nedkvitne 2014, 161), by which time English fishermen were fully established in Icelandic waters (Chapters 7–8). Subsequently the main market for Norwegian cod was the continent. Germany continued to be a destination for stockfish from Bergen until *c.* 1700, albeit supplemented with fish from direct Hanseatic trade to Iceland, the Faroe Islands and Shetland, which first began in the fifteenth century (Friedland 1973; Mehler and Gardiner 2013; Nedkvitne 2014; Chapters 7–8). Holland also emerged as a significant market for Bergen stockfish from the fifteenth century (Wubs-Mrozewicz 2009, 195). Continued international demand meant that Arctic Norway had an economic 'parachute' in the wake of the crises of the fourteenth century. In the hundred years following the Black Death, fishermen received three times as much grain for their stockfish (Chapter 5). Settlement archaeology suggests that long-established fishing settlements, such as Borgvær in Lofoten, survived post-plague depopulation of the wider region (Wickler and Narmo 2014). The new, specialised fishing villages of Arctic Norway also prospered, and continued to do so beyond the end of the Middle Ages. They were abandoned in the seventeenth century, in favour of more diversified farming and fishing settlements, after a fall in stockfish prices that began around AD 1500 (Chapters 4–5).

The Scanian herring fishery also continued to prosper in the late Middle Ages. Systematic customs accounts regarding Baltic fisheries first become available in the late fourteenth century. As we have seen above, however, whatever the deleterious impact of late medieval depression, the sale of barrelled Scanian herring was booming in the early decades of the fifteenth century. Annual exports were in the tens of thousands of tonnes, and *c.* 17,000 men participated directly in the fishery (Chapter 2). The value of barrelled herring *vis-à-vis* stockfish was also rising in the years between 1259 and 1550, presumably due to increasingly rigorous quality controls regarding the production of the former (Chapter 5). As a product, it reached as far inland as Switzerland (Jahnke 2009, 180). Barrelled herring of lower value were also made on the Baltic island of Bornholm, and from the fifteenth century, a commercial fishery serving local markets developed in the Limfjord of northern Jutland (Chapter 2).

The dominance of Scanian herring in international trade was ultimately superseded by Dutch catches from the North Sea, as far afield as the Northern Isles of Scotland, which were cured on-board ship. This practice seems to have begun in the late fourteenth or early fifteenth centuries (Unger 1978). It expanded in the sixteenth century (Bennema and Rijnsdorp 2015), with a shift from Scanian to Dutch barrelled herring in European shipping being evident by the time of the earliest Øresund toll registers of the 1560s (Chapter 2). The fish-bone evidence from late medieval Denmark is not abundant, but a decline in herring fishing may be evident in the zooarchaeological record from sites such as Tårnby, on the Øresund (Chapter 13). Here herring bones dropped to only 33% of the assemblage (by NISP) in the sixteenth century, from a high of 77% in the late tenth–thirteenth centuries.

In England, different sea fisheries expanded and contracted in the late Middle Ages. In the course of the late fourteenth and fifteenth centuries, West Country fishermen (especially from Devon and Cornwall) developed large-scale local and long-distance fisheries for such species as cod, hake and ling (Kowaleski 2000; 2003). The coast of Ireland was a major destination, but Iceland was also visited (see further below). On one occasion in 1438, fishermen of Saltash (in Cornwall) were fined for going to Finnmark (Arctic Norway) without licence (Childs 1997, 285). Temporary fishing stations (some operated by Gascon merchants) were established already in the early thirteenth century, but permanent fishing villages developed in the late Middle Ages, during the late fifteenth and early sixteenth centuries (Fox 2001; Kowaleski 2014, 47; Littler 1970, 203). Although the Black Death of the mid-fourteenth century reduced the number of fishermen and boats in some communities, the long-term trend was towards expansion in both the quantity of trade and the distances travelled (Kowaleski 2014). After John Cabot sailed from Bristol and 'discovered' Newfoundland in 1497, West Country fishermen played a major role in England's migratory cod fishery in North America, plying waters that were initially most frequented by ships from France, Portugal and Spain (Candow 2009b; Kowaleski 2003; Pope 2004).

In contrast, the English east coast herring fishery experienced a serious and long-lasting decline from the

second half of the fourteenth century (Kowaleski 2003, 191–8). Saul's (1982; 1983) study of the fate of Great Yarmouth assembles diverse and persuasive evidence. To provide one striking example, in 1334 it was the fourth-highest taxed of England's provincial towns (after Bristol, York and Newcastle), whereas by 1377 it was eighteenth (Saul 1982, 76). Crucially, its trade of preserved herring for Gascon wine (which had engaged 65 ships from Yarmouth in 1303–11) was in decline by the 1360s at the latest.

Great Yarmouth's fate was not unique; the decline affected all of eastern England. The causes are open to debate given that the pan-European market for preserved herring remained strong. A variety of environmental and socio-economic challenges probably converged (Kowaleski 2003, 193–8). The harbours of many ports experienced siltation and/or destruction by storm surges in the fourteenth century. Another key variable was the decline in continental boats landing fish in England, in the face of political and military instability. Increasingly, fishermen from the Low Countries salted their catch on board and finished curing it in home ports – a trend that culminated in the practice of barrelling herring at sea which underpinned the post-medieval Dutch fishery mentioned above. Other factors influencing the decline of the English herring trade include major military requisitions of the fishing fleet and the superior quality of competing products. As noted above, imported barrelled herring, initially from the Baltic, began to find a market in England during the fourteenth century (Nedkvitne 2014, 517–18).

Another variable is the fact that emerging late medieval fisheries of the English West Country were better placed for the French wine trade, in which Yarmouth had played a leading role during the thirteenth and early fourteenth centuries. Most important of all, however, may have been the rising cost of labour after the Black Death, a critical variable in the decline of formerly large-scale salt production (by boiling sea water) in eastern England (Bridbury 1955, 26). The competitive advantage of the eastern English herring fishery must itself have evaporated in the absence of a cheap local preservative. Imported solar salt from Atlantic Europe was available to all from the late fourteenth century, but it was most accessible to the emerging fisheries of the southwest (Kowaleski 2003, 226).

As England's fishery for North Sea herring ceased to be profitable, a reduced number of fishermen from diverse locations along the east coast shifted their emphasis to other species, especially cod caught in increasingly distant waters – in the North Sea (e.g. the Dogger Bank), off the coasts of Denmark and Norway and around Iceland (Childs and Kowaleski 2000). Direct English involvement in Iceland began in the first two decades of the fifteenth century (Childs 1995, 12). It entailed both trade for stockfish (preserved without salt by local fishermen) and fishing by

English vessels to make a product that was salted and dried (Chapter 8). These activities were not without controversy and political intervention, given that Hanseatic and royal Danish interests were best served if Icelandic exports were funnelled through Bergen. Yet the 1400s, nevertheless, became known as Iceland's 'English century' (Chapters 7–8). Many settlements of varying scales participated, but the merchants of Hull (and of Bristol, in the southwest) were particularly involved (Childs 1995). Iceland remained an important source of dried cod for the English market into the seventeenth century (Jones 2000). Based on stable isotope and ancient DNA evidence, it is thought to have probably provided some of the provisions on the Tudor warship *Mary Rose*, which sank in 1545 (Hutchinson *et al.* 2015).

The decline of England's east coast herring fishery saw the emergence of multiple competitors, only some of which have been considered thus far. Wales, a nation that has not yet figured in this chapter, has important evidence of early fish traps – in estuarine settings and on the western seaboard (e.g. Godbold and Turner 1994; James and James 2003; Turner 2002) – but not yet an abundant zooarchaeological record of medieval fish bone finds. Nevertheless, textual evidence reveals the existence of a coastal herring fishery before the thirteenth-century English conquest of Gwynedd (Chapter 3). It developed into a major commercial enterprise attracting migrant labour in the late fifteenth and sixteenth centuries (Kowaleski 2003, 219–20).

Scotland's sea fisheries also prospered in the fifteenth and sixteenth centuries. Royal customs on herring were introduced in 1424 (Rorke 2001, 197), and the volume of exports rose dramatically in the sixteenth century (Rorke 2005, 153). Salmon were the target of initial customs charges in 1398 and of systematic taxation from 1425 (Rorke 2001, 182–3). Exports of cod were explicitly added to the Scottish customs system in the 1460s, having previously been evident from anecdotal sources (Rorke 2001, 197; see also above). The availability of solar salt from western France, which increased from the 1360s (Bridbury 1955, 67), must have partly fuelled this increase in trade, particularly of barrelled herring and barrelled salmon. The Scottish picture was not, however, a uniform one. As noted above, local dried cod production declined in late medieval Orkney (governed as a Scandinavian earldom until 1468) in the wake of political interventions by the expanding kingdoms of Scotland and Norway that temporarily weakened the archipelago's international connections (Barrett 2012).

As in so many of the regions discussed, Ireland's marine fisheries boomed in the late Middle Ages, before a decline in the seventeenth century following the suppression of independent Gaelic lordships. By the fifteenth century, fish were one of the island's main exports, and Irish waters also attracted fishermen from distant shores (Chapter 9). Trips from England's West Country have been mentioned above, but in the late fifteenth and sixteenth centuries, Breton and Spanish vessels were also very active along the Atlantic coasts of Ireland, trading with Gaelic lordships and reaching agreements that allowed them to make their own catches. Cod, ling and hake were the main targets of these migrant fisheries. Cod and ling were also caught in the Irish Sea; some were processed for later consumption at Arran Quay in Dublin in the fifteenth–early sixteenth centuries. But herring constituted the major catch along Ireland's eastern and southern coasts (Chapter 9; Chapter 19). Many of these herring were cured for export. The evidence is mainly historical, probably due to limited use of sieving during the recovery of archaeological fish bone assemblages, but a consignment of barrelled herring from the wreck of the sixteenth-century Drogheda boat provides a clear example of Ireland's trade in this commodity (Harland 2009).

Fishing and fish trade also increased in the eastern Baltic Sea (east of Bornholm) in the late Middle Ages. Stable isotope evidence suggests a shift in the origin of cod bones found in towns such as Gdańsk and Uppsala. Many thirteenth–fourteenth-century specimens may represent imports from beyond the Baltic Sea (e.g. Norwegian stockfish), whereas most fifteenth–sixteenth-century finds have comparatively local isotopic signatures (Orton *et al.* 2011). This trend is supported by zooarchaeological evidence regarding butchery patterns, with later medieval assemblages being more likely to include skull bones and cod of small estimated TL (Jonsson 1986; Chapters 11–12). Of note is a distinctive method of preparing fish for drying that was applied to both cod and pike in the eastern Baltic Sea region: anterior vertebrae were removed, but cranial bones and caudal vertebrae were both left in the finished product. This pattern was first recognised in fifteenth–sixteenth-century Uppsala (Jonsson 1986), but it has also been identified tentatively in a fifteenth-century assemblage from Mała Nieszawka, a castle of the Teutonic Knights near Toruń. It can be equated speculatively with the Baltic dried fish product known as *strekfusz*, which seems to have been variously manufactured from pike and cod, and which begins to appear in historical sources during the fourteenth century (Hoffmann 2009; Chapter 12; see above). Baltic stockfish were also manufactured in the late Middle Ages, with Hel (near Gdańsk) and Curonia being two known centres of production (Nedkvitne 2014, 130). The late medieval growth of fishing and fish trade in the eastern Baltic occurred against a backdrop of major urban expansion and cultural influences from Scandinavia, the Teutonic Order and the German Hansa (Hoffmann 2009; Orton *et al.* 2011; forthcoming; Chapters 11–12).

Conclusions

What then are the answers to the eight questions posed at the outset of this book (Chapter 1)? First, was the growth of early medieval sea fishing a correlate of state formation and urbanisation? Urbanism was a key variable in many contexts, ranging from the Baltic and North Sea *wics* of the 'long eighth century', to the urban boom associated with the turn of the first and second millennia AD, to the growth of medieval towns in the Baltic region between the twelfth and fourteenth centuries. Sometimes urbanisation, state formation and fish trade went hand in hand, as in the conquest of Polish coastal towns by the Piast dynasty in the tenth century or the royal Norwegian investment in both the Lofoten fisheries and in Bergen in the years around AD 1100. In other cases (such as Truso in Poland) incipient urbanisation and its appetite for sea-fish could precede state formation, and of course urban growth continued long after the initiation of political centralization in all regions of northern Europe.

Secondly, were sea fisheries developed in response to social drivers, such as Christian fasting practices, elite demand and/or the cultural foodways of migrant communities? The role of elite demand (and investment) was probably critical in stimulating many of northern Europe's sea fisheries, from the provisioning of estate centres in Anglo-Saxon England to Hanseatic and royal Danish involvement in the Øresund herring fishery. Elite demand was direct, through the requirements of supplying households, and indirect, through the need for wealth from taxation and other renders. Medieval migrants also played a role in spreading traditions of sea fishing and/or sea-fish consumption. Examples range from Viking Age Scandinavian settlement in parts of Scotland to western influence in the eastern Baltic region during the thirteenth and fourteenth centuries. The role of Christian fasting has been more controversial. During the High and late Middle Ages it clearly helped foster a market for stored fish products suitable for consumption at times such as Lent. In earlier centuries, however, the evidence has proven ambiguous (cf. Barrett *et al.* 2004, 629–30; 2011, 1523; Frantzen 2014, 232–45; Lampen 2000, 210 11). Moreover, demand for sea-fish existed even among non-Christian communities (e.g. Viking Age Scandinavians). Nevertheless, in some contexts Christianisation converged with state formation as a probable catalyst for growing fish trade, as in tenth-century Poland.

Thirdly, was there an unprecedented sea-fishing revolution at the turn of the first–second millennium AD? There are exceptions to the hypothesised rapid 'fish event horizon' in England and elsewhere around the North and Baltic seas *c.* AD 1000. Earlier examples of sea-fish consumption exist, even in England, and the chronology of expanding fishing activity can be tracked over centuries rather than decades in contexts such as Flanders. Nevertheless, sea fishing did increase in importance across most of the geography covered by this book in the years between AD 850 and 1050. More sea-fish bones are found at more sites and/or farther from the coast. Even in cases where earlier finds of marine fish bones have been made (at Bishopstone in southern England, for example), stable isotope analysis of human bones suggests that marine

protein was not yet a major contributor to diet. With specific exceptions (e.g. from Scandinavia and the coast of eastern England), isotopic evidence for substantial consumption of the sea's harvests dates to the Viking Age and later.

Fourthly, did sea fishing expand as a result of human impacts on freshwater ecosystems? The authors of the chapters in this book are divided on the issue. What is clear is that the consumption of freshwater fish declined in quantitative importance through time, both in absolute numbers and in relative terms *vis-à-vis* marine fish. Where the evidence is available, as at York, it is suggested that this decline may have been associated with a decrease in fish size. However, it is difficult to know how representative this case is of wider patterns. In Belgium, for example, there are differences in the representation of freshwater fish between urban and rural assemblages, suggesting variable access and/or differing degrees of human impact on aquatic ecosystems. This issue clearly merits more research. Nevertheless, the fact that freshwater fish were increasingly monopolised by the elite (often via aquaculture using fish ponds) – and protected by legislation – in the High Middle Ages (Dyer 1988; Hoffmann 2014, 272; Serjeantson and Woolgar 2006) may imply that they had become rare in the wild.

Fifthly, could marine fisheries of the High Middle Ages have overfished formerly superabundant species, such as cod and herring? Human impacts on premodern marine ecosystems have been demonstrated in other contexts (Rick and Erlandson 2008; Roberts 2007). Moreover, catches to fuel the major trades of stockfish and barrelled herring were large in scale and of great economic importance. Nevertheless, based on the present evidence and on comparison with twentieth- and twenty-first-century data, it seems that fishing pressure on the Norwegian (northeastern Arctic) cod used for stockfish production was probably sustainable (Chapter 5). Moreover, although herring catches in the late medieval Øresund fishery did exceed those of around AD 1900, the fishery continued to be successful into the sixteenth century, and its demise is best explained in terms of political change (Chapter 2). We cannot yet quantify the scale or impact of fishing pressure in the North Sea during the critical years of expansion around AD 1000, and medieval Irish Sea catches are also unmeasured. Despite promising efforts, it is also not yet possible to disentangle the impacts of human predation and natural processes (such as climate change) on the growth, chemistry and genetics of marine species as inferred from scientific analyses of skeletal remains (e.g. Bailey *et al.* 2008; Bolle *et al.* 2004; Ólafsdóttir *et al.* 2014). In cases where there were clear reductions in sea fishing and/or the size of marine fish caught during the late Middle Ages – such as cod fishing in Orkney and Hordaland, or herring fishing in East Anglia – socio-economic changes are the most likely explanation. Overall, it is too soon to make generalisations regarding the possible impact of medieval sea fishing on cod and herring. As with human impacts on freshwater fish, much important work remains for the future.

Turning to the sixth and seventh questions, when did the long-range trade of high-bulk and low-value staples, such as salted herring and dried cod, really begin? When did such trade expand to a pan-European scale? The answers are closely tied to trends in the emergence and growth of urbanism and elite demand noted above. Around both the Baltic and North seas, herring were an important staple in the *wics* of the 'long eighth century'. Many must have been cured for transport, because herring spoil quickly and some of the sites, such as Dorestad and York, are inland river ports. The butchery evidence from Truso in Poland may even suggest early experimentation with the gutted cure later made famous by the Scanian and Dutch fisheries. By the eleventh century, the large-scale herring fisheries of eastern and southern England may already have been exchanging salted and smoked fish for continental wine, a pattern better evidenced from the twelfth century and, particularly, the thirteenth century. Gascony (and briefly La Rochelle) were the focus of this trade by the time that historical records became abundant, but earlier sources of wine, along the Seine and the Rhine, may have been the first destinations of English herring exports. In the eastern Baltic, herring were being transported to inland strongholds of the Piast dynasty by the tenth–eleventh centuries. In the western Baltic, an export fishery of barrelled herring developed in Rügen during the twelfth century, partly under the influence of the growing town of Lübeck. In the thirteenth century, the focus of production shifted to Scania, culminating in the largest known medieval fishery.

The cod trade has a different chronology and geography. In Norway, there was a lengthy tradition (that stretches into prehistory) of producing stockfish. On present evidence, we can suggest that its transport to Scandinavian towns as part of a commercial trade is likely to have begun by the eleventh century at the latest. Export of stockfish beyond Scandinavia, in exchange for cloth, grain and wine (beer was imported only after 1200), probably also began on a small scale in the eleventh to twelfth centuries. However, it is in the thirteenth–fourteenth centuries that this product becomes most visible in the fish bone record of consumer settlements around the North and Baltic seas. By this date some small-scale producers (in Orkney, for example) were being excluded from the market, whereas others, such as Icelanders, were incorporated into the international stockfish trade. Elsewhere, cured (probably dried and salted) gadid products were being manufactured in settlements of the Scottish, Irish and English coasts by the thirteenth–fourteenth centuries at the latest. These became major articles of trade in the course of the fourteenth and fifteenth centuries.

Lastly, to what degree did demand for sea-fish influence the increasingly 'global' destinations of late medieval mariners? The answer to this question has occupied a central place in the popular perception of fisheries history (e.g. Kurlansky 1997; Fagan 2006). Many goods were widely exchanged in medieval Europe, including the cloth, grain, wine and beer for which cod and herring were sometimes traded (Berggren *et al.* 2002). Moreover, the transport of luxuries such as pepper entailed pan-Eurasian networks throughout the chronology of present concern (e.g. Loveluck 2013, 123, 315). Nevertheless, the combination of high-bulk cargoes, long-distance transport and seafaring competence involved in fishing and fish trade means that these activities made a particularly significant contribution to emerging globalisation at the transition from the Middle Ages to the Renaissance. Long-range interconnections and geographically differentiated specialisations of labour connected every axis of the compass rose. Stockfish from Arctic Norway found their way to Tallinn and Tartu. Herring from the North Sea were enjoyed in Bordeaux. Herring from the Baltic Sea were consumed as far inland as Switzerland. Spanish, French and English fishermen caught Irish hake. Icelandic cod, probably transported by English fishermen, provisioned the crew of the *Mary Rose*. Thus the migratory Newfoundland fishery that began in the sixteenth century – which exceeded the North American fur trade in economic importance (Candow 2009a, 411) – was only an incremental step from late medieval distant-water activities. Yet the complex consequences of this step, and all that preceded it, were not trivial. Medieval fish trade entailed a self-perpetuating drive for more fish from more distant waters. It created opportunities for wealth creation on an unprecedented scale. It contributed to extensive long-range trade networks. It made regional economies susceptible to contingent access to products and markets. It made even 'remote' communities prosperous, but also exposed them to conquest, economic domination and/or disease. Lastly, it created the preconditions for long-term impacts on marine ecosystems that continue to unfold.

Acknowledgements

I thank the contributors to this volume for making a preliminary synthesis of medieval fishing and fish trade possible. Inge van der Jagt and Annemarieke Willemsen assisted with information regarding the Netherlands. Sæbjørg Walaker Nordeide kindly provided a copy of Johannessen (1998). James Gerrard drew my attention to Dio's *Roman History*. Suzanne Needs-Howarth made helpful suggestions to improve the clarity of my argument. The research has been supported by the Leverhulme Trust and this article is also based upon work from the COST Action Oceans Past Platform, supported by COST (European Cooperation in Science and Technology).

References

Alheit, J. & E. Hagen. 1997. Long-term climate forcing of European herring and sardine populations. *Fisheries Oceanography* 6(2): 130–9.

Bailey, G., J. Barrett, O. Craig & N. Milner. 2008. Historical ecology of the North Sea basin: an archaeological perspective and some problems of methodology, in T. C. Rick & J. M. Erlandson (eds) *Human Impacts on Ancient Marine Ecosystems: a global perspective*: 216–42. Berkeley: University of California Press.

Bailey, M. (ed.). 1992. *The Bailiffs' Minute Book of Dunwich, 1404–1430*. Woodbridge: Boydell Press/Suffolk Records Society.

Barrett, J. H. 1997. Fish trade in Norse Orkney and Caithness: a zooarchaeological approach. *Antiquity* 71: 616–38.

Barrett, J. H. 2003. Culture contact in Viking Age Scotland, in J. H. Barrett (ed.) *Contact, Continuity and Collapse: the Norse colonization of the North Atlantic*: 73–111. (Studies in the Early Middle Ages). Turnhout: Brepols.

Barrett, J. H. 2012. Being an islander, in J. H. Barrett (ed.) *Being an Islander: production and identity at Quoygrew, Orkney, AD 900–1600*: 275–92. (McDonald Institute Monograph). Cambridge: McDonald Institute for Archaeological Research.

Barrett, J. H. & M. P. Richards. 2004. Identity, gender, religion and economy: new isotope and radiocarbon evidence for marine resource intensification in early historic Orkney, Scotland, UK. *European Journal of Archaeology* 7: 249–71.

Barrett, J. H., R. P. Beukens & R. A. Nicholson. 2001. Diet and ethnicity during the Viking colonization of northern Scotland: evidence from fish bones and stable carbon isotopes. *Antiquity* 75: 145–54.

Barrett, J. H., A. K. Hufthammer & O. Bratbak. 2015. Animals and animal products at the Late Iron Age settlement of Bjørkum, Lærdal: the zooarchaeological evidence. Unpublished report for the University Museum of Bergen.

Barrett, J. H., A. M. Locker & C. M. Roberts. 2004a. 'Dark Age Economics' revisited: the English fish bone evidence AD 600–1600. *Antiquity* 78: 618–36.

Barrett, J. H., A. M. Locker & C. M. Roberts. 2004b. The origin of intensive marine fishing in medieval Europe: the English evidence. *Proceedings of the Royal Society B* 271(1556): 2417–21. doi: 10.1098/rspb.2004.2885

Barrett, J. H., R. A. Nicholson & R. Cerón-Carrasco. 1999. Archaeo-ichthyological evidence for long-term socioeconomic trends in northern Scotland: 3500 BC to AD 1500. *Journal of Archaeological Science* 26. 353–88.

Barrett, J., A. Hall, C. Johnstone, H. Kenward, T. O'Connor & S. Ashby. 2007. Interpreting the plant and animal remains from Viking-age Kaupang, in D. Skre (ed.) *Kaupang in Skiringssal*: 283–319. Aarhus: Aarhus University Press.

Barrett, J. H., C. Johnstone, J. Harland, W. Van Neer, A. Ervynck, D. Makowiecki, D. Heinrich, A. K. Hufthammer, I. B. Enghoff, C. Amundsen, J. S. Christiansen, A. K. G. Jones, A. Locker, S. Hamilton-Dyer, L. Jonsson, L. Lõugas, C. Roberts & M. Richards. 2008. Detecting the medieval cod trade: a new method and first results. *Journal of Archaeological Science* 35(4): 850–61.

Barrett, J. H., D. Orton, C. Johnstone, J. Harland, W. Van Neer, A. Ervynck, C. Roberts, A. Locker, C. Amundsen, I. B. Enghoff, S. Hamilton-Dyer, D. Heinrich, A. K. Hufthammer, A. K. G. Jones, L. Jonsson, D. Makowiecki, P. Pope, T. C. O'Connell, T. de Roo & M. Richards. 2011. Interpreting the expansion of sea fishing in medieval Europe using stable isotope analysis of archaeological cod bones. *Journal of Archaeological Science* 38: 1516–24.

Becker, C. & G. Grupe. 2012. Archaeometry meets archaeozoology: Viking Haithabu and medieval Schleswig reconsidered. *Archaeological and Anthropological Sciences* 4: 241–62.

Beerenhout, B. 2009. Archeozoölogie – vissen, in M. Nokkert, A. C. Aarts & H. L. Wynia (eds) *Vroegmiddeleeuwse bewoning langs de A2: een nederzetting uit de zevende en achtste eeuw in Leidsche Rijn: Deel 2*: 335–46. Utrecht: Stadsontwikkeling Gemeente Utrecht.

Benecke, N. 1982. Zur frühmittelalterlichen Heringsfischerei im südlichen Ostseeraum – ein archäozoologischer beitrag. *Zeitschrift für Archäologie* 16: 283–90.

Benecke, N. 1987. Die Fischreste aus einer frühmittelalterlichen Siedlung bei Menzlin, Kreis Anklam. *Bodendenkmalpflege in Mecklenburg Jahrbuch* 34: 225–39.

Bennema, F. P. & A. D. Rijnsdorp. 2015. Fish abundance, fisheries, fish trade and consumption in sixteenth-century Netherlands as described by Adriaen Coenen. *Fisheries Research* 161: 384–99.

Beresford, M. W. 1988. *New Towns of the Middle Ages: town plantation in England, Wales, and Gascony*. Gloucester: Sutton.

Berggren, L., N. Hybel & A. Landen (eds). 2002. *Cogs, Cargoes, and Commerce: maritime bulk trade in northern Europe, 1150–1400*, Toronto: Pontifical Institute of Mediaeval Studies.

Bolle, L. J., A. D. Rijnsdorp, W. Van Neer, R. S. Millner, P. I. Van Leeuwen, A. Ervynck, R. Ayers & E. Ongenae. 2004. Growth changes in plaice, cod, haddock and saithe in the North Sea: a comparison of (post-)medieval and present-day growth rates based on otolith measurements. *Journal of Sea Research* 51: 313–28.

BoneInfo nd http://livelink.archis.nl/livelink/livelink.exe?func=ll&objId=3118333&objAction=browse&sort=name

Bridbury, A. R. 1955. *England and the Salt Trade in the Later Middle Ages*. Oxford: Clarendon Press.

Buitenhuis, H. 2011. Faunaresten, in A. Ufkes (ed.) *Een archeologische opgraving in de vroegmiddeleeuwse ringwalburg van Domburg, gem. Veere (Z.)*: 149–60. Groningen: ARC Archaeological Research and Consultancy.

Buitenhuis, H., D. C. Brinkhuizen, C. Van Loon & T. D. Ridder. 2006. *Gat in de Markt 1.101: het dierlijk botmateriaal* (VLAK-verslag 15.5). Vlaardingen: Vlaardings Archeologisch Kantoor (VLAK).

Campbell, J. 2002. Domesday herrings, in C. Harper-Bill, C. Rawcliffe & R. G. Wilson (eds) *East Anglia's History: studies in honour of Norman Scarfe*: 5–17. Woodbridge: Boydell Press.

Campbell, B. M. S. 2010. Nature as historical protagonist: environment and society in pre-industrial England. *Economic History Review* 63: 281–314.

Candow, J. E. 2009a. The organisation and conduct of European and domestic fisheries in northeast North America, 1502–1854, in D. J. Starkey, J. T. Thór & I. Heidbrink (ed.) *A History of the North Atlantic Fisheries. Volume 1: from early times to the mid-nineteenth century*: 387–415. Bremen: Verlag H. M. Hauschild.

Candow, J. E. 2009b. Migrants and residents: the interplay between European and domestic fisheries in northeast North America, 1502–1854, in D. J. Starkey, J. T. Thór & I. Heidbrink (eds) *A History of the North Atlantic Fisheries. Volume 1: from early times to the mid-nineteenth century*: 416–52. Bremen: Verlag H. M. Hauschild.

Carus-Wilson, E. M. 1962–63. The medieval trade of the ports of the Wash. *Medieval Archaeology* 6–7: 182–201.

Cary, E. (ed.) 1969. *Dio's Roman History*. Vol. 9. Cambridge, MA: Harvard University Press.

Cerón-Carrasco, R. 2005. *Of fish and men ('De iasg agus dhaoine'): a study of the utilization of marine resources as recovered from selected Hebridean archaeological sites.* (British Archaeological Report 400). Oxford: Archaeopress.

Cerón-Carrasco, R. 2011. The ethnography of fishing in Scotland and its contribution to icthyoarchaeological analysis in this region, in U. Albarella & A. Trentacoste (eds) *Ethnozooarchaeology: the present and past of human–animal relationships*: 58–72. Oxford: Oxbow Books.

Childs, W. R. 1995. England's Icelandic trade in the fifteenth century: the role of Hull, in P. Holm, O. U. Janzen, O. Uwe & J. Thor (eds) *Northern Seas Yearbook, 1995*: 11–31. (Fiskeri- og søfartsmuseets studieserie 5). Esbjerg: Fiskeri- og Søfartsmuseet.

Childs, W. R. 1997. The commercial shipping of south-western England in the later fifteenth century. *Mariner's Mirror* 83: 272–92.

Childs, W. & M. Kowaleski. 2000. Fishing and fisheries in the Middle Ages, in D. J. Starkey, C. Reid & N. Ashcroft (eds) *England's Sea Fisheries: the commercial sea fisheries of England and Wales since 1300*: 19–28. London: Chatham.

Clavel, B. & J.-H. Yvinec. 2010. L'archéozoologie du Moyen Âge au début de la période moderne dans la moitié nord de la France, in J. Chapelot (ed.) *Trente ans d'archéologie médiévale en France: un bilan pour un avenir*: 71–87. (IXe Congrès international de la Société d'archéologie médiévale, Vincennes, 16–18 juin 2006; Publication du CRAHM). Caen: Presses Universitaires.

Colley, S. 1984. *Fish Remains from Hamwic (Saxon Southampton): Six Dials variability study.* (Ancient Monuments Laboratory Report 4579). London: English Heritage.

Colson, J. 2014. London's forgotten company? Fishmongers, their trade and their networks in later medieval London, in C. Barron & A. F. Sutton (eds) *Proceedings of the 2012 Harlaxton Medieval Symposium: The Medieval Merchant*: 20–40. Donington: Shaun Tyas.

Coy, J. 1977. *Small Mammal, Bird, Amphibian and Fish Bones from Soil Samples Taken from Site VIII, Hamwih, Southampton.* (Ancient Monuments Laboratory Report 2324). London: English Heritage.

Crabtree, P. J. 2010. Agricultural innovation and socio-economic change in early medieval Europe: evidence from Britain and France. *World Archaeology* 42: 122–36.

Crumlin-Pedersen, O. 2010. Aspects of the origin of Atlantic and Baltic seafaring, in A. Anderson, J. H. Barrett & K. V. Boyle (eds) *The Global Origins and Development of Seafaring*: 109–27. Cambridge: McDonald Institute for Archaeological Research.

Dixon, P. 1986. *Excavations in the Fishing Town of Eyemouth 1982–1984.* Edinburgh: Borders Architects Group.

DN Diplomatarium Norvegicum 1847–2011. *Diplomatarium Norvegicum.* Vol. 5. Oslo: Dokumentasjonsprosjektet. Available at: http://www.dokpro.uio.no/dipl_norv/diplom_felt.html.

Dobney, K. & A. Ervynck. 2007. To fish or not to fish? Evidence for the possible avoidance of fish consumption during the Iron Age around the North Sea, in C. Haselgrove & T. Moore (eds) *The Later Iron Age in Britain and Beyond*: 403–18. Oxford: Oxbow Books.

Dobney, K., D. Jaques, J. H. Barrett & C. Johnstone. 2007. *Farmers, Monks and Aristocrats: the environmental archaeology of an Anglo-Saxon estate centre at Flixborough, north Lincolnshire, UK.* Oxford: Oxbow Books.

Dyer, C. C. 1988. The consumption of fresh-water fish in medieval England, in M. Aston (ed.) *Medieval Fish, Fisheries and Fishponds in England.* (British Archaeological Report 182): 27–38. Oxford: British Archaeological Reports.

Edvardsson, R. 2010. The Role of Marine Resources in the Medieval Economy of Vestfirðir, Iceland. Unpublished PhD dissertation, City University of New York.

Enghoff, I. B. 1999. Fishing in the Baltic region from the 5th century BC to the 16th century AD: evidence from fish bones. *Archaeofauna* 8: 41–85

Enghoff, I. B. 2000. Fishing in the southern North Sea region from the 1st to the 16th century AD: evidence from fish bones. *Archaeofauna* 9: 59–132.

Ervynck, A. & W. Van Neer. 1998. Het archeologisch onderzoek van de voedseleconomie van laatmiddeleeuwse steden: mogelijkheden en eerste resultaten voor Leuven, in L. Bessemans & Museum Vander Kelen-Mertens (ed.) *Leven te Leuven in de late Middeleeuwen*: 79–94. Leuven: Peeters.

Ervynck, A., W. Van Neer & M. Pieters. 2004. How the North was won (and lost again): historical and archaeological data on the exploitation of the North Atlantic by the Flemish fishery, in R. A. Housley & G. Coles (eds) *Atlantic Connections and Adaptations: economies, environments and subsistence in lands bordering the North Atlantic*: 230–9. (Symposia of the Association for Environmental Archaeology 21). Oxford: Oxbow Books.

Fagan, B. 2006. *Fish on Friday: feasting, fasting and the discovery of the New World.* New York: Basic Books.

Fox, H. 2001. *The Evolution of the Fishing Village: landscape and society along the south Devon coast.* Oxford: Leopard's Head.

Frantzen, A. J. 2014. *Food, Eating and Identity in Early Medieval England.* Woodbridge: Boydell Press.

Friedland, K. 1973. Der Hansische Shetlandhandel, in K. Friedland (ed.) *Stadt und Land in der Geschichte des Ostseeraums*: 66–79. Lübeck: Max Schmidt-Romhild.

Geraghty, S. 1996. *Viking Dublin: botanical evidence from Fishamble Street.* Dublin: Royal Irish Academy.

Godbold, S. & R. C. Turner 1994. Medieval fishtraps in the Severn estuary. *Medieval Archaeology* 38: 19–54.

Guðmundsson, F. 1965. *Orkneyinga saga.* Reykjavík: Hið Islenzka Fornritafélag.

Hamerow, H. 2007. Agrarian production and the emporia of mid Saxon England, ca. AD 650–850, in J. Henning (ed.) *Post-Roman towns, trade and settlement in Europe and Byzantium. Volume 1: the heirs of the Roman West.* Berlin: Walter de Gruyter.

Hamilton-Dyer, S. 1997 The Lower High Street Project, Southampton: the faunal remains: Unpublished Report.

Hamilton-Dyer, S. 2001. Bird and fish remains, in M. Gardiner, R. Cross, N. Macpherson-Grant & I. Riddler, Continental trade and non-urban ports in middle Anglo-Saxon England: excavations at Sandtun, West Hythe, Kent. *Archaeological Journal* 158: 255–61.

Hansen, G. 2005. *Bergen c. 800–c.1170: the emergence of a town.* (Bryggen Papers Main Series 6). Bergen: Fagbokforlaget.

Hansen, G. 2015. Bergen AD 1020/30–1170: between plans and reality, in J. H. Barrett & S. J. Gibbon (eds) *Maritime Societies of the Viking and Medieval World*: 182–97. Leeds: Maney/ Society for Medieval Archaeology.

Hardy, K. 2015. Variable use of coastal resources in prehistoric and historic periods in western Scotland. *Journal of Island and Coastal Archaeology*. doi: 10.1080/15564894.2015.1103336

Harland, J. 2006. Zooarchaeology in the Viking Age to medieval Northern Isles, Scotland: an investigation of spatial and temporal patterning. Unpublished PhD dissertation, University of York.

Harland, J. 2007. Status and space in the 'fish event horizon': initial results from Quoygrew and Earl's Bu, Viking Age and medieval sites in Orkney, Scotland, in H. Hüster Plogmann (ed.) *The Role of Fish in Ancient Time*: 63–8. (Proceedings of the 13th Meeting of the ICAZ Fish Remains Working Group). Rahden: Marie Leidorf.

Harland, J. 2009. Technical report: fish remains from the Drogheda boat, Ireland. Report prepared by Centre for Human Palaeoecology, Department of Archaeology, University of York.

Harland, J. 2010. Technical report: the fish bone from Bon Accord, Aberdeen (site code 20215). Report 2010/02. Report prepared by Centre for Human Palaeoecology, Department of Archaeology, University of York.

Harland, J. F. & J. H. Barrett. 2012. The maritime economy: fish bone, in J. H. Barrett (ed.) *Being an Islander: production and identity at Quoygrew, Orkney, AD 900–1600*: 115–38. (McDonald Institute Monograph). Cambridge: McDonald Institute for Archaeological Research.

Heinrich, D. 1987. *Untersuchungen an mittelalterlichen Fischresten aus Schleswig: Ausgrabung Schild 1971–1975*. Neumünster: Wachholtz.

Heinrich, D. 2006. Die Fischreste aus dem Hafen von Haithabu – handaufgelesene funde, in U. Schmölcke, K. Schietzel, D. Heinrich, H. Hüster-Plogmann & K. J. Hüser, *Untersuchungen an Skelettresten von Tieren aus dem Hafen von Haithabu*: 157–93. (Berichte über die Ausgrabungen in Haithabu 35). Neumünster: Wachholtz.

Helle, K. 1968. Anglo-Norman relations in the reign of Håkon Håkonsson (1217–63). *Medieval Scandinavia* 1: 101–14.

Hodgson, G. W. I., C. Smith, A. Jones, M. Fraser, A. K. G. Jones, D. Heppel, R. Cerón-Carrasco, A. S. Clarke, I. H. M. Smart, R. B. Longmore & D. McKay. 2011. *Perth High Street Archaeological Excavation 1975–1977 Fascicule 4: living and working in a Medieval Scottish burgh, environmental remains and miscellaneous finds*. Perth: Tayside & Fife Archaeological Committee.

Hoffmann, R. C. 1996. Economic development and aquatic ecosystems in medieval Europe. *American Historical Review* 101: 631–69.

Hoffmann, R. C. 2002. Carp, cods and connections: new fisheries in the medieval European economy and environment, in M. J. Henninger-Voss (ed.) *Animals in Human Histories: the mirror of nature and culture*: 3–55. Rochester, NY: University of Rochester Press.

Hoffmann, R. C. 2009. *Strekfusz*: a fish dish links Jagiellonian Cracow to distant waters, in P. Górecki & N. van Dusen (ed.) *Central and Eastern Europe in the Middle Ages: a cultural history*: 116–24. London: I. B. Tauris.

Hoffmann, R.C. 2014. *An Environmental History of Medieval Europe*. Cambridge: Cambridge University Press.

Holmes, M. 2014. *Animals in Saxon and Scandinavian England: backbones of economy and society*. Leiden: Sidestone Press.

Hull, B. D. & T. C. O'Connell. 2011. Diet: recent evidence from analytical chemical techniques, in H. Hamerow, D. A. Hinton & S. Crawford (eds) *The Oxford Handbook of Anglo-Saxon Archaeology*: 667–87. Oxford: Oxford University Press.

Hutchinson, W. F., M. Culling, D. C. Orton, B. Hänfling, L. Lawson Handley, S. Hamilton-Dyer, T. C. O'Connell, M. P. Richards & J. H. Barrett. 2015. The globalization of naval provisioning: ancient DNA and stable isotope analyses of stored cod from the wreck of the *Mary Rose, AD* 1545. *Royal Society Open Science* 2: 150199.

IJzereef, G. F. & Laarman, F. 1986. The animal remains from Deventer (8th–19th centuries AD). *Proceedings of the State Service for Archaeological Investigations in the Netherlands* 1986: 405–43.

Jahnke, C. 2009. The medieval herring fishery in the western Baltic, in L. Sicking & D. Abreu-Ferreira (eds) *Beyond the Catch: fisheries of the North Atlantic, the North Sea and the Baltic, 900–1850*: 157–86. Leiden: Brill.

James, H. & T. James 2003. Fish weirs on the Taf, Towy and Gwendraeth estuaries, Carmarthenshire. *Carmarthenshire Antiquary* 39: 1–27.

Johannessen, L. 1998. *Fiskevær og fiskebuer i vestnorsk jernalder: en analyse av strandtufter i Hordaland*. Bergen: Arkeologisk institutt, Universitetet i Bergen.

Jones, E. T. 2000. England's Icelandic fishery in the early modern period, in D. J. Starkey, C. Reid & N. Ashcroft (eds) *England's Sea Fisheries: the commercial fisheries of England and Wales since 1300*: 105–10. London: Chatham.

Jonsson, L. 1986. Finska gäddor och Bergenfisk – ett försök att belysa Uppsalas fiskimport under medeltid och yngre Vasatid, in N. Cnattingius & T. Neréus (eds) *Uppsala stads historia. Volume 7: från Östra Aros till Uppsala: en samling uppsatser kring det medeltida Uppsala*: 122–39. Uppsala: Historiekomm.

Kerr, T. R., G. T. Swindles & G. Plunkett. 2009. Making hay while the sun shines? Socio-economic change, cereal production and climatic deterioration in early medieval Ireland. *Journal of Archaeological Science* 36: 2868–74.

Keyser, R. & P. A. Munch (eds) 1849. *Norges gamle love indtil 1387*. Vol. 3. Christiania: Chr. Gröndahl.

Kowaleski, M. 2000. The expansion of the south-western fisheries in late medieval England. *Economic History Review* New Series 53: 429–54.

Kowaleski, M. 2003. The commercialization of the sea fisheries in medieval England and Wales. *International Journal of Maritime History* 15(2): 177–231.

Kowaleski, M. 2014. Coastal communities in medieval Cornwall, in H. Doe, A. Kennerly & P. Payton (ed.) *A maritime history of Cornwall*: 43–59. Exeter: University of Exeter Press.

Kowaleski, M. & W. Childs. 2000. The internal and international fish trades of medieval England and Wales, in D. J. Starkey, C. Reid & N. Ashcroft (eds) *England's Sea Fisheries: the commercial sea fisheries of England and Wales since 1300*: 29–35. London: Chatham.

Küchelmann, H. -C. 2015. Matjeshering or fish soup? Animal remains from a Hanseatic merchants' site in Bremen, in A. Körösi & Á. Szotyori-Nagy (eds) *Hungarian Grey, Racka, Mangalitsa*: 263–70. Budapest: Museum and Library of Hungarian Agriculture.

Kurlansky, M. 1997. *Cod: a biography of the fish that changed the world*. Toronto: Alfred A. Knopf.

Lampen, A. 2000. *Fischerei und Fischhandel im Mittelalter: Wirtschafts- und sozialgeschichtliche Untersuchungen nach urkundlichen und archäologischen Quellen des 6. bis 14. Jahrhunderts im Gebiet des Deutschen Reiches*. Husum: Matthiesen.

Lauwerier, R. C. G. M. & L. S. de Vries. 2004. Lifting the iceberg – BoneInfo and the battle to save archaeological information, in R. C. G. M. Lauwerier & I. Plug (eds) *The Future from the Past: archaeozoology in wildlife conservation and heritage management*: 167–75. Oxford: Oxbow Books.

Lepiksaar, J. & D. Heinrich. 1977. *Untersuchungen an Fischresten aus der frühmittelalterlichen Siedlung, Haithabu*. (Berichte über die Ausgrabungen in Haithabu 10). Neumünster: Wachholtz.

Lie, R. W. & T. Fredriksen. 2007. Beinmaterialet fra Vesle Hjerkinn, in B. Weber (ed.) *Vesle Hjerkinn – kongens gård og sælehus*: 141–67. Oslo: Universitetets Kulturhistoriske Museer.

Littler, A. S. 1970. Fish in English Economy and Society Down to the Reformation. Unpublished PhD dissertation, University College of Swansea.

Locker, A. 1997. The fish bones, in P. Mills *Excavations at the Dorter Undercroft, Westminster Abbey: 111–3. Transactions of the London and Middlesex Archaeological Society* 46.

Locker, A. 2012. The fish assemblage, in P. Cope-Faulkner *Clampgate Road, Fishtoft: archaeology of a middle Saxon island settlement in the Lincolnshire fens*. Sleaford: Heritage Trust of Lincolnshire.

Lødøen, T. & G. Mandt. 2010. *The Rock Art of Norway*. Oxford: Windgather Press.

Loveluck, C. 2013. *Northwest Europe in the Early Middle Ages c. AD 600–1150*. Cambridge: Cambridge University Press.

Magendans, J. R. & J. A. Waasdorp. 1989. *Franken aan de Frankenslag: een vroeg-middeleeuwse nederzetting in 's-Gravenhage*. (VOM-reeks 1989/2). Den Haag: Gemeente 's-Gravenhage.

Mägi, M. 2015. Bound for the eastern Baltic: trade and centres AD 800–1200, in J. H. Barrett & S. J. Gibbon (eds) *Maritime Societies of the Viking and Medieval World*: 41–61. Leeds: Maney/Society for Medieval Archaeology.

Maitland, P. S. & R. N. Campbell. 1992. *Freshwater Fishes of the British Isles*. London: Harper Collins.

Makowiecki, D. 2012. Badania archeoichtiologiczne szczątków ze stanowiska Janów Pomorski 1, in M. Bogucki & B. Jurkiewicz (eds) *Janów Pomorski: wyniki ratowniczych badań archeologicznych w latach 2007–2008*: 302–23. Elbląg: Muzeum Archeologiczno-Historyczne.

McCarthy, M. 1998. Archaeozoological studies and early medieval Munster, in M. A. Monk & J. Sheehan (eds) *Early Medieval Munster*: 59–64. Cork: Cork University Press.

McCormick, F. 2008. The decline of the cow: agricultural and settlement change in early medieval Ireland. *Peritia* 20: 210–25.

McCormick, F. 2014. Agriculture, settlement and society in early medieval Ireland. *Quaternary International* 346: 119–30.

Mehler, N. & M. Gardiner. 2013. On the verge of colonialism: English and Hanseatic trade in the North Atlantic islands, in P. E. Pope & S. Lewis-Simpson (ed.) *Exploring Atlantic Transitions: archaeologies of transience and permanence in new found lands*: 1–14. Woodbridge: Boydell Press.

Müldner, G. 2009. Investigating medieval diet and society by stable isotope analysis of human bone, in R. Gilchrist & A. Reynolds (eds) *Fifty Years of Medieval Archaeology*: 327–46. Leeds: Maney.

Müller-Wille, M. & A. Tummuscheit. 2004. Viking-age proto-urban centres and their hinterlands: some examples from the Baltic area, in J. Hines, A. Lane & M. Redknap (eds) *Land, Sea and Home: proceedings of a conference on Viking-period settlement*: 27–39. (Society for Medieval Archaeology Monograph 20). Leeds: Maney.

Mulville, J. 2002. The role of Cetacea in prehistoric and historic Atlantic Scotland. *International Journal of Osteoarchaeology* 12: 34–48.

Murphy, P. 2009. *The English Coast: a history and prospect*. London: Continuum.

Näsman, U. & E. Roesdahl. 2003. Scandinavian and European perspectives – Borg I:1, in G. S. Munch, O. S. Johansen & E. Roesdahl (eds) *Borg in Lofoten: a chieftain's farm in north Norway*: 283–99. Trondheim: Tapir Academic.

Naumann, E., T. D. Price & M. P. Richards. 2014. Changes in dietary practices and social organization during the pivotal late Iron Age period in Norway (AD 550–1030): isotope analyses of Merovingian and Viking Age human remains. *American Journal of Physical Anthropology* 155: 322–31.

Nedkvitne, A. 2014. *The German Hansa and Bergen*. (Quellen und Darstellungen zur Hansischen Geschichte, Neue Folge 70). (Revised version of PhD). Vienna and Cologne: Böhlau.

Nicholson, R. A. 2007. The fish remains, in J. Hunter, J. M. Bond & A. N. Smith (eds) *Investigations in Sanday, Orkney, Vol 1: Excavations at Pool, Sanday. Multi-period settlement from Neolithic to Late Norse times*: 263–79. Kirkwall: Orcadian Ltd/Historic Scotland.

Nicholson, R. A. 2010. Fish and fishing from the Pictish to the Norse centuries, in S. J. Dockrill, J. M. Bond, V. E. Turner, L. D. Brown, D. J. Bashford, J. E. Cussans & R. A. Nicholson (eds) *Excavations at Old Scatness, Shetland: Volume 1: the Pictish and Viking settlement*: 156–67. Lerwick: Shetland Heritage Publications.

O'Connor, T. P. 1989. *Bones from Anglo-Scandinavian Levels at 16–22 Coppergate*. London: Council for British Archaeology.

O'Connor, T. P. 2001. On the interpretation of animal bone assemblages from *Wics*, in D. Hill & R. Cowie (ed.) *Wics: the early medieval trading centres of northern Europe*: 54–60. Sheffield: Sheffield Academic Press.

Oksanen, E. 2013. Economic relations between East Anglia and Flanders in the Anglo-Norman period, in D. Bates & R. Liddiard (eds) *East Anglia and its North Sea World in the Middle Ages*: 174–87. Woodbridge: Boydell Press.

Ólafsdóttir, G. Á., K. M. Westfall, R. Edvardsson & S. Pálsson. 2014. Historical DNA reveals the demographic history of Atlantic cod (*Gadus morhua*) in medieval and early modern Iceland. *Proceedings of the Royal Society B* 281: 20132976.

Orton, D. C., D. Makowiecki, T. de Roo, C. Johnstone, J. Harland, L. Jonsson, D. Heinrich, I. B. Enghoff, L. Lõugas, W. Van Neer, A. Ervynck, A. K. Hufthammer, C. Amundsen, A. K. G. Jones, A. Locker, S. Hamilton-Dyer, P. Pope, B. R. MacKenzie, M. Richards, T. C. O'Connell & J. H. Barrett. 2011. Stable isotope evidence for late medieval (14th–15th C) origins of the eastern Baltic cod (*Gadus morhua*) fishery. *PLoS ONE* 6(11): e27568. doi: 10.1371/journal.pone.0027568

Orton, D. C., J. Morris, A. Locker & J. H. Barrett. 2014. Fish for the city: meta-analysis of archaeological cod remains and the growth of London's northern trade. *Antiquity* 88: 516–30.

Orton, D., E. Rannamäe, L. Lõugas, D. Makowiecki, S. Hamilton-Dyer, A. G. Pluskowski, T. O'Connell & J. H. Barrett. forthcoming. The Teutonic Order's role in the development of a medieval eastern Baltic cod fishery, in A. G. Pluskowski (ed.) *The Ecology of Crusading, Colonisation and Religious Conversion in the Medieval Eastern Baltic: Terra Sacra II.* Leiden: Brepols.

O'Sullivan, A. 2001. *Foragers, Farmers and Fishers in a Coastal Landscape: an intertidal archaeological survey of the Shannon estuary.* Dublin: Royal Irish Academy.

O'Sullivan, A., F. McCormick, T. R. Kerr & L. Harney. 2014. *Early Medieval Ireland AD 400–1100: the evidence from archaeological excavations.* Dublin: Royal Irish Academy.

Parks, R. L. & J. Barrett. 2009. Fish bone, in S. Lucy, J. Tipper & A. Dickens (eds) *The Anglo-Saxon settlement and cemetery at Bloodmore Hill, Carlton Colville, Suffolk*: 304–6. (East Anglian Archaeology 131). Ipswich: Suffolk County Council.

Perdikaris, S. & T. H. McGovern. 2008. Codfish and kings, seals and subsistence: Norse marine resource use in the North Atlantic, in T. C. Rick & J. M. Erlandson (eds) *Human Impacts on Ancient Marine Ecosystems: a global perspective*: 187–214. Berkeley: University of California Press.

Perdikaris, S. & T. H. McGovern. 2009. Viking Age economics and the origins of commercial cod fisheries in the North Atlantic, in L. Sicking & D. Abreu-Ferreira (eds) *Beyond the Catch: fisheries of the North Atlantic, the North Sea and the Baltic, 900–1850*: 61–90. Leiden: Brill.

Pollard, A. M., P. Ditchfield, E. Piva, S. Wallis, C. Falys & S. Ford. 2012. 'Sprouting like cockle amongst the wheat': the St Brice's Day Massacre and the isotopic analysis of human bones from St John's College, Oxford. *Oxford Journal of Archaeology* 31(1): 83–102.

Pope, P. E. 2004. *Fish into Wine: the Newfoundland plantation in the seventeenth century.* Chapel Hill: University of North Carolina Press.

Prummel, W. 1983. *Early Medieval Dorestad, an Archaeozoological Study.* (Excavations at Dorestad 2, Nederlandse Oudheden 11, Kromme Rijn Projekt 2). Amersfoort: Rijksdienst voor het Oudheidkundig Bodemonderzoek.

Prummel, W. & D. Heinrich. 2005. Archaeological evidence of former occurrence and changes in fishes, amphibians, birds, mammals and molluscs in the Wadden Sea area. *Helgoland Marine Research* 59: 55–70.

Radtke, C. 1977. Bemerkungen zum mittelalterlichen Fischfang mit Heringszäunen in der Schlei, in K. Schietzel (ed.) *Berichte über die Ausgrabungen in Haithabu* 10: 123–40. Neumünster: Karl Wachholtz.

Rannamäe, E. & L. Lõugas. forthcoming. Animal exploitation in Karksi and Viljandi during the late prehistoric times and the Middle Ages, in A. G. Pluskowski (ed.) *The Ecology of Crusading, Colonisation and Religious Conversion in the Medieval Eastern Baltic: Terra Sacra II.* Leiden: Brepols.

Rick, T. C. & J. M. Erlandson (eds) 2008. *Human Impacts on Ancient Marine Ecosystems: a global perspective*, Berkeley: University of California Press.

Rigby, S. H. 1993. *Medieval Grimsby: growth and decline.* Hull: University of Hull Press.

Roberts, C. 2007. *The Unnatural History of the Sea.* Washington, D.C.: Island Press.

Rorke, M. 2001. Scottish Overseas Trade, 1275/86–1597. Unpublished PhD dissertation, University of Edinburgh.

Rorke, M. 2005. The Scottish herring trade, 1470–1600. *Scottish Historical Review* 84: 149–65.

Rose, S. 2011. *The Wine Trade in Medieval Europe 1000–1500* London: Continuum.

Russ, H. 2011. Fish remains, in M. Rowe (ed.) Land to the East of Cartergate Grimsby, North East Lincolnshire. Report prepared for Preconstruct Archaeological Services, Lincoln.

Russ, H., I. Armit, J. McKenzie & A. K. G. Jones. 2012. Deep-sea fishing in the Iron Age? New evidence from Broxmouth hillfort, south-east Scotland. *Environmental Archaeology* 17: 177–84.

Russow, E., L. Lõugas, L. Maldre, S. Hiie, K. Kihno, H. Luik, V. Kadakas, K. Sarv, U. Kadakas, A. Kalm & M. Reppo. 2013. Medieval and early modern suburban site in Tallinn, Tartu Road 1: artefacts and ecofacts. *Archaeological Fieldwork in Estonia* 2012: 149–70.

Saul, A. 1982. English towns in the late middle ages: the case of Great Yarmouth. *Journal of Medieval History* 8: 75–88.

Saul, A. 1983. The herring industry in Great Yarmouth. *Norfolk Archaeology* 38: 33–43.

Sawyer, P. 2013. *The Wealth of Anglo-Saxon England.* Oxford: Oxford University Press.

Schmölcke, U. 2004. *Nutztierhaltung, Jagd und Fischfang: zur Nahrungsmittelwirtschaft des frühgeschichtlichen Handelsplatzes von Groß Strömkendorf, Landkreis Nordwestmecklenburg.* Lübstorf: Archäologisches Landesmuseum Mecklenburg-Vorpommern.

Schmölcke, U. & D. Heinrich. 2006. Die Tierknochen aus dem Hafen von Haithabu – Schlämmfunde, in C. Radtke (ed.) *Untersuchungen an Skelettresten von Tieren aus dem Hafen von Haithabu*: 195–239. (Berichte über die Ausgrabungen in Haithabu 35). Neumünster: Wachholtz.

Sear, D., S. R. Bacon, A. Murdock, G. Doneghan, P. Baggaley, C. Serra & T. P. LeBas. 2011. Cartographic, geophysical and diver surveys of the medieval town site at Dunwich, Suffolk, England. *International Journal of Nautical Archaeology* 40: 113–32.

Serjeantson, D. 2013. *Farming and Fishing in the Outer Hebrides AD 600 to 1700: the Udal, North Uist.* Southampton: Highfield Press.

Serjeantson, D. & C. M. Woolgar. 2006. Fish consumption in medieval England, in C. M. Woolgar, D. Serjeantson & T. Waldron (eds) *Food in Medieval England: diet and nutrition*: 102–30. Oxford: Oxford University Press.

Sharples, N. M., C. Ingrem, P. Marshall, J. Mulville, A. Powell & K. Reed. 2015. The Viking occupation of the Hebrides: evidence from the excavations at Bornais, South Uist, in J. H. Barrett & S. J. Gibbon (ed.) *Maritime Societies of the Viking and Medieval World*: 237–58. Leeds: Maney/Society for Medieval Archaeology.

Stronach, S. 2002. The medieval development of South Leith and the creation of Rotten Row. *Proceedings of the Society of Antiquaries of Scotland* 132: 383–423.

Thór, J. Th. 2008. Icelandic fisheries, *c.* 900–1900, in D. J. Starkey, J. Th. Thór & I. Heidbrink (eds) *A History of the North Atlantic Fisheries. Volume 1: from early times to the mid-nineteenth century*: 323–49. (Deutsche maritime Studien/German Maritime Studies 6). Bremen: Hauschild.

Tsurushima, H. 2006. The eleventh century in England through fish-eyes: salmon, herring, oysters, and 1066, in C. P. Lewis (ed.) *Anglo-Norman Studies* 29: 193–213. (Proceedings of the Battle Conference). Woodbridge: Boydell Press.

Turner, R. 2002. Fish weirs and fish traps, in A. Davidson (ed.) *The Coastal Archaeology of Wales*: 95–108. York: Council for British Archaeology.

Tys, D. 2015. Maritime environment and social identities in medieval coastal Flanders: the management of water and environment and its consequences for the local community and the landscape, in J. H. Barrett & S. J. Gibbon (eds) *Maritime Societies of the Viking and Medieval World*: 122–37. Leeds: Maney/Society for Medieval Archaeology.

Unger, R. W. 1978. The Netherlands herring fishery in the late Middle Ages: the false legend of Willem Beukelszoon of Biervliet. *Viator* 9: 335–56.

Unger, R. W. 2002. Beer: a new bulk good, in L. Berggren, N. Hybel & A. Landen (eds) *Cogs, Cargoes, and Commerce: maritime bulk trade in northern Europe, 1150–1400*: 113–27. Toronto: Pontifical Institute of Mediaeval Studies.

Wickham, C. 2005. *Framing the Early Middle Ages: Europe and the Mediterranean, 400–800*. Oxford: Oxford University Press.

Wickler, S. & L. Narmo. 2014. Tracing the development of fishing settlement from the Iron Age to the Modern Period in northern Norway: a case study from Borgvær in the Lofoten Islands. *Journal of Island & Coastal Archaeology* 9: 72–87.

Wouters, W., L. Muylaert & W. Van Neer. 2007. The distinction of isolated bones from plaice (*Pleuronectes platessa*), flounder (*Platichthys flesus*) and dab (*Limanda limanda*): a description of the diagnostic characters. *Archaeofauna* 16: 33–95.

Wubs-Mrozewicz, J. 2009. Fish, stock and barrel. Changes in the stockfish trade in northern Europe, *c.* 1360–1560, in L. Sicking & D. Abreu-Ferreira (eds) *Beyond the Catch*: 187–208. (The Northern World 41). Leiden: Brill.